BUILDING FIRE PERFORMANCE ANALYSIS

BUILDING FIRE PERFORMANCE ANALYSIS

Robert W. Fitzgerald
Worcester Polytechnic Institute, Worcester, Massachusetts, USA

John Wiley & Sons, Ltd

Telephone (+44) 1243 779777

Email (for orders and customer service enquiries): cs-books@wiley.co.uk
Visit our Home Page on www.wileyeurope.com or www.wiley.com

Other Wiley Editorial Offices

John Wiley & Sons Inc., 111 River Street, Hoboken, NJ 07030, USA

Jossey-Bass, 989 Market Street, San Francisco, CA 94103-1741, USA

Wiley-VCH Verlag GmbH, Boschstr. 12, D-69469 Weinheim, Germany

John Wiley & Sons Australia Ltd, 33 Park Road, Milton, Queensland 4064, Australia

John Wiley & Sons (Asia) Pte Ltd, 2 Clementi Loop #02-01, Jin Xing Distripark, Singapore 129809

John Wiley & Sons Canada Ltd, 22 Worcester Road, Etobicoke, Ontario, Canada M9W 1L1

Wiley also publishes its books in a variety of electronic formats. Some content that appears in print may not be available in electronic books.

Library of Congress Cataloging-in-Publication Data

Fitzgerald, Robert W.
Building fire performance analysis / Robert W. Fitzgerald.
p. cm.
ISBN 0-470-86326-9 (Cloth : alk. paper)
1. Building, Fireproof. 2. Fire prevention–Inspection. I. Title.
TH1065. F574 2004
693.8'2–dc22
2003025150

British Library Cataloguing in Publication Data

A catalogue record for this book is available from the British Library

ISBN 0-470-86326-9

Typeset in 9.5/11.5pt Times by Laserwords Private Limited, Chennai, India
Printed and bound in Great Britain by TJ International, Padstow, Cornwall
This book is printed on acid-free paper responsibly manufactured from sustainable forestry in which at least two trees are planted for each one used for paper production.

CONTENTS

PREFACE

The building/fire system is enormously complex. Dynamic fire changes interact with active and passive fire defenses and human activities. The fire and site-specific building features influence a variety of time-related interactions and interdependencies. A century ago, this complexity combined with the rudimentary status of fire science and engineering led conventional wisdom to solve the fire problem through the regulatory process rather than by performance solutions.

About 40 years ago, fire scientists and engineers began to address fire technology through comprehensive and cooperative international efforts. During the past 15 or 20 years, a knowledge explosion of fire sciences and technical understanding has occurred. A comparison of the content during the past 20 years in the *SFPE Handbook of Fire Protection Engineering*, the *NFPA Fire Protection Handbook*, and published fire standards clearly indicates this information explosion. This technological growth, combined with historical knowledge and experience enables building defenses and fire behavior to be predicted with greater understanding than at any previous time.

This book describes a way of thinking to analyze building fire performance using contemporary fire knowledge and experience. The analytical framework focuses on the functions of fire and fire defenses to understand how a building will behave during a fire. The process integrates micro component behavior with macro building performance. Evaluations blend state-of-the-art fire science and engineering with experience and knowledge. Logical steps enable the gap between what is known and what is needed to be filled by judgment. Ways of dealing with variability and uncertainty are integral to the analysis.

A comprehensive understanding of performance enables an expanded variety of applications to be addressed. Applications may range from resource management (How can I do better with what I now have?) to technical decisions (How can I improve the performance quality?). Because the focus is analytical, one can compare alternatives on a rational basis. Thus, risks between a proposed performance-based design may be compared with prescriptive code requirements. A fire department may tailor its operational planning to avoid placing fire-fighting forces in unusual jeopardy. Corporations may portray implications of alternate mixes between accepting losses, insuring risks, and reducing risks with better protection measures. Proposed alternatives may be logically evaluated for cost and effectiveness. The understanding that unfolds from a performance evaluation enables an individual to make better routine, day-to-day decisions and to communicate more effectively with others.

Language and culture are very different among architects, engineers, the fire service, owners, code officials, product suppliers, and the public. Communication between groups on decision-making rationale is often difficult. The knowledge explosion of recent years has enabled some individuals to become experts within selective specialties. While this creates increased knowledge of certain components, it may not adequately address the holistic building behavior or improve the dialogue between interested groups. The goal of this book is to provide an enhanced understanding of the holistic process so that communication among different groups will enable better decisions to be made more easily.

ACKNOWLEDGEMENT

On April 27, 1972, the General Services Administration (GSA) published PBS 5920 *Building FireSafety Criteria.* Appendix D was entitled "Interim Guide to Goal Oriented Systems Approach to Building FireSafety." It is the seed from which this book has grown.

Building Fire Performance Analysis: A Way of Thinking may be viewed as a way station along the road from GSA Appendix D toward performance-based design procedures. Whatever may eventually happen at the final destination, my participation has been a thirty-year adventure filled with wonderful people and fond memories. The journey has been very rewarding personally and professionally. I feel fortunate to have participated in this part of the expedition.

Neither this book nor the adventure was envisioned when a middle-aged structural engineer decided to shift professional focus from design and construction to the building regulatory system with emphasis on the building code. Although fire was a major part of the building code, it was naively assumed that fire regulations had a technical foundation similar to other engineering disciplines. It was anticipated that technical justification for fire requirements merely required an extension of normal scientific and engineering principles.

Although these early assumptions were quickly dispelled, a new world began to emerge to an individual who initially believed he had a broad understanding of the building process. The professional community involved with fire safety was dedicated and passionate about its role to protect society against the ravages of fire. Although a technical basis for fire was very deficient when compared to other mature engineering disciplines, the experience, knowledge, and technical judgment of its practitioners was remarkable. Individuals who practice in the enormously complex fire safety field seem to acquire an understanding, thought process, and decision-making ability that rivals the best talent of other disciplines. Over the years, these individuals have been very generous of time, talent, and resources to help a struggling novice understand how they think and make decisions in the complex system of fire and buildings.

This book is a compilation of the talents of others in the fire and academic communities. My role has been that of a scribe who borrowed the thoughts and experiences of my colleagues, tested them, organized the process, and put their ideas into words. I owe a great debt of gratitude to those many individuals who have made this book possible.

It is impossible to acknowledge all of the individuals who have contributed to the formation of this book. Certain individuals who made specific contributions should be recognized, and I attempt to acknowledge those individuals here. The list is very long, and human imperfection may have caused me to inadvertently miss some deserving individuals. I hope that those many individuals who should have been recognized here will forgive me.

Of the many individuals who are responsible for this document, three deserve special recognition, and many others distinct acknowledgment. I treasure their friendship and help.

Rolf Jensen invested enormous time and resources into the development of this method and had a major role with this book. Rolf could understand and correct not only what I was saying, but what I was thinking. He had tremendous insight and knowledge of fire protection engineering and its people. This book improved over the years because of Rolf Jensen. I am eternally grateful for the enormous time that he spent in reading and influencing this publication. More importantly, I am indebted for his advice, guidance, and friendship over these many years. Rolf Jensen was a giant in fire protection engineering. His recent passing is a major loss for the profession and for me personally.

Harold E. "Bud" Nelson established the original basis for this book with the publication of what was affectionately known as Appendix D. This method should be properly called the *Nelson Method* after the man who created it over 30 years ago. Appendix D is, in my opinion, the most significant single document in modern fire safety literature. Bud's insight and creative mind started this way of thinking and his direction during the early years of its history was critical.

Rexford Wilson was a key figure in the early development of this method. He was an important teacher and practitioner during the early developmental years. Rexford was an early mentor who taught me much about the fire protection business. Rexford is an enormously creative individual who provided major direction to me and this method during our formative periods. I am grateful for his guidance and friendship over the years.

Lieutenant John Carlson, Training Division, Worcester Fire Department was my good friend and teacher for many years. John was my tutor in fire fighting, the fire service, and life. John affected the way this method has unfolded. His patient guidance and friendship are treasured, and he is greatly missed.

The Worcester, Massachusetts, Fire Department has had a major influence on my understanding of the fire service, fire, and buildings. It is impossible to adequately express my gratitude for the guidance, tolerance, and friendship that I have received during a thirty-year association with this organization. They took a raw academic novice under their wing and taught him rudimentary fire fighting and so much more about building fires. For many years, I was called for all multiple-alarm fires and ran with many other incidents. The burning building was my laboratory, and the Worcester Fire Department provided my education. A building looks very different during a fire than after it is out. After the fire, we often reviewed decisions and reasoning and discussed what was done well and the mistakes that were made. Their mentoring, my participation on the fire ground, and the plain enjoyment of hanging out with a great group will be with me forever. During the early years in the development of this method, nearly 60 officers of the Worcester Fire Department attended my courses, became an important part of my education, and made improvements in these evaluation methods. Building details that affected the fire or fire fighting, even in routine fires, were called to my attention, and we analyzed their influences carefully. So many of these individuals have been a part of my life, it is difficult and perhaps unfair to identify only a few. Nevertheless, Ed Hackett, Jim Nally, George Beringer, Bill Hobbs, Ken Henderson, Joe Hennigan, Jim Callery, Walter Giard, and Dennis Budd deserve special note. Of recognitions I have received over the years, the most appreciated has been my badge and helmet as honorary chief of the Worcester Fire Department and the feeling of belonging to that organization.

Dick Stevens was my first mentor in the fire protection business. His kind and gentle guidance and instruction set me on this marvelous adventure.

Jim Shields has made major contributions to this method as it was becoming more clearly defined. Jim's insight, probing questions, and advice have been responsible for significant directional changes and improvements in recent years.

Cliff Harvey has been a passionate and dedicated advocate of this method. He proved that the method can be applied by the fire service in a community. As chief fire marshal for the Boulder, Colorado, Fire Department, Cliff made it work and work well. During and since that time, Cliff has given valuable advice and support to this book, the method, and me.

Wayne Moore has been a steady, reliable friend, supporter, reviewer, and advisor over the many years of this adventure. Wayne has made many contributions to the organization and evaluation techniques. No matter how busy Wayne may be, he is always available to help, advise, and talk.

Rob Richards and the US Coast Guard Research & Development Center have been responsible for much of the progress on this method. His leadership, questions, support, and applications testing for ships were responsible for many improvements. His use of computer analysis for ship applications led to many innovations in the procedures. Many technical advancements can be attributed to Rob and the many individuals in Coast Guard fire research.

Brian Meacham has given advice, guidance, ideas, help for many years. Brian's understanding of risk analysis and management in this area involving state-of-the-art fire protection have been very helpful in making decisions and identifying directional changes.

Len Albano has made important contributions to this book and its applications. In recent years, Len has provided critical thinking and responses. His insight, availability for discussion, and advice have been important to recent directions. His knowledge of structural design for fire and his understanding of building design and construction practices are a valuable resource to me and to the fire profession.

The Fire Service College at Morton-in-Marsh, England has been a major proving ground for risk analysis and risk management using this method. ACO Goeff Winkworth, Tony Barnes, Mike Robins, and the many dedicated fire service officers who have been applying and teaching the method for risk assessment and management deserve clear recognition. They have educated me with their experiences and feed back, and some of their recommendations appear in this volume. My association and friendship with the college and the many individuals associated with it has been very enjoyable and rewarding.

Over the years, many students *at Worcester Polytechnic Institute (WPI)*, have contributed to the development of this method in a variety of ways. Theses, projects, class exercises, and other activities have created, tested, and modified many techniques. Sometimes they showed that an idea should be abandoned. Other times they confirmed practicality or established a new procedure. All activities advanced the knowledge and understanding of this method. District Chief Jim Callery, Ken Menke, Kurt Ruchala, Tim Rodrique, and Glenn Corbett made major contributions to the M curve analysis. Other individuals developed components or performed comprehensive testing. Special recognition should be given to Dave Demers, Mike McGreal, Tom Capaul, Mike Wojcik, Doug Nadeau, Wally Pizzano, Peter Johnson, Craig Van Anne, Greg Ghosh, Mike Ferreira, Beth Newton Tubbs, Elham Falsafi, Bob Till, Dick Chutoransky, Nick Cricenti, Mike Isner, Ali Saffar, Tom Barry, John Titus, Jeff LaSalle, Bernie Till, Gene Cable, Johannes Almas, John Mahoney, Tzusheng Shen, Casey Grant, Eddie Mui, Li Fan, Matt Johann, Hossein Davoodi, Hamid Bahadori, John Farley, Jonathan Hill, John Fitzgerald, Doug Beller, Mario Ley, Suprapto, Scott Deal, Chris Marrion, Rolando Roque, Rich Pehrson, Mark Crompton, Graham Marsh, Ken Bland, Tom Klem, Bill Pucci, Brian Dolph, Paul Donga, P.J. McGuire, Tom Bedard, Renee Lantz, Colleen Wade, Ales Jug, Brian Gilda, and Amanda Greaney.

Current and former WPI faculty members have helped in a variety of ways over the years. They include Dave Lucht, Bob Zalosh, Jonathan Barnett, Nick Dembsey, John Woycheese, Craig Beyler, Ed Clougherty, and Dick Custer of the Fire Protection Engineering Department. Ray Scott was a major force in the computer-based analysis. Frank Noonan, Art Heinricher, Brian Savalonis, Bob Wagner, and Guillermo Salazar have all contributed to the evolution in ways that could not have been made without their help. In some cases their tangible efforts advanced techniques or validated theory and practices. In other cases their advice, criticism, and counsel have contributed to quality improvements.

A large group of individuals have contributed or supported activities at various times over the years. Dave Bouchard, Anne Phillips, Hank Roux, Roger Wildt, Jack Watts, Bob Thompson, Terry Mills, John DeRis, Ron Alpert, Ed Jerger, Hank Wakabayashi, Chuck House, Narindra Gunaji, Bob Schifiliti, Dayse Duarte, Ove Pettersson, Mike Sprague, Donovan Ryan, Alan Beard, Yngve Anderberg, G. Ramachandran, Jonathan Sime, Jake Pauls, Ralph Transue, Ann Spets, and Vidar Stenstad are in this group.

Participants in a variety of courses over the past 30 years deserve special appreciation. The number is literally in the thousands. The courses ranged from two-, three-, or five-day short courses to seven- and fourteen-week undergraduate, graduate, and professional courses. Because these courses were mechanisms to advance and test the knowledge, no two courses were the same. Most of the time, a course was used to test new theories or practices. Each course provided a new opportunity to advance knowledge and test results. Participants included graduate and undergraduate students, fire fighters, plant engineers, college faculty, code officials, insurance inspectors, fire protection engineers, architects, and a range of others relating to ship and building activities. The thought process and evaluation procedures were under rapid development, particularly during the first 20 years. They still are, although on a different level. Enough recent changes have occurred that a student of only five years ago would note significant differences. Except for vocabulary and general organization, early students would not recognize the materials. Although instructors did the best they could at the time, many things that were taught eventually proved to be incorrect, whereas others were correct and useful. Nevertheless, the way of thinking and the goal to understand fire and buildings as an integrated system remained constant. Methods improved in part because the participants were tolerant of the state of the art at the time. Most of the progress occurred because the participants joined in a united effort to make it better. They asked "Why" or said, "I don't agree with you, it should be . . ." During the early years, participant questions were recorded so that an appropriate response could be developed. Sometimes the questions took several years to reach a satisfactory answer.

The evolution of this work could not have happened without a loving and tranquil home life. I have been blessed with very tolerant and understanding wives. Ann had to contend with my learning curve. Peg had to endure the writing of this book. This book is here now because of Peggy's active encouragement and support.

It is with gratitude that I recognize the individuals who have made such long-term and significant contributions to the evolution of this method. It is with regret that I am unable to find the space to acknowledge the literally thousands of others who have made incremental, yet no less significant contributions. This method is the result of many individuals who have helped to move it bit by bit along its path. I have been privileged to act as the scribe to them.

1 UNDERSTANDING, DECIDING, COMMUNICATING

1.1 The destination

The goal of this book is to organize the system of fire and buildings in a way that explains fire performance. The understanding that evolves during a performance evaluation will help professionals who work in the building industry make better decisions and communicate more effectively with others.

The procedures described in this book can be used for any existing or proposed new building involving any code, standard, or regulatory requirements. In addition to being applicable to any building size or use, the framework for analysis and evaluation procedures may also be adapted to ships, tunnels, and mass transport systems.

Before describing fire performance evaluations, we shall briefly discuss practices relating to building fire safety today. These practices are a basis for the way decisions are made. They affect ways of thinking about design, regulatory compliance, performance expectations, and fire risk management. If performance analysis requires a different way of thinking, a brief review of the current way of thinking enables the distinctions to be put into context.

1.2 Codes and standards

Fire safety decisions today are dominated by building codes, associated standards of practice, and insurance considerations. This is natural because they have been a part of the building process for nearly a century and have demonstrated success in reducing fire losses. The thought process for fire-related decisions is heavily influenced by experiences and interpretations of codes and standards.

Modern building code and standards development can be associated with the first few decades of the twentieth century. At the turn of the century, the second industrial revolution and the population growth in cities was well established. Fire was clearly recognized as a major threat to business, commerce, and society. For example, on March 21 and 22, 1916 three cities in the states of Georgia, Tennessee, and Texas had conflagrations that destroyed over 2700 buildings during that single period. The economic and social impact was devastating. A long-standing recognition of the fire threat, combined with regular and severe fire disasters provided a clear realization that something had to be done.

Building Fire Performance Analysis R. W. Fitzgerald
 ISBN: 0-470-86326-9 (HB)

Under the leadership of the insurance industry, it was decided that the most effective way of dealing with this major problem was through the building code system. Although building codes did exist at that time, they were not effective in dealing with the fire problem. A restructuring of the codes to deal with fire safety became a major focus of attention. Fire issues during the period around World War I evoked active debate and major decisions culminating with the publication in 1927 of the first edition of the Uniform Building Code (UBC) of the International Conference of Building Officials. This code established a framework from which the organization and practices of succeeding codes have been derived.

Structural, mechanical, and electrical parts of a modern building code are generally performance oriented. However, fire regulations are prescriptive for a number of reasons. The fire safety system is far more complex than other building design disciplines, and a century ago the knowledge base of fire technology was very rudimentary. Individuals with a professional focus on fire protection were not a part of the design team because fire was an "abnormal" rather than a "normal" design function. Existing practices in the building industry combined with historical precedents in the wording of building codes provided a way to structure and enforce fire safety requirements. This environment and a recognized urgency to solve the fire problem became driving forces toward the expediency of developing prescriptive rather than performance codes for fire.

One may look back with admiration at the prescriptive code system that was implemented eight decades ago. The influence of that code and standard system can clearly be identified with major improvements in fire safety and the preservation of economic values. Efforts by enforcement officials in code administration were a major factor in this success. During this time period, advances in fire equipment and knowledge, code improvements, better code administration, the fire service, and insurance protection have greatly diminished the fear of fire among citizens and reduced the concern of fire losses in the business community.

Nevertheless, the existing prescriptive code and standards system is not without its shortcomings. Prescriptive fire regulations developed by consensus committees may be described as a compilation of good practices that have a weak technical basis. They are "easy" to administer, yet difficult to control. Modern regulatory requirements are overwhelming, and the designer is often placed in an adversarial position with conflicts between regulatory conformance and building functionality needs. Questions often arise regarding the justification of certain requirements for a site-specific building condition. Cost and effectiveness are viewed differently by code officials and designers. Differences of opinion are rarely resolved by rational, analytical procedures to predict performance, because an analytical framework has not been established by the fire community.

It is important to recognize that a prescriptive code assumes the responsibility for fire safety rather than a design professional. The code official is a "policeman" obliged to administer the law. Consequently, in a prescriptive code environment, a building is required to "meet the code" rather than to be a fire-safe building. Insofar as the code specifies safety features, the building is presumed to provide a level of safety consistent with those requirements. When codes change, that base level of safety changes. Preexisting buildings are rarely required to be upgraded to the latest requirements. Over a period of time, any city will contain buildings that have been designed in conformance with widely varying sets of requirements. Which are correct? When should a building be upgraded to more modern requirements? How much is enough? How do we know?

The level of safety provided in a building code is indeterminate. Although a perception exists that a code-complying building is a safe building, that is not necessarily the case. Regulatory practices provide no way to measure the level of fire safety for a prescriptive code complying building. We do recognize that it is possible to configure architecture and to satisfy modern code requirements in a way that a building will perform poorly in a fire and pose a great risk to its occupants and mission. It is also possible to have a building of the same occupancy and size constructed in the same city under the same code to perform very well and have a minimal risk to its occupants and functions. The difference has more to do with designing fire defenses using understanding and sensitivity than in spending more money for fire-related features.

With all of the strengths and defects in building codes, accepted standards, and their administration, it is important to recognize that today's codes and the fire service provide whatever public fire safety is available to the citizen. And the code is the law.

1.3 Routine practices

In the design and approval of buildings for fire, routine practices are rarely routine. In building design and construction, compliance with the local building code is the usual focus. A thoughtful harmonizing of fire defenses by a trained and skilled fire safety professional to provide identified performance objectives and goals is rarely done. Both extremes can obtain necessary construction permits and certificates of occupancy. One is not necessarily more expensive than the other. Sometimes greater design investment can be offset by lower construction costs.

Regulatory approval practices and inspections by authorities having jurisdiction (AHJs) also can exhibit a wide range of attention. Some AHJs may require only two statements from the designer. One is that the building design complies with all applicable codes and standards, and the second is that construction followed the plans and specifications. Building permits and certificates of occupancy may be given upon receipt of these letters with little or no design review and construction inspection. At the other extreme, a code authority may meticulously review plans, approve equipment designs, conduct acceptance testing, and provide regular field inspections. The quality and extent of the services provided by AHJs depends on the workload of individuals, their qualifications and experience, and community management attitudes. A wide range of enforcement experiences can exist among jurisdictions.

Questions that arise during regulatory approvals can be complex and difficult. For example, how does an AHJ decide if a proposed design alternative is equivalent to a regulatory requirement? These are often known as trade-offs and they may have a substantial impact on building fire performance. Costs are important. How much is enough and what is excessive? Decisions of this type can have a major impact on building functionality, costs, and performance.

Practices of building and fire departments are nonuniform. In some jurisdictions the fire department may never be consulted or involved in the process. In other locations the entire approval process is administered within the fire department. Sometimes the fire department is consulted for advisory opinion only. In other cases the fire department may have approval authority only for certain requirements. A wide range of procedures exist, although a trend of greater fire department participation is emerging.

Differences of opinion about certain code interpretations, uneven enforcement, variability of participation by local fire departments, and the inadequacies of many architectural and engineering design teams in understanding fire and fire defense behavior (as opposed to understanding code requirements) can lead to very different performance among buildings. The vast inventory of existing buildings that have been constructed under different codes and conditions enables one to recognize the fire performance variability that can exist within any community.

The theme of this discussion is that the building code and its administration may produce a legal building, but that credential does not assure safety from fire. The emergence of the modern building code and its enforcement have contributed to an enormous improvement of fire safety. However, it is not difficult to identify legal buildings in which a fire caused extensive destruction or loss of life. All buildings and building operations have risk. Some buildings have much greater risk than others. Building code compliance as evidenced by a certificate of occupancy indicates that the building provides some level of safety, although that level is neither defined nor measured.

1.4 A way of thinking

Building performance evaluations require a different way of thinking from traditional practices. Because the emphasis is on understanding and describing performance, regulations and standards of practice offer little help. Instead, one needs an integrated framework to structure the process and behavior estimates to calibrate performance. Dynamic value estimates are based on available information and knowledge for the specific building being evaluated.

THE FOCUS IS ANALYSIS

Fire safety professionals are usually absorbed with demands of design. Design often involves obtaining information about code requirements, applicable standards, and AHJ expectations for obtaining necessary building permits and certificates of occupancy. Inherent to the process is the way in which fire defenses work and cost comparisons of alternate proposals. Design makes decisions about the best actions to take within existing constraints.

This book does not address building design. It does not recommend appropriate ways of designing, suitable values for design, or what to do to produce a better design. The design will have been completed as a prerequisite to evaluating how that design will perform.

The focus of this book is the analysis (evaluation) of building fire performance and the associated risk characterizations to people, property, operational continuity, neighbors, and the environment. An analysis starts with conditions that exist or have been selected for a proposed design. An analysis is structured around understanding what is rather than what should be.

ANALYTICAL FRAMEWORK

Performance evaluations are structured by an analytical framework that examines the functional performance of fire defenses and building features. The systems framework integrates the micro behavior of individual components with the macro building performance.

The framework organizes the complete building/fire system into discrete components that are based on functional performance. These components incorporate traditional active and passive fire defenses, such as sprinkler suppression, manual fire fighting, detection and alarm, compartmentation, and structural frame behavior. Life safety analysis is an integral part of the risk characterizations and is based on the performance expectations for the specific building being evaluated.

Performance evaluations enable one to compare alternatives in a disciplined manner. Thus, one can compare on a consistent, rational basis the impact of specific building features such as sprinklers, early detection and alarm, and smoke management systems on performance that affects risks to life safety, property damage, or operational continuity.

The framework is based on a thought process that tracks functional component performance. Because the thought process and component behavior are so intertwined, each chapter on component evaluation includes a description of the functional operation of the component. This description establishes a logic for the analytical structure.

PERFORMANCE MEASURES

The integration of performance measures into the analytical framework provides the basis for understanding building behavior. The function of this book is to organize a way of thinking that enables one to incorporate building observations with state-of-the-art fire science and engineering, traditional practices, standards, experience, and judgment.

There has been an explosion in knowledge about fire and fire defense behavior in recent years. In particular, the *SFPE Handbook of Fire Protection Engineering* [1], the *NFPA Fire Protection Handbook* [2], published standards of practice [3], computer programs [4] and a variety of books provide a wealth of information that may be incorporated into performance evaluations. This book describes how to organize that knowledge into measures of performance.

Fire performance is dynamic. For one period of time, one may be certain that a component will not act. During another time period one may be confident that it will act. Between these positions, a window of uncertainty exists. Performance within this window of uncertainty is estimated and expressed in descriptive terms or as a subjective probability. Performance is based on information and time available for the evaluation. The framework organizes the analytical needs to enable appropriate information to be selected for event evaluations. The way of thinking can remain constant while confidence in performance estimates grows as knowledge increases.

We work in a very imperfect world. Dynamic behavioral estimates almost always involve poorly understood and uncertain situations. Sometimes the insufficient knowledge is due to our own inadequacies. Sometimes it is because the state-of-the-art fire science and engineering are inadequate. Sometimes it is because important information is not available. Often, sufficient time and resources are not available to acquire the necessary information. Normally it is a combination of all these factors.

Even with all the difficulties encountered in performance evaluations, it is possible to make credible estimates with confidence. The process requires an organized, disciplined framework for thinking as well as a willingness to incorporate whatever knowledge is available to make the best estimates possible under the circumstances that exist at the time of evaluation. Judgment is an integral part of the process. Judgment is the glue that helps to blend available information and encode expected behavior into performance estimates that can be documented if necessary.

TRANSPARENT LOGIC

The understanding that evolves from performance evaluations enhances communication with others. Decisions affecting building fire performance often are made by individuals who have relatively little knowledge of buildings and fire or fire defense behavior. It is important to be able to describe clearly and easily the significance of site-specific building features on building fire performance. Documenting the logic for building performance becomes easier with the understanding that evolves from a complete systems evaluation.

1.5 Evaluation levels

In the world of fire resource and risk management applications, decisions for routine day-to-day decisions involve uncertainty. Koen [5] defines an engineering method as the strategy for causing the best change in a poorly understood or uncertain situation within the available resources. This concept is useful for fire performance evaluations and risk management applications that routinely involve uncertainty, a need or desire for additional information, and limited available time. Although much information and knowledge exists in fire safety, technical knowledge and standardized design guidelines have yet to reach the level of other mature technical disciplines. Consequently, an important aspect of applications is to make appropriate and efficient use of the available resources of time, money, knowledge, information, available procedures, equipment, and confidence.

The idea that one size does not fit all is an accurate concept when reading this book. Practical applications that involve fire performance can have a wide range of decision-making needs and cost limitations. To accommodate this range of needs and costs, three levels of performance evaluation are described:

- level 1: basic understanding;
- level 2: detailed understanding;
- level 3: sensitivity investigation.

A level 1 evaluation develops a basic understanding of the fire performance of a building and its risk characterizations. The goal is to understand the macro performance of a site-specific building. The key word is *understand*. A level 1 evaluation enables an individual to define the problem, identify the important building features that influence its performance, describe the behavior and the basis for those expectations, and characterize the risk. A focus of a level 1 evaluation involves the ability to recognize the ingredients that are important to performance and risk and gain a sense of proportion for relative magnitudes. Another aspect is time. Level 1 evaluations are rapid. The organization of the process, selection of key details, and ways of estimating performance are important.

Because a level 1 evaluation provides the basic understanding of how the building works, it is a part of all evaluations. Future actions are based on this knowledge. Perhaps the information is sufficient to make management decisions and no additional evaluations are necessary. Perhaps a comprehensive risk management program may be needed. Perhaps the results indicate that a level 2 or level 3 evaluation is needed only for one or a few components. A level 1 evaluation gives information about the building to make informed decisions on future courses of action.

The level 1 evaluation provides the basic information for a level 2 evaluation. A level 2 evaluation focuses on details of component quality and building behavior. While the organization, general structure, and thought process are the same as for level 1, details that affect performance receive much greater attention. The level 2 framework organizes the process in greater detail and the evaluation describes their influence more accurately.

Site-specific building features, fire science, engineering, and experiential knowledge are the main sources of information for performance estimates. Because the framework forces evaluations for relatively specific and narrow conditions, judgment can be used with greater confidence to integrate knowledge and information. When important "what if" questions arise about different conditions or materials, a level 3 sensitivity analysis may be appropriate. A level 3 evaluation investigates the significance of variations on performance.

1.6 Applications

Fire safety design decisions are made with specificity. That is, the water density of a sprinkler system is xx gpm/ft^2; type yy photoelectric smoke detectors will be installed in the following locations; fire department connections to the standpipes are located zz feet to the west of the main entrance; or the local fire department will respond with two pumpers, one ladder, a chief, and a staff of twelve. Someone during the design and construction process or in the local community government will have made those decisions. Decisions such as these can have a significant influence on the building's fire performance, or they may be relatively unimportant to the outcome of a fire. An evaluation can assess the significance of any detail with respect to the building's fire performance. It uses the performance knowledge to characterize risks and to document building or building alternative comparisons in a technically credible manner.

The primary objectives of a performance evaluation are

- to *understand* the building fire performance and risk characterizations;
- to *use* that understanding to make day-to-day work decisions easier; and
- to *communicate* more effectively with others.

The organized, structured procedures will enable decision makers to understand the problem, examine details, develop ways of strategic thinking, identify and evaluate alternatives, and recognize implications of decisions. The insight that is gained by a thoughtful performance analysis enables one to discuss and evaluate alternatives and recognize clearly why one course of action may be more desirable than another.

There are two major types of application that benefit from the understanding that follows a disciplined fire performance and risk characterization analysis. One addresses the management question, How can I do better with my available resources? The other considers technical decisions: Are the fire defenses appropriate for my objectives and how can I improve their performance quality?

Some examples of resource management are:

- *Operational planning*: local fire service officers can plan appropriate and safe fire department responses to specific potential fire incidents.
- *Performance expectations*: local fire service officers can communicate the fire suppression and life safety performance expectations more effectively to owners and occupants.
- *Fire risk management*: consultants and corporate managers can formulate fire risk management programs that integrate appropriate fire defense measures, insurance selection, and loss expectations into cost-effective plans that address specific needs.
- *Emergency planning*: consultants and corporate operations personnel can formulate emergency operational procedures that are tailored to the needs and activities of the specific building being studied.
- *Resource allocation*: business managers can make more informed decisions relating to allocations between accepting losses, transferring risks through the purchase of insurance, and reducing the risk by improving the building fire performance.
- *Risk discrimination*: insurance companies can evaluate the risk characterizations for different site-specific buildings more objectively to decide if they will insure and at what price.

Examples of technical decisions may include

- *Equivalency acceptance*: local authorities having jurisdiction (AHJ) can compare building code equivalency proposals on a consistent basis.
- *Code interpretation*: local AHJ examiners can interpret the functional basis for prescriptive regulatory requirements.
- *Performance-based design approval*: local AHJ examiners can have a basis for comparing a performance-based design with the equivalent performance that would be expected from prescriptive code compliance.
- *Impairment planning*: local fire departments can document a rationale for identifying acceptable equivalency alternatives during temporary impairment of fire protection systems for servicing.
- *Cost-effectiveness comparisons*: fire safety engineers can compare cost and effectiveness among fire defense alternatives more rationally.
- *Fire reconstruction*: fire investigators can plan and conduct fire reconstructions to compare the manner in which a building actually performed with the manner in which it had been expected to perform.

- *Design alternative comparisons*: an architectural design team can compare the fire performance and associated costs that could be expected from prescribed building regulations with those which could be expected using a desired alternative design.

1.7 Road map

Some fire professionals are knowledgeable about the complete systems performance of buildings. Others may specialize in certain parts of the complete process and have less detailed knowledge about other parts. Many individuals may understand codes well, but are not as comfortable with details of equipment. The fire service may be very knowledgeable of fire ground operations whereas others who have no experience will feel uncomfortable making the estimates. Some individuals may be starting on the road of fire protection. A variety of reader backgrounds are anticipated. The materials are organized to accommodate those differences.

The destination is a way of thinking about fire and buildings. This requires a general functional understanding of the major components and the critical events that influence their behavior. It also includes an acceptance that perfection is not attainable, but we do the best we can with what we have. The more one knows, the greater the comfort level in making estimates. Often a specialist may be very comfortable with estimating performance within his or her competence, but very uncomfortable when confronted with an area outside that scope of knowledge. We cannot rely on codes and standards to replace our understanding.

The way of thinking moves back and forth between microanalysis of individual components and the macrobehavior of the site-specific building. The intent is to relate the pieces to the whole. To do this, topics are organized into two packages. Chapters 2 through 6 describe basic concepts and tools for analysis. Chapter 4 describes a building that provides a unified thread for performance analysis. If one is interested in background, Appendix A describes the relationship of this framework to traditional risk assessment tools. Appendix B discusses the rationale for using subjective probability as the basis for performance measures, and Appendix C illustrates levels 1, 2, and 3 for a single component example.

Chapters 8 through 16 focus on the individual components. Chapters 18, 19, and 20 look at risk characterizations and risk management. Specialists in component design will be very knowledgeable in individual topics. However, the purpose is to describe the functional operations that form the basis for the thought process described by the framework. A basis for estimating performance is also provided. As with all performance estimates, the more one knows, the more narrow the window of uncertainty. Information for the major factors that influence performance provide a base for evaluations. In-depth references can provide more detailed information, if needed.

The goal of this book is to describe a way of thinking. An integral part of this thinking is decision making involving numbers and a sense of proportion. This gives some difficulty in dealing with measurements and units. SI units are generally used with fire science calculations, and there is no difficulty with direct conversion between English units and SI units. However, sometimes for convenience and often because the author doesn't know appropriate standard dimensions, US fire service or construction practices have been used. Practices in different countries use local standards. For example, the usual hose length in the United States is 50 ft; this is about 16 m. A visibility of 3 m is about 10 ft and a length of 13 ft is about 4 m. When the goal is conceptual association rather than precise correlation, a sloppiness has been used in conversions to convey a sense of proportion. Whether proper or not, this sloppiness is intentional. I hope that readers will forgive these shortcomings. Bon voyage.

References

1. *SFPE Handbook of Fire Protection Engineering*, 3d ed. Copyright by SFPE 2002, and published by NFPA.
2. *Fire Protection Handbook*, 19th ed. Published by NFPA, 2003.

3. NFPA Standards. The National Fire Protection Association publishes a large number of consensus fire standards. Many of these standards provide excellent background information to evaluate component performance and risk management.
4. Olenick, S. M. and Carpenter, D. J. An updated survey of computer models for fire and smoke. *Journal of Fire Protection Engineering*, Vol. **13**, No. 2, May 2003.
5. Koen, B. V. *Definition of the Engineering Method.* American Society for Engineering Education, Washington, D.C., 1985.

2 FIRE DEFENSES

2.1 Introduction

Traditional fire defenses provide a tool kit for protecting buildings. Their selection and location as well as the quality of installation and maintenance has an important influence on building performance. Evaluations involve understanding individual component (micro) behavior within an interactive (macro) behavior of the complete building system.

This chapter lists the fire defenses and gives a brief description of their function and use. The objective is to provide a terminology and basic understanding of their use. Then, when we describe the analytical framework in later chapters, the physical relationships between the fire and the quality and performance of a specific defense can be put into perspective.

2.2 Building fire defenses

The organizational framework for a fire performance evaluation incorporates all of the site-specific building and fire protection features that contribute to the building's behavior. The active and passive fire defenses provide a structure for that framework. Some terminology may be slightly different from common usage to establish an unambiguous representation of the function.

ACTIVE FIRE DEFENSES

Definition: A device or action that must receive a stimulus to act in a real or a perceived fire condition.

Traditional elements

1. Automatic detection and alarm system
   a. Fire detection
   b. Alarm process
   c. Occupant-alerting process
   d. Fire department notification process
   e. Equipment activation

Building Fire Performance Analysis R. W. Fitzgerald
 ISBN: 0-470-86326-9 (HB)

2. The automatic sprinkler system
3. Fire department operations
   a. Fire suppression
   b. Search and rescue
   c. Property protection and salvage
4. Trained building fire brigade
5. Special hazard, automatic suppression systems
   a. Halon systems (being phased out)
   b. Carbon dioxide systems
   c. Dry chemical systems
   d. Foam agent systems
   e. Water-spray systems
6. Occupant activities
   a. Detection, occupant alerting, and fire department notification
   b. Extinguishment
   c. Property protection
   d. Information receiving or transmitting
   e. Help others
   f. Escape
7. Special features
   a. Smoke management
   b. Automatic elevator recall
   c. Automatic closing of selected doors or ducts
   d. Emergency lighting and signage
   e. Communication devices
   f. Emergency power

PASSIVE FIRE DEFENSES

Definition: A building component that remains fixed in the building whether or not a fire emergency exists.

Traditional elements

1. Insulation of structural elements to prevent failure.
2. Barriers to prevent extension of the flame/heat or smoke/gas from one space to another.
3. Opening protectives in barriers, such as doors or dampers, to inhibit the movement of flame, heat, smoke, or gases into the adjacent space.
4. The egress system.
5. The fire attack routes.

These defenses are discussed briefly here. Subsequent chapters that describe the analytical framework and evaluations provide more detailed and functional descriptions of the defenses.

2.3 Active fire defenses

The preceding list of common active fire defenses is an inventory for building analysis or design. A description of their characteristics provides a useful base for understanding building fire performance.

FIRE DETECTION AND ALARM SYSTEMS

Fire detectors are small devices that sense the intrusion of combustion products. Detectors can be designed to sense heat, flame, smoke, or other products of combustion. Performance effectiveness depends on their location relative to the fire, the movement of the combustion products, and detector sensitivity. These instruments may detect fires ranging in size from very small to relatively large.

A significant potential failure mode is a high unwanted nuisance alarm rate (often called false alarms) that may lead to an inadequate or inappropriate response. These nuisance alarms may be due to inadequate maintenance, a sensitivity that is not appropriate to the environment, or a malfunction of circuitry or signals. A second type of failure occurs when a detector fails to actuate when enough products of combustion reach the sensing area. A third type of failure involves locating the instruments in positions where the combustion products are delayed or prevented from reaching the detector as rapidly as desired. When this occurs, the fire becomes much larger than may be intended for the expected fire defenses.

Detection and alarm services have several uses. The most common is to sound an alarm or other signal intended to alert occupants of a fire emergency in the building. This is called a local alarm system. A second type of activity transmits a signal to a location within or outside the building to summon help. A third type of operation occurs when the alarm triggers activation of other fire defenses such as closing doors, automatic fire suppression, or smoke control systems.

Notification of the fire department using alarm systems can take place in one of four ways. One is with a *proprietary, supervisory station system* that transmits a trouble supervisory or alarm signal to a central location. The location may be within the building being protected or another building controlled by the property owner. Proprietary supervising station fire alarm systems are common in large buildings or with owners of multiple buildings. Owners of these properties normally have a security force and desire to monitor their own security, HVAC (heating, ventilating, air conditioning), fire alarm systems, or other building operation systems.

The second and third ways to notify a fire department are by transmitting a trouble supervisory or alarm signal to a *central station* or a *remote servicing station* location. A central station is a facility staffed 24 hours a day and operated by a commercial company that provides trouble supervisory and alarm monitoring and service for buildings. A central station will retransmit an alarm to the fire service communication center and also provide repair services for the equipment. A remote servicing station is normally an alarm-monitoring facility at the fire service communication center. When located here, the fire department often transmits trouble and supervisory signals to someone else who will respond to those signals. When the fire department does not wish to directly monitor the fire alarm systems in its jurisdiction, it may designate a private company to provide the service.

An *auxiliary fire alarm system* is a fourth type of fire department notification. An auxiliary system has direct connections from the building's fire alarm system to the fire service communication center. This type of service provides the fastest notification to the fire department.

In building evaluations it is useful to distinguish the different functions that occur after detection by terms associated with specific performance activities. Therefore the evaluation functions are to *alert* occupants, *notify* the fire department, and *release* other fire protection devices. Although these terms are not a part of present codes or the vernacular of the trade, they provide an unambiguous identification of functional performance.

THE AUTOMATIC SPRINKLER SYSTEM

An automatic sprinkler system consists of a water supply, control valves, a piping distribution system, sprinklers, and a monitoring system for trouble and alerting alarms. The objective of the sprinkler system is to extinguish or control a fire within a size that will not be a threat to people, property, or operational continuity. Sprinkler systems have a broad range of design alternatives that allow one to tailor the hardware to meet the needs of a variety of hazards and environmental conditions.

FIRE DEPARTMENT OPERATIONS

The fire department is an important part of the active fire defense system for a building. Many buildings rely exclusively on a combination of barrier effectiveness to delay fire propagation and the fire department to extinguish a hostile fire.

A building evaluation integrates the building systems and the local fire department in the analysis. The function of an analysis is to understand and identify the building's site, architectural, and fire suppression features that influence local fire department operations. On the building fire ground, a local fire department has three major roles. One is fire extinguishment, the second is search and rescue, and the third is property protection and salvage. The combination of local community resources and the building itself forms the basis of understanding and evaluation of the performance.

Community fire emergency resources can vary broadly within a geographic region. Some communities may send four engines, two ladders, a rescue squad, and a chief to a first alarm. A neighboring community may have a standard first alarm response of two engines, one ladder truck, and a chief. Still other communities may respond with a single engine. The number of rooms and the fire size that a fire department can extinguish with an aggressive interior attack varies with the number of fire suppression companies that are available. The community water supply is another important factor. Most urban communities in North America and Europe have underground water distribution systems and hydrants for water supply. Many suburban and rural communities rely on ponds or streams, tank truck water supplies, or private water supplies provided at the building site.

Building and site features have a major influence on fire extinguishment capabilities. Access of fire apparatus to the building is an important consideration. Landscaping, terrain, or adjacent buildings can influence fire-fighting operations significantly. The number and location of fire hydrants and the water availability are important.

Building features influence fire fighting by helping or hindering fire ground operations. A clear and ready access to the building from several possible locations is one factor. Another is an ability to locate the fire quickly and accurately. To help in this task, annunciator panels aid in locating fire detector or sprinkler actuation.

The ability to apply water to the fire quickly is critical to interior fire fighting. Interior standpipes, which are vertical water pipes usually located in stairwells or corridors, reduce the need for laying hose lines great distances. Standpipes may be wet or dry. Wet standpipes are filled with water for rapid availability whereas dry standpipes must be charged by the fire department from an exterior water source. Standpipes have hose connections at each floor level.

The architectural layout influences fire fighting. Fire fighters must pull hose lines through or around architectural obstacles, such as doors, stairs, lengthy corridors, and corners. The number and condition of these architectural obstacles influences the time required to stretch a hose line. The time to apply water to a fire is important because as time increases, the increased fire growth makes extinguishment more difficult. Fire department staffing available for suppression has an important influence on the duration for applying water to the fire.

Another important activity of fire department operations is search and rescue. Life safety is the most important concern of the fire fighters. If it appears that lives may be at risk, a fire chief will direct available fire fighters to search the rooms and rescue any trapped occupants. This activity requires many of the available fire fighters.

A final fire ground activity is property protection and salvage. Most commonly, this involves protecting exposed zones within the fire building or neighboring exposed buildings from fire extension. Another activity may entail protecting property from water damage by moving the contents to a central location and covering them with tarpaulins. Although salvage after a fire was a routine operation in earlier times, this activity is not usually done today.

TRAINED BUILDING FIRE BRIGADE

Some buildings have a trained building fire brigade that is available to provide a more rapid initial fire attack. This early fire attack supplements the actions of the local fire department. Building fire brigades are common in larger industrial plants and rare in most other occupancies. In recent years their use has been diminishing. The capability of building fire brigades varies enormously. Some building fire brigades are mini fire departments with proper equipment and training. Other building fire brigades provide an inappropriate sense of security because they devote few resources to staffing, training, experience, and equipment. This contributes to ineffective fire fighting.

SPECIAL HAZARD AUTOMATIC SUPPRESSION SYSTEMS

Some buildings house materials that are easily ignitable and fast burning. Flammable liquids, some gases, and certain chemicals fall into this category. Usually the building's functional operations use these materials. Dip tanks, machining processes, spray booths, storage tanks for flammable liquids, and deep fat fryers for cooking may present special hazards in manufacturing or commercial buildings. Often these hazards are located in smaller areas within larger operational spaces, therefore automatic protection for these locations is called spot protection or specific hazard protection.

Automatic extinguishing systems tailored to the hazard provide quick extinguishing action and keep operational downtime to a minimum. The most common types of special hazard extinguishing systems are carbon dioxide (CO_2), dry chemical, foam, and water-spray systems. Halon systems had been popular for many years, but have been phased out because of the effect of halon on the atmospheric ozone layer. Nevertheless, many halon systems still exist today. Special hazard systems supplement the fire protection and are not considered as a substitute for the automatic sprinkler system. A major factor in the evaluation of special hazard systems involves the definition of the hazard and its characteristics. Assessment of the extinguishing agent capability and its design effectiveness to suppress this hazard requires a good knowledge of the hazard, extinguishing agent, and storage and delivery system.

OCCUPANT ACTIVITIES

The occupant can contribute to some active fire defenses. While escape is the advice most often given when a fire occurs, the occupant may frequently provide assistance to limit the size of the fire or protect other people and property.

Humans are sensitive in recognizing and discriminating a fire's combustion products. Occupants often become a critical link in the detection and notification process. The human interface is the most common means to notify the fire department of a fire. In addition, occupants may alert other occupants of the existence of a fire emergency and also assist in evacuation of others.

Building occupants often provide information to the fire department on the fire's location and potential dangers for others remaining in the building. When an individual can give detailed information about the existence and location of critical business operations, the fire ground officer may be able to provide defensive protection for those important spaces exposed to the fire.

Occupants often extinguish small fires early in the fire growth process. An occupant can be very effective in extinguishing small smoldering and early flaming fires. Fire extinguishers are intended for occupant use to put out small fires.

SPECIAL FEATURES

Buildings can provide many special features that help in fire suppression, egress, and emergency operations. Changing an air-handling system into an emergency smoke management mode is becoming a common feature in many building designs. The smoke management system often attempts to exhaust smoke from fire floors and to pressurize other rooms, such as stairwells and corridors, preventing smoke from migrating into those spaces.

A design may integrate a variety of other features that can enhance life safety features. For example, emergency lighting and signage can help occupants and fire fighters to move through the building with a better sense of direction and confidence. Communication systems for alerting occupants and providing emergency information are available. Many buildings have an automatic elevator recall so that the fire department can control the use of elevators to reduce damage and life loss. Detection instruments can trigger the closing of selected doors held open by magnetic devices and the shutting of dampers in air ducts. Reliability can be enhanced by using emergency electrical systems that operate when the building's normal power supply is interrupted.

2.4 Passive fire defenses

Passive fire defenses do not change their mode of operation after a fire occurs or is perceived. Often, passive fire protection features provide additional time during which the active defenses can become effective. They also provide movement routes for occupants and emergency forces.

STRUCTURAL FIRE PROTECTION

All building materials deteriorate under the high temperatures produced by a fire. Besides high temperatures, the duration of heat energy application causes progressive deterioration to the point of collapse. The objective of this type of passive fire protection is to delay collapse of the structural system beyond the point where the fire has burned itself out.

Structural steel, reinforced concrete, timber, and prestressed concrete are the most common structural materials used in construction. Techniques for providing protection against the heat energy transfer enable the structural frame to resist excessive deformation or premature collapse. The type of protection depends on the type of the structural framing system.

Unprotected (i.e., exposed) structural steel loses strength very rapidly in a fire. Methods are available to delay the heat energy transfer from the fire gases to the structural frame. Mineral spray-on materials or gypsum board coverings usually protect steel beams and girders. Concrete encasement was once a common method of providing structural fire protection but it is not common today. Intumescent coatings have value for specialized applications.

A common way to provide protection for joist construction uses a suspended membrane ceiling. A membrane ceiling delays excessive heat accumulation in the space between the ceiling and the floor above. Membrane systems should be constructed to allow thermal expansion of the ceiling framing without collapse. Lighting fixtures and air diffusers must be compatible with the ceiling design and construction. The entire system of structural framing, suspended ceiling, and lighting and air diffusers is described as a floor–ceiling assembly or a roof–ceiling assembly.

Reinforced concrete uses a minimum thickness of concrete cover to protect the steel reinforcing bars. Prestressed concrete is protected in the same way.

Timber can lose strength upon heating, and the material also augments the fire if it burns. When wood burns, a char forms to provide some insulation to continued pyrolysis and delay collapse for a time. Encasing the wood with gypsum boards is a common way to protect wood members.

BARRIERS

We define a barrier as any surface that will delay or prevent products of combustion from propagating into an adjacent space. This definition is broader than traditional descriptions because we are interested in the full range of barrier behavior rather than the fire resistance requirements of a standard test. Barrier performance evaluates the prevention or time delay of flame/heat or smoke/gas movement into an adjacent space.

Typical barriers from a code viewpoint are floor–ceiling and roof–ceiling assemblies, interior partitions that separate building fire area, fire walls, and enclosed corridors and shafts. Fire walls are specially constructed barriers intended to prevent any extension of fire from one side to the other. Designers attempt to ensure that a fire wall remains stable if a fire destroys the structural framing on one side of the wall.

Code-defined partitions and fire walls specify a fire resistance rating based on the ASTM E 119 (or ISO 834) standard fire endurance test. The test classifies construction systems according to their relative ability to prevent fire propagation from one space to another. Classifications are based on a standardized fire and expressed as a time in hours. Many in the construction business are under the misconception that the hourly rating represents the time a barrier assembly will last in a building fire. The assembly may fail in a considerably shorter time or it may be successful for a considerably longer time. Construction and field conditions influence barrier performance.

A standard test ends when a barrier fails by cracking, collapse, or excessive thermal transmission. If one of those conditions occurs during a building fire, ignition can occur in an adjacent space. However, the type of ignition and the continuum of barrier behavior are important to building performance. When failure occurs by a small crack or excessive thermal transmission, the fire growth in the next room is slower, and emergency forces can extinguish the fire more easily. On the other hand, when ignition into the next room occurs through a large hole or partial barrier collapse, fire growth is rapid and emergency forces are unlikely to extinguish the fire before it fully involves the room. Consequently, performance evaluations use the full range of barrier behavior.

OPENING PROTECTIVES

Many barriers have openings to allow the passage of people, light, heat, air, plumbing, electricity, or other building services from one room to another. The fire resistance will change whenever a barrier is opened or penetrated. Opening protectives, such as doors, windows, shutters, dampers, and fire stopping restore some fire resistance to barriers.

THE EGRESS SYSTEM

A building is expected to provide a safe path for occupants to leave. A means of egress is defined as a continuous and unobstructed path from any location in a building to a public way. A means of egress consists of three separate and distinct parts. The first is the *exit access*. The exit access is the path that connects any location in the building with an exit. Typically, corridors become the exit access for most buildings. The second component is the *exit*. An exit is a specially protected space separated from other parts of the building by construction that can provide a temporary area of refuge for occupant movement to the exit discharge. Stairwells and exterior doors are often the building exits. An *exit discharge* is the third component that terminates the means of egress. The exit discharge leads to a public way, usually outside the building.

A means of egress is a designed path to the outside that incorporates additional safety features. While no building code identifies a level of safety, all codes give rules for constructing egress paths. Many buildings do not have all the features of a code-complying means of egress. These nonconforming routes may be described as "ways out." While a layman's perception that any way out is an "exit," this is not the technical definition of the term.

FIRE ATTACK ROUTE

For most practical situations, a fire attack route is the inverse of the egress path. Fire fighters must transport themselves and their equipment to the fire location. This often involves movement in a direction opposite to the occupants who may be leaving the building.

An interior fire attack is usually launched from the stairwells that are usually exits. Stairwells house standpipes and provide some protection to fire fighters from heat and smoke. This enables fire fighters to pause, secure their self-contained breathing apparatus (SCBA), and prepare to place hose lines for water application.

2.5 Closure

The building architecture and construction are the basis for functional performance evaluations. The fire defenses are just one part of the complete building/fire system. A fire performance analysis moves back and forth between understanding the quality and operation of individual (micro) components and the (macro) performance of the building as a whole. A systems way of thinking integrates component functions and the complete building performance. The goal of a performance analysis is to understand the building. A systems understanding of performance enables one to identify creative alternative solutions, make better decisions for courses of action, and communicate more effectively. The fire defenses described in this chapter provide one set of tools to understand building fire performance.

3 BASIC CONCEPTS

3.1 Introduction

The goal of a performance evaluation is to understand the building/fire system behavior during a fire. The understanding evolves from a methodical examination of the fire and its effect on the active and passive fire defenses. An analysis examines the components in isolation and then combines related parts to provide an interactive picture of performance. The dynamic nature of the system causes different components to phase in and out of action over time.

The process systematically organizes the functional behavior of the building/fire system to describe individual and interactive performance. It involves an analytical structure to guide the thought process and encodes relevant information to estimate behavior. This chapter identifies concepts and definitions needed to structure consistent performance evaluations.

3.2 Concepts and definitions

SPACE/BARRIER ORGANIZATION

All buildings are assemblies of spaces and barriers. The functions and uses of the building are the major influences in establishing the locations and orientations of the barriers that define the volumetric spaces. A *space* is the volume enclosed by barriers. A *barrier* is any surface that can delay or stop the movement of flame/heat or smoke/gas through a building.

Normally spaces are the rooms of a building and barriers are the walls, floors, and ceilings. Figure 3.1 shows this pattern. However, we need some flexibility in defining spaces and barriers to adapt the evaluation process to a variety of applications. For example, Figure 3.2 shows a section of a building with suspended ceilings and partitions terminated at the ceiling level. Fire can propagate from room 1 to room 2 through the separating partition (path A). Alternatively, the fire can enter room 2 along the path through the suspended ceiling (barrier) into the plenum space and back through the ceiling into room 2 (path B).

To use a consistent analysis, we define any potential fire propagation path as a sequential grouping of spaces and barriers regardless of architectural classification. For example, in fire reconstructions, propagation may occur through concealed spaces such as between studs of a wall. In this situation an initial space is a room. Then the barrier is the gypsum wallboard surface

Building Fire Performance Analysis R. W. Fitzgerald
 ISBN: 0-470-86326-9 (HB)

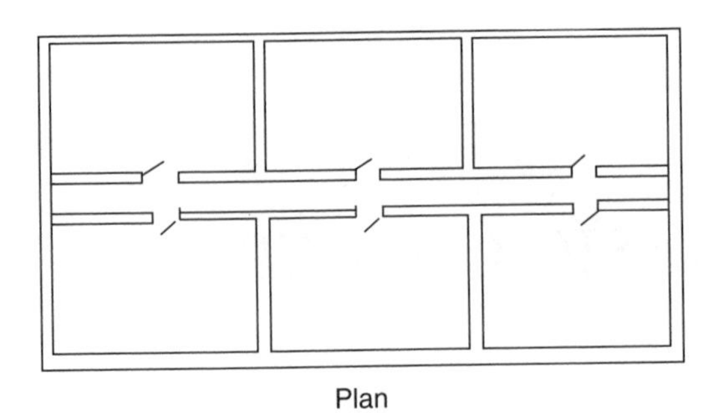

Figure 3.1

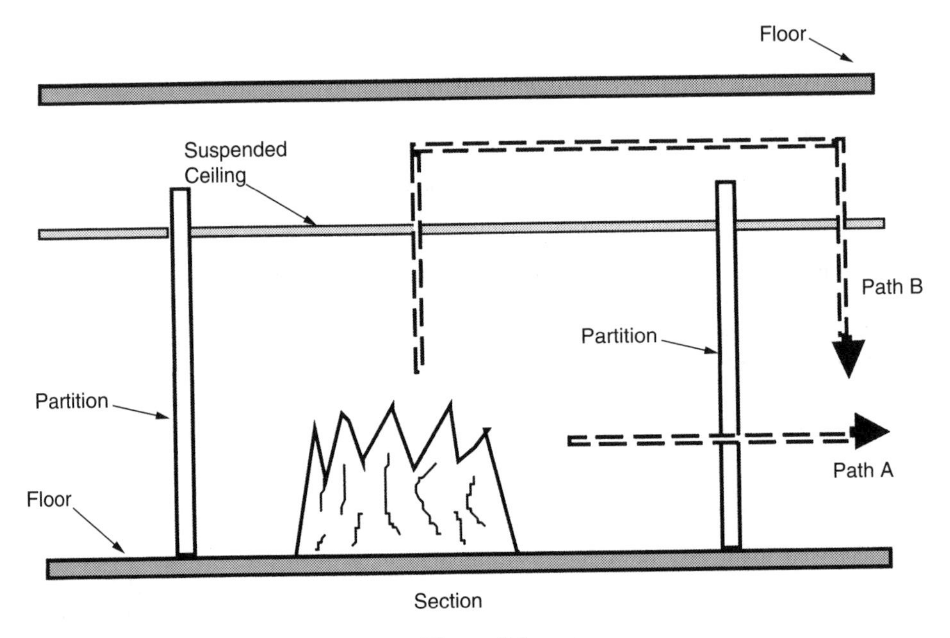

Figure 3.2

and the next space is the volume between the studs. Fire stopping becomes the next barrier along the propagation path.

Another example of the space/barrier concept is illustrated by a large, open-volume room such as a warehouse, department store, or manufacturing facility. A large, unconfined space may be broken into a group of smaller zones, as shown in Figure 3.3. This zoning allows one to analyze the effect of fire origins in different locations and the ability of a fire department or a sprinkler system to confine a fire to a specified zone. The boundaries of the zones are defined by *zero-strength* barriers. A zero-strength barrier is a fictitious plane that has no resistance to the passage of heat or smoke. This technique enables one to use the space/barrier concept for analytical consistency and time sequencing in complicated situations.

FIRE

For analytical purposes, fire is separated into the combustion products of flame/heat and smoke/gas. Each impacts the building, its occupants, and the contents at different speeds and

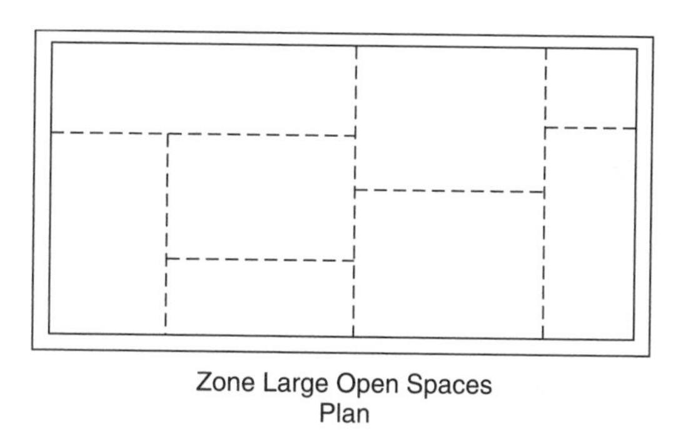

Zone Large Open Spaces
Plan

Figure 3.3

in different ways. A building analysis evaluates the impact of each component as separate but related parts.

IGNITION

We define ignition (IG) as the self-sustained burning of the target fuel. To describe this event more specifically, we define ignition as the appearance of the first fragile flame on the fuel surface. For our analyses, overheat and smoldering occur before ignition and are not included in this definition of ignition.

ESTABLISHED BURNING

Established burning (EB) is defined as the fire size that starts the building analysis. EB is selected to reflect the needs and conditions of a specific space or zone for the building under consideration. Examples of EB might include

- a preselected fire size that marks the beginning of a more predictable fire growth scenario (e.g., one may envision a wastepaper basket fire 10 in (250 mm) in height or 20 kW as EB);
- a selected fire condition (e.g., an oil-spray fire size);
- a defined smoldering or smoke production condition (e.g., a cable tray fire);
- a spark in an explosive atmosphere.

Established burning identifies the demarcation between fire prevention and the building fire design.

Established burning is an important concept. For compartmented buildings, a flame size of about 20 kW (about 250 mm in height) seems most appropriate for EB. It is difficult to forecast fire behavior for sizes below 20 kW because so many factors influence the initial combustion characteristics. However, when the combustion process has become established, its future behavior can usually be predicted with more confidence.

FULL ROOM INVOLVEMENT

Full room involvement (FRI) is a condition where the exposed combustible surfaces in a room are burning. Often this happens suddenly due to a very rapid ignition of hot fire gases that simultaneously ignite all combustible surfaces. In fire literature this condition is usually described as *flashover*. When flashover occurs, the fire in the room is rapidly transformed into a fully involved room fire.

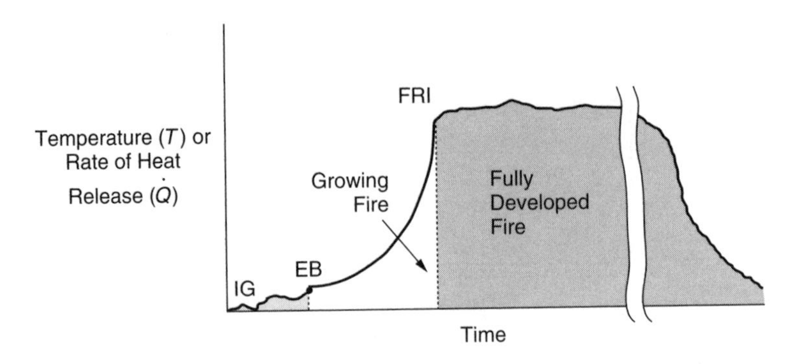

Figure 3.4

Full room involvement can also occur by spreadover. Spreadover involves successive ignitions of adjacent fuel packages. The fire moves through a space by progressive flame movement rather than by a rapid flashover. In some configurations of fuel and room geometry, both may occur at about the same time. For our purposes, we shall consider full room involvement (FRI) and flashover to be synonymous and to represent the same condition.

FULLY DEVELOPED FIRE

After full room involvement (FRI), the fire will burn for an extended time until the fuel is nearly consumed or until it is extinguished. The fully developed fire, sometimes called the postflashover fire, burns at a generally uniform rate for an extended time. This segment of a fire history can affect the structural frame and the barriers significantly. The fire endurance of assemblies and structural members is based on the heat energy produced by a fully developed fire.

Figure 3.4 illustrates a time–temperature (or time–heat release rate, $\dot{Q}$) history for a representative room fire. It shows the positions of ignition (IG), established burning (EB), full room involvement (FRI), and the fully developed fire in the process.

BARRIER PERFORMANCE

A barrier is any surface that can delay or prevent the movement of flames or smoke from one space to the adjacent space. A barrier can be combustible or noncombustible, complete or incomplete, penetrated or unpenetrated, load-bearing or non-load-bearing, fire rated or not fire rated.

At any instant of time, a barrier can be in only one of three states. Considering the ignition potential on the unexposed side, these states are as follows:

- The barrier is successful (B). Ignition does not occur in the adjacent space.
- The barrier experiences a small, localized failure ($\overline{T}$ failure) that can cause an ignition in the adjacent space. If the potential for ignition exists, even if fuels are not present, a $\overline{T}$ failure is defined to exist. Examples of a $\overline{T}$ failure might include the following: one or more cracks may open; a small hole develops; the hole may have existed before the fire; or sufficient heat can be transmitted through the barrier that an ignition could occur on the unexposed side. If a $\overline{T}$ failure occurs, the hot-spot ignition would result in a normal fire development in the adjacent space.
- The barrier experiences a large failure ($\overline{D}$ failure) that can cause a massive influx of fire gases into the adjacent space. Here a large opening (such as an open door, a broken window, or a partial collapse) will allow a substantial influx of heat energy into the adjacent room. This

massive influx of heat energy from a $\overline{D}$ failure would result in full room involvement within a short time.

Barrier performance depends on construction, the heat energy application (fire severity), and the fire duration. Barriers deteriorate over time as a fire continues to burn. Consequently, the particular status of a barrier can change as the fire continues. A common progression might start with a successful barrier performance (B) that withstands the fire heat energy for a time. After continued exposure to the heat energy, the barrier may develop a small, hot-spot ($\overline{T}$) failure. As the fire continues to burn, the small failure may enlarge until it becomes a massive ($\overline{D}$) failure as the heat energy continues to be applied against the barrier surface.

A building performance evaluation incorporates the conditions that exist or are likely to exist in the field. This includes all of the construction features that may be present. Thus, features such as an open door, unpacked through construction, or a window become part of the barrier analysis. The barrier has a significant influence on the fire performance of spaces beyond the fully involved room. Therefore an evaluation must be able to accommodate differences in condition relatively easily to gain a good understanding of building performance. Fortunately, a methodical barrier analysis can easily incorporate differences in barrier conditions into field performance predictions.

LIMIT OF FLAME MOVEMENT

The limit of flame movement is the extent to which the fire spreads before it is terminated. The term "limit" (L) may be applied to the extent of fire spread in a space or to the extent of fire spread in a building. The context will make the meaning clear.

Within a space, the fire may be terminated (limited) in several ways:

- the fire goes out itself (I);
- automatic sprinkler suppression (A);
- fire department extinguishment (M).

Within a building, the fire may be limited in several ways:

- termination within the space of origin (L_1);
- successful barrier performance that prevents movement to an adjacent space (B);
- termination within a barrier/space module beyond the space of origin (L_n).

After the constituent parts are evaluated, one can combine related outcomes to produce a composite picture of the extent of expected flame/heat damage.

STRUCTURAL FRAME BEHAVIOR

Knowledge of structural frame stability is important in risk characterizations for occupants and fire fighters as well as for building property and operational continuity after the fire. The structural frame may support or be a part of barriers that delay or prevent fire propagation to additional spaces. A separate structural analysis for the expected fire conditions provides an understanding of how the supporting framework will perform.

Failure of the structural frame may be defined as either collapse or excessive deformation. The structural frame evaluation (Fr) describes the time duration for which a structural member can support a specified load without failure.

SMOKE/GAS TENABILITY

Smoke and fire gases are risks for humans. Smoke and gas contamination may also be a risk to electronic data storage and in the operation of sensitive equipment. Measures for these combustion products must be defined to evaluate the tenability performance (Sm) for a building.

TARGET SPACE

A target space is a room where people may be located during a fire. For example, a target space may be a corridor segment or a building exit through which an individual may transit. A target space could also be a room in which an individual may be confined (e.g., hospital or prison), or a designated area of refuge to protect an individual during the fire emergency. With equipment or information protection, the target space is the room that houses those items.

3.3 Performance evaluations

Fires are dynamic events. Time durations and the sequencing of different components are critical to understanding a building's performance during a fire. Performance evaluations require us to order and understand time line sequencing for the components.

BUILDING PERFORMANCE EVALUATIONS

A performance evaluation involves an organized procedure to understand and describe

1. The limit (extent) of flame/heat damage (the L curve). This includes
   a. The fire
   b. Barrier effectiveness to limit fire propagation
   c. Sprinkler system effectiveness in controlling or extinguishing a fire
   d. Fire department extinguishment
2. Times during which selected target spaces remain tenable (Sm curve)
3. The structural frame's ability to avoid failure (Fr curve)

All of the active and passive fire defenses and other building features that affect fire performance are incorporated into one or more of these parts.

RISK CHARACTERIZATION

Understanding the building performance for flame movement, smoke movement and contamination of target spaces, and structural frame behavior enables risk characterizations to be made for

- people, during egress and defend in place;
- property;
- operational continuity (mission);
- threat to neighbors;
- threat to the environment.

FIRE PREVENTION

Fire prevention includes ignition prevention and the prevention of established burning after ignition. Occupant extinguishment is incorporated into the prevent EB analysis.

Fire prevention effectiveness has an important role in the total system of fire and buildings. However, we uncouple prevention from the building's performance given EB for two reasons. The first is that prevention practices can mask significant shortcomings in a building design. A clear understanding of the manner in which a building is likely to perform and the risk characterization for occupants and operations given ignition and established burning should be recognized and clearly understood. The second reason for uncoupling the prevent EB component from a building performance evaluation is that remedial practices and temporary recommendations can involve fire prevention actions that differ substantially from routine activities. A clear understanding of the impact of recommendations on the development of a special-purpose prevent EB plan often enables a better risk management program to be developed.

4 THE ANATOMY OF BUILDING FIRES

4.1 Introduction

Buildings and fire form a complex system involving dynamic changes in fire behavior, coupled with the actions of active and passive fire defenses as well as interdependencies and interactions with building systems and human activities. Every fire is different, and each building poses a different set of conditions that affect fire performance. Nevertheless, one may develop an organizational framework that examines the parts in a consistent, methodical manner. Viewing each building with the same systematic framework contributes to the development of a process that enhances fire performance understanding and allows one to make comparisons.

This chapter introduces an organization for evaluating building fire safety. The description of a fire in a small building will serve as a model for identifying the analytical framework. This fire scenario is a composite of observed buildings and fire case studies. Some buildings will perform better in a fire, others will perform worse; different fire locations will pose different problems. Nevertheless, the fundamental analytical process remains the same. Only the performance measures change. This example will illustrate a variety of applications throughout the book.

4.2 The building

As an introduction to performance evaluations, we will describe a building and fire scenario. This scenario is used to structure a framework that may be used for all performance evaluations.

Figure 4.1 shows a site plan for a small, four-story apartment building. Each living unit houses one or two individuals ranging in age from 52 to 95. The ground floor includes a lobby, office, storage and building services, and seven apartments. The upper three floors contain 12 apartments per floor. An elevator is in the lobby at the center of the building and stairwells are located at each end. Laundry rooms and trash rooms are on each upper floor opposite the elevator.

The ground floor is constructed with a concrete slab on grade and concrete masonry block walls. Floor–ceiling assemblies for the upper three floors are constructed with 2 in × 10 in wood joists at 16 in c/c supporting a 1/2 in layer of plywood. This plywood surface supports a 1 1/2 in layer of concrete, a carpet pad, and a carpet. A 1/2 in layer of fire-rated gypsum wallboard attached to metal runners forms the ceiling of the assembly.

Building Fire Performance Analysis R. W. Fitzgerald
 ISBN: 0-470-86326-9 (HB)

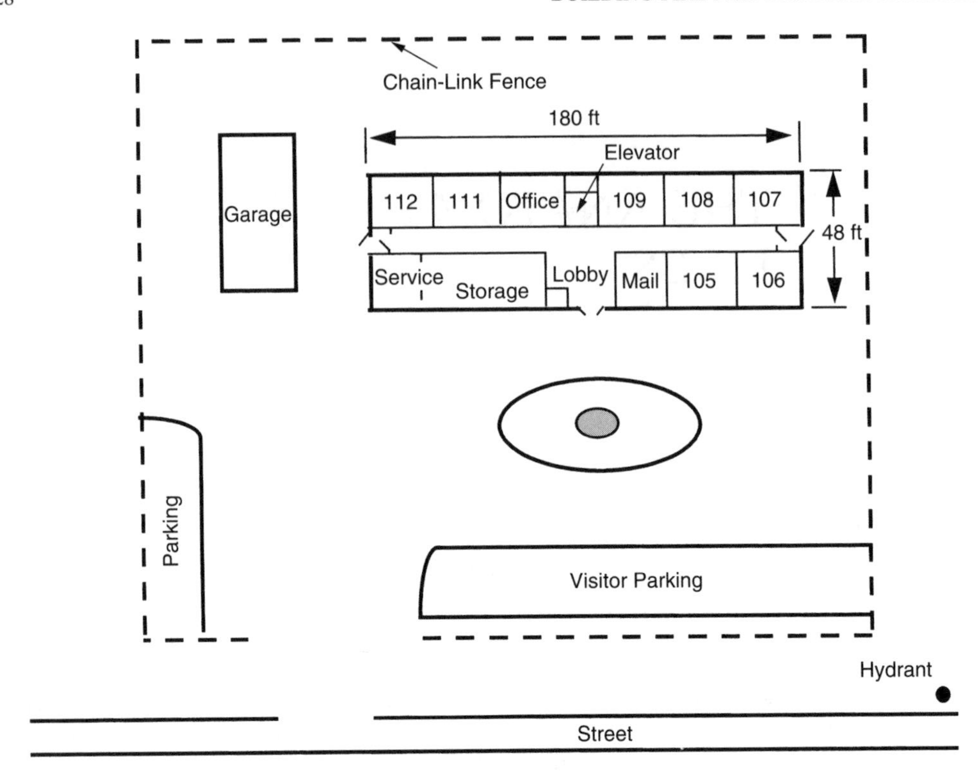

Figure 4.1

The exterior walls are constructed of 2 in × 4 in wood studs spaced 16 in c/c with one layer of 1/2 in gypsum wallboard on each side. Fiberglass batt insulation fills the voids between the studs. On the exterior of the building, the outer layer of gypsum wallboard is covered with a sheet of 3/4 in thick non-fire-rated plywood.

Interior walls between the apartment units and the corridors are constructed of 2 in × 4 in wood studs covered on each side with 1/2 in fire-rated gypsum wallboard. Fiberglass batt insulation fills the voids between the studs. Interior walls within the apartments are constructed of 2 in × 4 in wood studs covered on each side with 1/2 in gypsum wallboard. The wall finish within the apartments may be painted wallboard, wallpaper, or 1/4 in wood paneling, depending on the taste of the tenant. The corridor ceilings are painted, and the walls are covered with wallpaper. Entrance doors to the apartments are 1 5/8 in solid core wood doors.

The roof framing consists of pitched trusses having a 20 in overhang and spaced 24 in on center. The trusses have 2 in × 6 in top and bottom chords and 2 in × 4 in web members. Connections are toothed connector metal gusset plates. A 5/8 in plywood roof sheathing is covered with asphalt shingles. The soffit of the overhang is covered with 1/2 in plywood. Figure 4.2 illustrates the construction. The roof void space is insulated with blown-in cellulosic insulation. Void space air vents are placed on the gable ends of the building and at frequent intervals in the overhang soffit.

Figure 4.3 shows a typical upper floor plan and Figure 4.4 shows a typical apartment layout. Figure 4.5 shows a partial elevation. Each apartment has a small balcony off the living room, and each bedroom contains a 6 ft × 4 ft window in the exterior wall. A forced-air system from a central heating unit in the service area of the ground floor supplies heat to each apartment.

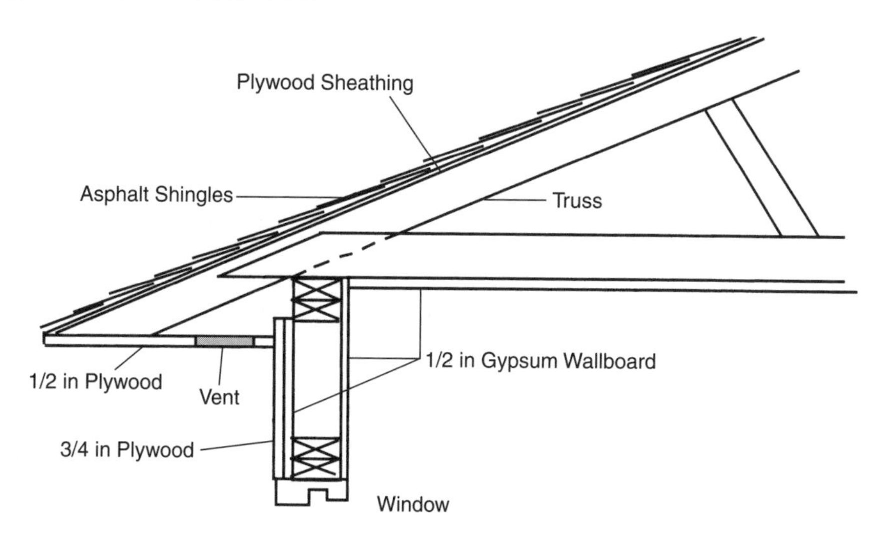

Figure 4.2

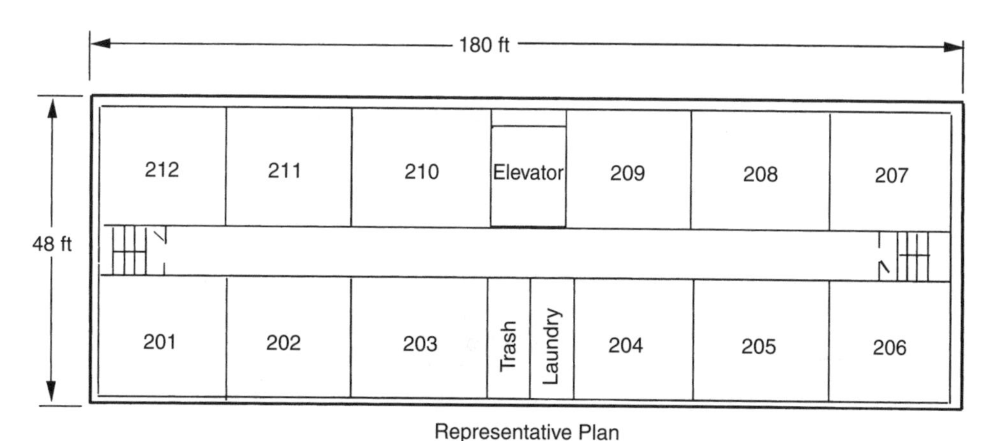

Figure 4.3

4.3 The fire protection systems

- Each apartment is equipped with a single-station, battery-powered smoke detector that will sound an alarm only within the apartment.
- Fire doors are installed at the exits and at each side of the elevator. These doors are held open by steel wires connected by fusible links.
- Three 3A-40BC dry chemical fire extinguishers are located in each of the two stairwells.
- No automatic detection and alarm system is present except the single-station smoke detectors in individual apartments.
- No manual fire alarm system is in the building.
- No closers are installed on the apartment doors to the corridor.
- No automatic sprinkler system is present.
- No fire-fighting standpipe is installed.

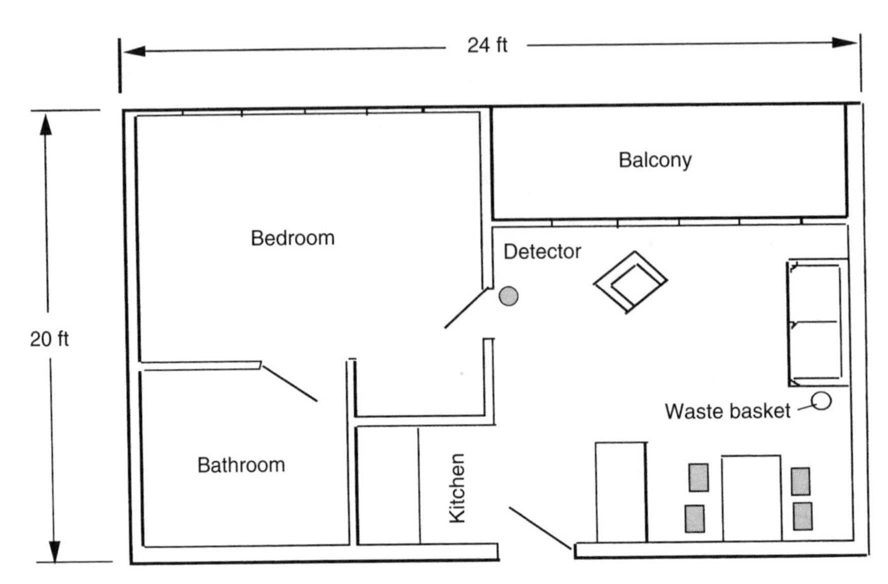

Representative Studio Apartment

Figure 4.4

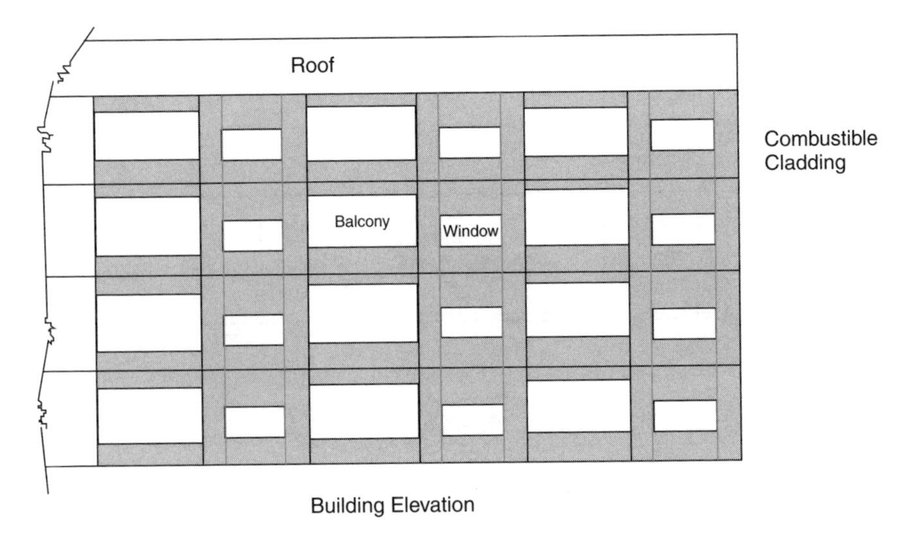

Building Elevation

Figure 4.5

- The local fire department has a complement of 62, and an on duty staffing of 13 people.
- The department runs two engines and one ladder truck and provides paramedic-level emergency medical services.

4.4 The fire

At approximately 5:00 A.M. the maintenance manager for the building was collecting rubbish from the trash rooms on each floor. Upon reaching the third floor, he heard a smoke detector sounding in apartment 310.

Upon hearing the smoke detector sound, the maintenance manager knocked on the door of apartment 310. Hearing no response, he used a master key to open the door and found the apartment filled with black, heavy smoke about 4 ft from the floor level. He crawled a few feet into the apartment to determine if anyone was there. Because of the heavy smoke and no indication of any occupants, he left the apartment to alert other occupants on the third floor and call the fire department. He knocked on a few doors near the apartment and shouted "fire," then took the elevator to the ground-floor office to call the fire department. He did not close the door to apartment 310 when leaving.

After calling 911 and notifying the communication center of a fire, the maintenance man returned to the third floor by the east stairwell to alert other residents. Unfortunately, at this time the corridors of the third and fourth floors were charged with dense, black smoke and he could not move along these corridors.

Upon receipt of the 911 call, the communication center dispatched engines 1 and 2, ladder 3, and a chief officer to the building. Upon arrival, the officer of engine 1 saw black smoke rising from the rear of the building and called a second alarm. Because of building access problems, engine 1 stopped in front of the building. The maintenance manager informed the officer of a fire on the third floor. At this time the building was charged with smoke, a fire was raging in apartment 310 and extended into the corridor. The bedroom window had broken and fire was extending along and up the exterior siding. Occupants, many elderly and infirm, were calling for help, and several were trapped. The staffing for engine 1 is one officer and two fire fighters.

The occupant of apartment 410, directly above the fire was awakened by his bedroom window exploding inward and fire and smoke entering from outside. A similar event occurred a few minutes later with the occupants of apartment 411. Most building occupants become aware of the fire upon hearing the sirens of the responding engines a few minutes later.

The fire officer and the other two companies that arrived within two minutes were faced with a dilemma. Is it better to extinguish the fire to prevent its extension and continued production of smoke, heat, and toxic products? Or is it better to attempt an immediate rescue of the occupants? Insufficient staffing and equipment, a complicated, continually changing fire scene, and inadequate information make decisions difficult.

4.5 The anatomy of a building fire

The previous sections have described a building and a fire scenario. Only general observations were provided, and the description stopped at a crucial point of the scenario. Let us pause for a moment to structure a way of looking at a fire and add a few additional details for this building.

A fire scene can be very complicated because many different actions and interactions occur within a relatively short time. To describe the different activities, let us pretend that a cameraman can hover above the building with a motion picture camera that has X-ray vision. That is, the camera lens can see through materials and record the actions of the fire, the building, and all human and other responses associated with the fire. This motion picture initially would be very confusing because so many things occur simultaneously. So we also assume that the camera has filters which edit out all information except the influences for a particular component of interest. This will allow a more leisurely examination of each component as a separate entity. After examining each component, we can systematically recombine them to understand the total building fire performance. The camera filters will enable us to view

- the fire in the room of origin;
- the fire, barrier effectiveness, and propagation into other spaces;
- the fire and the action of automatic sprinklers, if present;

- the fire, detection, and fire department notification;
- the fire, local fire department response, fire attack, and extinguishment.

This group of components defines the flame/heat analysis. A flame/heat evaluation provides a good understanding of time-related fire conditions.

Returning to the concept of the filtered motion picture, one could then view in isolation

- the structural frame behavior, and
- the smoke generation, smoke movement, and contamination of target spaces.

Finally, the filtered motion picture could record

- detection, alerting, and communication with occupants;
- egress routes and movement of people who are attempting to leave the building;
- the condition and well-being of people who stay in the building.

This case study describes the building performance after ignition and established burning (EB) had occurred. The building design dominates performance and we evaluate its performance in isolation. Although some traditional practices may start with a hazard and ignition source assessment, we intentionally separate this assessment from the building performance. Although fire prevention is an important part of a total fire risk management program, it is not relevant to building analysis. This way of thinking provides greater flexibility in developing complete risk management programs.

4.6 Fire in the room of origin

Returning to the fire scenario, consider the actions recorded on the filtered components of the motion picture. A brief discussion of each component is incorporated into the description.

FIRE GROWTH POTENTIAL

For the fire described in Section 4.4, assume that EB occurred near a bed in the bedroom of Figure 4.4. This fire did continue to burn and reach flashover. Some rooms have a greater propensity for reaching full room involvement than do others. It is possible to discriminate between the relative fire growth potential for different rooms by evaluating the amount and type of fuels, their arrangement, the room size, and the ventilation conditions. For now, one should recognize that different room interior designs exhibit different potential hazards for fire growth. Here the scenario led to FRI.

The fire near and in the bed grows and gives off heat, flame, smoke, and gases. The buoyancy of the products causes a hot layer to form at the ceiling. As the fire continues to grow, the layer gets hotter, deeper, and flows out into the adjacent room. Eventually the hot layer and the contents will reach a temperature at which flashover occurs.

AUTOMATIC SPRINKLER SUPPRESSION

Here no sprinkler system exists and automatic fire extinguishment did not occur.

FIRE DEPARTMENT EXTINGUISHMENT

Several sequential events must occur before fire department extinguishment can take place. These include detection, fire department notification, arrival at the scene, first water application, and extinguishment. The fire continues to grow and propagate in the time duration from established burning to initial water application. These events take time to accomplish. In this building, the fire speed and the time delay for detection and notification would make it impossible for fire department extinguishment before FRI within the room of origin or even the apartment of origin.

THE LIMIT IN THE ROOM OF ORIGIN

In this scenario, fire growth reached flashover relatively quickly. The automatic sprinkler system did not exist and suppression intervention by the fire department did not occur. Thus, the fire was not limited to the room of origin. Here it also was not limited to the apartment of origin.

4.7 Barrier effectiveness and fire propagation

Gypsum wallboard provides a fire endurance that normally prevents fire propagation to adjacent rooms. However, penetrations and weaknesses incorporated into construction can reduce the barrier effectiveness. In this example the door between the living room and the bedroom is open. This completely nullifies any barrier fire resistance and allows a massive amount of heat and fire gases to move into the adjacent room to flash over the entire apartment. Also, when the maintenance manager left the door to the corridor open after leaving the apartment, that barrier failed and a massive amount of heat and smoke was pushed into the corridor.

Besides the interior fire spread due to poor barrier effectiveness, fire propagated to additional rooms through the external walls. The fire caused the bedroom window to break and flames to emerge from the opening. These flames ignited the wood siding and propagated up the building. In turn, this flame movement broke the window of apartment 410 and ignited contents in this apartment. Some time after that, apartment 411 also became involved. As the flames moved up to the roof overhang soffit, they began to attack the wood panel cover and to enter the roof air vents. Whatever the actual mechanism for penetration, it would not take long for the fire to breach this barrier and involve the entire roof space.

The space/barrier modules of the building geometry define the fire propagation paths. Figure 4.6 shows a partial set of fire propagation paths for this scenario. For the external propagation, the barrier from the room of origin (bedroom) and the outside space failed by the window breaking. The next barrier failure was the window of apartment 410, and the next space was the bedroom in apartment 410. Similarly, the barrier between the outside and the roof space was the soffit under the overhang. When that barrier failed, the fire extended into the roof void. We describe flame/heat movement by space/barrier fire propagation paths, and a graphical network of potential paths such as Figure 4.6 becomes a valuable analytical tool.

4.8 Fire department operations

When the first-in engine company arrives, the apartment of origin (apartment 310) is fully involved and pumping flame, heat, and much smoke through the doorway into the corridor. Also, the fire has extended through the bedroom window up the exterior wall and into apartment 410 and later into 411. In addition, the wood soffit beneath the roof overhang is under attack, and the fire will soon enter the roof space. The fire officer knows little of this upon arrival.

Fire suppression is only one aspect of fire ground operations. Life safety and rescue are major priorities in a fire. The total management of a fire scene by the fire ground commander becomes a complex problem. All activities are time dependent with a fire that continues to grow until hose streams can prevent its extension and begin to knock it down.

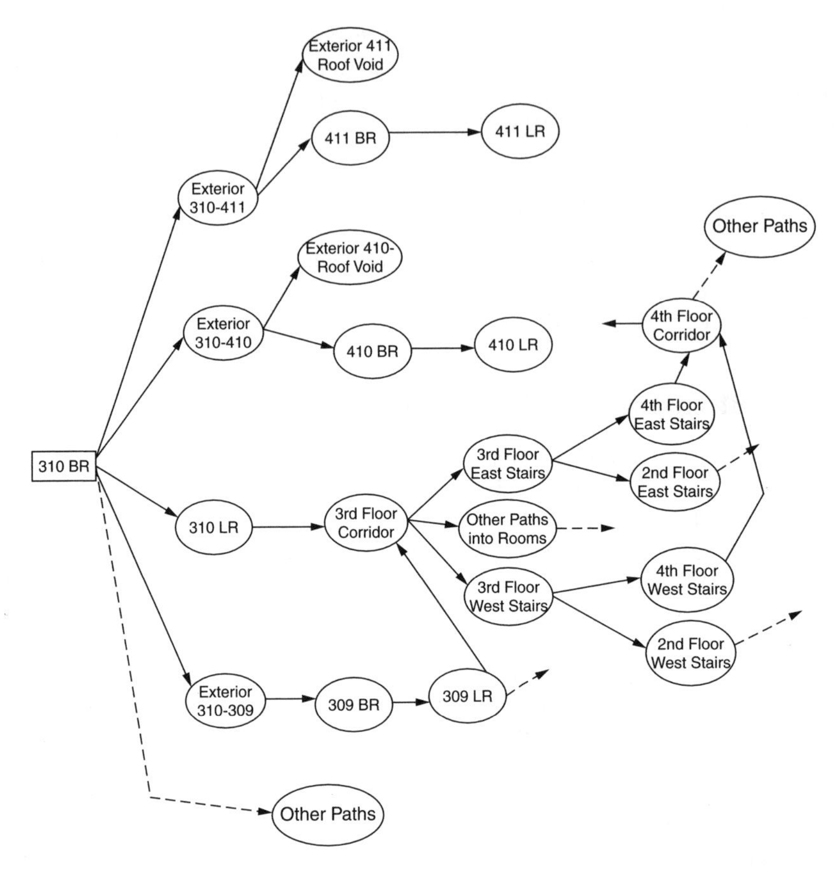

Figure 4.6

The initial response for this example is a chief officer, two engines, and one ladder truck. A total staffing of thirteen people is a very inadequate force with which to handle this incident. Calling off duty fire fighters and requesting mutual aid from neighboring communities will augment the initial staffing. However, the accumulation of these additional resources uses up time in a situation where time is the most precious commodity.

The limit of fire damage in this building is primarily dependent on the effectiveness of barriers and the actions of the local fire department. Although the gypsum wallboard between rooms and apartments performs well, the loss of barrier effectiveness because of open doors allows substantial fire extension. In addition, the combustible exterior siding provides a way for the fire to extend through windows into other apartments and up the building's exterior. This mechanism also enables the fire to extend into the attic space by breaching the overhang soffit. Installing automatic door closers and changing the exterior siding to a noncombustible material would have significantly altered the outcome of this fire. However, those features do not exist for the described situation, and the fire department at arrival faces a very large and complex fire ground operation.

4.9 The structural frame

The structural frame for this building is made of wood. When wood is protected, it can exhibit good fire resistance. However, all structural materials deteriorate when subjected to the heat energy of a fire. A structural frame analysis can identify whether a structure will collapse during a fire and the approximate time available before failure. After the fire enters the attic, that entire space will become involved quickly and the roof system will collapse shortly after that. In addition, drop-down ignitions into fourth-floor apartments can be expected shortly after the attic becomes involved.

4.10 Smoke movement

Smoke and a wide variety of airborne toxic gases are natural products of combustion. Building fires generate an enormous quantity of these products. They are pushed out of the building, within the building, or both. An early problem with smoke is a reduction in visibility. Later concerns arise because of physical debility to humans that can result from the other toxic products.

In this fire, large quantities of smoke are released outside the building through the open windows. This smoke, which is largely air, is replaced by fresher air that will enable more combustion to take place and produce more smoke and heat. The open door to the corridor will discharge large quantities of dense smoke and heat into the corridor. If wind blows into the room, the condition will worsen. Escape routes become impassable when visibility deteriorates substantially or too much heat is present. This untenable condition can occur quickly. Delay in fusible link operation will allow smoke to move into the exits. This smoke contamination will make upper-level corridors and stairs impassible to occupants. Here the target spaces along the egress paths remain tenable for only a brief time.

4.11 Life safety

Building occupants have three ways to avoid death or injury. The first is to leave the building before the egress paths are blocked. A second is to remain in their apartments until the fire goes out. This requires that the barriers are successful in preventing fire extension, the structural frame does not collapse, and too many toxic products of combustion do not reach the individual. A third is by fire department or building management rescue efforts.

Leaving the building by escaping along an egress route is the most commonly advocated way of preserving life. To do this, the fire must be detected and occupants alerted promptly. Here the fire was detected, but occupants were not alerted because the building could not sound an alarm to occupants. Consequently, a 5:00 A.M. alerting of occupants, some of whom may be hard of hearing, must be relegated to their awareness of cues from the fire. By the time cues are recognized, the fire will have become large and dangerous, and egress routes will be blocked.

After an occupant becomes aware of a fire, a preevacuation time interval occurs. Upon receiving a fire cue, an individual may take one or several actions before deciding to leave the building. Often, an individual will get dressed and investigate to determine whether a danger really exists. That individual may alert others, call the fire department, or take a variety of other actions before eventually deciding to leave the room. The absence of an occupant alert to residents on the third and fourth floor will delay the start of movement until conditions deteriorate and egress paths become impassible. Any additional preevacuation activities will exacerbate the time problem.

These occupants may decide to wait on the balcony until the fire department can rescue them. The chain-link fence and lack of access on three sides of the building will cause some delay in rescue. The first responding forces are clearly insufficient for a fire of this magnitude. The time delay for the fire department to obtain additional staffing and equipment may be relatively long. The occupants at the rear of the building are in more imminent danger than those at the front, because the wood siding will propagate the fire on the exterior. Anxiety of occupants and fire fighters will be a common emotion.

4.12 Performance evaluations and risk characterizations

The building and site design has a major influence on fire performance and the type of risk that can be expected for occupants, functions, and neighbors. The selection, quality, and location of fire defenses establishes fire performance. Later chapters will describe techniques for making performance evaluations.

The first step in evaluating building performance is to describe the fire that can be expected. This is a part of the flame/heat movement (L curve) analysis which involves the following parts:

1. Select a room of origin.
2. Select fire growth hazard categories for the room of origin and other rooms that can be involved in fire propagation.
3. Determine barrier weaknesses and the ease with which a fire can move into additional rooms.
4. If the building is nonsprinklered, fire damage and risk are dominated by barrier effectiveness (to keep the fire to a small number of rooms) and the building's ability to help the local fire department extinguish the fire. There are two parallel time lines:
   a. Design fire describing time durations for room-to-room fire propagation.
   b. Time duration for first water application to the fire. This time line includes four quantities:
      - Fire size at detection.
      - Time duration for fire department notification.
      - Time duration for apparatus response.
      - Time duration to find the fire, establish supply water, and lay attack lines.

Comparing these time lines enables one to estimate the fire size and conditions at the time of first water application. Knowing these fire conditions and the available community resources, one can estimate fire extinguishment performance.

5. If the building is fully sprinklered, the flame/heat movement and damage are dominated by the design fire, the sprinkler system reliability, and its operational effectiveness. An evaluation involves three aspects:
   a. Select design fire characteristics (fire size and associated heat release rates and fire plume momenta).
   b. Identify system reliability.
   c. Estimate its effectiveness in controlling or extinguishing the fire.

The local fire department is normally available to provide a mutual-aid support.

Having understood the building fire conditions (L curve), one may evaluate the smoke tenability (Sm curves) and structural frame stability (Fr curves). This package of evaluations identifies time durations and environmental conditions that may be expected. This performance information provides the knowledge base from which to characterize the risk to occupants and functions housed in the building.

4.13 Summary

A building performance evaluation and the associated risk characterizations can be described by a methodical, systematic process that analyzes individual components and integrates them into a complete systems understanding.

A building fire is a very dynamic process. Time is a major ingredient in all performance evaluations. The building, its fire defenses, and people can be integrated on a time basis to create a clear understanding of the risks associated with any existing or proposed new building.

Figure 4.7 provides a concise organization for the complete process. Each event can be evaluated independently. The time relationships show that when some events are taking place, other

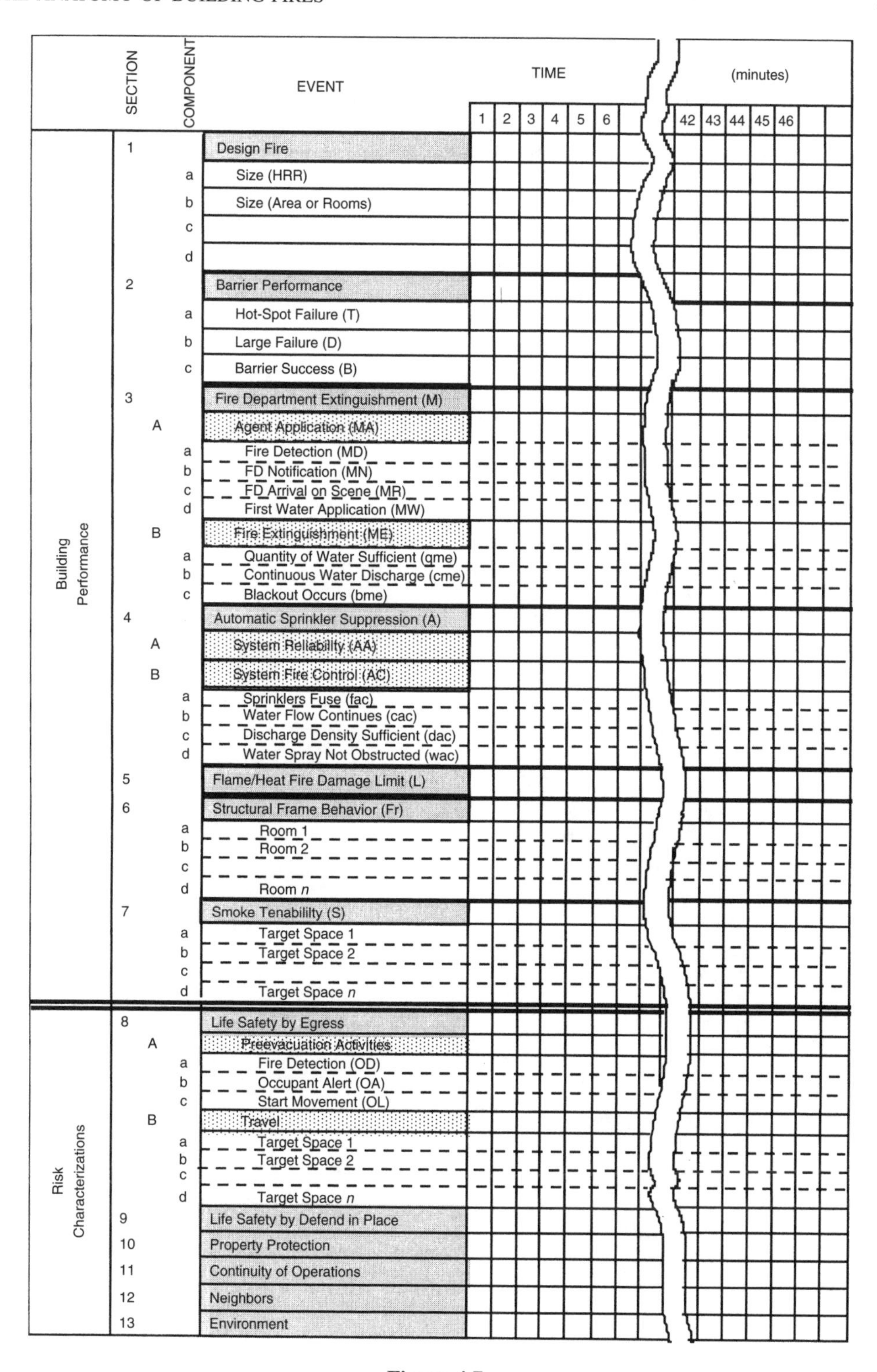
SECTION
COMPONENT
EVENT
TIME
(minutes)
1 2 3 4 5 6
42 43 44 45 46
Building Performance
1 Design Fire
a Size (HRR)
b Size (Area or Rooms)
c
d
2 Barrier Performance
a Hot-Spot Failure (T)
b Large Failure (D)
c Barrier Success (B)
3 Fire Department Extinguishment (M)
A Agent Application (MA)
a Fire Detection (MD)
b FD Notification (MN)
c FD Arrival on Scene (MR)
d First Water Application (MW)
B Fire Extinguishment (ME)
a Quantity of Water Sufficient (qme)
b Continuous Water Discharge (cme)
c Blackout Occurs (bme)
4 Automatic Sprinkler Suppression (A)
A System Reliability (AA)
B System Fire Control (AC)
a Sprinklers Fuse (fac)
b Water Flow Continues (cac)
c Discharge Density Sufficient (dac)
d Water Spray Not Obstructed (wac)
5 Flame/Heat Fire Damage Limit (L)
6 Structural Frame Behavior (Fr)
a Room 1
b Room 2
c
d Room n
7 Smoke Tenability (S)
a Target Space 1
b Target Space 2
c
d Target Space n
Risk Characterizations
8 Life Safety by Egress
A Preevacuation Activities
a Fire Detection (OD)
b Occupant Alert (OA)
c Start Movement (OL)
B Travel
a Target Space 1
b Target Space 2
c
d Target Space n
9 Life Safety by Defend in Place
10 Property Protection
11 Continuity of Operations
12 Neighbors
13 Environment

Figure 4.7

events have not yet started and still others may have been completed. Figure 4.7 provides a way to view the complex process to understand time frames and interrelationships. The rest of this book describes ways to evaluate, portray, and integrate these events to acquire understanding of a building's performance. By understanding building performance, we can communicate more effectively with people who may not be as knowledgeable about fire behavior.

5 A WAY OF THINKING

5.1 Introduction

A performance analysis provides information and understanding of expected behavior so that one can make better decisions. This knowledge gives an insight into planning and solutions to manage or change the risks. Clear understanding and logical judgment enhance the ability to communicate effectively with a broad range of individuals and professions that interface with the building. *Understanding – deciding – communicating* is the core of effective fire program management.

Fire performance evaluations require a different way of thinking. Fire science and engineering is not mature enough to provide adequate analytical tools to address all needs. Standard practices are not sensitive enough to discriminate expected behavior. The analytical framework and evaluation tools of this book organize a way of thinking into systematic, logical procedures that blend and use the best qualities of all available tools. These procedures apply methods and thought processes that are regularly used in professions associated with building design and fire applications.

5.2 The building/fire performance system

Fire performance consists of three parts: (1) the limit of flame/heat damage, (2) the time during which important target spaces remain tenable, and (3) the time during which the structural frame remains stable. Evaluation of these components produces the knowledge of how a building will perform during a fire.

The fire challenge (which defines the design fire) greatly affects fire defense performances. The design fire describes the time-related release of combustion products that are the loading against which the fire defenses must perform. Each fire defense is associated with the specific fire products that affect the functional operations. For example, the design fire for a sprinkler system analysis (or performance design) provides continuous graphs of the heat release rate, area of fire spread, fire plume momenta, and speed of growth (αt^2). All these factors affect sprinkler system performance. Similarly, the design fire for fire department suppression describes the time durations for fire growth both in area and number of rooms involved. In all functional analyses, we base the design fire on continuous fire development with no suppression intervention.

The design fire is an essential part of every fire component evaluation. Figure 5.1, Section 1 identifies the design fire characteristics with which to compare defense functions. Rows 1 a,

Building Fire Performance Analysis R. W. Fitzgerald
© 2004 John Wiley & Sons, Ltd ISBN: 0-470-86326-9 (HB)

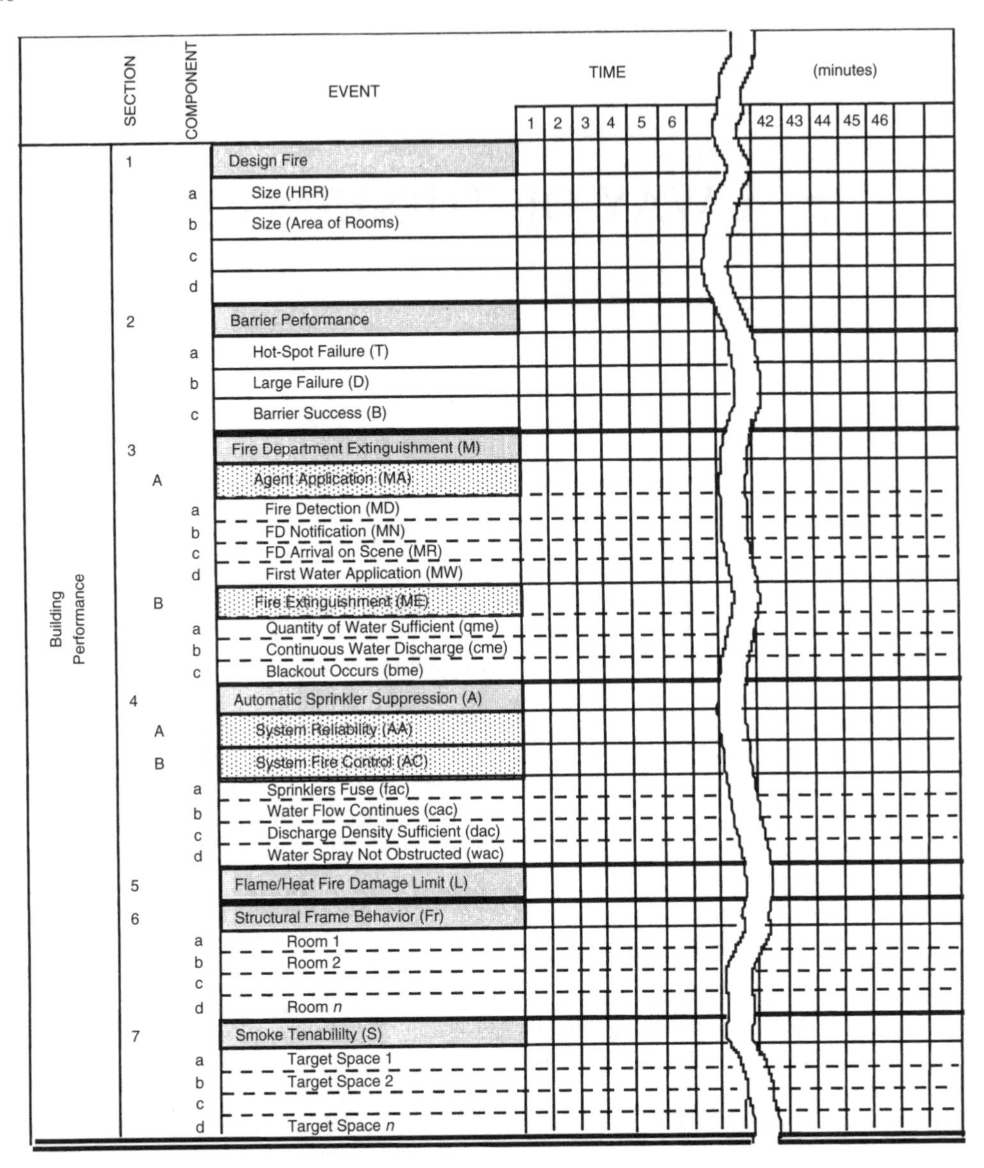

Figure 5.1

b, c, ... identify the specific products of combustion quantities with which to analyze the fire defense functions. Because Figure 5.1 shows the complete performance system, individual component evaluations use only those combustion products of Section 1 that apply to the defense being analyzed.

It may be useful to pause and look at Figure 5.1 more carefully. The horizontal rows identify the components that make up a performance evaluation. Because fire performance is dynamic, the vertical columns can show time sequences during which the fire defenses are operational. As the subject evolves, Figure 5.1 will describe other concepts for thinking about fire performance evaluations.

Section 4.5 introduced the concept of an X-ray camera with filters to isolate individual component performance. Each horizontal row of Figure 5.1 becomes a motion picture of a filtered component during the fire. Sequencing the different components allows one to understand their interactions better and to describe interactive systems performance.

Each vertical column of Figure 5.1 provides an opportunity to "stop the world" and study the status of the building fire with a "photograph" at that instant of time. Even further, one can isolate an individual cell to study component performance at that instant. This enables one to analyze the component responses of the design fire and the influence of other components on its behavior.

An analysis of appropriate components and time periods gives an initial knowledge of building fire performance. One can extend this initial knowledge to develop greater understanding and a more comprehensive picture of the systems performance. The questions of what to select and how much is enough become central to an effective analysis.

5.3 Performance evaluations

A performance evaluation involves two ingredients. One is a framework that tracks the analytical thought process, and the other is a way to discriminate quality and operational effectiveness. Although we can use the framework in isolation, we greatly magnify our understanding when we describe performance with numbers. We further enhance communication with visual descriptors of component and building behavior. Visual graphics help an individual develop a better personal understanding of behavior and describe expectations more effectively to others.

Note that the objective for an analysis is to estimate the performance of a fire defense component for existing or defined conditions. The question is not, Was the component designed and installed in accordance with applicable codes and standards? Rather, the question is, Will the component perform the function for which it was intended? And when? The focus is on understanding performance, not compliance approvals.

We will introduce the process of making component evaluations with an example. Consider the actuation of a fire detector within a building. Here the question is not, Will the detector actuate? Rather, the question is, Will the detector actuate within 1, 2, 3, ..., n minutes?

Every fire defense component has three regions of behavior across the time line of a fire. The first is an interval where one is certain that it will not operate or act. Another is a segment of time where one is certain that it will operate or act. Between them is a window of uncertainty in which one is not able to describe the behavior with certainty. To illustrate these conditions, consider Figure 5.2, which describes fire detector actuation at a specific time or fire size.

An initial goal of detection analysis is to estimate the time durations of each of these segments. The factors that influence these time durations are

- the design fire location and the time relationship of combustion products release;
- the location, type, and sensitivity of the detection instrument;
- the transport of combustion products between the fire and the detector.

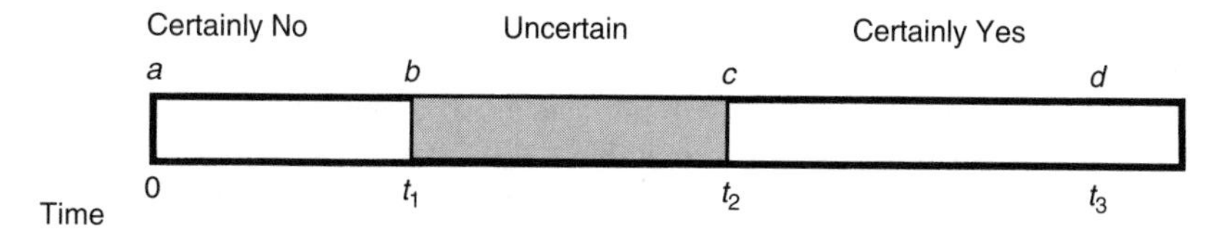

Figure 5.2

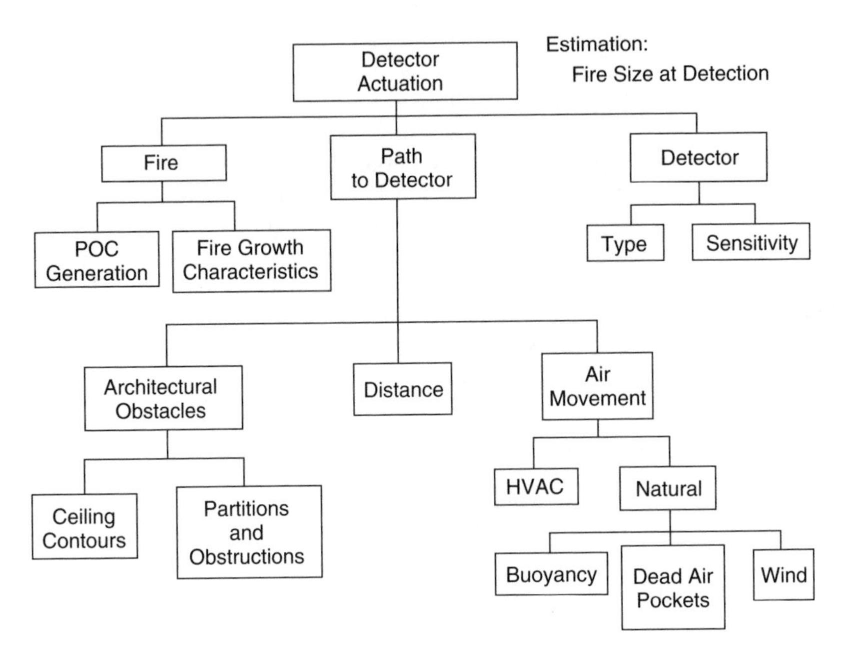

Figure 5.3

Figure 5.3 identifies factors associated with these groups. A performance analysis estimates whether a specific detector will actuate *n* minutes after a unique (i.e., defined) fire ignition. This involves answering two questions:

- Given the design fire development for *n* minutes, will enough combustion products reach the detector to cause actuation for the most sensitive setting for this type of detector?
- At *n* minutes, will the amount of combustion products (POC) cause actuation for the detector sensitivity?

One can ask these questions for each specified fire size. The answers provide the basis for the performance estimate. Each answer will be "certainly yes," "certainly no," or "I am not certain." Thus, for the series of questions "Will the detector actuate at 1, 2, 3, ..., *n* minutes?" the individual is estimating the performance in each cell along the horizontal row of *fire detection (MD)* of Figure 5.1. The results become the basis for constructing the bar chart of Figure 5.2.

5.4 The window of uncertainty

The way in which we deal with uncertainty has a major influence on our understanding of performance and the ability to communicate with others. For example, at a position within the window of uncertainty close to "certainly no" (point *b* of Figure 5.2), the uncertainty is near a condition where we believe the detector will not actuate. The opposite occurs near point *c*. It is useful to calibrate the performance estimate within the window of uncertainty.

Estimates of uncertainty can be expressed in verbal terms (e.g., very unlikely, a good chance, a high probability) or in numerical measures (e.g., a probability of 10%, 60%, or 80%). Both ways have value in understanding and communicating fire performance.

During the evolution of the analytical procedures described in this book, no topic has received more attention and scrutiny than the way uncertainty is handled. Probability has had a dual meaning from the time it was first invented nearly three and a half centuries ago. There are two extreme definitions (perhaps we can call them philosophies) and many positions between these extremes. One definition, called classical or objective, is that probability is a measure of the frequency of occurrence of one event out of a possible set of the same events. The second definition of probability is called personal or subjective, and describes an expected likelihood of performance. This interpretation is more commonly used for unique situations where experimental repetition under the same conditions is impossible or unfeasible. Objective probability is viewed as a physical property of the function being analyzed whereas subjective probability is interpreted as an intellectual process. Appendix B discusses interpretations and applications of probability in greater detail.

When this book describes the calibration of an estimate, it uses subjective probability. Subjective probability is a measure of the degree of belief that an individual has in a judgment or an opinion. The literature describes this process as encoding a state of knowledge. Assessments are based on acquired knowledge, such as observational facts, calculated values, computer models, physical relationships, experimental information, failure analyses, and personal experience. Assessments use the full spectrum of information that is available and seems relevant to the problem. Uncertainties are clearly recognized.

Here we interpret probability assessments as an estimate of the expected performance for a site-specific condition. For example, one should not interpret a probability of 0.4 or 0.9 literally as a statistical frequency of 40 times or 90 times out of 100. Rather, one should interpret the numbers as a calibration estimate by an individual making the assessment. In this illustration we recognize that a condition having a value of 0.9 is much more likely to occur than a condition having a value of 0.4.

We base a probability estimate on a deterministic evaluation of the situation involved. For example, assume that an individual evaluates a sprinkler system and estimates it will control a fire within a floor area of $25\,ft^2$ as $P(A) = 0.80$. This means that based on the information that was available for the evaluation, the individual believes that the likelihood the sprinkler system will control the fire before it extends beyond $25\,ft^2$ is 80%. The numerical value expresses a best judgment for the analysis of that specific fire size.

Return now to the initial performance estimates of Section 5.3. One can expand the original bar chart (Figure 5.2) to incorporate calibrated performance estimates, as shown in Figure 5.4. Here, performance is expressed as a probability having values between 0.0 and 1.0. Both extremes show certainty. A probability of 0.0 means there is a certainty the event will not occur. A value of 1.0 is a certainty that it will occur. The origin of this graph is at the top, and probability is shown positive downward.

Figure 5.4 typifies almost all of the graphs of component performance. The vertical axis represents the success probability of an event. The horizontal axis shows either time or fire conditions related to time. The graph shows the dynamic changes in expected performance of the detector as the fire (and time) continue to grow. Relative positions, as well as the span of the window of uncertainty, will vary depending on the fire situation being described.

Graphs of the component's dynamic performance can be related to Figure 5.1. The design fire time (or size) becomes the scale for the abscissa. If one were to show probabilistic values in each cell across any row, those values become the coordinates for the performance graph. Thus, each row of Figure 5.1 represents a continuous graphical description of that component's performance. If numbers are not used, one can construct a bar chart similar to Figure 5.2. When numerical probability estimates are shown in the cells, we obtain a continuous graph as in Figure 5.4.

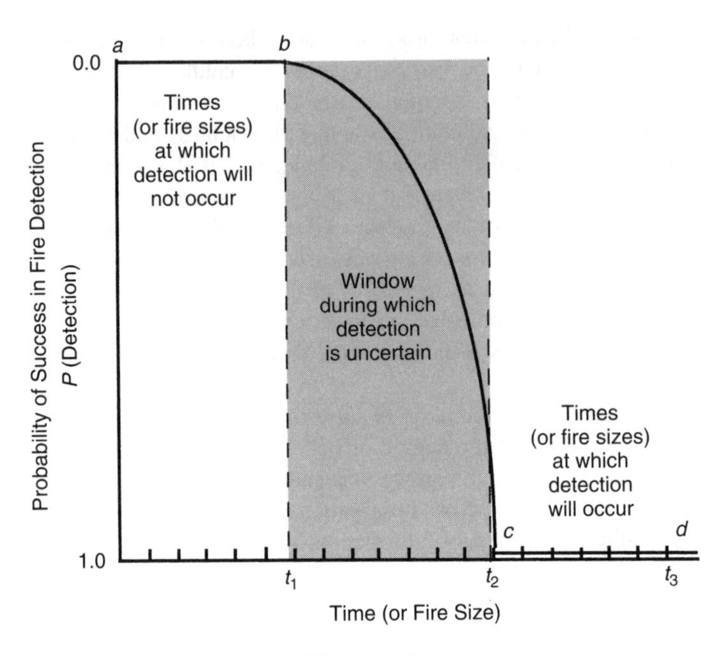

Figure 5.4

5.5 Estimating performance

An estimate is defined in Webster's Dictionary as

- to form an approximate judgment or opinion regarding the amount, size, weight, etc., of;
- to determine roughly the size, extent, or nature of;
- an approximate value or judgment of something as value, time, weight, etc.;
- a judgment or opinion of ... the qualities of something.

Performance evaluations in all technical disciplines predict a condition or outcome of functional events. Performance questions involve expectations of future conditions that have not yet been experienced. The answers can be described by an estimate. Confidence in the estimate is related to the understanding obtained from a combination of information from the study, system interactions, conditions under which performance is being evaluated, and available time. Information may include theoretical and empirical knowledge about the behavior of the system and its parts, relevant calculations, and observed details. Uncertainties associated with the loading, dimensions, properties, professional calculations, and other recognized conditions may be incorporated into the estimate. The estimate value is based on technical information and knowledge, experience, and judgment.

Let us pause for a moment to consider the meaning of "certainty." An absolute certainty means that there is no possibility. For example, in a room with no sprinkler system there is an absolute certainty that automatic suppression is zero. However, a functional absolute certainty is commonly viewed to mean "about one in twenty." Because the specific location of absolute certainty is difficult to decide, values of 0.0 and 1.0 are taken to mean "nearly 0 or 1." Thus, we recognize that the edges of certainty (points *b* and *c* in Figure 5.4) are near those values, and not absolutely the values.

The quality of an estimate depends on the amount of information that is used, the skill in translating that knowledge into performance estimates, and the time available to evaluate relevant details and conditions.

Different types of evaluation may be appropriate for a particular project. Although different evaluations use the same framework of thinking, they involve different amounts of information, types of detail, and time requirements. Sometimes it may be sufficient for an estimate to be an experienced top-of-the-head opinion to give a sense of proportion. At the other extreme, an estimate may be the result of a very detailed and rigorous study. One must tailor the level of effort to the needs and resources of the project.

Estimating involves an integration of technology, experience, judgment, and creativity. Information and knowledge form the basis of judgments and decisions. There is never enough information and knowledge of details to eliminate uncertainty. Consequently, uncertainty is an integral part of all estimates.

5.6 Evaluation levels

Professional time is a valuable commodity. The professional must conserve that commodity by reducing the time necessary to arrive at better decisions. Organizational planning and tailoring the details of an analysis to needs of the problem is one way of reducing analytical time.

We identify three different levels of evaluation that can associate a value added for efforts expended. These evaluation levels relate time and resources to the conditions and needs of the situation. The application of all three different levels uses the same framework, concepts, and thought process. Differences lie in the purpose of the study, time available, attention to detail, use of technology, and confidence in the results.

A *level 1 evaluation* develops a basic understanding of performance. It provides a rapid and inexpensive assessment for a specific building. This evaluation uses visual observations and mental estimations as the primary basis for making assessments. A level 1 evaluation gives a basic understanding of the building and a sense of proportion for performance expectations and risk characterizations. A level 1 evaluation is useful to screen buildings for priority identification; to identify and order the relative significance of different aspects of fire performance; or to establish a preliminary understanding of performance so that one may structure a detailed analysis more efficiently. A level 1 evaluation enables a trained professional to recognize abnormalities of condition that could lead to serious consequences that may require more careful analytical attention. Level 1 evaluations may be used for an individual component or for an entire building, a level 1 evaluation is a prerequisite to all level 2 and level 3 evaluations.

A *level 2 evaluation* develops a better technical understanding of the building performance by considering details more carefully and calculation procedures more generally. A level 2 analysis is useful to examine selected component performance in greater depth. Level 2 evaluations may help in approval decisions by AHJs for performance-based engineering designs or for code equivalency requests.

A *level 3 evaluation* provides an understanding of performance sensitivity. This level describes expected bounds of performance involving calculation variables for individual components and major system segments. Alternatively, this type of evaluation improves the understanding of the potential impact on performance of reasonable expected variability. For example, the influence ranges for different detector locations or site conditions can be described by a level 3 sensitivity analysis.

Evaluation levels 1, 2, and 3 each have different objectives and information resources. Although the organizational framework and basic thought process is the same for each level, the details of analysis are different. The distinctions are associated with the breadth and character of acquired knowledge as well as uncertainty of component performance descriptions. Additional information reduces the window of uncertainty and provides more technically defensible documentation.

5.7 Visual thinking

Level 1 evaluations give a good understanding of the important factors that influence fire performance of individual components and their interactions. These evaluations provide a sense of proportion for the way in which different fire defenses and risk characterizations behave with regard to time and fire growth. A graphical description of dynamic component performance enables one to understand the building better. It can also help to focus discussions with colleagues and to communicate better with others not in the fire business.

We will make a few observations about graphical performance descriptors. Other chapters will discuss these sketches more completely, and Chapter 6 will describe how to construct these graphs for level 2 evaluations. For now, we are interested only in reading level 1 performance sketches.

Figure 5.5 shows sketches of automatic sprinkler performance estimates for three buildings. The ordinate describes the probability of success for the sprinkler system to control different size fires. The origin is at the top. Thus, when the probability of success is 0.0, the sprinkler system certainly will not control the fire. When the probability is 1.0, the sprinkler system certainly will control the fire. Numbers between 0.0 and 1.0 calibrate the estimate in controlling the design fire. The abscissa shows design fire sizes.

Quality comparisons and reliability assessments are an important part of performance evaluations. Later chapters on component evaluations discuss ways of addressing these assessments. For now, we only wish to note that reliability is a part of the evaluation process that can be reflected, in part, by the vertical gap between the termination of the diagram and certainty of performance.

Level 1 sketches intentionally do not provide intermediate numerical values for the ordinate because the objective is to sketch relative component behavior. Although specific numbers are rarely used in level 1 evaluations (except the abscissa scale), details that affect performance are noted. A sense of proportion about the fire and fire defense details helps to order general relationships. This enables one to think visually about dynamic performance and to describe one's thoughts to others, if necessary.

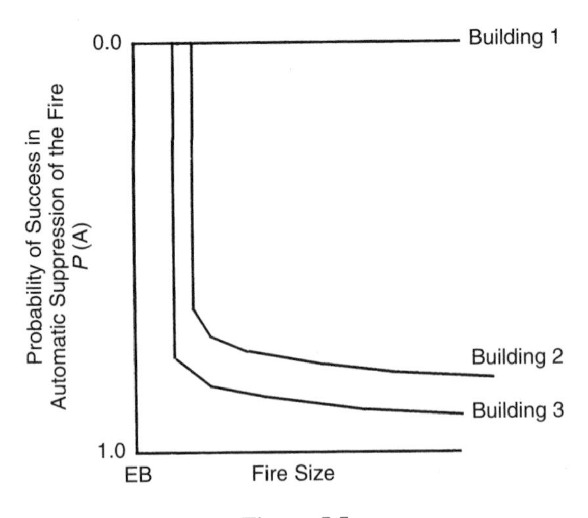

Figure 5.5

Example 5.1

Let's pretend. Assume that you are a supervisor and one of your inspectors has done a level 1 evaluation of sprinkler systems for three buildings to compare relative performance expectations. You have not yet seen the written report or spoken with the inspector. What observations can you make regarding the sprinkler systems of these three buildings shown in Figure 5.5?

Solution

Building 1 shows with certainty that the sprinkler system will not extinguish a fire of any size. Therefore the building does not have a sprinkler system. Perhaps one might argue that the inspector discovered an existing sprinkler system was impaired to the extent that it was not functional. As we will discuss in a later chapter, this type of condition normally would be assigned a very low value (e.g., 3% or 5%) to call attention to the condition of serious and clear impairment. Here we attribute the zero system success to the lack of a sprinkler system.

Both Buildings 2 and 3 have sprinkler systems. The system of Building 3 is better than that of Building 2. The graphs do not describe whether the differences are due to reliability or operational effectiveness. A report could document reasons for the performance differences.

The sketches for Buildings 2 and 3 show the characteristic shape for an A curve (for automatic sprinkler control). An initial horizontal segment describes the fire size at initial sprinkler fusing. The steep drop shows the success of the first sprinkler. As the fire extends, more sprinklers will open and success in fire control improves, although not as effectively as with the first few sprinklers. When the fire grows beyond the capability of the sprinkler system and cannot be controlled, the line becomes horizontal.

Example 5.2

Figure 5.6 shows M curves (manual fire department extinguishment) for three buildings. Again, you have not yet seen the written report or spoken with the inspector. Nevertheless, read these diagrams and make observations of what your inspector is trying to express.

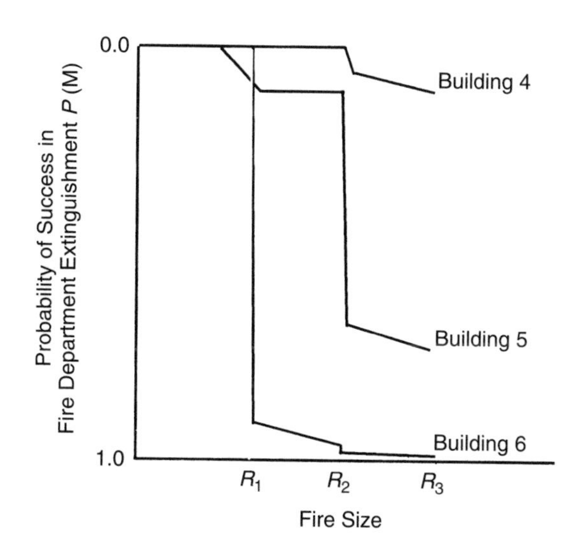

Figure 5.6

Solution

Performance questions for fire department extinguishment do not ask, Will the fire department extinguish the fire? But they do ask, Will the fire department extinguish the fire within 1, 2, 3, ..., n rooms? An M curve evaluates a building to show the likelihood that the local fire department can extinguish a fire. Indirectly, it describes the flame/heat destruction that can be expected before extinguishment. Here the building performance for fire department extinguishment is shown only for the first three rooms.

Certainly, the fire cannot be extinguished before initial water is applied to the fire. The initial horizontal segment of an M curve is based on time line comparisons of expected fire growth with events of detection, fire department notification, fire department arrival, and initial water application. The fire will certainly involve two rooms in Building 4 before initial water can be applied. Then barrier weakness, building conditions, or insufficient fire suppression resources (all or some of these factors) suggest there is a very low likelihood of extinguishment in the next room. This building is very weak with respect to fire department extinguishment performance.

The M curve for Building 5 shows that there is a chance of initial water application shortly before room 1 is fully involved. However, if the fire is not extinguished in room 1 (and there is a very low likelihood of this occurring), there is a near certainty that room 2 will become fully involved. A massive barrier failure ($\overline{\text{D}}$ failure), such as an open door, often causes this condition. Improved barrier performance between rooms 2 and 3, combined with building features and available fire-fighting resources, suggests a better likelihood of limiting the fire to two rooms. Nevertheless, if the fire is not stopped in the first two rooms, there is a chance that it will extend to additional rooms.

The expected performance of Building 6 says that the room of origin will be lost. However, one or a combination of a strong barrier between rooms 1 and 2, other building features that assist fire fighting, and local suppression resources shows that there is little chance for the fire to extend beyond one room.

Example 5.3

Risk characterization is a major interest in building performance assessments. Assume that a building inspector has evaluated three buildings and uses Figure 5.7 to describe level 1 estimates of the preevacuation activity time available before an occupant experiences untenable conditions when leaving a building. Although you have not yet seen the written report or spoken with the inspector, what observations can you use to compare the performance of these buildings?

Solution

The egress process involves two parts. The first is the time delay from EB until an occupant leaves the room of occupant origin. This time delay includes fire growth to detection, alerting the occupant, and time for the occupant decision to open the door and start to leave the building. The second part involves the time duration to move through the escape route to an exit discharge. Time line comparisons of untenable conditions in target spaces and an occupant in transit through those spaces identify situations when an occupant and untenable conditions may be in the same space at the same time.

The ordinate of Figure 5.7 shows the probability that an occupant will be successful in leaving a building with tenable conditions expected throughout the egress path. The abscissa shows the time delay between established burning (EB) and the occupant vacating a room to start the egress process. The initial horizontal segment $P(\text{BL}) = 1.0$ shows a certainty of success in leaving the building without experiencing untenable conditions. The final horizontal segment shows a certainty that untenable conditions $P(\text{BL}) = 0.0$ will be experienced. The window of uncertainty

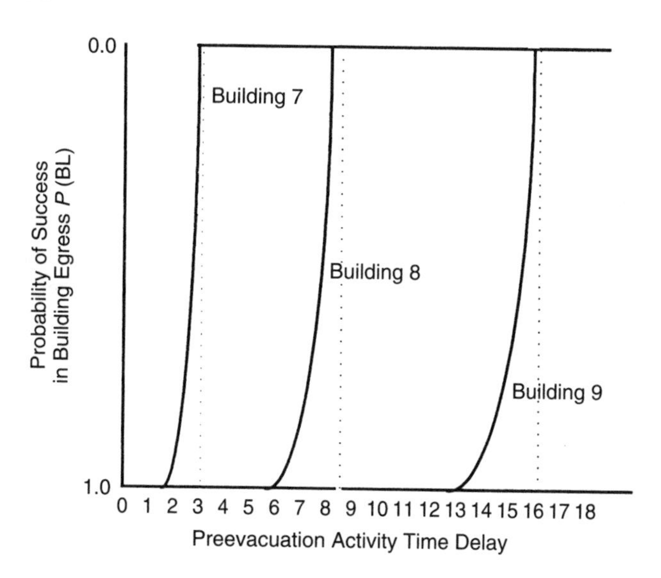

Figure 5.7

is the inclined portion between the certain conditions. Thus, if an individual leaves the room of occupant origin in Building 8 within 6 min after EB, there is a certainty that he or she can reach the exit discharge without experiencing untenable conditions. However, if the delay extends to 8 min, there is a certainty that the occupant will encounter untenable conditions along the egress path. If an individual leaves the room 7 min after EB, there is a high probability that untenable conditions will be encountered along the egress path. Building 9 provides substantial time for an occupant to leave, whereas an occupant of Building 7 can expect to experience untenable conditions within 2 min of EB.

The visual descriptions of Examples 5.1 to 5.3 tell a story. The initial purpose of the story is to help an inspector understand how a building component will behave during a fire. This visual thinking may be sufficient for the problem being addressed, or it may be augmented with brief documentation.

5.8 Example of effective communication

An L curve integrates all building and site fire features to describe the limit of fire damage. Before moving on from the concept of visual communication, let us describe briefly a building in which the L curve was constructed eight months before the fire occurred. The objective of this building analysis was to understand the building performance and describe it so that management could evaluate different alternatives and decide the most cost-effective solution. The communication used L curves to describe performance of the building as it existed and to show the effect of different proposed alternative solutions. Unfortunately, the fire occurred before the selected fire defense changes could be implemented.

The Military Records Center in Overland, Missouri, was a six-story reinforced concrete structure that housed military service records of members of the United States army, navy, air force, and marines and the associated information processing involving those records. The building was 726 ft long by 286 ft wide. The floor-to-ceiling height was about 12 ft. At the time of the fire, 60% of the space housed records storage and 40% was used for office operations. Two thousand occupants worked in the building during the day, and six occupants were on duty at night.

The building was built in 1956, and at that time most records were stored in metal fire cabinets. In 1960 the General Services Administration (GSA) was given responsibility for the building that resulted in substantial expansion of activities and a rapid increase in records storage. Most of the added records were placed in cardboard boxes on open shelving that extended from floor to ceiling in large open spaces.

A building fire performance evaluation was conducted by the GSA in October 1972. The four-page report of their conclusions noted the relative ease and speed of potential fire growth and the lack of effective fire defenses. Business managers usually have limited knowledge of fire behavior and fire defense effectiveness. A visual descriptor was prepared to portray the performance of the building in its present condition as well as the changes that could be expected with selected alternative designs. Figure 5.8 reproduces this graphical descriptor (with the vertical scale reflecting failure, rather than success) which today we call the L curve.

Figure 5.8 shows that the existing conditions (shown by the solid lines) will result in a high likelihood of extensive loss during normal operating hours and potentially disastrous damage during nonworking hours. If the cardboard boxes containing the files were rotated 90° to prevent paper exfoliation, the speed of the fire would be slowed, thus allowing a more timely intervention by the fire department. However, this change shows only marginal improvement. Also, this change would result in less efficient routine operations and was not cost-effective.

Two additional alternatives involving both working and nonworking hours were portrayed in this graph. The most effective design alternative from a fire damage viewpoint was to store the records in metal file cabinets. This changed the fire development conditions and speed of propagation. However, the cost was greater than for installing sprinklers, which was the other

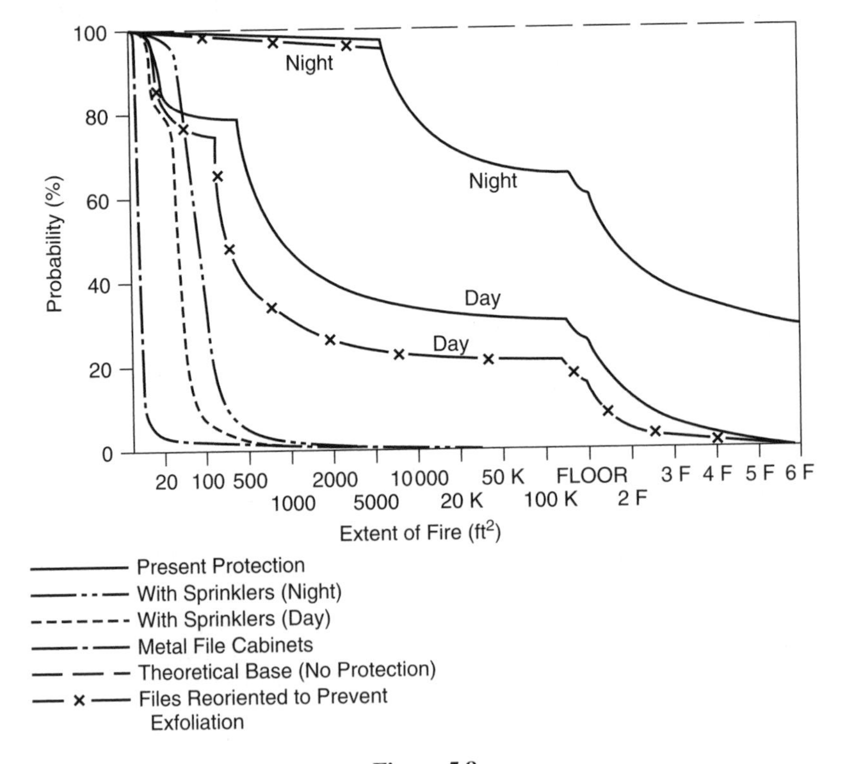

Figure 5.8

alternative. After considering this graphical story of potential damage under existing conditions and the effect of changes of the alternatives, management decided to install a sprinkler system. Unfortunately, the fire occurred before the sprinkler system was installed.

There are several significant aspects of this case study. One is that the engineers looked at the fire performance as an integrated system. This enabled the team to understand time-related interactions and focus on the issues that affected performance. Evaluations were expressed as the subjective judgment of the engineers using the technical knowledge base of 1970. Fire performance for the existing conditions and the alternatives were portrayed in a graphical, L curve format. The L curves became a powerful tool for communication. This tool enabled management to comprehend the problem, envision cost and effectiveness of alternatives, and quickly make a rational decision for a solution.

5.9 Summary

This chapter introduced a way of thinking for building performance evaluations. It may be used for any existing or proposed new building. It may be used with any set of building codes and standards, although applications do not focus on code and standards compliance. Rather, it answers questions about how a building will behave during a fire. The questions define a systematic organizational framework that structures the analysis. Performance estimates are based on expected behavior.

The primary goal of any analysis is to understand how a building will behave during a fire. After one develops the understanding, written documentation or visual communication can follow.

The technology of fire science and engineering has made significant progress during the past several decades. Nevertheless, compared with other more mature engineering disciplines, a substantial amount of uncertainty remains associated with calculations that predict performance. Consequently, complete confidence cannot be placed in the ability of calculations to accurately predict the behavior of a building over a broad range of fire conditions. One must supplement technological information with experience and judgment to understand fire defense performance. The function of the organizational framework is to structure the analytical process.

Fire performance evaluations require making decisions in an imperfect world of limited information, pressure for rapid responses, and uncertainty. Considering the present state of the art in fire science and engineering, decisions relating to acceptance or expected performance must reflect a value judgment of the individual making the assessment. That value judgment is calibrated as an estimate of the performance of the component. This estimate is a subjective judgment based on all the information available at the time of the evaluation.

Time is an important professional commodity. Three levels of evaluation may be used to satisfy different application needs. Level 1 is used with all evaluations. After completing a level 1 evaluation, decisions may be made regarding future work. If additional information is needed, one can target the specific factors that are important to the needs of a level 2 analysis. If information such as location or variable sensitivity is important, a level 3 analysis will bound ranges of component performance. Therefore one can tailor the building information needs with an analytical level that is appropriate.

Communication in fire safety is a dominant need in most performance studies. Sometimes one need only compare component time relationships to acquire a better personal understanding. Sometimes one wants to discuss estimates with colleagues in order to examine alternative courses of action. Often one needs to communicate with individuals who have little understanding of building fire behavior. The level 1 graphs described in this chapter are a useful visual thinking tool. When better understanding and discrimination are needed, use the level 2 procedures of Chapter 6.

6 FRAMEWORK FOR ANALYSIS

6.1 Introduction

The way of thinking described in Chapter 5 notes that the goal of a performance analysis is to understand the building's response to fire. This chapter introduces network diagrams as a tool to help the organization and analysis. The primary function of a network diagram is to guide an analytical thought process. A hierarchy of networks structures and integrates component assessments into a comprehensive framework of the entire system. This framework allows easy transition between microanalysis and macroperformance.

Network events and performance estimates are closely intertwined. The probability of a network event is a measure of its performance. Networks also function as an organization for calculations, the basis for constructing performance graphs, and an organization for technical documentation. This chapter describes techniques of network calculation and use. Later chapters describe techniques of evaluation and documentation and Appendix A discusses their development and theory.

6.2 Network diagrams

A network diagram is a semigraphical framework that describes a thought process. Two different types of network structure are used in this book. The first type, called a *continuous value network* (CVN), is analogous to a motion picture. In fire studies this is a scenario. A CVN starts with a specific event and identifies the future chain of events that provide a "script" (scenario) of possible outcomes. In Figure 5.1 the sequential cells along any horizontal row identify a continuous value network.

The second type of network, called a *single value network* (SVN), is analogous to a single frame of the motion picture. This would represent a single cell of Figure 5.1. A SVN enables one to "stop the world" and examine in detail the performance of the different system parts at that instant. This analysis can provide a relatively clear idea of what would happen at that moment in time.

Continuous value networks enable one to focus on the sequence of events. They help to identify conditionality and can incorporate time into the process. Single value networks allow one to stop at any desired instant or event and evaluate causes and conditions at that time.

Building Fire Performance Analysis R. W. Fitzgerald
 ISBN: 0-470-86326-9 (HB)

The coordination of these two types of networks gives a valuable insight into performance understanding. Astute selection of cells (i.e., times) for detailed study can produce an efficient and confident understanding of building performance.

6.3 Continuous value networks: concepts

A continuous value network traces the thought process for a component's performance. To illustrate this concept, consider the detector performance curve shown in Figure 6.1. The coordinates describe the likelihood of detection at fire sizes, $S_1, S_2, S_3, \ldots, S_n$. These sizes represent the design fire of Figure 5.1. The term $P(\text{Det}_n)$ describes the probability of detecting the fire within size S_n. The term $P(\text{Det}_n|\overline{\text{Det}}_{n-1})$ indicates a conditional probability. That is, the fire is detected within size S_n, *given* that it was not detected within smaller size S_{n-1}. The overbar $^-$ means "not," and the vertical bar | means "given that."

Figure 6.2(a) shows a CVN that describes the thought process to construct this performance curve. Figure 6.2(b) shows the same CVN using a concise notation for the events. One reads the CVN diagram and the associated performance curve as follows:

> Given, IG:
>
> Path 0–2–10 says that given ignition (IG), the fire is detected before reaching a fire size of S_1. The probability of this event is $P(\text{Det}_1)$. In Figure 6.1 this is associated with point c.
>
> Path 0–1–4–10 says that given IG AND the fire is not detected before size S_1 is reached, the fire is detected before size S_2 is reached. Symbolically, this conditional probability is expressed

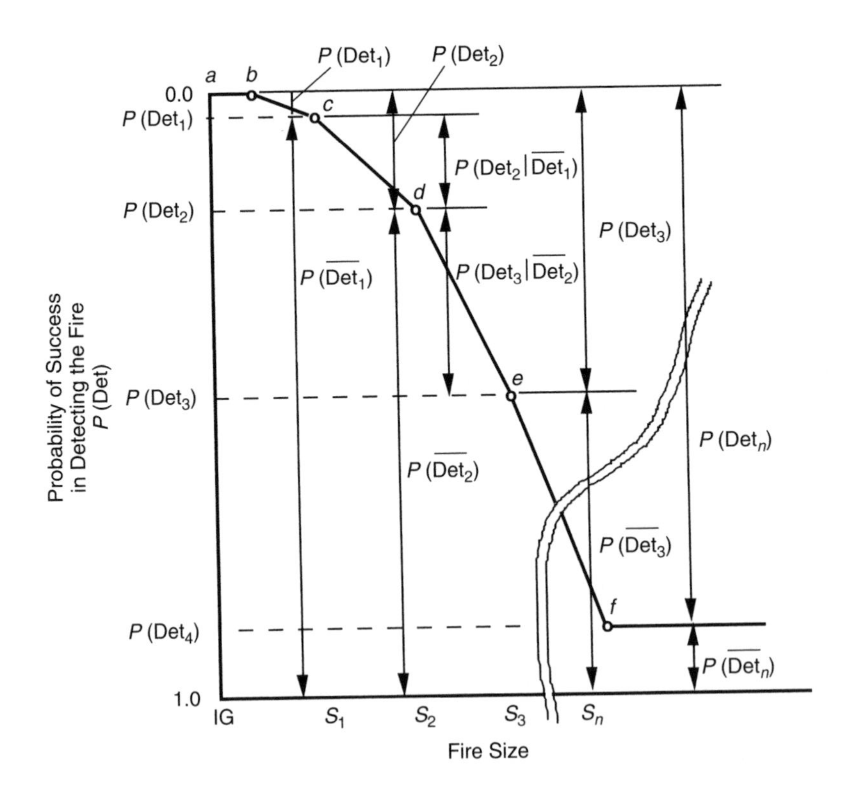

Figure 6.1

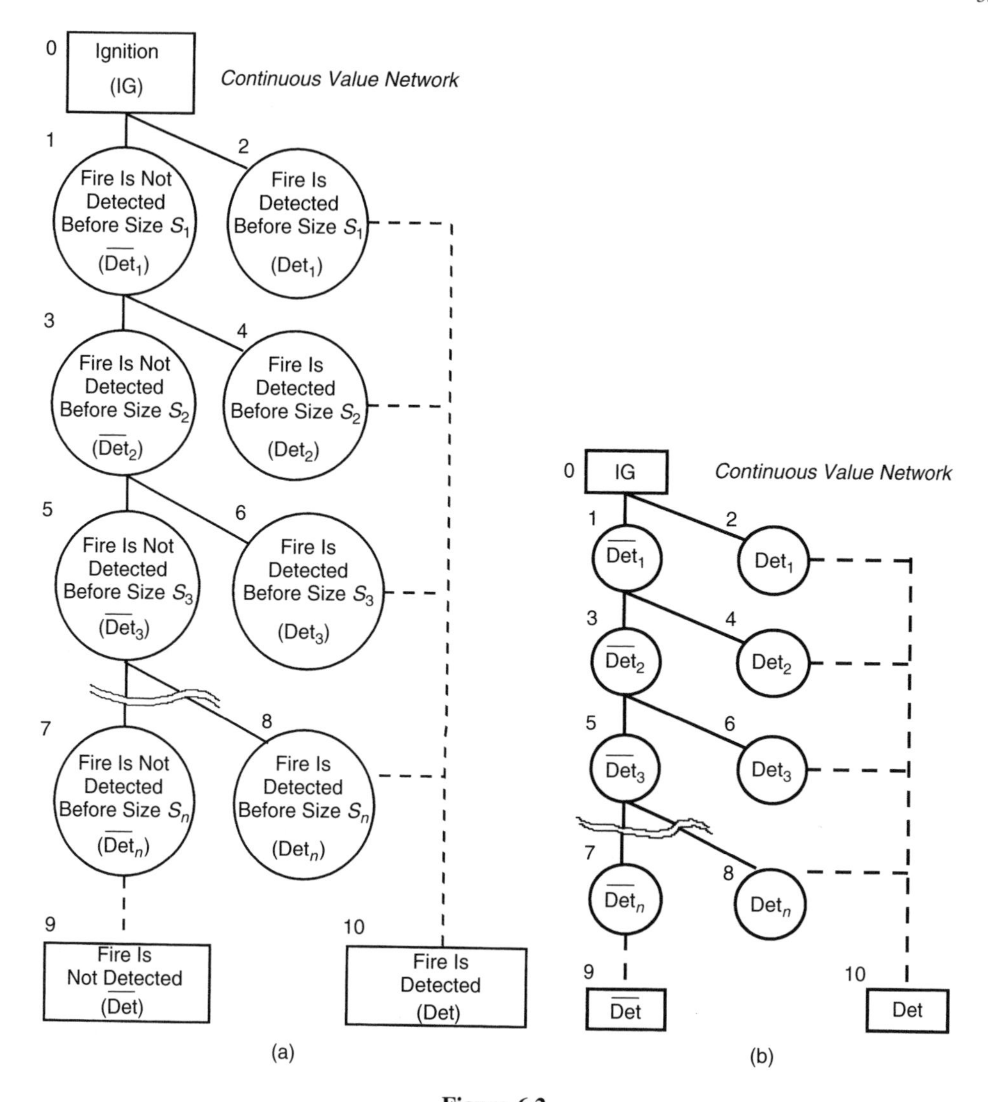

Figure 6.2

as $P(\text{Det}_2|\overline{\text{Det}}_1)$. The probability at size S_2 is $P(\text{Det}_2)$, indicated by point d. The likelihood that the fire will not be detected at this size is $P(\overline{\text{Det}}_2)$.

Path 0–1–3–5–8–10 says that given IG AND the fire is not detected within sizes 1, 2, 3, ..., $n-1$, the fire is detected within size n. This conditional probability is expressed as $P(\text{Det}_n|\overline{\text{Det}}_{n-1})$. At this fire size, the probability that the fire will be detected is $P(\text{Det}_n)$ and the probability that it will not be detected is $P(\overline{\text{Det}}_n)$; they are shown by point f in Figure 6.1.

Path 0–1–3–5–7–9 says that the fire is not detected, $P(\overline{\text{Det}})$.

We define CVN events in terms of performance. Each event and its binary companion constitute all possible outcomes for any given fire size or time. In theory a continuous performance curve may be described by a CVN that consists of all events along a horizontal row of Figure 5.1. In practice only a few events are chosen to avoid unnecessary complexity. The purpose of a CVN is to describe the analytical thought process that is the basis for a performance graph.

6.4 Continuous value networks: calculations and graphing

Continuous value networks describe a component's performance. When one selects probabilistic values for the events, the same framework structures calculations and performance graphs.

One should distinguish between a performance evaluation and the calculation of results. Evaluating performance is the art of selecting probability estimates that express a value judgment for an outcome. Calculations, on the other hand, are the mathematical operations on the probabilistic values. This chapter describes only the process of calculation and graphing results. Other parts of the book discuss evaluations.

Networks provide a form on which to organize calculations. When probability values are shown on the lines separating events, only two rules are needed to calculate the probability of an outcome:

- Multiply the probabilities along a continuous path to determine the outcome for that path.
- Add like outcomes to obtain a result.

Figure 6.3(a) to (c) show the three forms of graph paper. Each plays a different role in developing an understanding of performance, comparing alternatives, and communicating expected behavior. In each graph, the vertical scale shows the probability of success of the component being evaluated. The origin 0.0 (no success) is at the top and 1.0 (complete success) is at the bottom. The user identifies the dynamic measure (e.g., fire size, time) and value scale for the horizontal axis.

Evaluator A is used as a thinking tool for describing level 1 performance estimates. One does not need intermediate numerical values for the ordinate because only general performance characteristics, trends, and relative orders of magnitude are of interest. However, a measurement scale on the abscissa maintains consistency with comparisons.

Level 2 and level 3 evaluations use evaluator B to show component performance. Both axes show numerical values to discriminate differences in performance more distinctly. We use evaluator C only to integrate two or more components into a composite performance description. The expanded scale allows one to portray and visually compare substantial differences.

These evaluators are tools to help in thinking, discriminating, and communicating. The vertical scale for these evaluators is unusual. The paper modifies a probability scale so that one can visualize and discriminate orders of magnitude. For example, the expanded scale of evaluator C shows values of $0.99 = 99/100$, $0.999 = 999/1000$, and $0.9999 = 9999/10\,000$. This scale allows one to easily discriminate the effect of limit (L curve) performance over different extended space/barrier paths. Evaluator B does not use this expanded scale for individual component performance.

Example 6.1 illustrates the process of calculation using continuous value networks and graphing the results on evaluator B.

Example 6.1

Figure 6.4 shows a generic CVN that describes the likelihood for an automatic sprinkler system to control a fire. The subscripts refer to continuously increasing fire sizes. Assume that an individual has completed a performance evaluation and, given a specific room and design fire, the following values have been estimated:

- Fire size when the first sprinkler is expected to fuse: 600 kW. Therefore the probability of sprinkler control at a fire size less than 600 kW is $P(A_{600-}) = 0.0$.
- Probability of fire control when the first sprinkler fuses at 600 kW is $P(A_{600+}) = 0.6$.

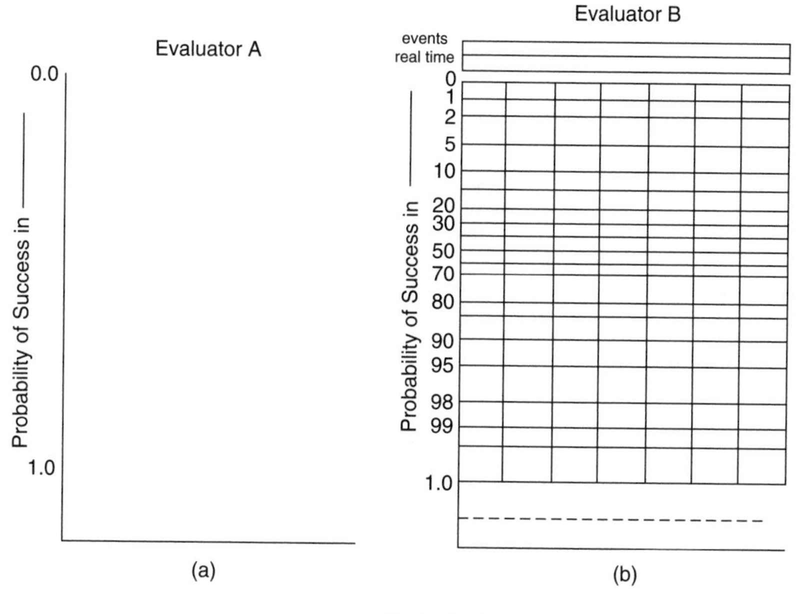
Evaluator A
0.0
1.0
Probability of Success in
(a)
Evaluator B
events
real time
0
1
2
5
10
20
30
50
70
80
90
95
98
99
1.0
Probability of Success in
(b)

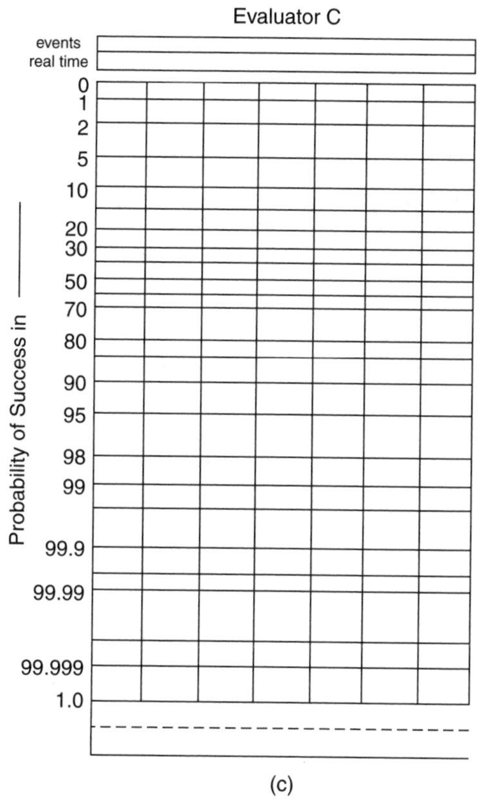
Evaluator C
events
real time
0
1
2
5
10
20
30
50
70
80
90
95
98
99
99.9
99.99
99.999
1.0
Probability of Success in
(c)

Figure 6.3

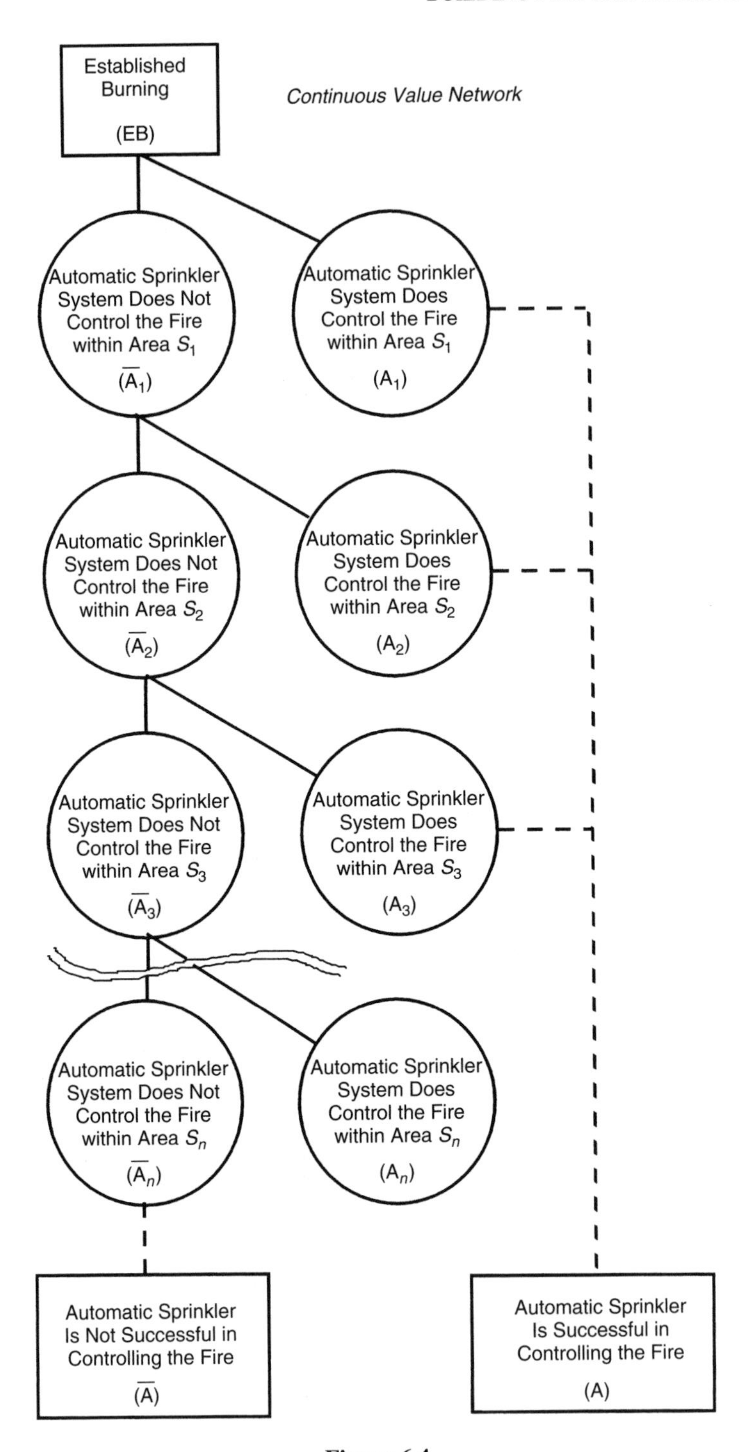
Established
Burning
(EB)
Continuous Value Network
Automatic Sprinkler System Does Not Control the Fire within Area S_1
($\overline{A}_1$)
Automatic Sprinkler System Does Control the Fire within Area S_1
(A_1)
Automatic Sprinkler System Does Not Control the Fire within Area S_2
($\overline{A}_2$)
Automatic Sprinkler System Does Control the Fire within Area S_2
(A_2)
Automatic Sprinkler System Does Not Control the Fire within Area S_3
($\overline{A}_3$)
Automatic Sprinkler System Does Control the Fire within Area S_3
(A_3)
Automatic Sprinkler System Does Not Control the Fire within Area S_n
($\overline{A}_n$)
Automatic Sprinkler System Does Control the Fire within Area S_n
(A_n)
Automatic Sprinkler Is Not Successful in Controlling the Fire
($\overline{A}$)
Automatic Sprinkler Is Successful in Controlling the Fire
(A)

Figure 6.4

- Probability that the sprinklers will control a 900 kW fire given that it was not controlled at 600 kW is $P(A_{900}|\overline{A}_{600}) = 0.75$.
- Probability that the sprinklers will control a 1600 kW fire given that it was not controlled at 900 kW is $P(A_{1600}|\overline{A}_{900}) = 0.4$.
- Probability that the sprinklers will control a 3 MW fire given that it was not controlled at 1.6 MW $= P(A_3|\overline{A}_{1.6}) = 0.4$.
- Probability that the sprinklers will control a 5 MW fire given that it was not controlled at 3 MW $= P(A_5|\overline{A}_3) = 0.1$.
- Probability that the sprinklers will control an 8 MW fire, given that it was not controlled at 5 MW $= P(A_8|\overline{A}_5) = 0.0$.

Construct an A curve to represent the cumulative sprinkler system success in controlling the fire.

Solution

Figure 6.5(a) shows a CVN with concise notation for the fire sizes and probability values. Although the problem statement gave only one set of values, the sum of each set of lines (here each binary pair) represents all possibilities that can occur from that event. That sum must always add to 1.0. Consequently, knowing the value for one event enables one to calculate the other by subtraction. Thus, when $P(A_{1600}|\overline{A}_{900}) = 0.4$, $P(\overline{A}_{1600}|\overline{A}_{900}) = 0.6$, as shown in Figure 6.5(a).

Using the calculation rules described above, the probability multiplication along each path is shown on Figure 6.5(b) and noted as outcomes (a) to (g) and (n). The cumulative limit values are noted as outcomes (h) to (n). The cumulative fire size is noted in the last column. Cumulative limit values are determined as follows:

$$P(A_{600-}) = \text{(a) and (h)} = 0.0,$$

$$P(A_{600+}) = (h + b = i) = 0.0 + 0.6 = 0.6,$$

$$P(A_{900}) = (i + c = j) = 0.6 + 0.3 = 0.9,$$

$$P(A_{1600}) = (j + d = k) = 0.9 + 0.04 = 0.94,$$

$$P(A_{3000}) = (k + e = l) = 0.94 + 0.024 = 0.964,$$

$$P(A_{5000}) = (l + f = m) = 0.964 + 0.0036 = 0.9676$$

$$P(A_{8000}) = (m + g = n) = 0.9676 + 0.0 = 0.9676.$$

These numbers may be rounded to the desired number of digits. Figure 6.5(b) shows the CVN graph of these numerical results. Note the following in this pictorial representation of sprinkler system performance:

- The origin of the graph is in the upper left corner.
- The probability that the fire will be controlled within 600– kW is plotted as 0.0, corresponding to outcome (h).
- The next coordinate, point (i), shows the estimate when sprinklers initially actuate. We plot the value of $P(A_{600+}) = 0.6$ directly under $P(A_{600-})$.
- The next coordinate, point (j), is the sum of the previous probability $P(A_{600+}) = 0.6$ and the conditional probability $P(A_{900}|\overline{A}_{600}) = (0.75)(0.4) = 0.3$. Thus, the value for point (j) is $0.6 + 0.3 = 0.9$.

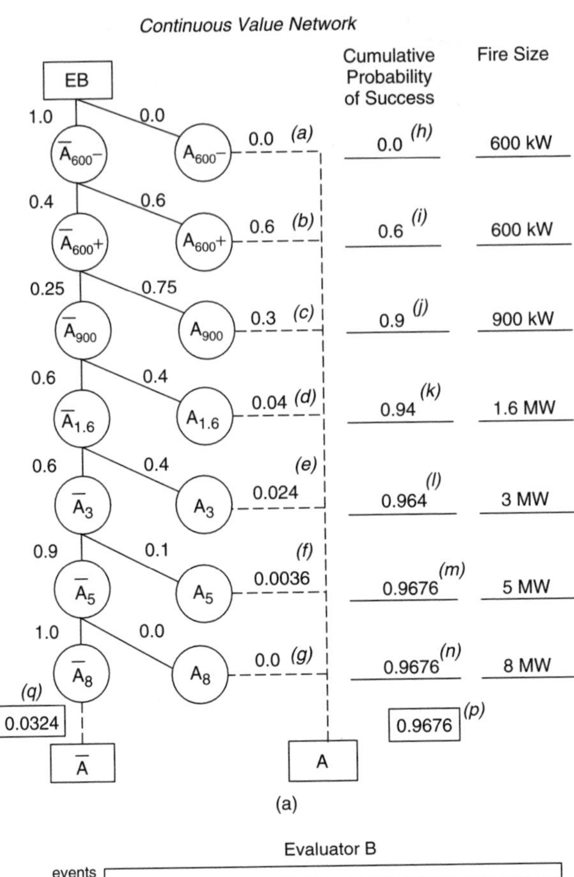

Continuous Value Network
Cumulative Probability of Success
Fire Size
EB
1.0
0.0
0.0 (a)
0.0 (h)
600 kW
0.4
0.6
0.6 (b)
0.6 (i)
600 kW
0.25
0.75
0.3 (c)
0.9 (j)
900 kW
0.6
0.4
0.04 (d)
0.94 (k)
1.6 MW
0.6
0.4
(e) 0.024
0.964 (l)
3 MW
0.9
0.1
(f) 0.0036
0.9676 (m)
5 MW
1.0
0.0
0.0 (g)
0.9676 (n)
8 MW
(q) 0.0324
0.9676 (p)
A

(a)

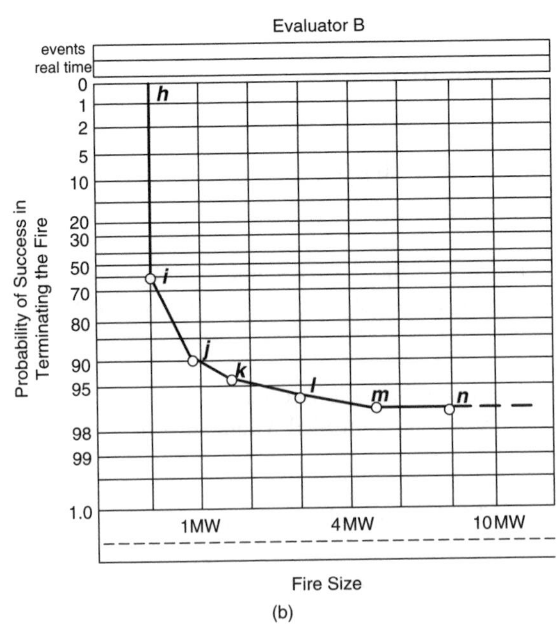

Evaluator B
events
real time
Probability of Success in Terminating the Fire
0
1
2
5
10
20
30
50
70
80
90
95
98
99
1.0
h
i
j
k
l
m
n
1MW
4MW
10MW
Fire Size

(b)

Figure 6.5

- Similarly, the other coordinates are obtained by adding the conditional change to the previous value. The conditional change is always the product of the probabilistic values along the chain from the start to the terminal event of the chain. Thus, point (k) is $P(\mathrm{A}_{900}) + P(\mathrm{A}_{1600}|\overline{\mathrm{A}}_{900})$. Or, $P(\mathrm{A}_{1600}) = 0.9 + 0.04 = 0.94$.
- The probability that a sprinkler system will not extinguish for a fire larger than 8000 kW is obtained by multiplying the probabilities along the chain from EB to $\overline{\mathrm{A}}$. Thus, $\mathrm{P} = (1.0)(0.4)(0.25)(0.6)(0.6)(0.9)(1.0) = 0.0324$.
- The sum $P(\mathrm{A}) + P(\overline{\mathrm{A}})$ is 1.0 always. Thus, $0.9676 + 0.0324 = 1.0$.

6.5 Single value networks: concepts

A single value network (SVN) is used to determine each coordinate of a continuous value graph. A SVN enables one to stop a scenario and analyze the conditions at that instant of time. The SVN organization allows one to evaluate related functions as a group and then combine them into a hierarchical package that structures a complete performance unit. The events are selected to enable individuals to make judgments in relatively comfortable increments – neither too large to cause discomfort in making estimates nor too small to require unnecessary repetition. The complete framework integrates a mathematical logic with an analytical thought process.

To illustrate SVN organization, Figure 6.6 shows the generic network hierarchy for evaluating sprinkler system performance in controlling a specific fire size. The evaluation incorporates two main components. The first evaluates the reliability of the system to deliver water to a fused sprinkler. We call this AA (for agent application). The second estimates the operational effectiveness of the sprinkler system to control a fire of specified size, given that water will issue from a sprinkler. We call this AC (for automatic sprinkler control). The reliability (AA) is evaluated once for the entire system or zone, whereas the operational effectiveness (AC) is evaluated for each fire size selected for analysis.

This example illustrates the hierarchical form of most networks. The factors that influence system reliability are independent of the operational effectiveness. Consequently, the evaluation of "Will the first drop of water (AA) come out of the fused sprinkler?" Agent application (reliability) is a precondition to "Will enough water come out of the sprinklers and be distributed effectively (AC) to control the fire being analyzed?".

Independence, dependence, and conditionality of the events have been incorporated into all networks. These considerations are important to evaluations, although they are not a problem in calculations. For example, events vaa (all valves are open) and waa (water will reach the fused sprinkler) are independent. Because each event is evaluated independently of the other, their order in Figure 6.6(b) is not important. The result (both in understanding and in numerical values) will be identical regardless of their order.

The starting event of Figure 6.6(c) is agent application (AA). That is, given AA, we evaluate the operational effectiveness (AC) as a separate entity. Figure 6.6(d) shows their combination. The event sequence in Figure 6.6(c) is not independent, and the functional order that guides judgmental thinking is important. For example, control (AC) is dependent first upon the determination that the sprinklers will fuse before the fire size increases to a larger size (fac). Then, based on the number of sprinklers that fuse, we sequentially estimate the water density (dac) and the water flow (cac) to supply that density. Finally, obstructions to effective water distribution (wac) are evaluated. Event conditionality or independence is usually provided in the discussion of component evaluations. We have taken care in structuring network order so that evaluations logically track a functional operations sequence.

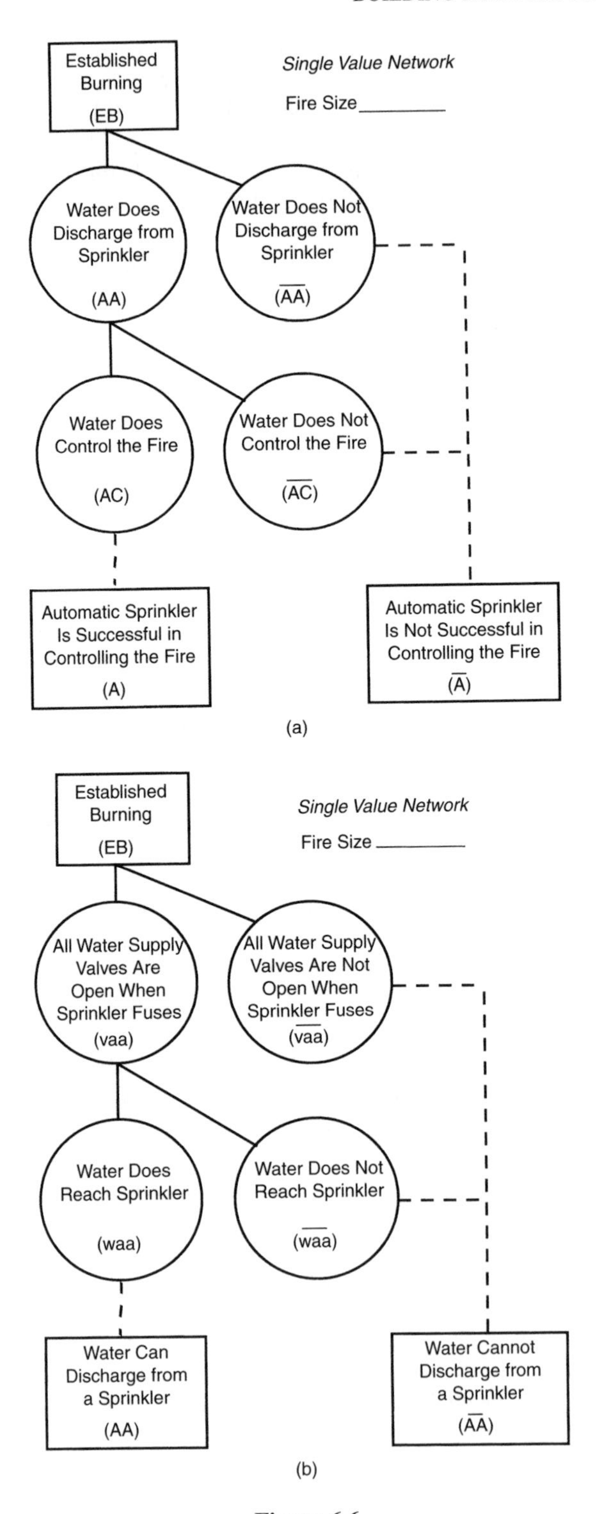
Established Burning (EB)
Single Value Network
Fire Size
Water Does Discharge from Sprinkler (AA)
Water Does Not Discharge from Sprinkler ($\overline{AA}$)
Water Does Control the Fire (AC)
Water Does Not Control the Fire ($\overline{AC}$)
Automatic Sprinkler Is Successful in Controlling the Fire (A)
Automatic Sprinkler Is Not Successful in Controlling the Fire ($\overline{A}$)
(a)
Established Burning (EB)
Single Value Network
Fire Size
All Water Supply Valves Are Open When Sprinkler Fuses (vaa)
All Water Supply Valves Are Not Open When Sprinkler Fuses ($\overline{vaa}$)
Water Does Reach Sprinkler (waa)
Water Does Not Reach Sprinkler ($\overline{waa}$)
Water Can Discharge from a Sprinkler (AA)
Water Cannot Discharge from a Sprinkler ($\overline{AA}$)
(b)

Figure 6.6

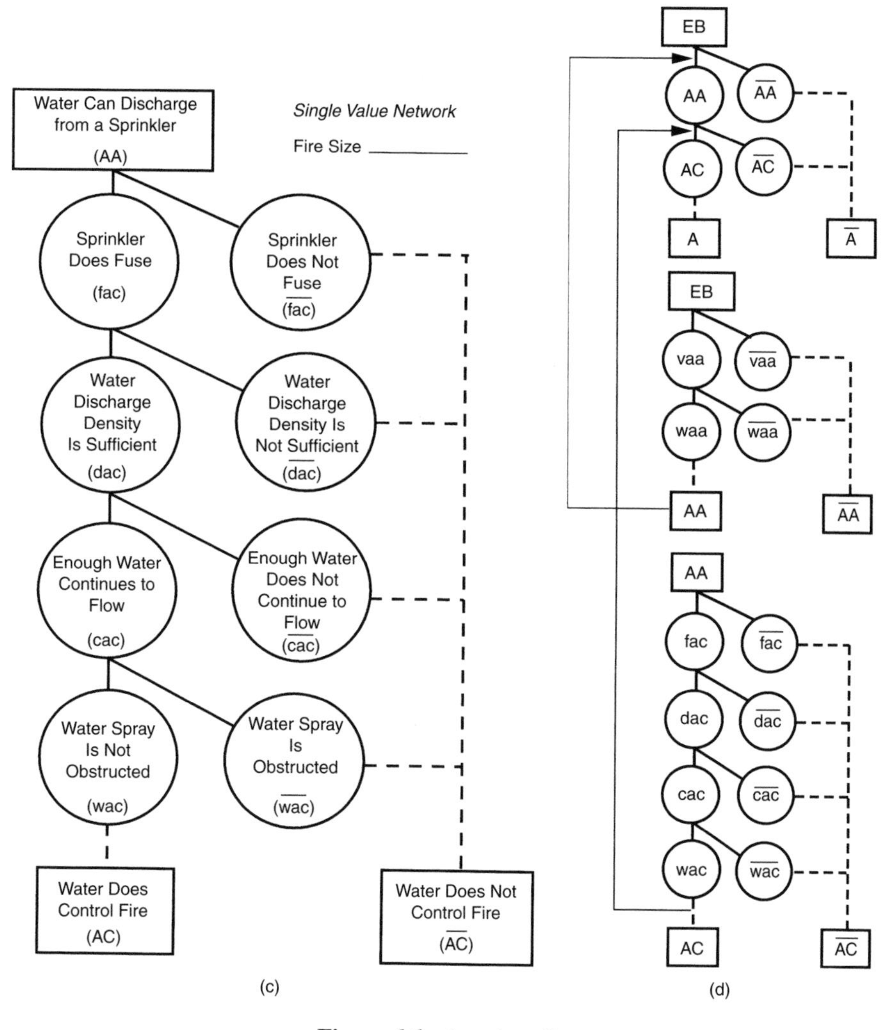

Figure 6.6 (*continued*)

6.6 Single value networks: calculations

Level 2 evaluations use single value networks to provide an opportunity to study details that affect performance. After one estimates probabilistic values for each event, the SVN can structure the calculations. The results provide a single coordinate for a continuous value graph.

Single value networks have the same appearance as continuous value networks. The rules for calculating with a network are the same as described in Section 6.4 except that one calculates only a single coordinate with an SVN. Cumulative probability values are neither necessary nor computed. Examples 6.2 and 6.3 illustrate the process.

Example 6.2

An individual has evaluated an automatic sprinkler system in a building. Calculate the probability that an automatic sprinkler system will control the design fire for the following estimates.

Fire size: 900 kW

Reliability assessment (AA)	Operational effectiveness (AC)
$P(\text{vaa}) = 0.98$	$P(\text{fac}) = 0.99$
$P(\text{waa}) = 0.999$	$P(\text{dac}) = 0.98$
	$P(\text{cac}) = 1.0$
	$P(\text{wac}) = 0.95$

Solution

This exercise demonstrates calculation procedures using network diagrams. The problem statement gives values for success, and we determine their complements by recognizing that each

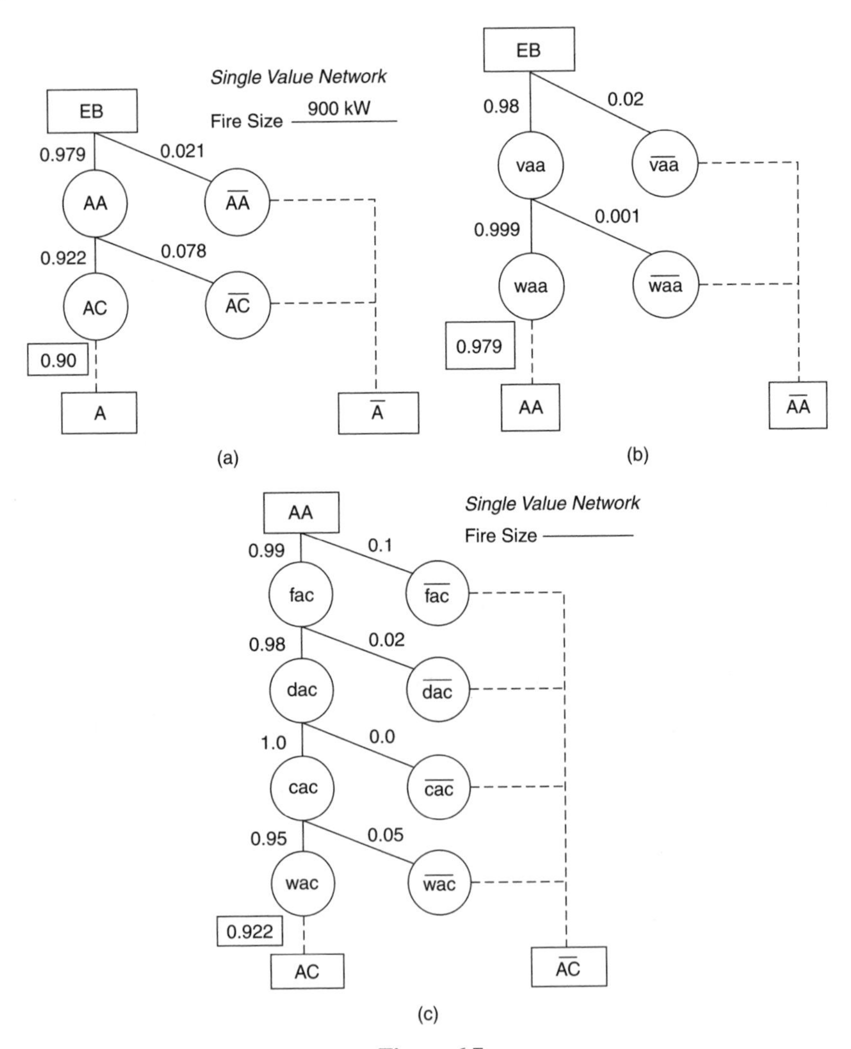

Figure 6.7

binary set must add to 1.0. Figure 6.7(b) shows the reliability calculation to be $P(\text{AA}) = 0.979$, and Figure 6.7(c) structures calculations for sprinkler system control to be $P(\text{AC}) = 0.922$. The control SVN of Figure 6.7(a) incorporates these results to calculate $P(\text{A}) = 0.9$. These calculations show a likelihood of 90% that the sprinkler system will control this fire of 900 kW.

Example 6.3

The individual who estimated the sprinkler system of Example 6.2 repeated the evaluation for a fire size of 1600 kW. The following estimates were made.

Fire size = 1600 kW

Reliability assessment (AA)	Operational effectiveness (AC)
Same as Example 6.2 because it is the same room	$P(\text{fac}) = 1.0$ $P(\text{dac}) = 0.96$ $P(\text{cac}) = 1.0$ $P(\text{wac}) = 0.999$

Solution

Figure 6.8(b) shows the reliability to be 0.979. Figure 6.8(c) shows the likelihood of sprinkler system control as $P(\text{AC}) = 0.959$. Figure 6.8(a) combines these results to calculate $P(\text{A}) = 0.94$. The calculations show the sprinkler system control to be $P(\text{A}) = 0.94$.

6.7 Networks and performance curves: discussion

The continuous value networks discussed in Section 6.5 describe the thought process that defines performance curves. Unfortunately, a major difficulty arises in evaluating the conditional probabilities. Except for the ST (self-termination) curve described in Chapter 8, there is no practical way to estimate the conditional probabilities for CVN evaluations. Fortunately, the SVN evaluations can provide a way around this difficulty. Let us look at this apparent inconvenience more carefully.

The SVN evaluations of Examples 6.2 and 6.3 describe performance estimates for the selected fire sizes of 900 kW and 1600 kW. These values correspond to the values of points j and k in Figure 6.5(a) and (b). The difference between these values is the increment (point $d = 0.04$) between these sizes. If SVN values were determined for a series of fire sizes, the CVN curve could be constructed, and conditional probabilities could be back-calculated from those SVN values. However, this mathematical exercise does not produce any new understanding about the sprinkler system performance.

The main attribute of a continuous value curve is to provide a visual portrayal of component performance. This useful attribute can be valuable in contributing to your own understanding or in communicating differences in performance conditions. One can sketch a CV curve by understanding the SVN performance for one or a few astutely selected fire conditions. The SVN evaluations provide the foundation for understanding performance. One can extrapolate this knowledge to sketch CV curves of dynamic performance.

6.8 The L curve

The L curve describes the limit (extent) of expected flame/heat damage for a building which provides the knowledge base on which smoke analysis, structural frame performance, and risk

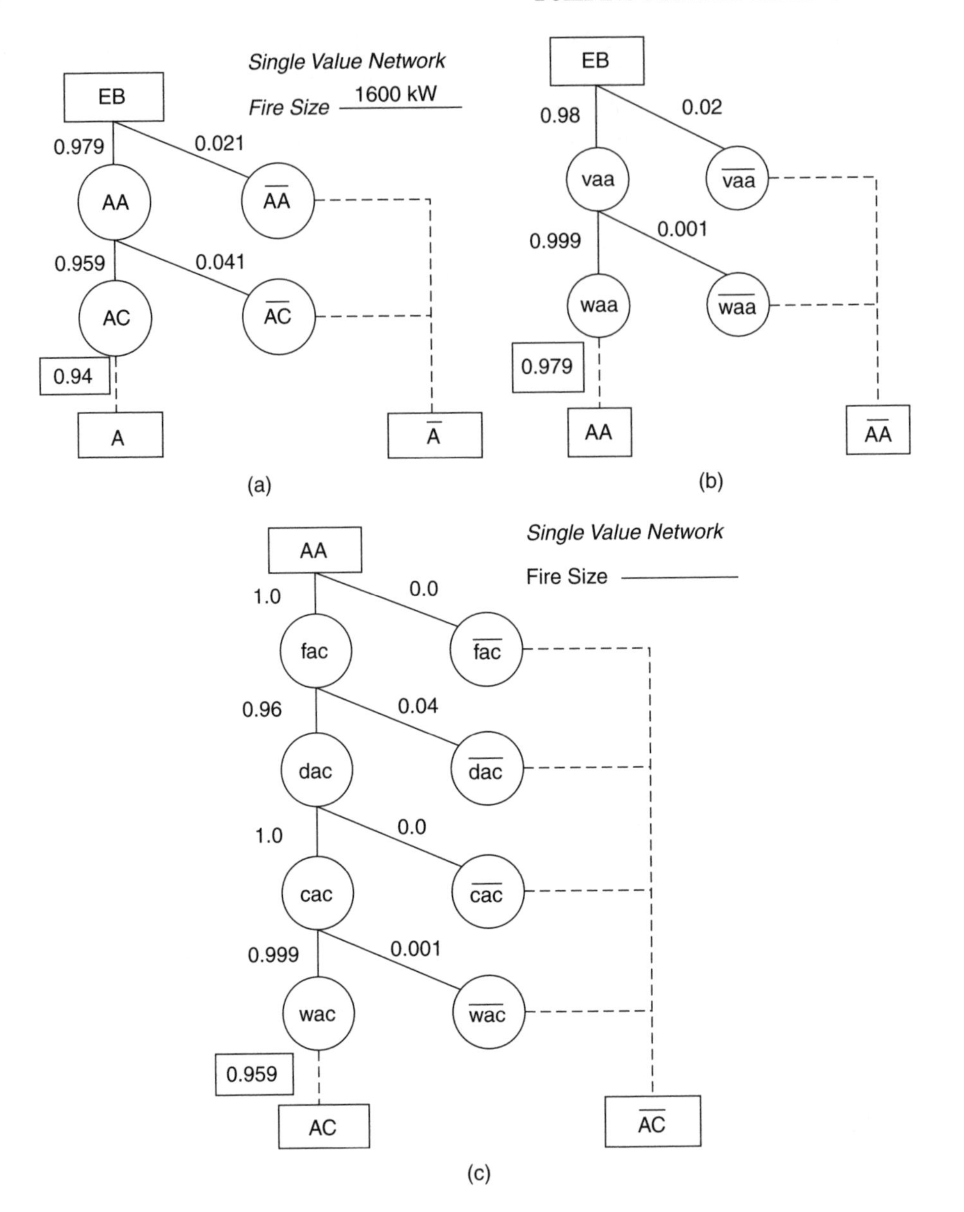

Figure 6.8

characterizations are founded. The L curve graph is also a convenient way to compare design alternatives.

The term "L curve" may signify a room of origin, a building zone, or an entire building. The context of use will make the meaning clear. Here we show L curve calculations to illustrate a way to integrate conditions and provide a unifying picture of performance. Chapter 14 describes additional applications.

6.9 The L curve for a room of origin

A fire can go out in a space in any of three ways. (1) A fire can go out itself; this is shown by an I curve and explained in Chapter 8. (2) An automatic sprinkler system can put out the fire; this is shown by an A curve and explained in Chapter 13. (3) The fire department can extinguish the fire; this is shown by an M curve and explained in Chapter 12.

Figure 6.9 shows a SVN that combines these components to obtain L curve coordinates. Because A and M are evaluated independently, their order after $\bar{I}$ is not important. The calculated

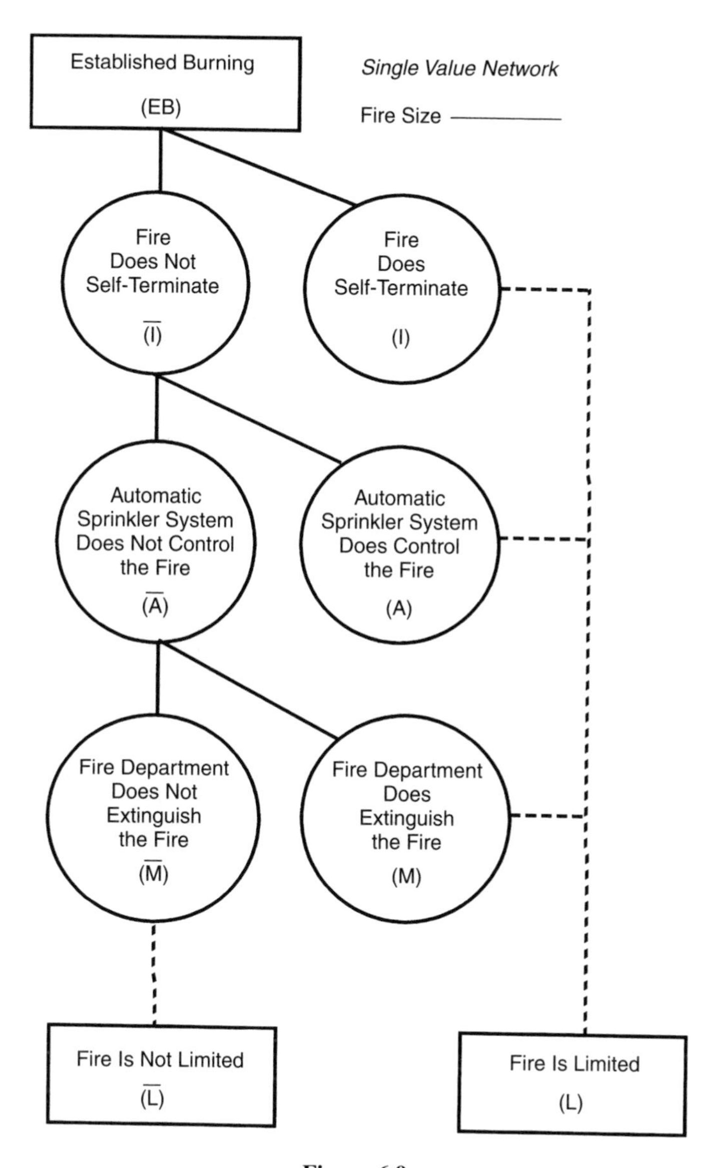

Figure 6.9

L value will remain the same regardless of their order. The order in Figure 6.9 is generally used for convenience. Examples 6.4 and 6.5 illustrate calculations to construct an L curve after the I, A, and M curves have been evaluated.

Example 6.4

The I, A, and M curves of Figure 6.10 represent CV performance estimates for the fire and suppression conditions for a room of origin. Calculate the coordinates for the limit (L) for fire sizes of 350 kW, 1 MW, and FRI. Sketch the L curve for the room.

Solution

Each of these three L curve coordinates represents an instant in time or, in this case, a specific fire size. The SVN of Figure 6.11(a) to (c) show the L curve calculations for each of the fire sizes. Note that the individual outcomes are added directly. The dashed line of Figure 6.10 shows the L curve sketch.

Discussion

Incorporating the I, A, M values and their complements into Figure 6.11 allows one to calculate the L values for these fire sizes. One need not calculate intermediate cumulative values because an SVN is used.

The calculations for any single value network will produce the same numerical result regardless of the order in which the events are positioned. In Figure 6.11 one can reverse the order of A

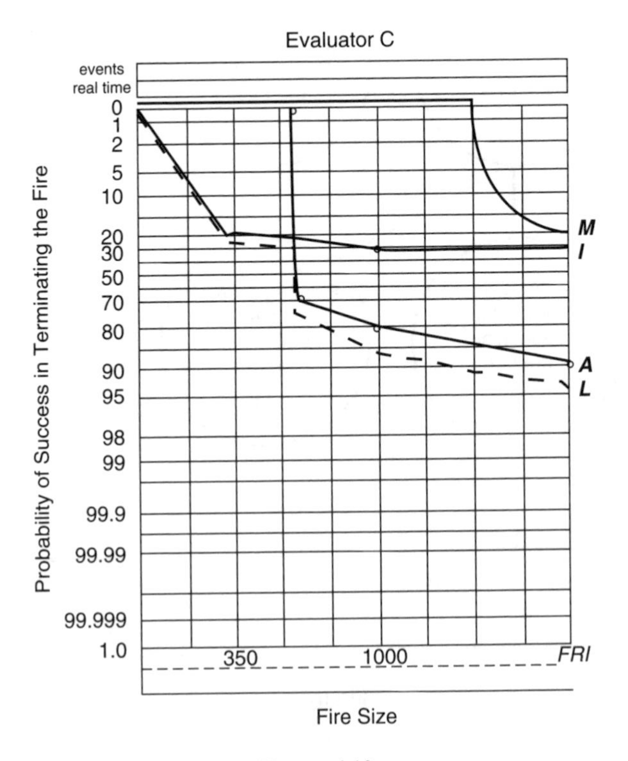

Figure 6.10

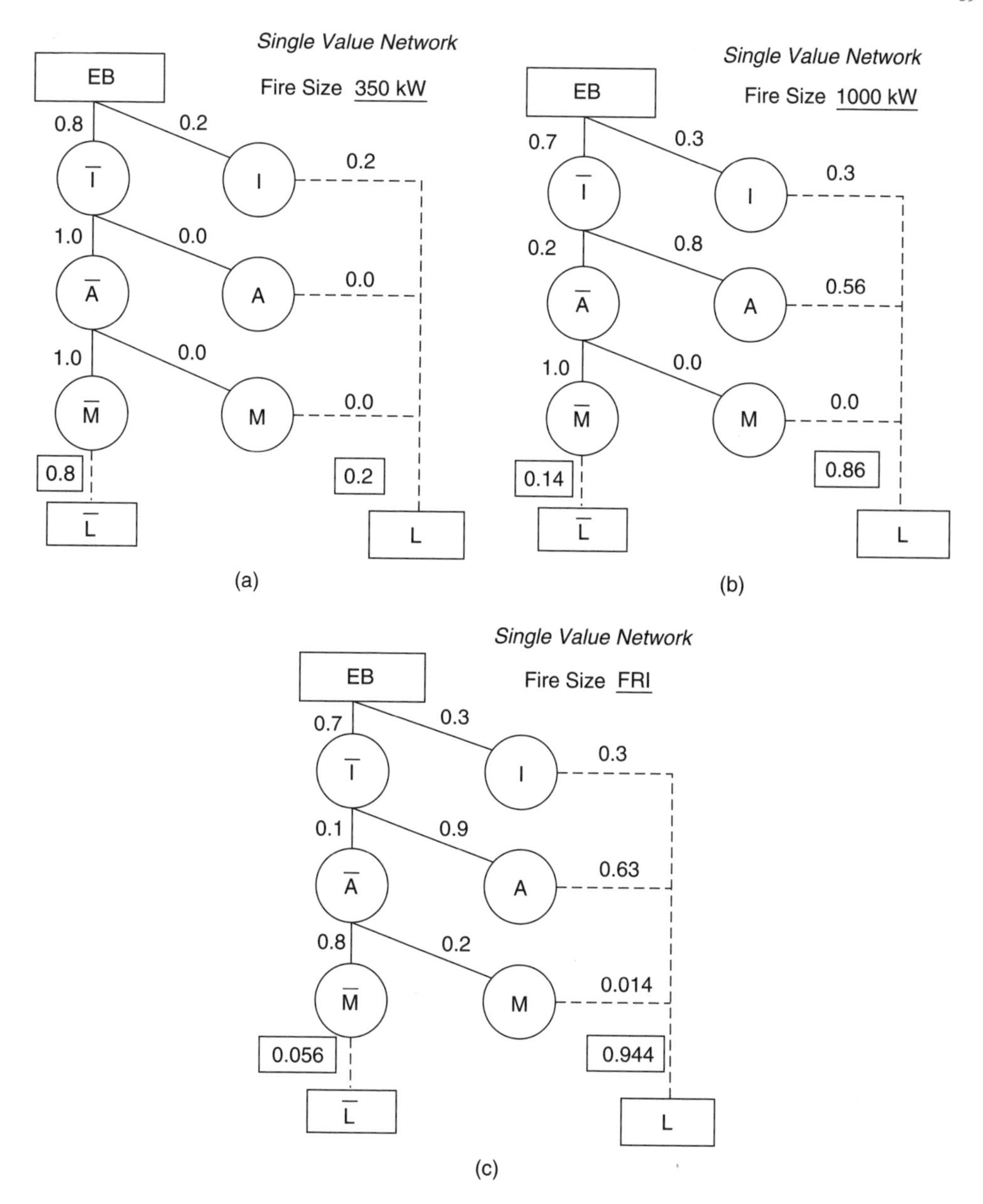

Figure 6.11

and M and obtain the same numerical L value. Note that this comment relates to calculations and not to evaluations (i.e., determining the values of probability). When making evaluations for probabilistic estimates, the dependency (and order) may be very important.

It is rarely necessary to calculate more than one coordinate because at any fire size, the L curve is always better than or equal to the best of the other curves. It is always below (better than) the best of the I, A, and M curves. Consequently, if one calculates a single point (usually at the full room size), the curve for the room can be sketched by observation.

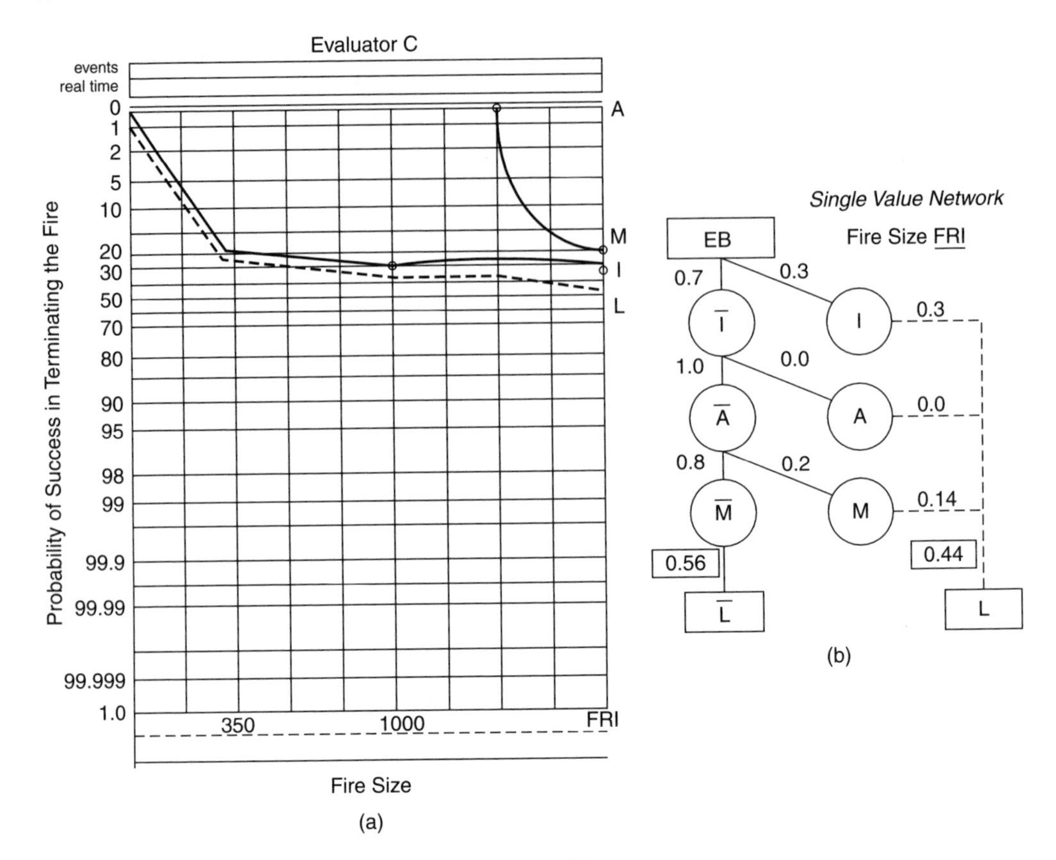

Figure 6.12

Example 6.5

Solve Example 6.4 if the automatic sprinkler were not in the room.

Solution

Figure 6.12(a) shows the I, M, and A curves for this condition. The probability of success of the nonexistent sprinkler system must be zero, as shown by the horizontal A curve.

Only the single coordinate at FRI will be calculated, as shown in Figure 6.12(b). The remainder of the L curve can be sketched by recognizing that the L curve is always better than the best of its component curves.

6.10 The L curve for a building fire path

The L curve can also describe the flame/heat performance for a sequence of rooms, a cluster of rooms, or for an entire building. To illustrate the concept of space/barrier fire propagation, consider the three-room building of Figure 6.13. Assume a fire starts in room 1 and, depending on the contents, construction, and suppression effectiveness, it may or may not extend to rooms 2 and 3.

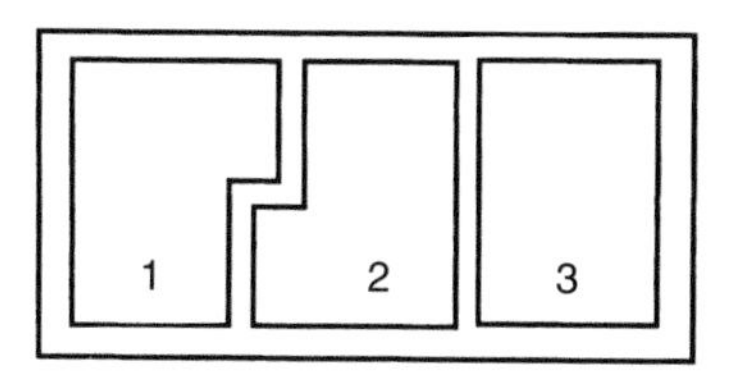

Figure 6.13

Let L signify the limit of flame spread in a room, and B signify barrier success in preventing an ignition in the adjacent space. Figure 6.14(a) shows the analytical network that describes all possible outcomes for the building, given an EB in room 1.

It is inconvenient to describe the events with the wording detail shown in Figure 6.14(a). Figure 6.14(b) shows the same network in symbolic form. As described earlier, the overbar $^{-}$ means "not."

To describe the meaning of the L curve network, each event is numbered for reference. The outcomes for the possible paths are as follows:

Path 0–1–3–5–7–9–11 shows the events that must occur for total building destruction.

Path 0–2 shows that the fire will be terminated before leaving room 1.

Path 0–1–4 shows that the fire fully involved room 1 and the barrier stopped the fire at that point.

Path 0–1–3–6 shows that the fire fully involved room 1, breached the barrier between rooms 1 and 2, and terminated somewhere in room 2.

Path 0–1–3–5–8 shows that the fire fully involved room 1, breached the barrier between rooms 1 and 2, fully involved room 2, and was stopped at that point by the barrier between rooms 2 and 3.

Path 0–1–3–5–7–10 shows that the fire fully involved room 1, breached the barrier between rooms 1 and 2, fully involved room 2, breached the barrier between rooms 2 and 3, and was terminated somewhere in room 3.

A single diagram can describe all possible magnitudes of fire damage within this building. One need only construct a network to identify the space/barrier elements and interpret the meaning of the events. One can structure a more complicated building similarly to describe the path or paths to represent the fire propagation of interest.

Example 6.6 illustrates L curve calculations and graphs.

Example 6.6

An inspector evaluated the performance of the three-room building in Figure 6.13 and determined the limit of flame movement for each space and the barrier effectiveness. The results are as follows:

$$\text{Limit of flame movement in room 1} = P(\text{L}_1) = 0.30,$$

$$\text{Success of the barrier between rooms 1 and 2} = P(\text{B}_{1-2}) = 0.20,$$

$$\text{Limit of flame movement in room 2} = P(\text{L}_2) = 0.40,$$

$$\text{Success of barrier between rooms 2 \& 3} = P(\text{B}_{2-3}) = 0.20,$$

$$\text{Limit of Flame movement in room 3} = P(\text{L}_3) = 0.15.$$

Calculate the limit of flame movement for the building and plot the results on evaluator 3.

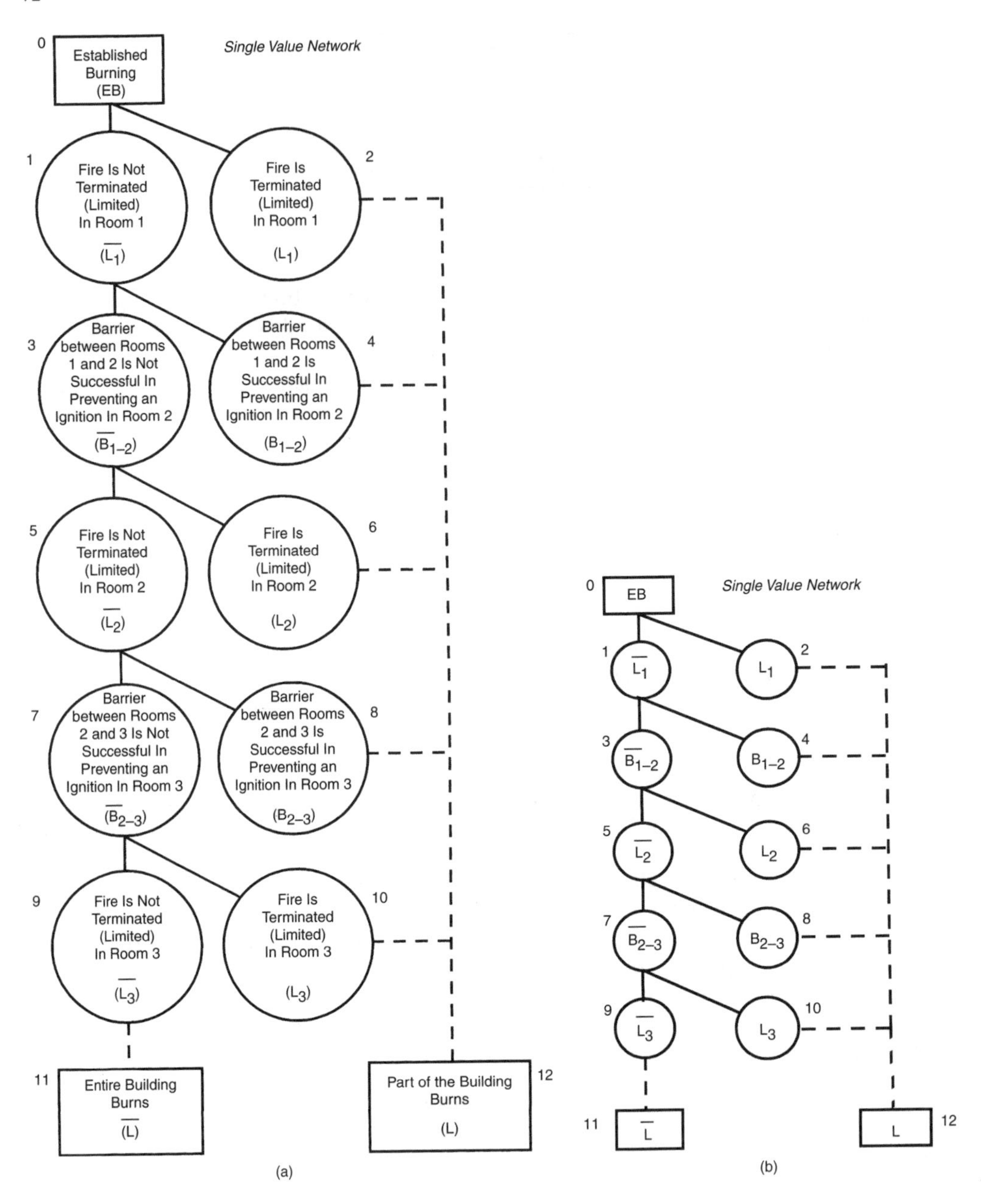

Figure 6.14

Solution

Figure 6.15(a) shows the CVN for all possible outcomes for this space/barrier analysis. Alphabetical symbols correspond to the event descriptions given above.

Using the calculation rules of Section 6.4, the probability along each path is shown on Figure 6.15(a) and noted as outcomes (a) to (e) and (l). The cumulative limit values are noted as

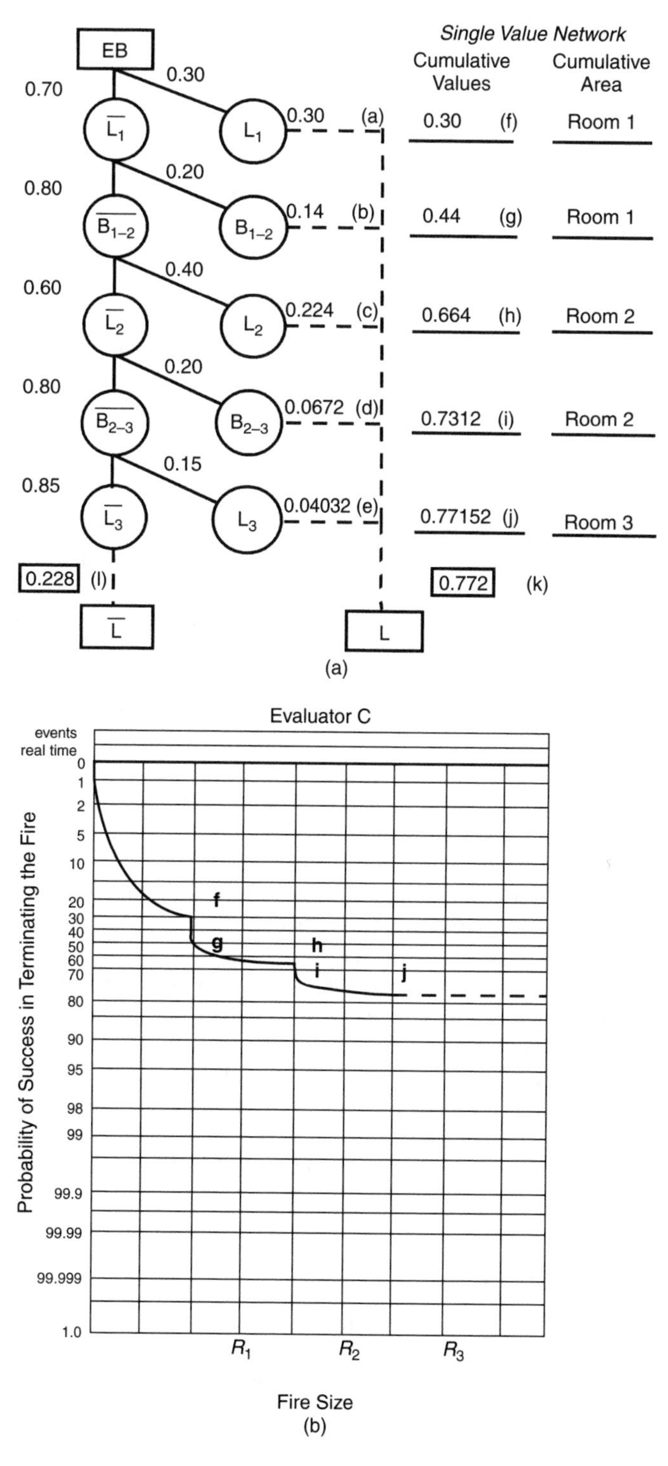

Single Value Network
Cumulative Values
Cumulative Area
EB
0.70
0.30
0.30 (a)
0.30 (f)
Room 1
0.80
0.20
0.14 (b)
0.44 (g)
Room 1
0.60
0.40
0.224 (c)
0.664 (h)
Room 2
0.80
0.20
0.0672 (d)
0.7312 (i)
Room 2
0.85
0.15
0.04032 (e)
0.77152 (j)
Room 3
0.228 (l)
0.772 (k)
L
(a)
Evaluator C
events
real time
Probability of Success in Terminating the Fire
f
g
h
i
j
R1
R2
R3
Fire Size
(b)

Figure 6.15

outcomes (f) to (k). The associated building sizes are shown in the last column. The cumulative limit values are

$$P(\mathrm{L}_1) = \text{(a) and (f)} = 0.30,$$

$$P(\mathrm{B}_{1-2}) = (\mathrm{f} + \mathrm{b} = \mathrm{g}) = 0.30 + 0.14 = 0.44,$$

$$P(\mathrm{L}_2) = (\mathrm{g} + \mathrm{c} = \mathrm{h}) = 0.44 + 0.224 = 0.664,$$

$$P(\mathrm{B}_{2-3}) = (\mathrm{h} + \mathrm{d} = \mathrm{i}) = 0.664 + 0.0672 = 0.7312,$$

$$P(\mathrm{L}_3) = (\mathrm{i} + \mathrm{e} = \mathrm{j}) = 0.7312 + 0.04032 = 0.77152.$$

Figure 6.15(b) shows the graph of these values. In this graph, note the following:

1. The origin of the graph is in the upper left corner.
2. The limit for room 1 is plotted as 0.30, corresponding to outcome (f).
3. The next coordinate, point (g), is $P(\mathrm{B}_{1-2})$ and is plotted directly under $P(\mathrm{L}_1)$.
4. Note that up to point (g) the graph shows five items:
   a. The probability of success in preventing FRI in room 1 is $P(\mathrm{L}_1) = 0.30$.
   b. A fire that started in room 1 has a probability of success in avoiding an ignition in room 2 of 0.44.
   c. The results are cumulative. The expectation that the barrier between rooms 1 and 2 will have a likelihood of success of 20% is applied only to the 70% likelihood that the fire has not terminated before involving the full room. Consequently, (0.70)(0.20) = 0.14. This is the vertical distance between (f) and (g). The probability that the fire *can cause* an ignition in room 2 is 1.00 – 0.44 = 0.56. Therefore a fire originating in room 1 will have a 56% likelihood of producing an ignition and EB in room 2.
   d. Similarly, point (h) applies the 40% chance of room 2 termination to the 56% likelihood of the fires to get beyond the barrier between rooms 1 and 2. Therefore the success of limiting the fire to two rooms is 0.44 + 0.224 = 0.664.
   e. The terminal point (j) says that an established burning (EB) started in room 1 will be 77% successful in avoiding full building involvement.

In Figure 6.15(a) the numerical value representing success in avoiding full building involvement is point (j), having a numerical value of 0.77152. Its complement is the likelihood that the entire building will be lost. This is obtained by multiplying the probabilistic values along the path from EB to L_3. This value is 0.22848. These corresponding numbers must always add to 1.00 to signify that all the possibilities have been included. One can round off the numbers to values that seem appropriate. Here we may say there is a 23% chance of full building involvement and a 77% likelihood that the fire will terminate somewhere before full building involvement. Figure 6.15(b) shows the building L curve.

6.11 L curve communication

The main purpose of a graph is to communicate information. In fire safety evaluations, these graphs show performance comparisons for different conditions. Let us pause to read some L curves and to interpret the different building performance characterizations they represent.

Figure 6.16 shows a representative L curve for three sequential rooms. The evaluation would have been based on acquired knowledge of the building. Performance is the probability of success in limiting the fire damage to the spaces shown. The location of the graph's origin means that any position near the top of the graph shows a low probability of termination that describes a

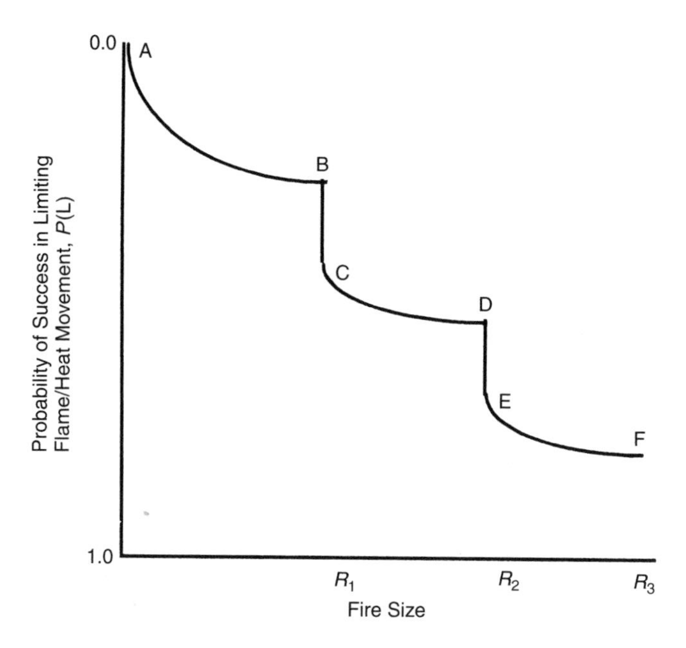

Figure 6.16

bad fire safety performance. Conversely, positions near the bottom of the paper indicate a good performance.

In Figure 6.16 segment AB describes a judgment of the relative likelihood that a fire will terminate within the room of origin. The vertical line from B to C describes the barrier effectiveness between room 1 and room 2. A short line indicates an ineffective barrier. A long line signifies a strong, effective barrier. If the vertical line had extended down to the value 1.0, the individual believes with certainty that the fire cannot propagate into the adjacent room. In Figure 6.16 the individual made a value judgment that there is some likelihood the fire can breach the barrier and extend into the adjacent room.

Continuing to read the L curve, segment CD shows the likelihood that the fire will terminate within room 2. Vertical line DE describes the effectiveness of the barrier between rooms 2 and 3, and segment EF indicates fire termination within room 3. Here the individual making the evaluation believes there is a chance that the fire could involve all three rooms because the certainty value of 1.0 was not reached.

Figure 6.17 shows L curves for two three-room buildings (or two alternative fire defenses within the same building). Building A is noticeably weaker than Building B. Within the room of origin, the two buildings are identical. However, the barriers between rooms 2 and 3 show that the barrier performance for Building B is somewhat better than that of Building A. Building B has a substantially more effective barrier between rooms 1 and 2. The barrier effectiveness provides a much better total performance for Building B.

Figure 6.18 shows L curves for two additional buildings (or alternatives). The barrier between rooms 1 and 2 in Building C is completely ineffective because there is no vertical segment. The barrier between rooms 2 and 3 is much more effective in preventing fire propagation into room 3. Building D has the same ineffective barrier between rooms 1 and 2. However, the L curve for room 1 is quite good. This might occur because different fire defenses, such as a sprinkler system, had been installed. Therefore, the cumulative L curve for the three-room fire path of Building D is much better than that for Building C.

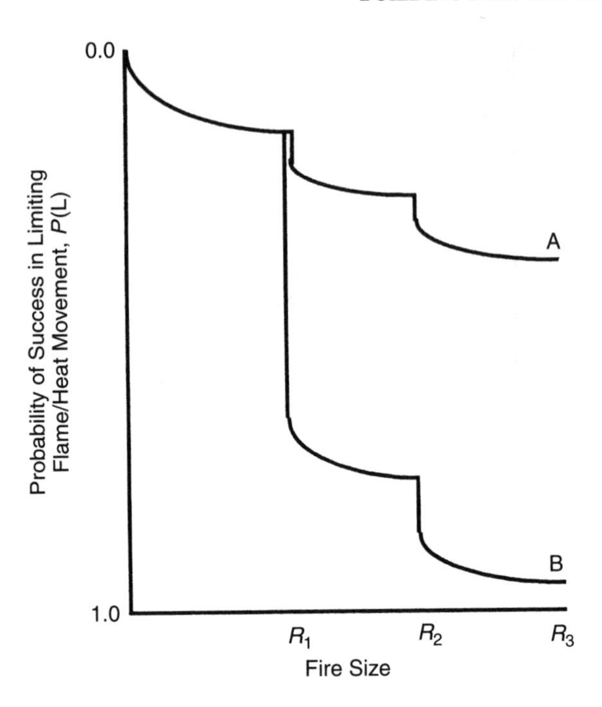

Figure 6.17

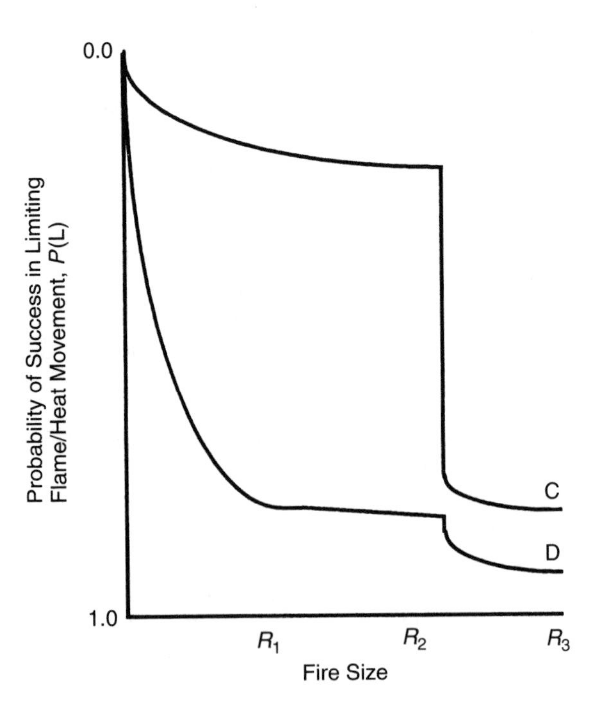

Figure 6.18

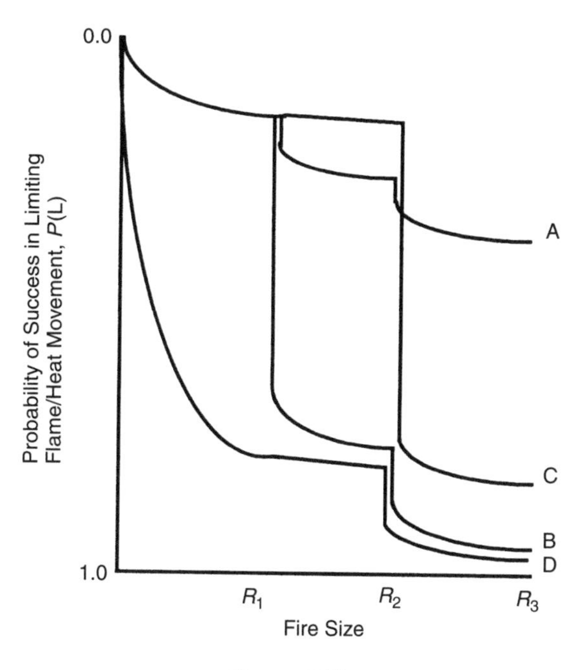

Figure 6.19

Figure 6.19 shows all four buildings on the same evaluator. The L curves enable one to compare the performances of each building. Alternatively, these L curves could represent the flame/heat damage for four design alternatives of the same building. This pictorial relationship enables decision makers to recognize anticipated damage impacts for different design alternatives.

This discussion uses L curves to illustrate comparative performance. Each major component has a curve associated with its performance. As one uses this way of thinking to describe performance, the ordering of the important components becomes evident and important interactions can become more easily recognized.

7 PROLOG TO APPLICATIONS

7.1 Introduction

People use buildings to perform a function, or perhaps a variety of functions. Fire is rarely a priority during the design process or its long-term use. Fire safety features are usually incorporated to satisfy a building code or an insurance recommendation. Although risk awareness is slowly changing these perceptions, decision makers often perceive fire safety costs to be greater than their value, until the fire.

Risk is associated with all building design and operations. Variations in building and site plans, fire defenses, occupant attributes, and functional operations make each building unique. Unfortunately, few individuals or corporations are aware of fire risks or think about them other than to purchase insurance. One function of this book is to help fire safety professionals tell a story about ways to manage fire risk to individuals in a way that a layperson can understand and use in making decisions.

It is important to realize that fire safety is not fire dynamics, or egress analysis, or smoke control, or sprinkler systems, or detection, or fire departments, or structural fire endurance, or fire prevention, or risk management. The fire safety system is all these. One develops a systems understanding with methodical procedures that enable an individual to move easily between micro component performance and macro building behavior. Although the procedures are not complicated, they do require a way of thinking that augments available information with judgment to bridge the gap between what is known and what can be logically deduced and predicted.

This chapter provides an overview of the complete process. When studying details of individual components, one can overlook that each component is only one part of a complete system. The overview identifies component descriptions so that the detailed microanalyses can be placed into a macrocontext of performance and risk characterization.

7.2 Tools of the trade

Let's pause to review the basic tools of building performance analysis. Chapter 2 provides an inventory of fire defenses for a building whose functional operation becomes the basis for evaluating performance. The synthesis of the building architecture and construction with the fire defenses defines the fire system for analysis.

Building Fire Performance Analysis R. W. Fitzgerald
 ISBN: 0-470-86326-9 (HB)

Several concepts simplify a fire performance analysis. Established burning (EB) identifies the demarcation between fire prevention activities and building fire performance. Although fire prevention is important to a risk management program, we consciously detach it from a building performance analysis.

The space/barrier arrangement defines the paths by which fire, smoke, and people can move through a building. A fire scenario selects a room of origin and the associated barrier/space paths. The fire growth from EB to full room involvement (FRI) is the segment of primary interest in performance analyses for all spaces. After FRI, the fully developed fire defines conditions for evaluating structural performance, smoke movement, and life safety. The limit of flame/heat movement (described as the L curve) describes the likelihood and extent of flame/heat fire damage. Knowledge gained in an L curve evaluation provides the basis for structural frame and smoke tenability evaluations and risk characterizations associated with the building's performance.

Figure 4.7 defines the anatomy of the building/fire system. This one figure summarizes the entire performance analysis of the design fire, major fire defense components, and risk characterizations. Because fire performance is a dynamic process, the status of each component changes with time. The columns showing elapsed time allow one to recognize the phased sequence of component actions during the fire scenario. Thus, one can discern the sequencing of interactions or the status of each component at a glance.

A more important function of Figure 4.7 is less obvious. It provides a conceptual basis for visualizing analytical context. Each cell provides a location for a performance measure. We can describe performance estimates as certainly yes, certainly no, or uncertain. One can also calibrate the window of uncertainty with probability measures based on the physical, operational, and fire conditions that exist for the building.

Horizontal rows identify continuous value network (CVN) thinking that can create a visual perception of component behavior. Level 1 evaluations base performance on observations and mental estimates. Level 1 estimates are the foundation for all evaluations that provide a sense of proportion for important performance influences.

Level 2 evaluations examine details more carefully with single value networks (SVN) to give greater confidence in estimates. Vertical columns of Figure 4.7 help one to order what is happening with the different components at any instant of time. The individual cells provide a repository for level 2 estimates. One can often extrapolate a few carefully selected SVN cells to develop continuous performance curves.

Level 3 evaluations use the same SVNs to assess performance sensitivity for variations in location, parameters, and quality. Level 3 evaluations provide a base for identifying *if-then* situations when the value of additional information justifies its cost.

Level 1, level 2, and level 3 evaluations use the same thought process. The differences are in the level of detail and information and the time available for the analysis. Also, because almost all detailed analysis is based on SVN evaluations, familiarity with the integration of the micro related events allows a smooth transition to macro performance.

7.3 Fire prevention

Although fire prevention is an important part of the risk management system, we disconnect it from the building's performance analysis. Traditionally, fire prevention is viewed as preventing ignition. We modify this and define fire prevention to include both ignition control and preventing EB. Fire prevention measures are effective in reducing the likelihood of unwanted ignitions and in extinguishing small fires, but there is no satisfactory strategy that will successfully prevent all ignitions.

Established burning is the fire size at which fire behavior becomes more predictable. For most building analyses, EB is a fire of approximately 250 mm (10 in) in height. One can also envision

EB as a small wastepaper basket size fire. Although building fire defenses (except for detection) will not become active with fires of this size, occupants often extinguish these small fires.

Occupant extinguishment is an important part of a total fire risk management program. However, some occupants cannot provide active extinguishment because they may not be near the fire or they may be unable to act. Analytically, we consider occupant extinguishment only within the segment between ignition and EB. After EB the occupant is assumed to be unable or unwilling to extinguish the fire manually. Active extinguishment after EB occurs by the sprinkler system or the local fire department. Although this decision may be conservative for some situations, it provides a clear understanding of the influence of building design on fire performance.

When we define fire prevention as prevent EB, the analysis has two components. One is the traditional prevent ignition. The second is, given ignition, prevent the fire from growing to the size of EB. This second component, in turn, is influenced by fuel conditions (e.g., fire retardant treatments) or by occupant emergency extinguishment. Chapter 20 discusses prevent EB analysis.

7.4 Building types

There are only two kinds of building in the world: sprinklered and not sprinklered. The way of thinking about performance and risk is very different for each. Here are a few characteristics relating to each building type.

NONSPRINKLERED BUILDINGS

- The local fire department is the primary active fire defense.
- Barrier effectiveness is an important part of the design fire.
- Building occupants have a major role in performance success and risk reduction.
- Architectural design decisions help or hinder on fire department suppression.
- Some buildings survive only on the effectiveness of fire prevention.
- Risk management programs can have an important role in improving performance.
- The vast majority of buildings in the world are nonsprinklered.

SPRINKLERED BUILDINGS

- The sprinkler system is the primary active fire defense.
- Design fire characteristics are important to sprinkler system performance.
- Sprinkler system quality can vary substantially.
- The local fire department functions as a mutual-aid active defense.

7.5 Selection of the room of origin

A performance evaluation starts with EB in a room of origin and describes the likelihood of reaching full room involvement. Then one examines the barrier/space modules to determine the fire propagation potential along paths beyond the room of origin.

The room of origin selection is an important decision for a performance analysis. One may look at the selection from two different perspectives. If a computer analysis were available, every room, one at a time, becomes a room of origin. The computer analyzes fire propagation along all possible paths and allows one to organize a variety of information. This analysis provides an opportunity to study in detail all possible paths of fire propagation considering each room as a room of origin.

When a computer program is not available, only a few rooms need be evaluated to develop a good understanding of building behavior. One can usually extrapolate building performance from the information derived from a detailed analysis of one or a few rooms. Consequently, the room of origin selection becomes a key ingredient in a building performance evaluation.

One carefully selected room of origin usually provides a good understanding for much of the building's performance. One may examine atypical rooms if they contribute to a better understanding of the building. All other rooms are considered for the value they add to understanding the performance. For example, the evaluation of a 408-compartment ship [1] required a detailed examination of only eight compartments to provide a comprehensive understanding of performance. Other rooms and conditions were correlated to those few rooms that dominated performance.

Here are the types of question that identify candidate rooms of origin for a building:

- If a fire were to occur in the building, which room (or rooms) would cause the greatest problems from a life safety perspective?
- Which rooms pose the greatest threat to property protection or operational continuity? Where would be the worst room location for a fire to threaten these vulnerable spaces?
- Which rooms pose the greatest fire growth hazard?
- Which rooms will cause the greatest problems for manual fire-fighting?
- Which rooms have barrier weaknesses that can cause a cluster of rooms to become involved quickly?
- Which rooms are representative for the building?

Normally, most buildings have one or only a few rooms that dominate fire performance understanding. A careful evaluation of this (or these) rooms will give a clear insight into the building's performance. One can use this information to develop risk characterizations and to compare performance with other buildings or design alternatives.

Example 7.1

Identify an appropriate room of origin to evaluate the building described in Figures 4.1 and 4.3.

Solution

Two candidate rooms seem appropriate. Apartment 210 is representative of the other apartments and has a greater fire threat to the building and the other residents than the other units (except on the ground floor). A fire in either the living room or the bedroom of this apartment could extend to all three floors by exterior propagation from the bedroom window. If the corridor door were open during the fire, three floors would be quickly at risk for substantial smoke contamination. Also, a fire on the second floor will give local fire fighters more difficulty in extinguishment than a fire on the ground floor.

A second candidate room is the ground-floor storage and service area. A fire in this space has the potential of substantial smoke logging of the entire building. In addition, the fire growth potential may be greater than for other rooms. The frequency of ignition may be greater because of the location of heating and electrical distribution systems. Also, it is less likely to have occupants in the space in the event of an ignition. However, the frequency of ignition is not part of the consideration for room of origin selection or building analysis. If a complete fire risk management program were developed, fire prevention would be a part of that program.

Barrier/space paths for fire propagation is a selection consideration. Here doors within the apartment would normally be open, leading to a rapid $\overline{\text{D}}$ failure of interior barriers. Also, a potential exists for the fire to break bedroom window glazing. The combustible exterior siding provides fuel continuity for rapid upward fire propagation. Therefore barrier/space paths that allow external flame movement to threaten apartments on upper floors becomes an important consideration. We note that if the combustible exterior siding were replaced by noncombustible materials, the analysis would still consider this potential barrier/space propagation path. However, the probability for fire spread would change substantially because the surface flame spread propagation would be eliminated, although the spandrel distance between exterior windows remains the same.

Gypsum wallboard construction between apartments would be expected to have integrity because the building has a residential occupancy. Nevertheless, one would look for potential holes that can propagate fires to adjacent spaces. Of particular concern would be pipe and wire penetrations and heat and air return ducts.

The other primary barrier/space path of interest involves corridor doors. If the apartment door is open during a fire, one can expect a substantial amount of heat and smoke transfer through the opening. If it is closed, a very different fire scenario will occur. It is not uncommon for an occupant to leave a door open during escape from a fire.

7.6 Design fire concepts

Building fire performance depends on the response of one or an integrated group of fire defense components to the fire's combustion products. After an actual fire incident has occurred and the various building fire defenses have operated, one can observe results and recognize the individual and collective performances for that fire. However, before the fire, a performance analysis must estimate the effectiveness of the individual and collective fire defenses. The fire defenses are evaluated against a design fire that defines the burning characteristics and time-related combustion products generation.

Performance evaluations compare two constituents that can be described as a load and a resistance. The loads are the dynamic combustion product relationships defined by a design fire. The resistance is the fire defense component being evaluated. For example, the design fire for a sprinkler system evaluation includes (1) a relationship between the time of fire growth and the rate of heat release (αt^2 fire); (2) a relationship between heat release rate and floor area of fire involvement; and (3) a relationship between the fire plume momentum and the floor area of fire involvement. The resistance is the sprinkler system's ability to control (or extinguish) those combustion products. Another illustration describes the detection system performance. The design fire (i.e., load) for a detector is the time-related generation of the products of combustion associated with the detector's sensitivity. The resistance is the actuation of the detector.

Although it does not affect the design fire selection, building features can influence the transport of combustion products. For example, Figure 7.1(a) and (b) show the effect of ceiling projections

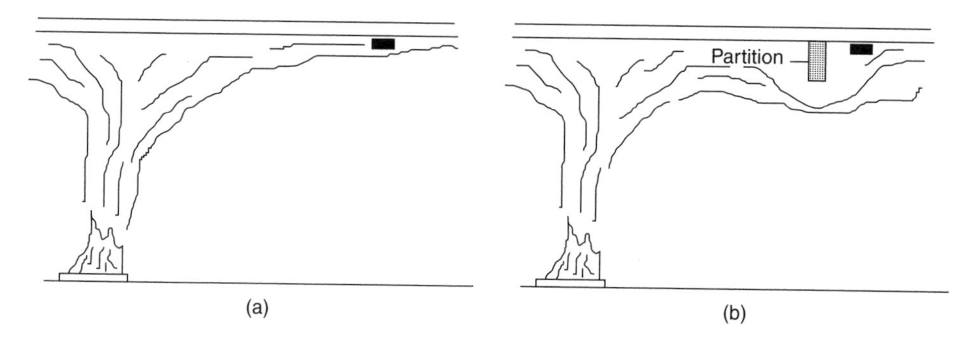

Figure 7.1

on combustion product movement to a detector or sprinkler. The fire size at actuation will be different for each situation, and a larger fire may affect the behavior of other fire defenses (individually or in combination).

Detailed single value (SV) studies for carefully selected conditions provide a clear understanding of component performance at a specific instant of time or condition. This is a major advantage of SVN analyses because the effect of the design fire and fire defense locations is incorporated into the estimates. Design fire characteristics do not change when building features affect combustion product transport. The analytical framework automatically incorporates building features into fire defense assessments.

7.7 The design fire

A design fire is the load against which to evaluate the fire defenses and the resultant risk to exposed people, property, and building operations. Building codes identify the design loading for other disciplines. The fire engineering profession and the codes have not yet established design fire characteristics to identify acceptable performance. Therefore individuals must identify the fire characteristics to evaluate performance.

A design fire for an existing building may be easier to define because uses, materials, quantities, and conditions may be observed and measured. The design fire for a proposed new building is often more complicated because less information is available and use variations are more difficult to envision. In either case the problem is more difficult because uses, architectural renovations, and contents regularly change as the building ages. How will these changes affect the fire performance? Should the resistance be increased to allow more adverse fire conditions to occur over time? How much is enough?

Whether the performance evaluation is for an existing or a proposed new building, one must envision and define a conceptual room model. This model is a room interior design consisting of the size and shape of the room, its interior finish, the contents, and their arrangement. Ventilation from doors, windows, and other openings has an important influence on fire development and products of combustion. However, performance applications separate ventilation assumptions from the other interior design conditions when developing design fire characteristics. The concept of defining a model room with default ventilation conditions simplifies performance evaluations greatly. If necessary, one can incorporate the effect of variations in ventilation to gain a better understanding of building behavior.

The design fire for a model room includes two distinct, but related parts:

- A fire growth hazard classification for the room interior design.
- Products of combustion characteristics associated with the fire growth hazard classification.

Chapter 8 describes room fire growth hazard classifications. Specific design fire characteristics are associated with the functional behavior of each fire defense. One may standardize the design fire characteristics for a model room. These standard characteristics may be adapted to atypical conditions if they make a difference in understanding performance. Design fire characteristics for component evaluations are as follows.

AUTOMATIC SPRINKLER SUPPRESSION

- A relationship between the time of fire growth and the rate of heat release.
- A relationship between heat release rate and floor area of fire involvement.
- A relationship between the fire plume momentum and the floor area of fire involvement.

FIRE DEPARTMENT EXTINGUISHMENT

- The time from EB to FRI for the room of origin and for fire propagation to a successively increasing cluster of rooms.

FIRE DETECTORS

- An identification of the combustion products that affect the detector.
- A relationship between the time from ignition and the generation of those combustion products.

BARRIER FIRE ENDURANCE

- A relationship between time and the cumulative heat energy release by the fire.

STRUCTURAL FRAME BEHAVIOR

- A relationship between time and the cumulative heat energy release by the fire.

SMOKE TENABILITY OF TARGET SPACES

- A relationship between time and smoke quantity generation.
- A definition of tenability for exposed people or objects.
- A relationship between smoke generation and the smoke characteristics.

FIRE GROWTH HAZARD CLASSIFICATION

- A classification to represent the relative ease for the model room to reach full room involvement.

A fire growth hazard classification is not essential to a performance evaluation, but its identification helps to understand expected fire behavior and conceptual room models. Guidelines and tools for rapid level 1 evaluations can be associated with classifications. Perhaps the most useful benefit of classifications is enhanced communication about the influence of interior design on fire conditions and performance.

Figure 7.2 shows the design fire section of Figure 4.7. The blank rows allow one to show other appropriate time-related combustion products for the specific fire defenses.

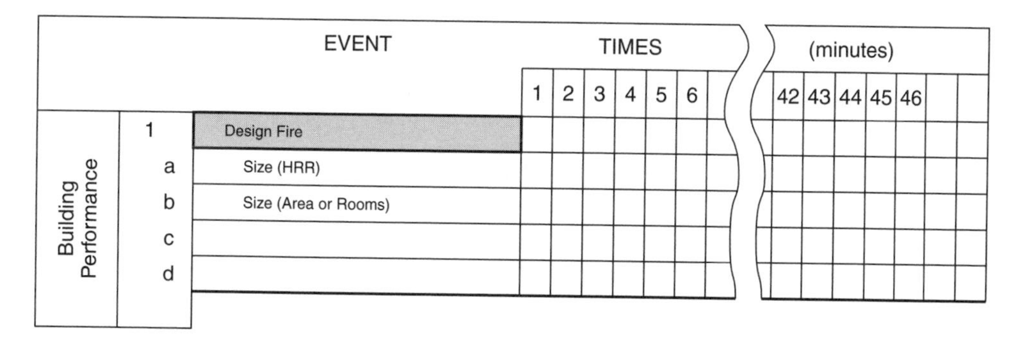

Figure 7.2

7.8 Performance analysis overview

The design fire is central to all performance evaluations. One should not view the design fire as a single fire, but as a composite of appropriate time–POC characteristics. If these relationships were standardized, one could evaluate fire defense components quickly and consistently. Because they are not standard, we must describe the fire products on an ad hoc basis.

We describe details of fire defense components and their evaluation in the next several chapters. A brief overview of the parts and their relationships will provide a road map for the complete process.

Figures 7.4 to 7.8 show the sections of Figure 4.7 that relate to the components.

7.9 Fire performance: M curve analysis

A building evaluation for fire department extinguishment (M curve) does not view the fire department in isolation. Instead it evaluates the building's ability to work with the local fire department. Evaluations examine the building and site features that influence fire ground operations and affect expected fire damage.

The performance measure for a building M curve is the fire area and number of rooms destroyed by the flame and heat before the fire is extinguished. Two questions are central to the building evaluation. The first asks, How large is the fire at the time of first water application? The second says, Given that fire size, the building architecture and construction, and local community resources, can the fire department extinguish this fire before it extends to additional spaces? If the answer is no, we continue analyzing until we find a fire size at which the fire can be extinguished.

We estimate the fire size at first water application by comparing the two time lines of Figure 7.3. The first shows the time-related fire propagation in the room of origin and the cluster of other rooms. The second estimates time durations for detection, fire department notification, apparatus response, and first water application.

Barrier effectiveness is an important part of the design fire. Extinguishment is rarely a problem when barriers to the room of origin delay fire propagation long enough to allow the fire department to respond and get water onto the fire. Except for large rooms, any trained fire company should extinguish a one-room fire. Difficulties escalate when the fire involves additional rooms before initial water application can occur.

Barrier fire endurance (i.e., fire rating) is rarely a factor in fire propagation to adjacent rooms within the time frame associated with fire department suppression. Holes in the barriers become the major focus of this analysis. Open doors, windows that can easily break, unprotected through

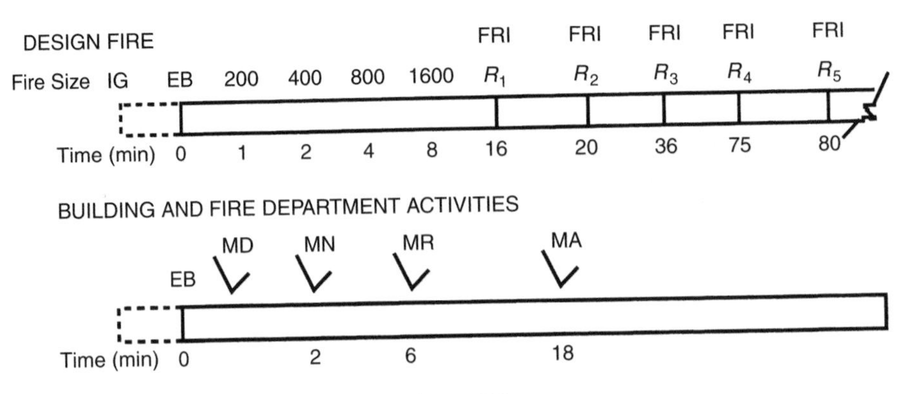

Figure 7.3

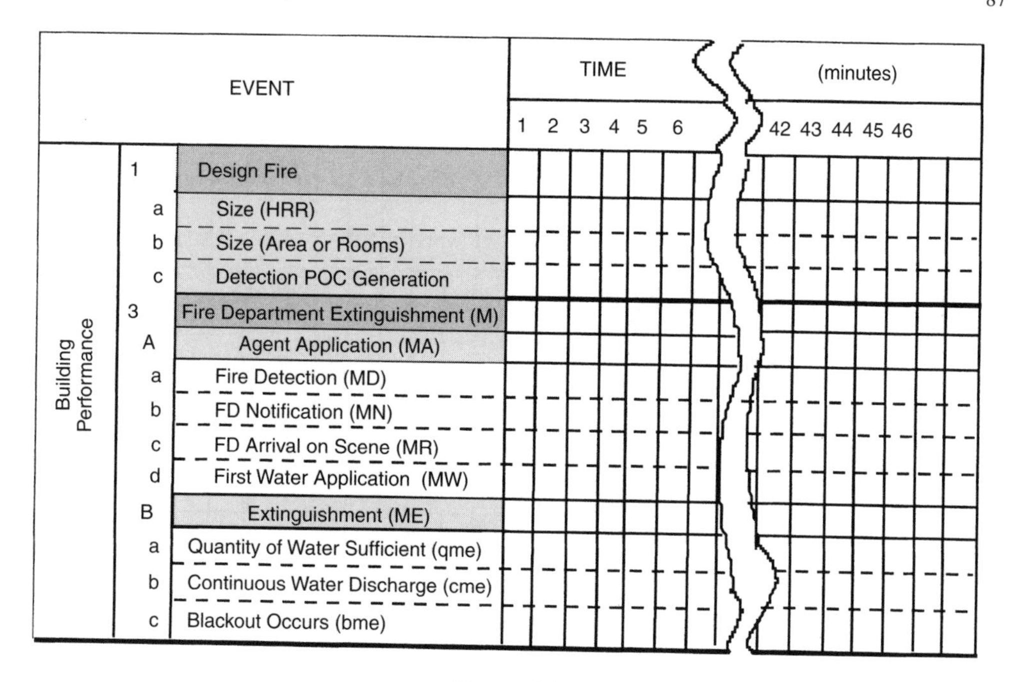

Figure 7.4

construction holes, and other openings with ineffective protectives allow a fire to propagate into adjacent rooms. Large fire areas, multiple rooms, or several floors of fire involvement cause difficulties in extinguishment.

The process of developing these time lines reveals much information about the building and its performance capabilities. With this analysis one gains knowledge of the approximate fire size, the effectiveness of building construction features, the number and locations of available hose lines, and water supply capabilities. This type of information helps to gauge whether the fire department can use offensive or defensive tactics to extinguish the fire.

7.10 Fire performance: A curve analysis

An A curve analysis estimates the likelihood that the sprinkler system will control a fire with 1, 2, 3, ..., n sprinkler heads. The A curve performance assesses the quality of the sprinkler system and the area of damage before the sprinklers control the fire. The analysis integrates three elements: (1) the design fire; (2) the reliability of the system to deliver water; and (3) the effectiveness of the sprinkler system in controlling or extinguishing the fire. When the analysis evaluates sprinkler control, some manual action is needed to extinguish the fire. When total extinguishment is assessed, only the quality and operational effectiveness are considered. Either type of evaluation may be selected, although performance expectations will differ.

The design fire for a sprinkler system evaluation involves the combustion products associated with sprinkler performance. These include relationships between

- time and fire growth (i.e., αt^2 fire);
- fire size (i.e., area) and heat release rate;
- fire size and fire plume momentum.

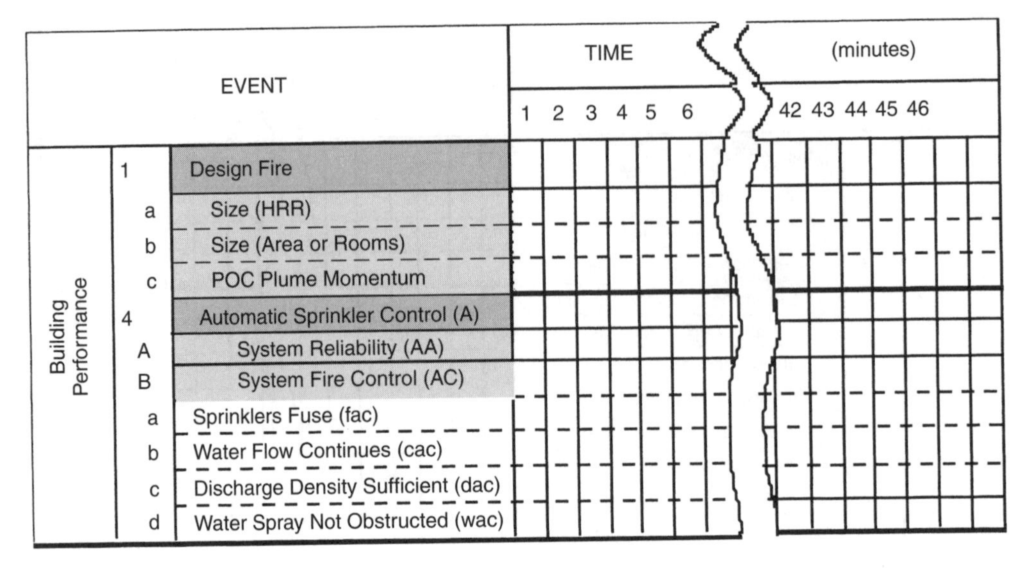

Figure 7.5

We define the reliability of a sprinkler system as the likelihood that water will issue from a fused sprinkler. The reliability is often associated with the valves being open at the time of the fire. Besides closed valves, a variety of other factors such as pipe scale, pump failure, and power failure may prevent water from reaching sprinkler. Maintenance of the sprinkler system is important to its reliability. System quality can vary substantially, and guidelines to appropriate practices help in estimating the installation's reliability.

An operational effectiveness evaluation estimates the likelihood that the sprinkler system will control (or extinguish) a design fire within *XX* ft^2 (m^2) in area. Single value estimates provide the information to construct a continuous value A curve. Operational effectiveness estimates are based on information from the following questions:

- Will the sprinkler fuse before the fire has extended beyond the fire size being evaluated?
- Is the water discharge density sufficient to control (or extinguish) the fire?
- Will enough water continue to flow until the fire is extinguished?
- Is the water-spray distribution obstructed?

7.11 Putting it together: the L curve

The L curve combines the fire defense performance evaluations. It combines the results of earlier analyses of

- the fire itself;
- automatic sprinkler suppression;
- fire department extinguishment;
- barrier effectiveness.

Although the L curve describes only the flame/heat damage of the fire, the process of making the estimates provides an understanding of building performance. The acquisition of knowledge about of building features and fire conditions furnishes much information useful to evaluate structural behavior and smoke contamination.

7.12 Structural frame behavior: the Fr curve

Structural collapse is a danger to occupants and fire fighters and also to property losses and operational continuity for the building. From a fire viewpoint, the structural frame supports the dead load of building components such as partitions, walls, floors, and roofs and any superimposed live loads.

The structural frame may support a barrier or it may be a part of a barrier construction. For example, a floor system illustrates a structural frame that is simultaneously a part of a barrier. Structural collapse will cause a barrier collapse while excessive deflection can cause cracks that allow flame movement penetration into the space above. Chapter 10 incorporates these differences into the barrier analysis and Chapter 15 discusses structural behavior.

Whatever the construction details, we evaluate structural frame performance as a separate component. As a part of the structural evaluation, we define failure as either collapse or excessive deformation. While collapse may cause a major catastrophe, deformations can cause cracking of barrier materials and allow flames to move into the next space even though collapse does not occur. It is possible for the heat from a fire to cause expansion of beams, girders, and joists. This expansion may move columns out of plumb and push walls toward collapse. These consequences are all a part of structural performance.

Applied loads influence the time at which the frame will begin to deform excessively or collapse. When dead and live loads are near to the design values, deformations will occur earlier than when the actual loads are below design values.

The insulation of structural members from a fire's heat is important to performance. Exposed structural steel, especially the lightweight members and trusses used in smaller buildings, are very susceptible to strength loss due to temperature rise in the member. Exposed light wood members can lose enough material by burning to be unable to support the applied loads.

When structural members are protected by insulating materials such as gypsum board, spray fireproofing or concrete, the strength can be maintained for much longer time durations. Therefore the presence or absence of insulating materials is a first observation. The quality and insulation thickness is another factor.

A third aspect in structural performance involves connection details and support conditions. Continuous flexural members and rigid connections provide a better resistance to deflection and collapse than does simply supported framing. Framing system observations of member sizes, materials, spans, and support conditions contribute to the structural frame evaluation. How the building is put together influences its structural performance.

If the fire is not extinguished directly and quickly, large amounts of fuel will cause a long-duration fire, and the adequacy of protection should be investigated. The design fire identifies time–heat generation and time–temperature relationships. The analysis adjusts this relationship to reflect the influence of heat losses. An SV estimation starts by identifying the applied load that the structure is expected to support. Then, for selected time intervals after EB, the following questions are evaluated:

- Will enough heat be generated and reach the surface of the structural member to reduce its strength to the point of failure?
- Will enough heat penetrate into the structural member to weaken it to the point of failure?

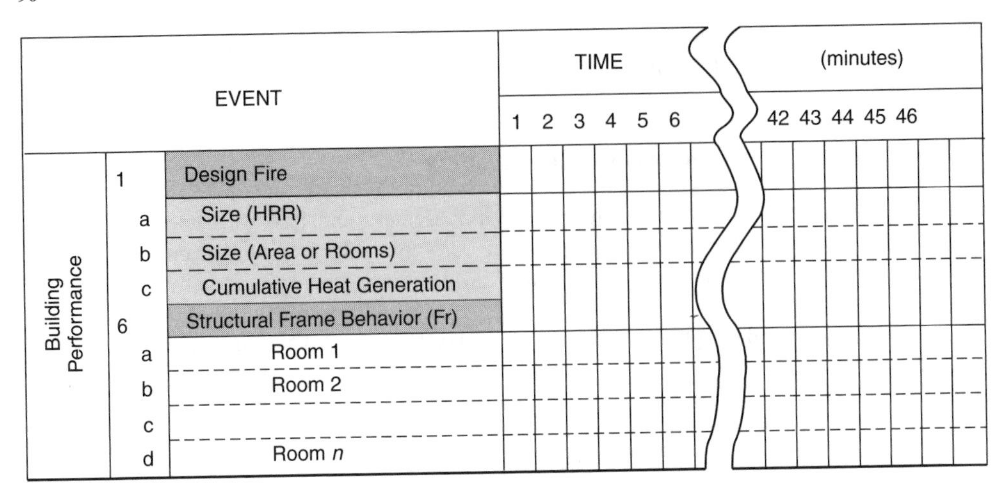

Figure 7.6

The initial question considers time and the amount of heat energy to reach the exterior of the structural member. Because insulation may range from nothing to substantial, the second question investigates whether the heat can weaken the structural member enough to cause failure.

This two-phase analysis allows one to separate time and heat movement to the structural member from the heat transfer into the member. This enables one to deal with the fire and the insulation and structural behavior as separate functions before combining them.

7.13 Smoke analysis

Flame/heat movement and structural frame analyses often provide a good understanding of building performance in a fire. During these studies, one may develop an awareness for potential smoke movement and life safety concerns. A performance analysis for smoke considers the smoke/gas generation the design fire loading. The resistance is the building's ability to prevent untenable conditions in selected target spaces.

A performance analysis starts by identifying a room of fire origin and one or more target spaces. A target space is a room critical to a life safety or property function. For example, a target space may be an important part of an egress path, such as a corridor segment or an exit. Alternatively, a target space might be a room that houses individuals who must be defended in place. Still further, a target space might contain equipment and data storage sensitive to the corrosive effects of combustion products. The analysis estimates smoke and gas movement from the room of origin to the target space and the accumulation of untenable conditions in the target space.

Target space tenability evaluations involve the following:

- identification of a room of origin and one or more target spaces important to performance;
- a definition for tenability;
- a design fire for smoke/gas generation and movement;
- identifying smoke movement paths between the room of origin and target spaces;
- barrier characteristics around the target space;
- venting and pressurization of the target space.

We define building performance as the time during which a target space remains tenable during a fire.

The design fire characteristics for a smoke movement analysis include

- a relationship of time and the volume of smoke generation;
- smoke temperatures;
- an obscuration measure to predict visibility;
- identification of corrosive or toxic gases for sensitive equipment.

Normally the smoke generation from a single room of origin will be sufficient to understand the building performance. Occasionally one may use a cluster of rooms that can become involved quickly to establish design fire smoke characteristics.

The tenability criterion uses any combustion product that can be calculated or estimated. Visibility is the most appropriate and useful tenability measure for life safety. We define tenability as the ability to see *XX* feet in smoke. The value of *XX* will depend on factors such as architectural plan complexity, familiarity of the occupant with the building, and characteristics of the occupants. The literature cites values ranging from 10 ft to 90 ft. Visibility distance selection is an important professional judgment and its value has a major influence on performance.

Normally, visibility will deteriorate before toxic gases become a life safety problem. Therefore visibility is the appropriate measure of performance for most situations. Defend-in-place evaluations for humans include toxic gas effects if they can be evaluated. However, even if another criterion (e.g., toxicity, radiation, heat) were the measure for tenability, the performance definition would still be the time duration that the target space remains tenable.

Some smoke will be pushed out of the building and some will move through available routes inside the building. The quantity that moves inside the building depends on fluid forces, the size and location of openings, distances, and the particulate and heat losses along the path to the target room. We estimate the fraction of the smoke that moves through the path of interest. Each single value estimate answers the question, What is the likelihood that the target space will remain tenable *XX* minutes after EB? Each SV estimate is based on examining the following performance questions:

- Will enough smoke reach the boundaries of the target room to make that room untenable?
- If enough smoke can reach the boundaries of the target room, will enough smoke accumulate in the target room to make it untenable?

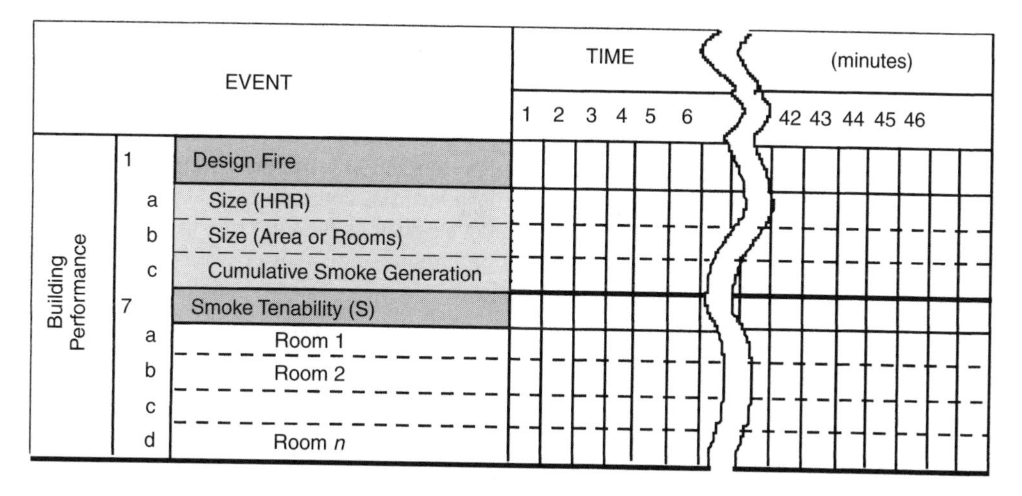

Figure 7.7

This two-phase analysis allows one to separate the smoke movement from its accumulation in the space. It enables one to examine the effectiveness of barrier integrity, space pressurization, and venting conditions in preventing smoke from entering into and accumulating in the target space.

7.14 Risk characterizations

After one gains an understanding of building performance, attention is directed to risk characterizations. Risk characterizations may be developed for

- people who are expected to leave the building;
- people who cannot (or choose not) to leave the building;
- property damage;
- operations interruption;
- threat to neighboring properties;
- threat to the environment.

Risk characterizations can tell the story for general and specific scenarios.

RISK TO PEOPLE

A risk characterization identifies only the likelihood that a person will encounter untenable conditions during a fire. We describe human risk as the likelihood that people and untenable conditions will occupy the same space at the same time. Although the likelihood of death or injury is not a part of the analysis, the story may describe possible outcomes.

RISK TO PROPERTY

The process of characterizing property losses uses a defend-in-place analysis. One identifies the property of value and its sensitivity to the flame/heat or smoke/gas products of combustion. The target room contains the property of interest. The risk characterization is based on the building performance evaluations and describes potential losses for different conditions and fire locations.

RISK TO OPERATIONAL CONTINUITY

Many buildings house operations that are sensitive to disruption and downtime due to fire damage. In some businesses a single piece of equipment may be essential to the functional operations and perhaps even the survival of the enterprise. The characterization starts with the identification of equipment or information important to the mission and business continuity. The target space is its location. The sensitivity to specific combustion products such as heat or corrosive gases defines the tenability condition. The performance evaluation provides a risk awareness for the effect on operational continuity of fire conditions and locations.

RISK TO NEIGHBORING PROPERTIES

The term "exposure" commonly describes neighboring units in or near the building being evaluated. The exposed property may be external, as with detached buildings, or it may be internal, as in apartments, business units, or connecting buildings. The macroanalysis provides a recognition for potential fire threats to others.

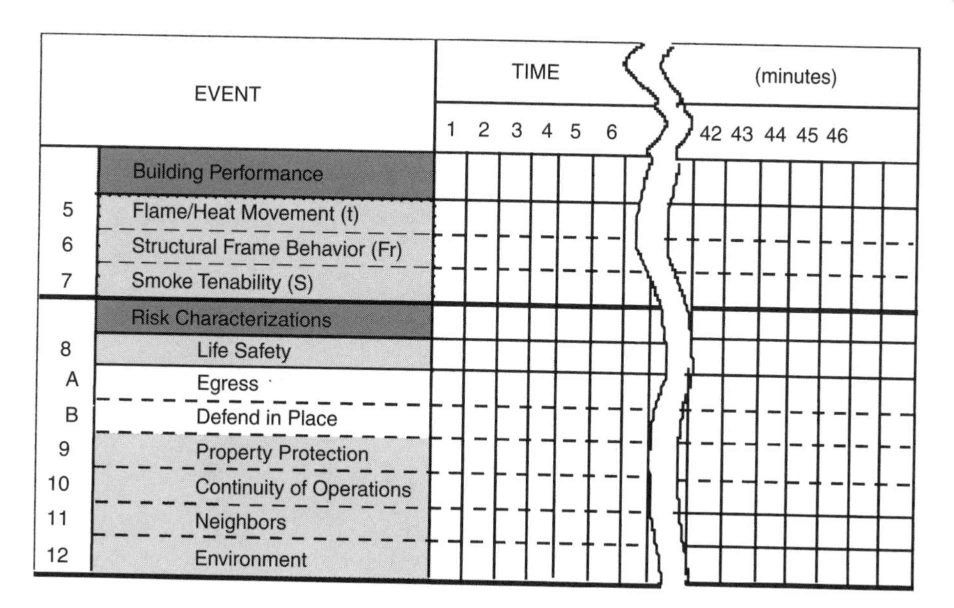

Figure 7.8

RISK TO THE ENVIRONMENT

Some buildings contain chemicals or other materials that produce hazards to the environment when they burn or mix with fire suppression water. These hazards may be airborne and transported to remote areas or they may seep into the ground and enter the water supply. Fire runoff may flow into sewers, rivers, or lakes to contaminate communities distant from the fire incident location.

An environmental risk characterization identifies building contents that can pose a serious or unusual problem when burned or combined with water from fire suppression operations. It also notes geographic or terrain conditions that influence or contribute hazardous by-product transport. The risk characterization describes the likelihood of hazardous by-products reaching locations to cause environmental damage.

RISK MANAGEMENT

Building performance and risk characterization evaluations enable one to adapt the information to a variety of applications. Risk management programs can assume a wide range of forms. Their organization and structure are discussed in Chapter 19.

7.15 Summary

The goal of a building evaluation is to understand the building sufficiently well to describe clearly its performance and the associated risk characterizations. This chapter provides an overview of the building/fire system that defines performance and risk. One component's performance affects other parts of the system and contributes to the overall building performance.

A performance evaluation provides the information to make better decisions and to communicate with others. Communication tells a story about the building and what is likely to occur with different fire conditions or scenarios. The story must be credible and relate to the specific building of interest. One must be able to adapt the story for a layperson or a technical expert and justify estimates with credible technical documentation. This requires an ability to move between a microanalysis and a macroevaluation.

Reference

1. Fitzgerald, R., Richards, R., and Beyler, C. Firesafety analysis of the polar icebreaker replacement design. *Journal of Fire Protection Engineering*, Vol. 3, No. 4, October 1991.

8 DESIGN FIRES

8.1 The need

The central ingredient of all performance evaluations is the fire. Each fire defense component requires a different time-related set of combustion characteristics to estimate its performance expectations.

A design fire is the starting point for performance applications. A design fire includes a hazard classification for the room's fire growth potential along with combustion products to analyze each fire defense. This poses several practical problems. The first and most critical is the time available for the application. One must quickly identify appropriate fire characteristics so that enough time remains available to analyze all of the system's fire defenses. A second is how to select the model room that establishes a mental concept and physical parameters on which to base the design fire. A third involves selection of the specific combustion products and their timed release. Another problem is finding a way to deal with variability in physical parameters and future unknown uses for the room. In spite of these difficulties, we will find a way to select a design fire.

The goal of this chapter is to provide a practical way to establish design fire characteristics quickly and with some confidence. For example, experience has shown that inspectors can complete level 1 evaluations for two moderate-sized buildings in one day. An inspector must make rapid, competent assessments of room design fires so that all of the other fire defenses can receive adequate attention. The results must be reasonably accurate because they provide a foundation for the more detailed level 2 and level 3 evaluations.

The functional behavior of room fires establishes an organization for analysis and selection of design fires. Initially we will look at level 2 scenario analyses for room fires. Level 3 studies enable one to compare performance variations arising from differences in condition of room construction or contents. An awareness of the effect of interior design variations on differences in fire growth is fundamental to selecting appropriate level 1 classifications for a room interior design.

Level 1 evaluations classify the fire growth potential for a room. This classification provides a way to include default values for other design fire characteristics. We base room classifications and their design fire characteristics on the natural behavior of the fire in isolation. No fire suppression intervention is considered when evaluating design fires.

Building Fire Performance Analysis R. W. Fitzgerald
 ISBN: 0-470-86326-9 (HB)

8.2 Fire in a room

From a combustion viewpoint, a fuel is anything that will burn in a fire. Some fuels have higher ignition temperatures and different combustion characteristics than others. These characteristics may affect the time and certainty of fire growth.

When a fuel is heated, pyrolysis releases volatile gases. When these volatile gases mix with air within a specific range of proportions, a flammable mixture is produced. Flaming ignition occurs when this flammable mixture reaches a high enough temperature. Two types of ignition sources can initiate a fire. A *piloted ignition* is caused by a localized heat source such as an open flame or a spark. An *auto ignition* (or self-ignition) occurs when a sufficiently high temperature of the pyrolyzed gas and air mixture is reached. Piloted ignition temperatures are much lower than autoignition temperatures.

Fuel types and characteristics, the heat application time and intensity, and the separation and relative positions of the heat source and target fuels all influence the likelihood of ignition. After ignition and in any phase of its development a fire can

- continue to grow;
- remain steady within a fuel package until the fuels are consumed, and then go out;
- go out because not enough critical conditions exist to sustain the combustion process.

Figure 8.1 is a representative graph for a fire heat release rate, Q, or the temperature, T, with time.

For the procedures of this book, *ignition* (IG) will be defined as the appearance of the first small flame. After ignition the fuel may go through a period of uncertain development. The flame may go back and forth between flaming and smoldering, it may nearly die out and then flash again slightly, it may go out, or it may eventually gain in strength and grow to established burning (EB). A useful size for EB is a fire power about 20 kW. For mental estimates, EB may be visualized as a flame of almost knee height or a small wastepaper basket fire. When the fire reaches EB it becomes established or stabilized, and its future course of action is more predictable.

The segment from EB to full room involvement (FRI) represents fire growth in the room. After FRI all fuels in the room are surface burning, and the ventilation supplied to the room controls the rate of continued combustion. The segment beyond FRI represents the postflashover or fully developed fire condition. This relatively long time duration will last until most fuel has been consumed and combustion can no longer be maintained. The fire will die out leaving only ash and smaller pieces of unburned fuel. The time duration of the fully developed fire depends on the type, amount and location of combustibles in the room; the ventilation conditions; and room size and construction characteristics. The fully developed fire continues until the fuel burns out.

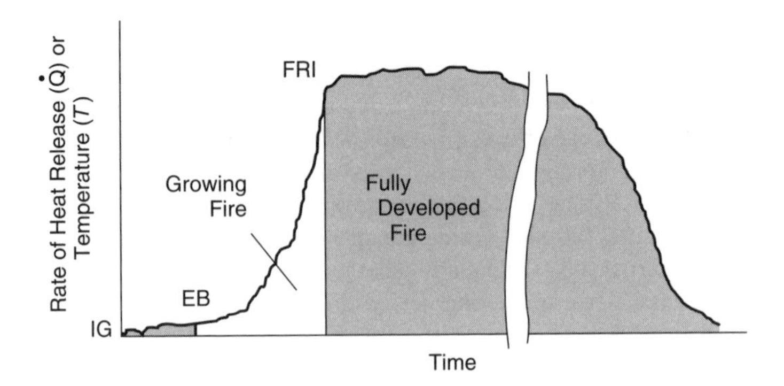

Figure 8.1

Room classifications consider only the fire growth segment from EB to FRI. The segment from IG to EB is analyzed as a part of fire prevention (Chapter 20). The fully developed fire beyond FRI is a part of barrier analysis (Chapters 9 and 10) and the structural frame behavior (Chapter 15). The segment from EB to FRI is a critical part in fire defense and risk characterization performance. This relatively short time duration in a room's fire history is a major constituent of performance evaluations.

The mechanisms by which fire grows from EB to FRI depend on the size, shape, and construction of the room, the amount and types of fuel and their arrangement, and the ventilation. For evaluation purposes, we will categorize rooms as *small rooms* or *large rooms*. The specific dimensions are not as critical as the means by which the fire develops and propagates. In small rooms, flashover is the dominant mechanism for reaching FRI. Spreadover is the chief mechanism for reaching FRI in large rooms.

8.3 Fire development in small rooms

Flashover is the dominant mechanism for reaching full room involvement in small and moderate sized rooms. Flashover is characterized by the following:

- Fire starts in a fuel package and continues to grow in size.
- A fire plume forms, and a hot layer of heated gases of combustion and soot develops at the ceiling.
- Ventilation provides sufficient air to support combustion.
- Hot-layer radiation and some radiation from the flames pyrolyze other unburned fuels in the room.
- Rapid ignition of all combustibles in the room occurs.

This fire development will take a period of time. For small and moderate-sized rooms, the time may vary from as little as 2–3 min to perhaps as long as 20 min. The rapid transition from fire in a room to full room involvement occurs within a matter of seconds. Let's describe the process in a little greater detail.

Fire heat is generated as a part of an exothermic chemical reaction. Some of that heat is radiated back to the unburned fuel, volatilizing more combustible gases and causing the fire to grow. Before describing fire development in a room, first consider a fire in the open air, unconfined by compartment boundaries, as shown in Figure 8.2(a). As the fire grows, air is drawn into the combustion reaction, and a fire plume forms. This fire plume is a column of hot gases that rises due to the buoyancy of the fluid. As the column of hot air and fire gases rises, additional air is entrained and the hot air and gases begin to cool. When enough heat has been lost to the entrained air, the gases lose their buoyancy effect and diffuse into the atmosphere.

Now consider the development of a similar fire within a room enclosure. During its early stages, the fire will behave much the same as if it were in the open air. However, as the fire continues to grow, its behavior changes. Room openings supply air and the fire plume again develops. The ceiling interrupts this plume, forming a hot gas layer as in Figure 8.2(b). As the hot gas layer descends below the top of the opening, fire gases that flow out are replaced by new air that flows into the room (Figure 8.2(c)). This new air provides fresh oxygen to the fire, allowing it to burn more vigorously and increasing the quantity of heat and fire gases.

The hot layer radiates heat to the other combustibles that have not yet ignited, causing them to volatilize more rapidly. When the hot combustible gases in the room reach their ignition temperature and enough oxygen is present, the fire spreads very rapidly. This phenomenon is called flashover.

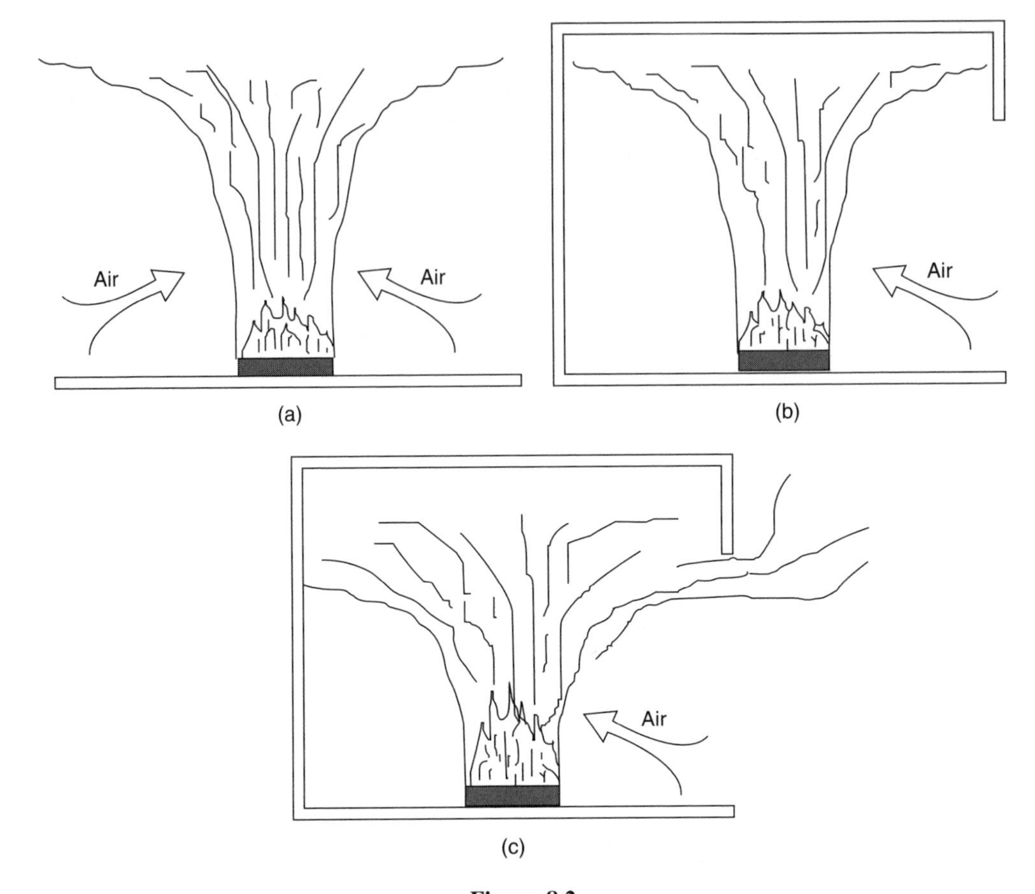

Figure 8.2

8.4 Interior design and model rooms

A room interior design for fire evaluation purposes consists of the

- size and shape of the room;
- interior finish;
- room contents;
- barrier construction.

A model room is the specific interior design on which the design fire is based. The model room enables one to envision the fire behavior and estimate fire characteristics to analyze fire defenses.

Fuels include the room contents and the interior finish. Interior finish is the interior surface of the room. It can have an important influence on the speed and ease of fire growth. Combustible wall and ceiling coverings, such as wood paneling, combustible fiberboard, or carpeting can contribute significantly to fire growth and to reaching full room involvement. Location and orientation of fuels become important to fire development. For example, carpeting on the floor rarely contributes to a fire development from EB to FRI. On the other hand, that same carpeting on a vertical surface can contribute significantly to conditions that encourage and accelerate flashover. Thin wall or ceiling coverings, such as paint, wallpaper, and the paper on gypsum wallboard contribute little to surface flame spread or to fire severity when applied directly to substrates such as plaster, concrete block, and reinforced concrete.

The room geometry can have an influence on fire development. Rooms with low ceilings contain the fire plume earlier and interrupt high flames, causing the flames to mushroom across the ceiling. This mushrooming effect radiates heat, pyrolyzing combustibles remote from the seat of the fire. Long, narrow rooms feed the heat back from the adjacent surfaces more easily. This causes an easier, more rapid fire growth than in rooms that have a squarer configuration. The insulation qualities of the barriers can effect fire development. Well insulated rooms contain heat better than poorly insulated rooms, increasing the ease of reaching flashover.

Ventilation is the size, height, and location of room openings. Ventilation can have a significant effect on fire development. When openings are small, inefficient combustion occurs, resulting in lower fire temperatures. However, comparatively less hot gas is released from the room, and most of the heat remains in the room. On the other hand, when the openings are large, more efficient combustion occurs, producing higher fire temperatures and a more rapid burnout. This condition causes large amounts of hot gases to leave the room.

Note that ventilation is not a part of the model room interior design. We assume that sufficient ventilation is available to allow efficient combustion to take place. Consequently, the room classification relates only to the interior design. We base classification categories on efficient combustion. Design fire characteristics for some fire defense analyses, such as structural performance and smoke movement, incorporate a default, standardized ventilation condition. On those rare occasions when differences in ventilation conditions become important to an application, one adjusts the design to reflect those differences.

8.5 Realms of fire growth

When flashover is the mechanism to reach FRI, fires develop through stages or realms of fire growth. Different factors dominate the fire behavior within each realm. When enough factors are present, the fire can grow and move into the next realm. When too many factors are weak or missing, the fire can burn for a time and decay without moving into the next realm. It is difficult to be more specific regarding factors or numerical quantities because of the synergistic nature of the components.

For small and moderate rooms, fire sizes can roughly describe the realms. Figure 8.3 shows a definition for realms of fire growth. The fire sizes describe the demarcation between realms. The

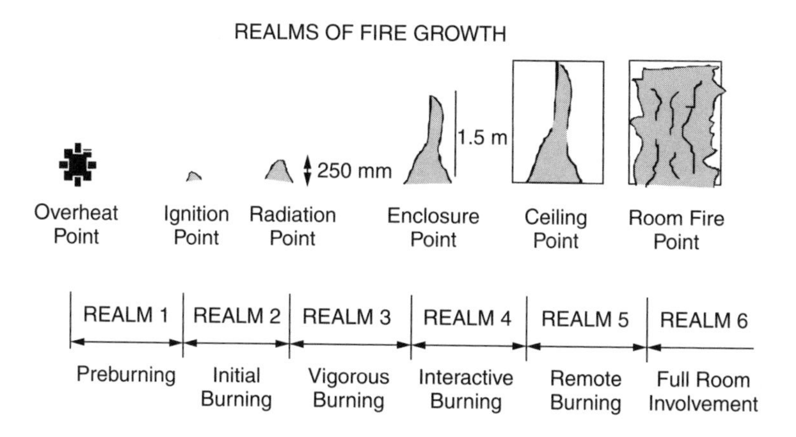

Figure 8.3

Table 8.1

Controlling factor	Realm 1	Realm 2	Realm 3	Realm 4	Realm 5	Realm 6
Fuel	Surface area receiving heat flux Ignitability	Continuity Ignitability Surface roughness $k\rho c$ of fuel Thickness Surface area	Continuity and feedback $k\rho c$ of fuel Surface area Ignitability Quantity	Arrangement and feedback Surface area Tallness Quantity	Quantity Arrangement	
Room container			Proximity of flames to walls Room insulation	Proximity of flames to walls Room insulation Ceiling height	Room insulation Ceiling height L/W ratio	
Ventilation				Size and location of openings HVAC operation	Size and location of openings HVAC operation	

overheat point — ignition point — radiation point — enclosure point — ceiling point — FRI

fire size names attempt to reflect conditions or events of the process. The transitions between the fire sizes are described as realms of fire growth. Realm names attempt to describe the burning conditions during these stages of fire growth. Table 8.1 identifies significant factors in each realm.

OVERHEAT POINT

A noticeable temperature increase begins accelerated volatilization. One may recognize discoloring or smoldering in some materials.

IGNITION POINT

The instant when the first small, fragile flame of burning appears.

RADIATION POINT

This is established burning (EB) for most room situations. The radiation point describes the fire size when fire growth becomes more heavily dominated by radiation feedback rather than by convection and conduction. For many common fire scenarios, this is a flame of almost knee height. The fire power is about 20 kW.

ENCLOSURE POINT

The fire size when the room enclosure first begins to influence and augment the fire growth. For small to moderate rooms, this is a flame height of about 1.5 m (4 1/2 to 5 ft), or nearly the height of a short person. The fire power is about 300 to 400 kW.

CEILING POINT

The fire size when sustained flames reach the ceiling. For an 8 ft ceiling, the fire power is about 800 to 1000 kW.

ROOM FIRE POINT

The condition of full room involvement.

REALM 1: PREBURNING

The period of heating and volatilization from the condition of overheat until ignition occurs. We define ignition to be the condition when flaming first occurs. Self-sustained smoldering may be present. This condition is not yet ignition, by definition.

REALM 2: INITIAL BURNING

The period during which the first small, fragile flames attempt to generate some substance and strength and grow to established burning (EB).

REALM 3: VIGOROUS BURNING

The domain in which an established burning grows within the fuel package and develops a strength and stamina of its own. The fire size grows from knee high (20 kW) to nearly the height of a person (400 kW) during this realm.

REALM 4: INTERACTIVE BURNING

The fire continues to grow beyond the enclosure point. This is usually a result of interactive burning between fuel packages or within one large fuel package. The fire power when the flames touch the ceiling is normally about 800 kW to 1 MW.

REALM 5: REMOTE BURNING

Beyond the ceiling point, the fire mushrooms along the ceiling and radiates heat energy to other fuels, causing an increased rate of volatilization. Some fuels remote from the initial ignition may experience autoignition, and additional fires may start in the room.

REALM 6: FULL ROOM INVOLVEMENT

In small to moderate-sized rooms, flashover will occur as a rapid fire involvement of all of the exposed combustibles.

8.6 Level 2 concepts

The fire growth potential describes the relative hazard of a room interior design. Some rooms have contents and conditions that are difficult to burn or they are arranged so that FRI is unlikely. In other rooms the interior design makes FRI almost a certainty. One can describe the reasons for expected fire behavior in a way that lay individuals can understand.

A level 2 analysis increases your understanding of how the model room will behave in a fire. One obtains this increased knowledge from studying details for selected fire scenarios. Because many potential scenarios can exist, judgment guides which location for EB will produce the most additional information about the model room fire behavior.

Each fire scenario involves a different combination of conditions that influences the fire behavior. Behavior is expressed in terms of success or failure in fire self-termination. An ST (self-termination) curve describes the probability that a fire will go out itself. Each scenario analysis produces a different ST curve. Graphing the results on evaluator 2 enables one to compare different scenarios and conditions.

8.7 Level 2 framework

Room fire development guides level 2 analyses. The fire size events that define the realm demarcations of Figure 8.3 become the focus of probability estimates. These fire sizes are

- enclosure point (E);
- ceiling point (C);
- room fire point (R).

Figure 8.4 structures the thought process. The questions for each event are as follows:

- Given EB, what is the probability that the fire will self-terminate before reaching the enclosure point, ST_E?
- Given that the fire grows to the enclosure point, what is the probability that it will self-terminate before reaching the ceiling point, ST_C?
- Given that a fire grows to the ceiling point, what is the probability that it will self-terminate before reaching full room involvement, ST_R?

Probability is a calibration of the belief that an event will occur. The results may be expressed as certainly no, certainly yes, or uncertain. Alternatively, they may use a numerical probability measure. Fire phenomena and expected fire behavior are the basis for the probability estimates. One can identify and document the technical basis on which a judgment is founded. Uncertainties and information gaps are clearly recognized. Triangular symbols (tics) on a network show the usual factors that influence event evaluations. Figures 8.5, 8.6, and 8.7 show factors that influence events of Figure 8.4.

Examples 8.1 to 8.4 illustrate calculations and graphs that describe expected outcomes for different hypothetical fire scenarios.

Example 8.1

A room has several fuel packages. Assume that a fire scenario has been evaluated for an EB in a specific location. Plot an ST curve for the following probability estimates.

Given EB, the likelihood that a fire will grow to reach the enclosure point is not a certainty, but very high. The value assigned to this event is $P(\overline{ST_E}) = 0.90$. The fire size at the enclosure

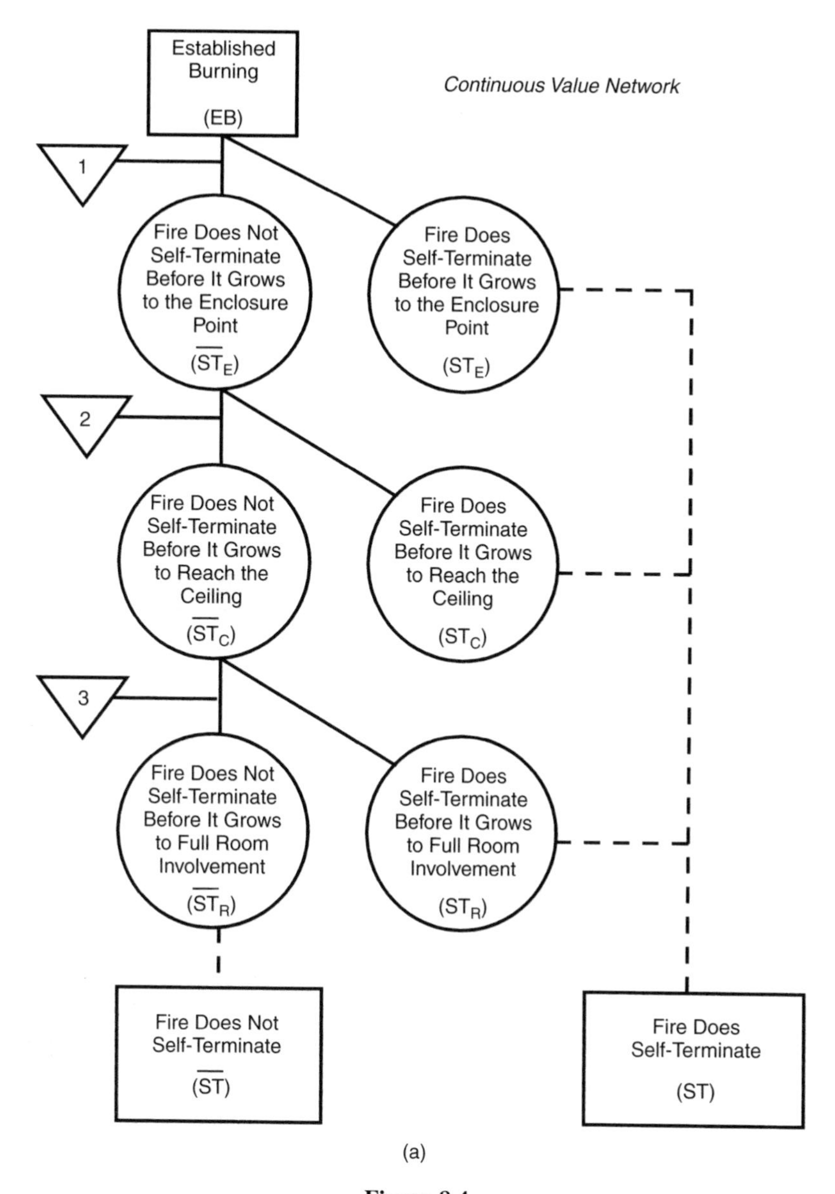

Figure 8.4

point was estimated as 300 kW. The likelihood of self-termination is the complement of $P(\overline{ST_E})$, which is $P(ST_E) = 0.10$.

Given that the fire reaches the enclosure point $(\overline{ST_E})$, the likelihood that it will grow to reach the ceiling point is even greater. The value assigned to this belief is $P(\overline{ST_C}) = 0.95$. Its complement is the probability that the fire will self-terminate, $P(ST_C) = 0.05$. The estimated fire size at the ceiling point is 900 kW.

Given that the fire reaches the ceiling point $P(\overline{ST_C})$, the likelihood that it will continue to grow and reach FRI is almost certain. Therefore the value for this event is $P(\overline{ST_R}) = 1.0$ and the value for self-termination of the fire is $P(ST_R) = 0.0$. The estimated fire size at FRI is about 5 MW.

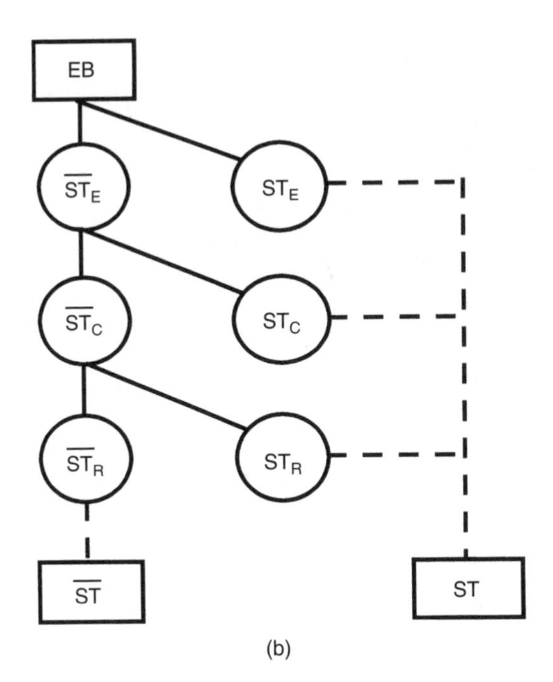

Figure 8.4 *(continued)*

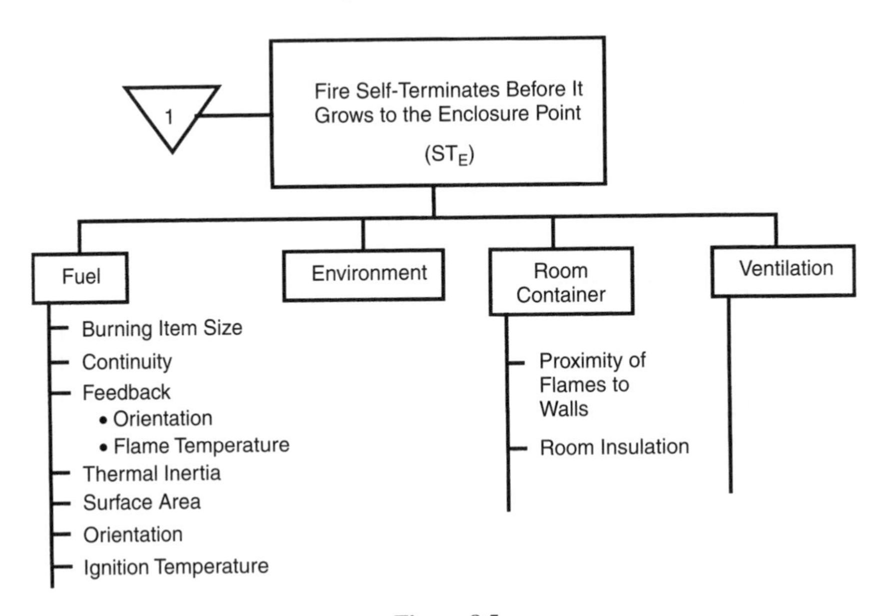

Figure 8.5

Solution

The scenario estimates express the judgment of the analyst and are incorporated into the CVN of Figure 8.8. The calculations follow the procedures described in Chapter 6. The graph of this fire scenario is ST curve A in Figure 8.12.

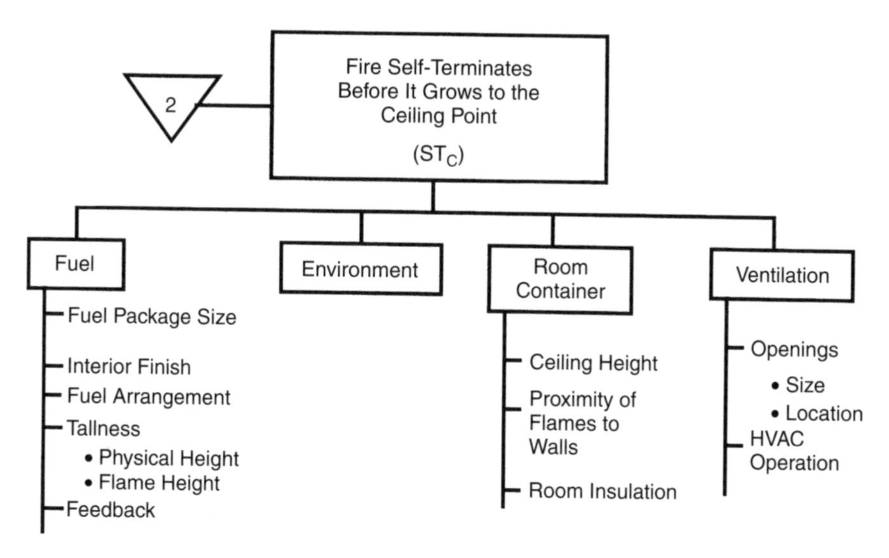

Figure 8.6

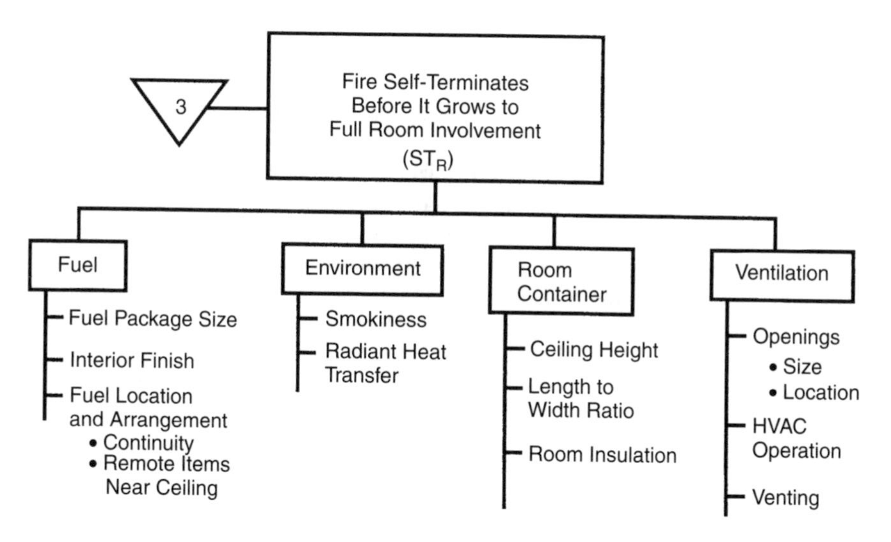

Figure 8.7

Discussion

The corresponding values from columns marked Cumulative Value and Fire Size are the coordinates for ST curve A of Figure 8.12. Theoretically the segments between coordinates are curved lines. However, because the general shape and relative positions are the important characteristics, straight lines are usually satisfactory.

Example 8.2

A different location for ignition and EB is selected for the room of Example 8.1. Plot an ST curve for the scenario using the following estimations.

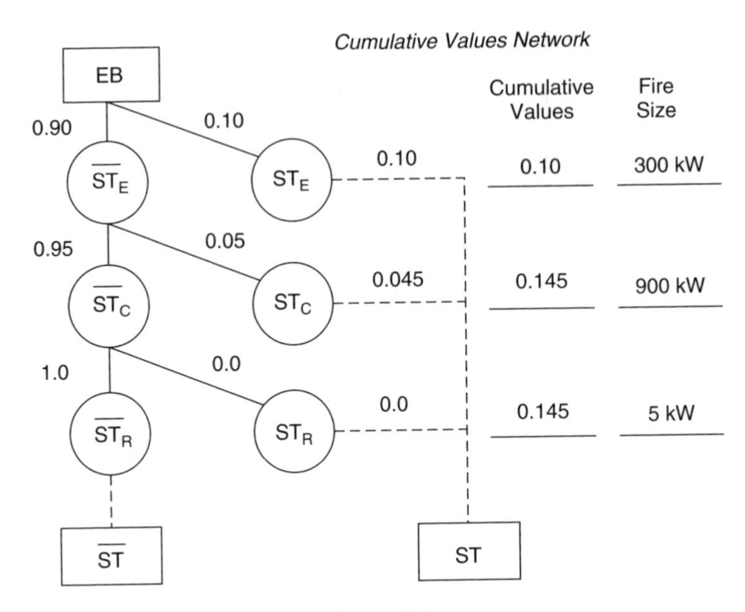

Figure 8.8

Given EB, the fuel package for this scenario will resist fire growth during the early stages of a fire. The likelihood that this fire will grow to reach the enclosure point was estimated as $P(\overline{ST}_E) = 0.40$. The fire size is about 400 kW.

Larger fires tend to continue increasing in size unless other factors are present to inhibit natural fire growth. For these scenario conditions the inspector feels that if the enclosure point is reached, the likelihood that the fire can grow to the ceiling point is relatively high. A value of $P(\overline{ST}_C) = 0.75$ was assigned. This fire size was estimated as 800 kW.

If the fire does reach the ceiling point, there is a large likelihood that FRI will result, However, the inspector feels that it is still not a certainty and selects a value of $P(\overline{ST}_R) = 0.90$ to represent this assessment. The estimated fire size at FRI is again about 5 MW.

Solution

The CVN of Figure 8.9 shows the estimated values and their complements. ST curve B of Figure 8.12 shows the resulting graph.

Example 8.3

A third location for EB was selected for the room of Example 8.1. Plot an ST curve for the following scenario evaluations.

Given EB, the likelihood that a fire will grow to reach the enclosure point was judged to be nearly certain. Self-termination will not occur, so $P(ST_E) = 0.0$. The estimated fire size at the enclosure point is 400 kW.

When the fire reaches the enclosure point, the analyst believes that again there is a certainty that the fire will grow to the ceiling point. Therefore the probability of success in fire termination before it reaches the ceiling point must be $P(ST_C) = 0.0$. The estimated fire size at the ceiling point for this scenario is 1 MW.

For this scenario, if the fire reaches the ceiling point, there is a certainty that the fire will grow to FRI. This certain value must be $P(ST_R) = 0.0$. This fire size was 5 MW.

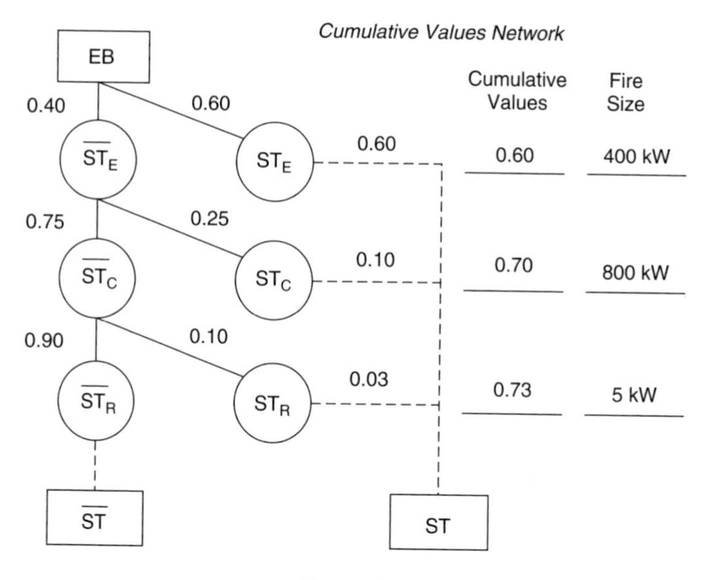

Figure 8.9

Solution

Here values were described in terms of a probability of self-termination. Figure 8.10 shows the calculations, and Figure 8.12 shows the results as ST curve C.

Discussion

Note that the values and ST curve express a judgment that an EB in this location will always result in FRI. Never will the fire self-terminate between EB and FRI. Therefore ST curve C

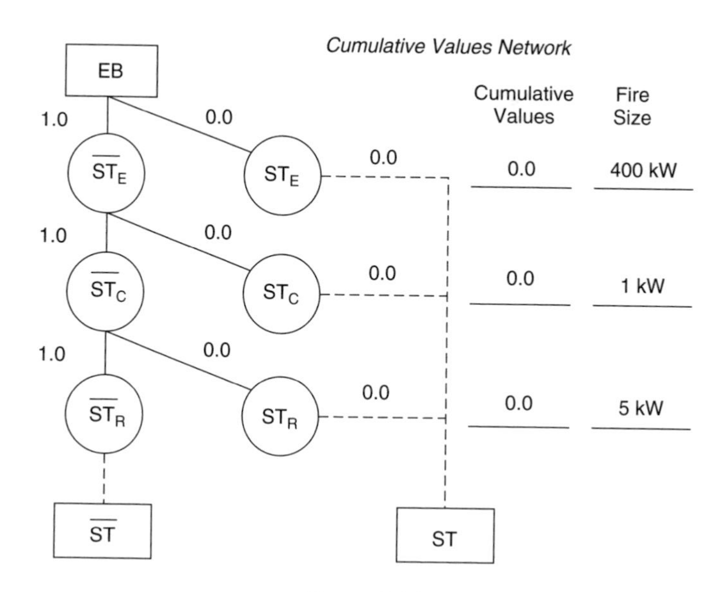

Figure 8.10

of Figure 8.12 is horizontal, depicting that the fire has no likelihood of self-termination during this scenario.

Example 8.4

A fourth EB location was selected for the room of Example 8.1. Plot an ST curve for a scenario that had the following evaluations.

Given EB, the inspector believes that the quantity and type of fuels in this fuel package and their arrangement make it impossible for a fire to grow to the enclosure point. Although the fire has a 60% likelihood of reaching 50 kW, conditions make it impossible to grow beyond that size and the fire will certainly go out without extension.

Solution

Figure 8.11 shows the values and calculations and curve D of Figure 8.12 the visual description. Note that Figure 8.11 modifies the fire size events for an additional event of ST_{50kW}. Generic events are convenient for rooms that routinely reach FRI by moving through the usual realms of fire growth. The purpose of a network is to guide a disciplined way of thinking and to express that thinking with others. One makes adjustments whenever it is appropriate to clarify the description. The only caution is that all network adjustments must be mathematically logical and cohesive.

8.8 Discussion

The event probabilities of Examples 8.1 to 8.4 describe judgments of fire behavior. To form an opinion, one systematically studies the relevant information. Networks force an individual to mentally place himself or herself in a position of an event condition and look forward only to the next event. The room and fire environment for that condition is envisioned and the analysis incorporates only information relevant to the event being evaluated. The probability measure is a judgment of the event's likely performance.

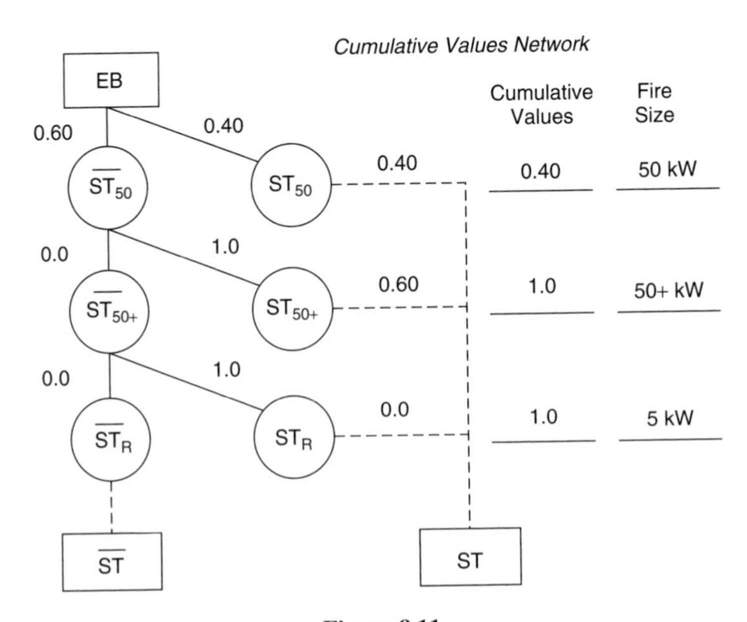

Figure 8.11

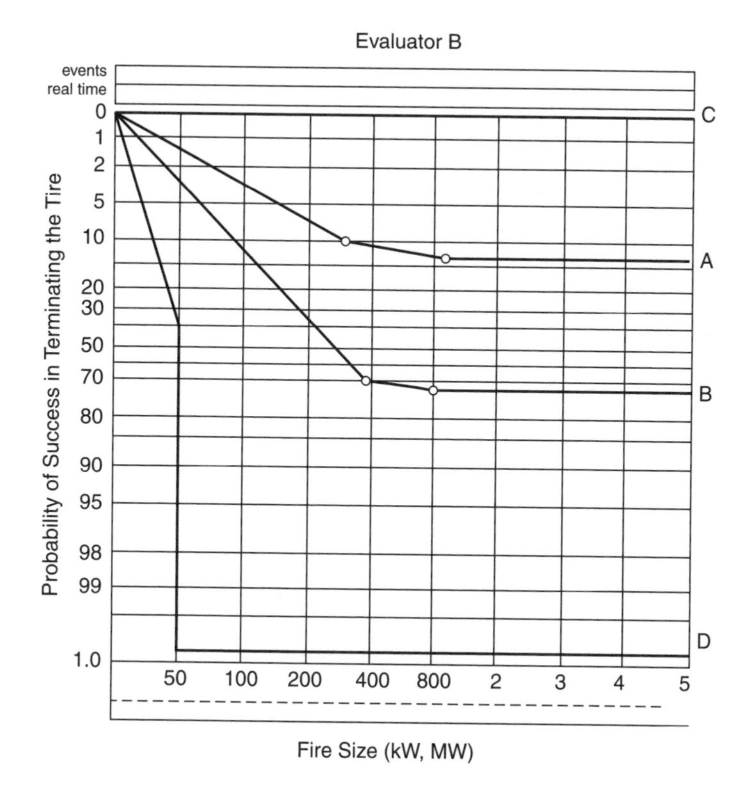

Figure 8.12

One bases an assessment on relevant information such as theoretical and empirical knowledge, calculations, computer analyses, previous experiences, and observed details. Uncertainties may exist regarding dimensions, properties, professional calculations, and knowledge. An estimate blends available information and knowledge with experience and judgment within the time constraints.

An analysis establishes the logic and technical rationale for performance estimates. Graphs tell a story of estimated scenario behavior. The ST curves of Figure 8.12 give a visual sense of expected outcomes when a fire starts in each of four different locations of a room. One can quickly discern the relative hazard and outcomes. If another individual disagrees with the estimates, specific events can be discussed to identify the technical bases for performance. Collectively, network probability estimates and their associated graphs establish a personal understanding of performance and enhance communication with others.

8.9 Large room concepts

Analysis for a large room follows a logic similar to that for a small room. The CVN is modified to reflect a thought process that traces the fire propagation mechanism. Large rooms are characterized by high ceilings, large volumes, and spaced fuel packages where radiation is the primary heat transfer mechanism that causes fire to propagate between fuel packages. Atria, malls, auditoria, and large manufacturing or storage spaces are examples of large rooms.

The distinction between large and small rooms is more than just a matter of size. The mechanism of fire growth and FRI guides the differentiation. A range of sizes and conditions blur differences between flashover and spreadover. Flashover correlations, computer programs, and theoretical and empirical relationships help one to understand a fire scenario. However, theory and applications must be appropriate for the dimensions and conditions being evaluated. Probability measures and graphs describe expected performance for conditions that do not match the constraints of available tools.

Full room involvement in large rooms starts with a fire in an initial fuel package. The fire conditions and proximity of adjacent fuel packages may cause radiation induced ignition and fire development in an adjacent fuel package. The fire continues spreading to other fuel packages until conditions reach a level where full room involvement or its equivalent occurs. While hot layers do form at the ceiling and back radiation does pyrolyze unburned fuels, the principal mechanism of fire growth is a radiation induced fire propagation throughout the arrangement of fuel packages. In moderate-sized rooms, spreadover may cause the initial fire growth, but room dimensions and ventilation characteristics may create flashover conditions as the fire grows.

The characteristics of a spreadover mechanism are

- A fire starts and grows in a fuel package.
- The burning fuel package radiates heat.
- The heat flux pyrolyzes fuels in adjacent fuel packages.
- Autoignition temperatures of adjacent fuels are reached.
- Fire in this adjacent fuel package grows.
- Heat radiates to additional fuel packages and the process continues.

Many of the factors in Table 8.1 remain valid. Although flame spread between fuel packages occurs regularly, the mechanism is difficult because radiation is the major factor in propagation. The most significant factors in the process are

- the size of the radiating item generating the heat flux;
- the spacing between the emitting heat source and the target surface;
- the heat flux received by the target fuel;
- the critical heat flux (irradiation) in the target fuel needed to cause ignition;
- the duration of heat flux.

8.10 Level 2 framework for large rooms

An ST network describes possible outcomes for a fire scenario involving a specific room, contents, and ignition location. Figure 8.13 describes the thought process for rooms in which spreadover is the mechanism for fire propagation.

Here are the questions that must be asked in evaluating the events:

- Given EB, what is the probability that the fire will self-terminate in fuel package 1?
- Given a fire in fuel package 1, what is the probability that this fire will ignite fuel package 2?
- Given fuel package 2 is ignited, what is the probability that the fire will self-terminate?

The process of assessing the path of ignition and fire growth continues through as many fuel packages as necessary to gain an understanding of the building and the fire.

Fire phenomena provide a technical basis for the judgmental estimates. Figures 8.14 and 8.15 show the major factors that influence fire propagation. The following examples of ST curve construction illustrate the process for different scenarios in a large room.

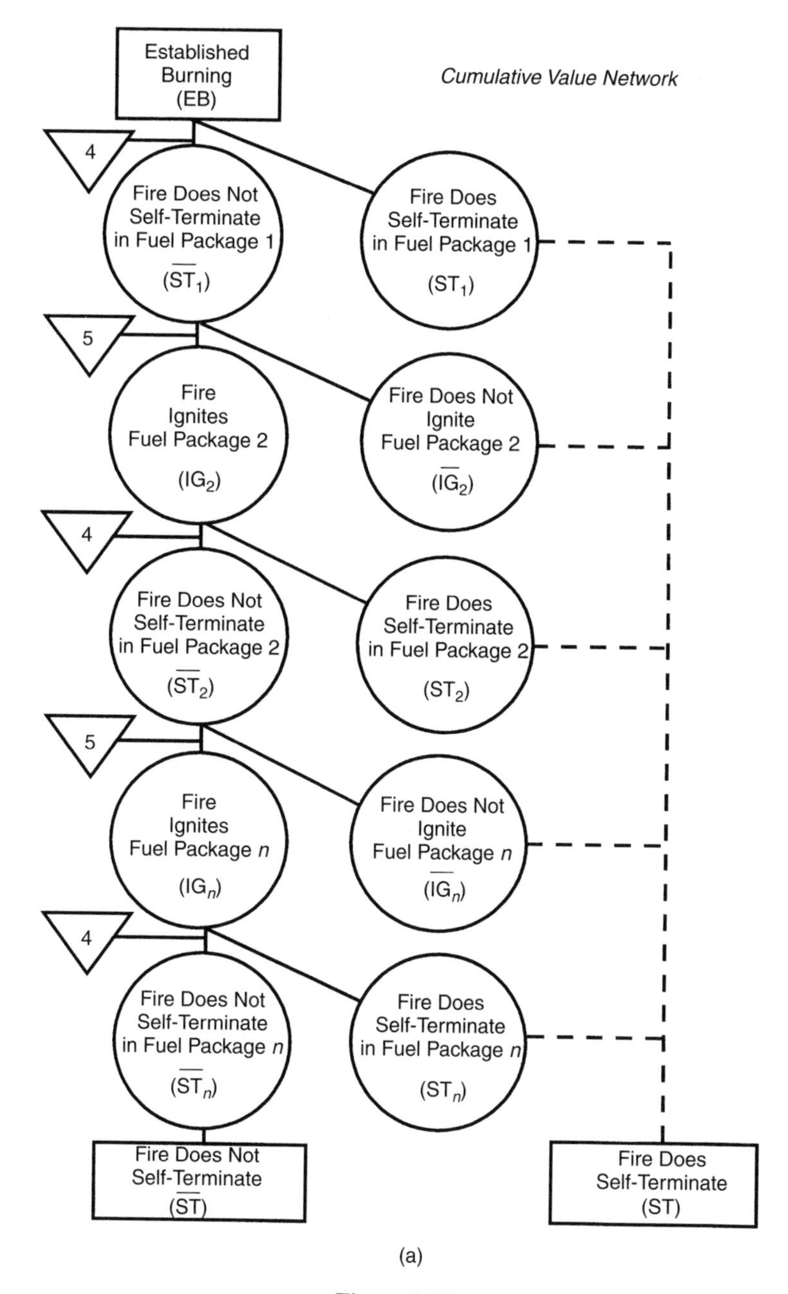

(a)

Figure 8.13

Example 8.5

A large room has several fuel packages. Assume that an inspector evaluates a fire scenario for EB in a specific location. Plot an ST curve for following estimates.

Given EB, the likelihood that a fire will grow to involve the entire fuel package is very high. A value of $P(\mathrm{ST}_1) = 0.15$ was assigned. The estimated fire size for fuel package 1 is 1.5 MW.

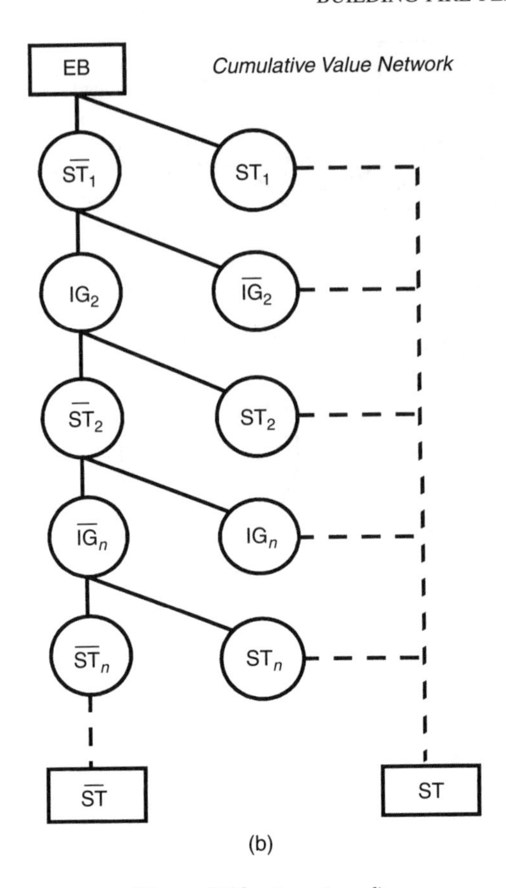

Figure 8.13 *(continued)*

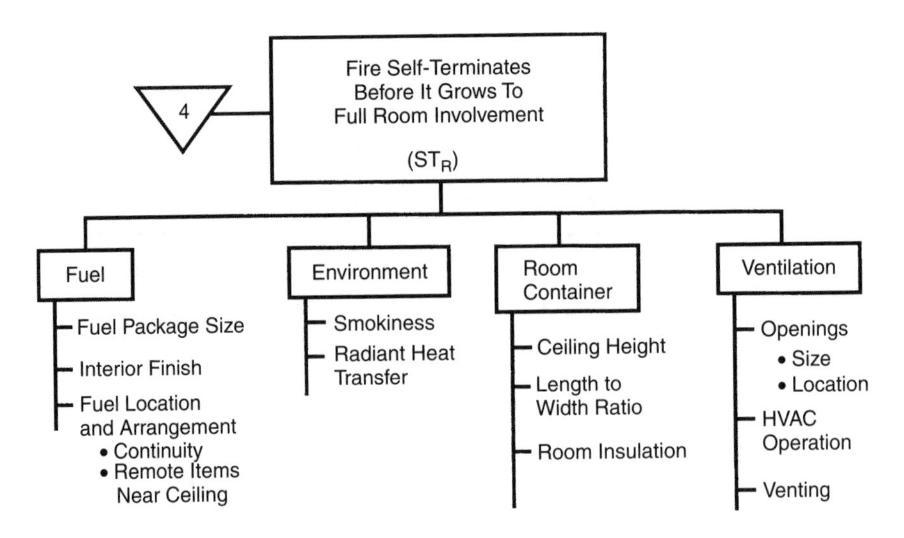

Figure 8.14

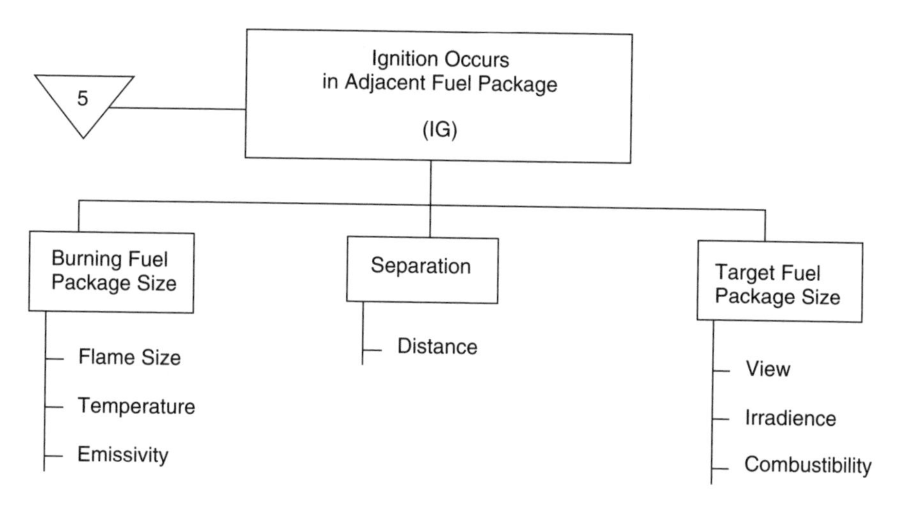

Figure 8.15

The likelihood that the fire will not self-terminate but continue to burn is the complement of $P(\mathrm{ST}_1)$, which is $P(\overline{\mathrm{ST}}_1) = 0.85$.

Fuel package spacing and radiant heat flux calculations indicate that the likelihood of ignition in fuel package 2 is relatively low. The value assigned to this event was $P(\mathrm{IG}_2) = 0.20$. Its complement is the probability that fuel package 2 will not ignite, $P(\overline{\mathrm{IG}}_2) = 0.80$.

Given that fuel package 2 ignites, the likelihood that the fire will self-terminate is low. A value for this event was estimated as $P(\mathrm{ST}_2) = 0.10$. The estimated fire size of fuel packages 1 and 2 is 5 MW.

Given involvement of fuel package 2, ignition of the next fuel package is extremely high, described by a value of $P(\mathrm{IG}_3) = 0.95$. Then, given ignition of fuel package 3, its full involvement is a certainty. Therefore $P(\mathrm{ST}_3) = 0.0$. The fire size of the three fuel packages was estimated at about 12 MW.

After these three fuel packages are fully involved, there is a certainty that the entire room will become fully involved.

Solution

The evaluations that express the fire behavior of the scenario described above are incorporated into the CVN in Figure 8.16. ST curve A in Figure 8.18 shows this scenario. Physical behavior is related to the numerical values and their locations on the networks.

Vertical discontinuities represent the likelihood of ignition in the adjacent fuel package. A short line signifies a relatively easy ignition potential, while a long line indicates difficulty in igniting the next fuel package.

Example 8.6

The inspector also evaluated another fire scenario for the room of Example 8.5. Plot an ST curve for the descriptions given below.

Given EB, the likelihood that a fire will grow to involve the entire fuel package is high. The value assigned is $P(\overline{\mathrm{ST}}_1) = 0.75$. The estimated fire size for a fire in fuel package 1 is 1.0 MW.

Given a fire in fuel package 1, the probability that fuel package 2 will ignite is very high. The value assigned is $P(\mathrm{IG}_2) = 0.90$.

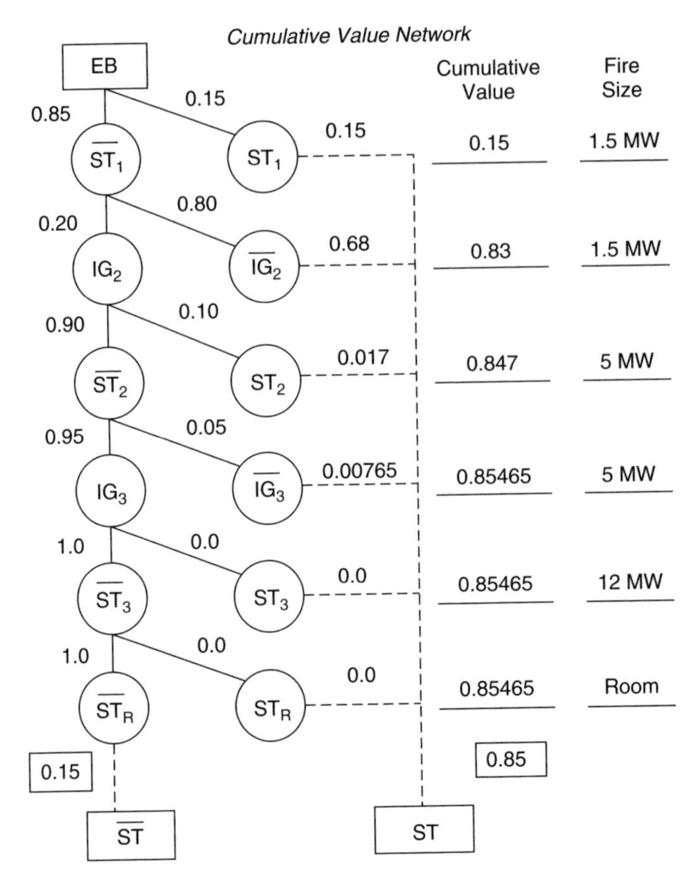

Figure 8.16

Given the ignition in fuel package 2, the probability that the fire will self-terminate is relatively low. An estimate for this event was $P(\mathrm{ST}_2) = 0.30$. The estimated fire size of fuel packages 1 and 2 is 3 MW.

Given a fire in fuel package 2, a large open space will make ignition of the next fuel package extremely difficult. A value of $P(\mathrm{IG}_3) = 0.05$ is selected.

Given that ignition of fuel package 3 occurs, estimated self-termination was $P(\mathrm{ST}_3) = 0.6$. The fire size for these three fuel packages is about 7 kW.

The individual feels there is no chance for this fire scenario to extend beyond the first three fuel packages.

Solution

The CVN of Figure 8.17 and ST curve B in Figure 8.18 describe this scenario. Curve B terminates at 7 MW. The dashed line indicates that the entire room is of a larger size. Dashed lines are useful to avoid ambiguity when expressing results to others. If only curve B were drawn and the line terminated at 7 MW, another individual might incorrectly interpret the entire room as of just that size.

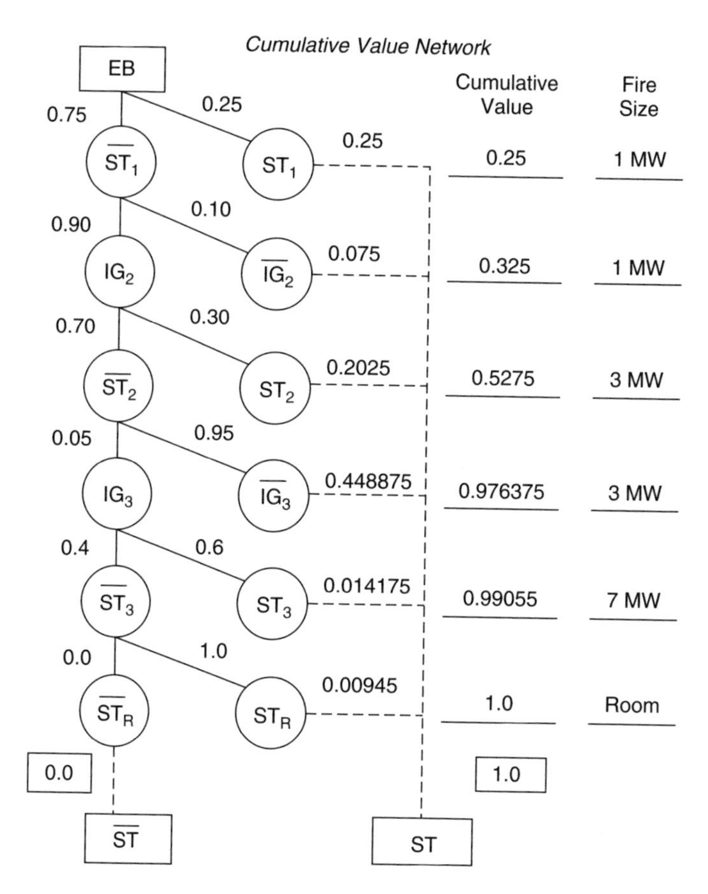

Figure 8.17

8.11 Level 3 evaluations

A level 2 evaluation enables one to understand performance and provide technical documentation, if necessary. It also enables one to discriminate which details affect performance and their relative influence.

A level 3 evaluation uses the same networks as a level 2 analysis. The difference is that level 3 develops a sensitivity to variations of parameter or detail. A level 3 evaluation will give a sense of proportion for an envelope of behavior ranges. For example, a level 2 analysis may estimate the likelihood of flashover for a carefully selected model room. Variations in contents, materials, sizes, and arrangements will cause different likelihoods for flashover. When the model room is chosen carefully, these variations seldom make an adverse difference in macroperformance. However, assume one wishes to understand time differences or flashover likelihood ranges for various types of upholstered furniture. A level 3 analysis will provide an understanding of the room behavior differences for different types of furniture.

Level 3 evaluations can examine influences on component behavior for variations in physical features or computational parameters, such as the effect of physical relationships or default values in computer programs. One can incorporate reasonable variations in network analyses to obtain *if-then* information. Normally, one uses level 3 analyses only for situations when the value of the added information is greater than its cost.

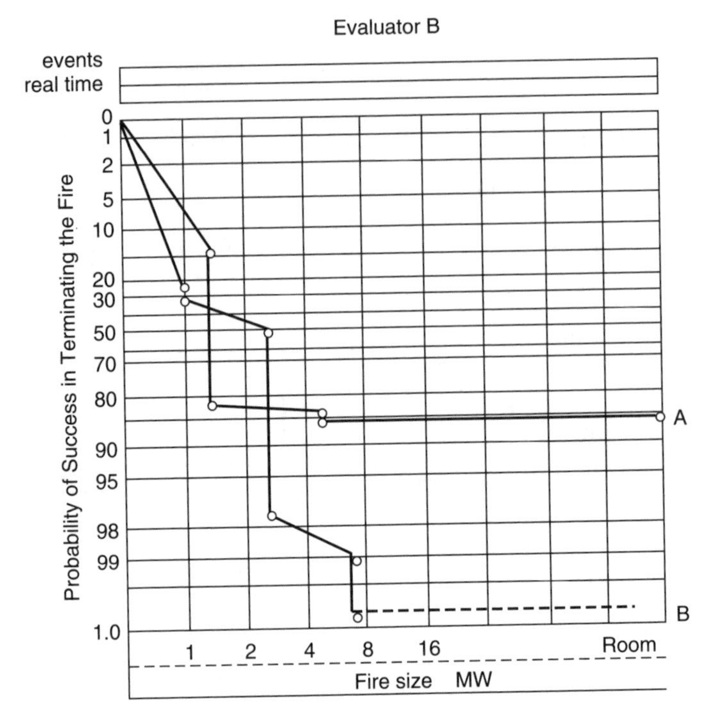

Figure 8.18

8.12 Level 1 concepts

The goal of a level 1 evaluation is to gain an understanding of the "big picture" for the way in which a component will work. It helps to distinguish between details that will have an important influence on performance and features that have little effect. Level 1 evaluations are relatively fast and comprehensive using predominantly observational information, mental estimates, and back-of-the-envelope guidelines.

A level 1 evaluation is the start for all performance analyses. It provides a sense of proportion for behavior and a base from which to make decisions on other actions. Often knowledge from a level 1 evaluation may be sufficient for project needs. If not, it identifies which topics should be examined in subsequent level 2 or level 3 analyses.

8.13 Concepts of fire growth potential classification

The interior design of a room defines the potential for fire growth. The classification of a room's fire growth potential is a valuable part of a design fire description. A fire growth potential classification describes the relative hazard of a room's interior design. A classification can also incorporate a variety of standardized fire characteristics to use with level 1 evaluations.

An interior design classification represents the relative ease and speed with which full room involvement (FRI) can occur. Some interior designs have attributes that make it difficult for FRI to occur. If FRI occurs at all, the time duration will be very long. On the other hand, other interior designs have conditions that produce FRI easily and quickly. Ranking room interior designs cultivates a sense of proportion for observational estimates of behavior. Also, one can more easily describe hazards posed by different features.

A fire growth potential classification is associated only with the room interior design. While ventilation does play an important part in room fire behavior, we assume that sufficient ventilation is available for efficient combustion. The classification considers only the room size and shape, interior finish, and the amount, type, characteristics, and distribution of contents. It considers only the segment of a room fire between EB and FRI. Ignition conditions, ventilation, and postflashover fire conditions are not considered in classifying the fire growth potential.

8.14 Fire growth potential classifications

Fire growth potential classifications are broad and rank the interior design as a composite entity. Although individual fire scenario outcomes are influenced by EB, the classification rank orders relative fire growth potentials for the composite room.

Broad descriptors enable one to discriminate room conditions and describe behavior clearly and easily. Figure 8.19(a) shows three groups of high, moderate, and low fire growth potential. This system is convenient because individuals can describe a room hazard quickly in terms such as red, yellow, green or bad, moderate, good. Figure 8.19(b) shows a five-group classification that takes the probability scale from certainly no (0.0) to certainly yes (1.0) and divides it into equal segments. Although there are situations where a three- or four-group ranking is useful, we will use the five-group classification here.

An established burning can occur at any location in a room. Each produces a different fire. Also, "tomorrow" the occupants may change the arrangement and materials of contents or introduce new sources for ignition. Ignition source locations and their frequency do not influence a room classification. We address the frequency, nature, and location of ignition in another part of the system. The fire growth potential classification is based only on the room's size and shape, its interior finish, the contents, and their arrangement. Figure 8.20 shows the factors that relate to the selection of an interior design classification selection.

8.15 Illustrations of the classification process

A level 1 classification ranks a room's fire growth potential. The rankings attempt to order the fire growth hazard by integrating the factors of Figure 8.20. Examples 8.7 to 8.10 describe the process. As one combines observational skills and experience with judgmental decisions the process becomes rapid and consistent.

Example 8.7

Figure 8.21 shows a model for room A and Table 8.2 describes its attributes. Select a fire growth potential classification for this room.

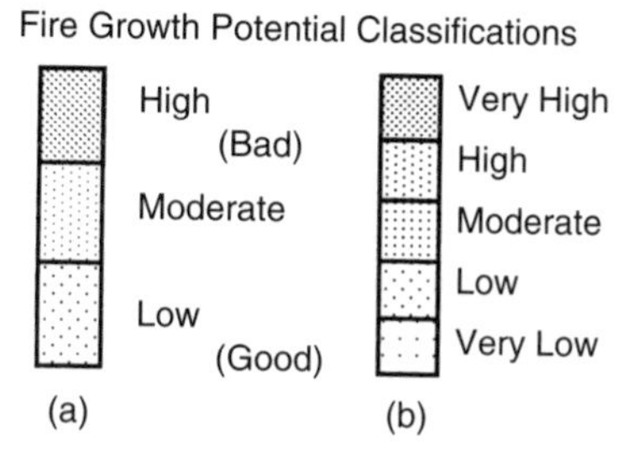

Figure 8.19

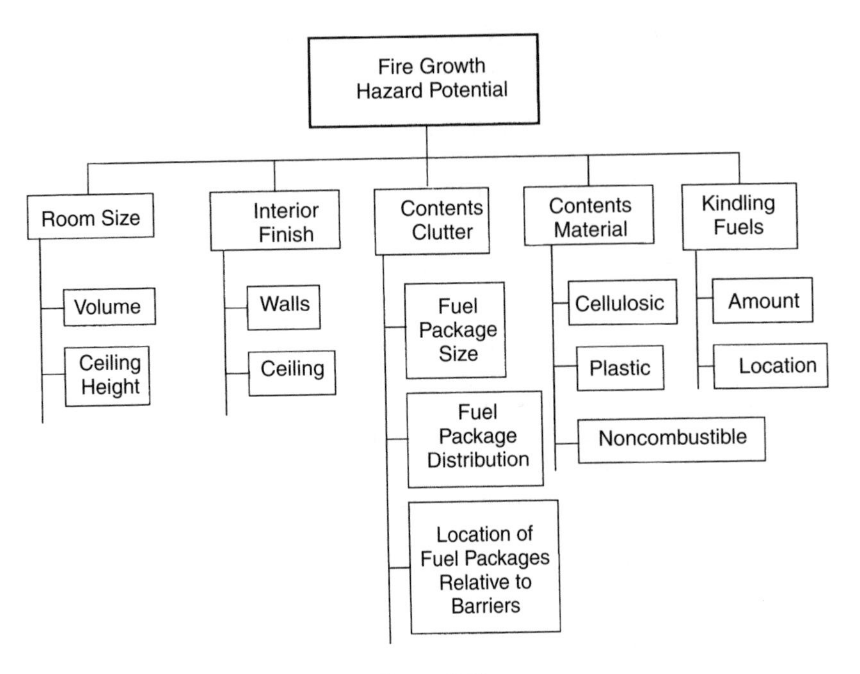

Figure 8.20

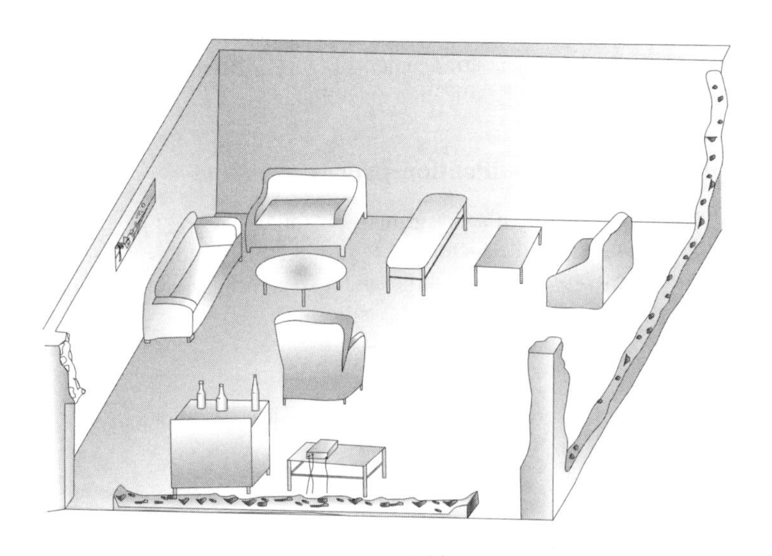

Figure 8.21

Solution

Classification: very Low potential for full room involvement.

Discussion

The factors of Figure 8.20 provide the rationale for Table 8.3. A fire in almost any location would have a low likelihood of causing FRI. The greatest potential difficulty is the combined

Table 8.2: Room A Description

Size	$L = 32$ ft, $W = 24$ ft, $H = 10$ ft
Interior finish	Walls: painted gypsum plaster Ceiling: painted gypsum plaster Floor: hardwood
Contents	Older cellulosic-based (wood, cotton) furniture; cotton batting padding in upholstered furniture
Arrangement	Approximately as shown; waste baskets and incidental occupant-related goods are not shown
Kindling fuels	Normal expectations for a living room in a home
Ventilation	Adequate

Table 8.3: Room A Classification Rationale

Room size	Floor area = 768 ft^2; volume = 7680 ft^3 Ceiling height 10 ft; the materials and sizes of the fuel packages will make it difficult for flames to touch the ceiling
Interior finish	Energy absorbing; will not contribute to flame spread
Contents clutter	Wide separation of furniture except at couch and love seat; estimate about 12% of floor area covered by furniture; gives a feeling of spaciousness
Contents material	Wood and cotton base of the upholstered furniture does not produce a high peak heat release rate; the major fire concern would be the coalescence of combined burning of the couch and love seat; perhaps flames can reach the ceiling; nevertheless, the likelihood of this fuel package causing FRI is low; likelihood of fire spread to other fuel packages is low
Kindling fuels	A few waste baskets, magazines, etc. These should not contribute much to fire growth or spread after EB, although they could be involved in the ignition to EB part of the process. Since the classification is based on a situation given EB, the kindling fuels will not be a major factor in the classification selection

fuel package of the couch and love seat. Even if FRI did occur from this fuel package, the time duration would be very long. The judgment is that this room and contents provide a very low likelihood of reaching FRI.

Example 8.8

Change the attributes for the room of Figure 8.21 to those of room B (Table 8.4) and select a fire growth potential classification.

Table 8.4: Room B Description

Size	$L = 16$ ft, $W = 12$ ft, $H = 10$ ft
Interior finish	Walls: painted gypsum plaster Ceiling: painted gypsum plaster Floor: hardwood
Contents	Older cellulosic-based (wood, cotton) furniture; cotton batting padding in upholstered furniture
Arrangement	Approximately as shown; waste baskets and incidental occupant-related goods are not shown
Kindling fuels	Normal expectations for a living room in a home
Ventilation	Adequate

Solution

Classification: moderate potential for full room involvement.

Discussion

This room is the same as room A except that the floor area is only 25% of room A and the volume is 20% of room A. The same furniture now occupies almost 50% of the floor area, contributing to a much greater sense of clutter and crowding. Although the room may be uncomfortable, it is not unlivable. The ceiling height is lower, making it more likely that a fire in the couch and love seat would produce flames to touch the ceiling. It is unlikely that a fire in other fuel packages would cause flames to touch the ceiling or spread to other items.

Although it may appear that there can be a large number of room interior designs and styles, most rooms use a relatively small group of common features. Dimensions, arrangements, content materials, and furnishings are quite manageable from a fire analysis viewpoint. The standards used by architectural and building and interior design professionals provide a basis for model room development by the fire protection community. A few selected model rooms provide standardized classifications that can be adjusted to reflect differences in condition.

A table such as Table 8.5 can be used to adjust factors from a basic model room. In Table 8.5 each factor can raise (i.e., make worse) or lower (i.e., make better) the likelihood that FRI will occur. Not all factors contribute equally to the fire hazard so we use a four-star rating system to describe importance. A rating of * provides little change, whereas a rating of **** indicates a significant importance level. Identifying these changes enables one to blend differing influences to provide a rationale for adjusting fire growth potential classifications.

The likelihood of FRI for Example 8.8 has increased and its time duration decreased because of the room dimensional changes. The major focus of attention continues to be the fuel package of the couch and love seat.

Table 8.5: Classification Adjustments Changing Room A to Room B

Factor	Direction of change from room A	Importance of change from room A	Discussion
Lower ceiling to 8 ft	↑ (worse)	**	Easier for the flames to reach ceiling and contain heat
Change floor area and volume	↑ (worse)	**	Based on smaller volume (easier to reach flashover) and closer spacing of the fuel packages, makes fire development more likely; incorporated into the room size change above
Interior finish	No change	–	
Contents clutter			Spacing is more conducive to fire development; this factor was incorporated into the room size change above
Contents materials	No change	–	
Kindling fuels	No change	–	Perhaps a slight increase in the likelihood of FRI because of room size reduction

Example 8.9

Change the attributes for the model room of Figure 8.21 to those of room C (Table 8.6) and select a fire growth potential classification.

Solution

Classification: high potential for full room involvement.

Discussion

This room is the same as room A except for the change in furniture construction and the addition of combustible interior finish (Table 8.7). While we do not know the burning characteristics, it is reasonable to anticipate that the heat release rate will be significantly higher for this modern upholstered furniture.

There is an uncertainty in whether to judge this room as "low" in the "high" classification or as high in the moderate classification. It seems more appropriate to place it somewhere near the boundary between these classifications. Here we will select a single classification; another way to describe this hazard will be introduced in Section 8.19. Here combustible interior finish

Table 8.6: Room C Description

Size	$L = 32\,\text{ft}$, $W = 24\,\text{ft}$, $H = 10\,\text{ft}$
Interior finish	Walls: mahogany plywood cover over gypsum plaster Ceiling: painted gypsum plaster Floor: hardwood
Contents	Modern furniture construction including foam plastic padding
Arrangement	Approximately as shown; waste baskets and incidental occupant-related goods are not shown
Kindling fuels	Normal expectations for a living room in a home
Ventilation	Adequate

Table 8.7: Classification Adjustments Changing Room A to Room C

Factor	Direction of change from room A	Importance of change from room A	Discussion
Ceiling height	No change	–	
Floor area and volume	No change	–	
Interior finish	↑ (worse)	**	The continuity provided by the combustible interior finish increases substantially the size of the fuel packages; it contributes to increased HRR and flame spread
Contents clutter	No Change	–	
Contents materials	↑ (worse)	***	High HRR and increased flame heights from the upholstered materials increase significantly the likelihood of FRI; the room size and ceiling height mitigate the increase somewhat
Kindling fuels	No change	–	

influenced the higher classification. If the interior finish were the same as in room A, a moderate fire growth potential would have been selected.

Many factors that have a bearing on the fire growth potential are unknown in these examples. For example, plywood attachment directly to the walls or with furring strips can influence fire growth. The flame spread rating is unknown. Similarly, details of furniture construction and its combustion properties can have a substantial influence on fire growth. How much detail is enough? One must balance the cost of information acquisition against its value for the application. Confidence improvement in performance estimates resulting from more information may have an important or an insignificant role within the scope of any specific composite building evaluation.

Although the analytical framework can use any design fire, the purpose of a performance evaluation is to understand the building. A design fire is not a statistical fire nor does it depict a specific scenario. Instead it represents a reasonable, yet severe challenge for the various building fire defenses. Consequently, if the model room is specified carefully, most potential concerns become irrelevant. Conditions that produce more severe fire conditions may be recognized and the fire characteristics adjusted accordingly. We may disregard conditions that produce less severe fires.

Example 8.10

Change the attributes for the room of Figure 8.21 to those of room D (Table 8.8) and select a fire growth potential classification.

Solution

Classification: very high potential for full room involvement.

Discussion

This room combines the smaller volume with combustible interior finish and modern furniture having expectations of a high heat release rate. Table 8.9 shows the factors to blend in deciding the classification.

The influence of a relatively small room, high heat release rate (HRR) of contents, and combustible interior finish all contribute to conditions in which flashover will be easy to achieve. The importance of each as a separate consideration in Table 8.9 is difficult to distinguish. However, one can recognize that each factor provides an additional contribution to the potential for FRI.

8.16 Discussion for room classifications

A suggested process for selecting a fire growth potential classification is as follows:

1. Identify the model room interior design.
2. Identify the factors that have the greatest influence on potential fire behavior.

Table 8.8: Room D Description

Size	$L = 16\,\text{ft}$, $W = 12\,\text{ft}$, $H = 8\,\text{ft}$
Interior finish	Walls: mahogany plywood cover over gypsum plaster Ceiling: painted gypsum plaster Floor: hardwood
Contents	Modern furniture construction including foam plastic padding
Arrangement	Approximately as shown; waste baskets and incidental occupant-related goods are not shown
Kindling fuels	Normal expectations for a living room in a home
Ventilation	Adequate

Table 8.9: Classification Adjustments Changing Room A to Room D

Factor	Direction of change from room A	Importance of change from room A	Discussion
Lower ceiling to 8 ft	↑ (worse)	**	Easier for the flames to reach ceiling; high HRR of armchairs could even reach the ceiling
Change floor area and volume	↑ (worse)	**	Based on smaller volume (easier to reach flashover) and closer spacing of the fuel packages, makes flashover more likely
Interior finish	↑ (worse)	***	The continuity provided by the combustible interior finish increases substantially the size of the fuel packages; it contributes to increased HRR and flame spread
Contents clutter	↑ (worse)	**	Spacing is more conducive to fire development
Contents materials	↑ (worse)	***	High HRR and increased flame heights from the upholstered materials increase significantly the likelihood of FRI
Kindling fuels	No change	–	

3. Consider different locations for established burning and associated fire development scenarios to get a "feel" for the importance of factors and the sensitivity of the model room to reaching FRI. Note that these locations should not be associated with potential ignition sources, but with different types of fires that could develop.
4. Mentally envision a synergistic fire development process to develop a sense for the relative ease of room fire development.
5. Gauge the room sensitivity to variations in arrangement and contents regarding the speed and likelihood for reaching FRI.
6. Select a classification for the room's fire growth potential.

The issue of accuracy and sensitivity of group distinction needs attention. The objective of classification is to rank interior designs for their fire growth potential, not to calculate outcomes of specific fire scenarios. Ranking classifications use judgment to integrate recognized field conditions with principles of fire dynamics. While calculations can give greater confidence in making selections, one should not replace skills in relating field observations to theory, calculation outputs, and performance. The focus is on recognizing fire development conditions to rank the relative fire growth hazard of rooms. This thought process is useful to describe likely outcomes to others.

A room classification is the starting point in evaluating how a building will behave in a fire. If a generic model room has been carefully selected, one may compare variations in other rooms to the model room. Five general calibration zones provide sensitivity and flexibility in the evaluation process.

The examples of Section 8.15 illustrate a process for selecting a fire growth potential classification for a room and its interior design. Figure 8.20 identifies the factors that dominate classification decisions. This figure provides guidance for identifying relative fire growth hazards quickly and consistently. For many applications this rapid assessment is sufficient. For others it becomes a base for a more careful additional analysis.

8.17 αt^2 fires

A design fire defines the time-related products of combustion that fire defenses are expected to resist. Although the fire growth potential classification is not a product of combustion, it is a convenient device with which to package design fire values.

The segment of fire growth from EB to FRI, as shown in Figure 8.1, can be approximated by a power law, $\dot{Q} = \alpha t^2$. The term α is a fire intensity coefficient that reflects the influence of fuel on the speed of fire growth. Although the power law is valid only for fire growth to the ceiling point, that condition is reasonably close to FRI for many small rooms. The detection industry uses specific values of α to define ultrafast, fast, medium, and slow fire types [1]. These classifications and their values for α are not convenient for building performance evaluations. Therefore we shall use slightly different values of α for building performance evaluations and different classification names to avoid confusion.

Figure 8.22(a) shows the five fire growth classifications, and Figure 8.22(b) shows a family of αt^2 growth curves that can be associated with those classifications. Roughly, one can relate the speed of fire growth to the likelihood of FRI. When we segment the time line of Figure 8.22(b) into 4 min intervals, the αt^2 fires can be associated with fire growth potential classifications. After a classification is selected, a value for α may be selected, as shown by the default values of Figure 8.22(a). These values for α were obtained by using a heat release rate of 1500 kW with the maximum time duration for each classification. This classification provides an initial approximation for a time–fire size relationship that is satisfactory for most building applications. If one recognizes differences in conditions or wishes to use a specific fire intensity coefficient to address a particular concern, the values should be modified.

8.18 The design fire

Figure 8.23 shows the design fire part (Section 1 a, b, c, ...) of composite Figure 4.7. One can insert design fire values for the different products of combustion into these cells. For example, the fire size (HRR) of line 1a may be estimated using the αt^2 fires up to about FRI. The intensity coefficient α represents room conditions rather than item conditions. After FRI the HRR for fire sizes must be estimated based on observation, empirical correlations, or computer analyses.

The fire size is also identified in terms of floor area or number of rooms. Line 1b of Figure 8.23 locates this information. The time periods after FRI are important to constructing design fires for the M curve. Chapter 9 describes techniques for making these estimates. The undesignated rows for Section 1 c, d, ... are used to estimate the combustion products for other fire defense evaluations. For example, sprinkler analysis would add fire plume momentum to the earlier information. Barriers and structural frame analysis use the fire's cumulative heat energy release and the fraction applied to the barrier or structural member. Smoke uses smoke generation and temperature characteristics. Section 1 shows the fire-generated information to analyze the fire defense components. Collectively they define the design fire characteristics.

8.19 The I curve

Chapter 6 discusses the limit of flame movement, described by the L curve. Section 6.8 notes that the fire can go out in only three ways: (1) self-termination (I curve), (2) automatic sprinkler suppression (A curve), and (3) fire department extinguishment (M curve).

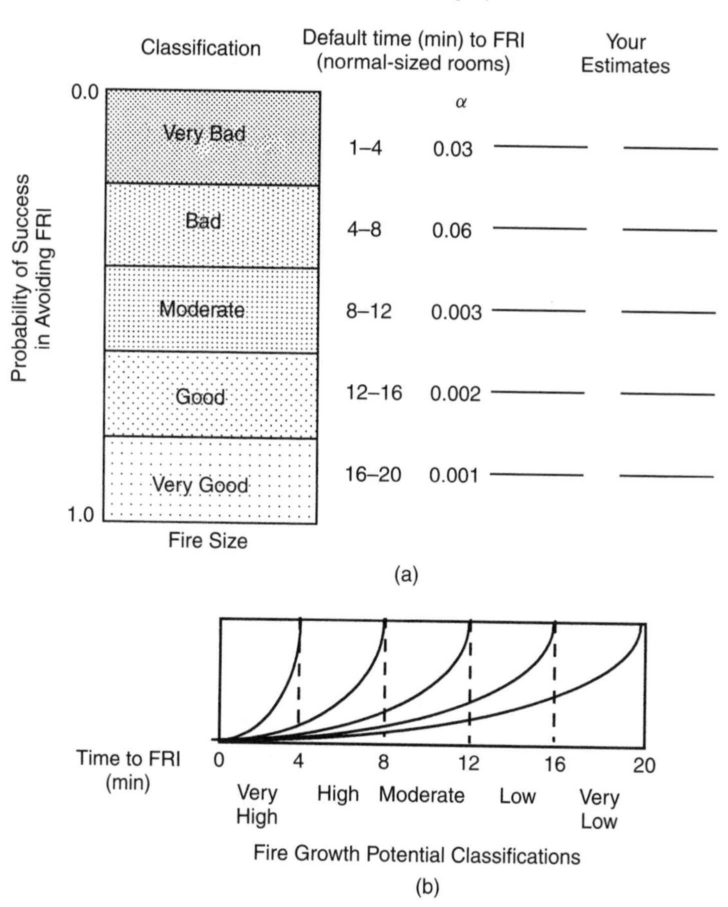

Figure 8.22

		EVENT	TIME							(minutes)						
			1	2	3	4	5	6		42	43	44	45	46		
Building Performance	1	Design Fire														
	a	Size (HRR)														
	b	Size (Area or Rooms)														
	c															
	d															

Figure 8.23

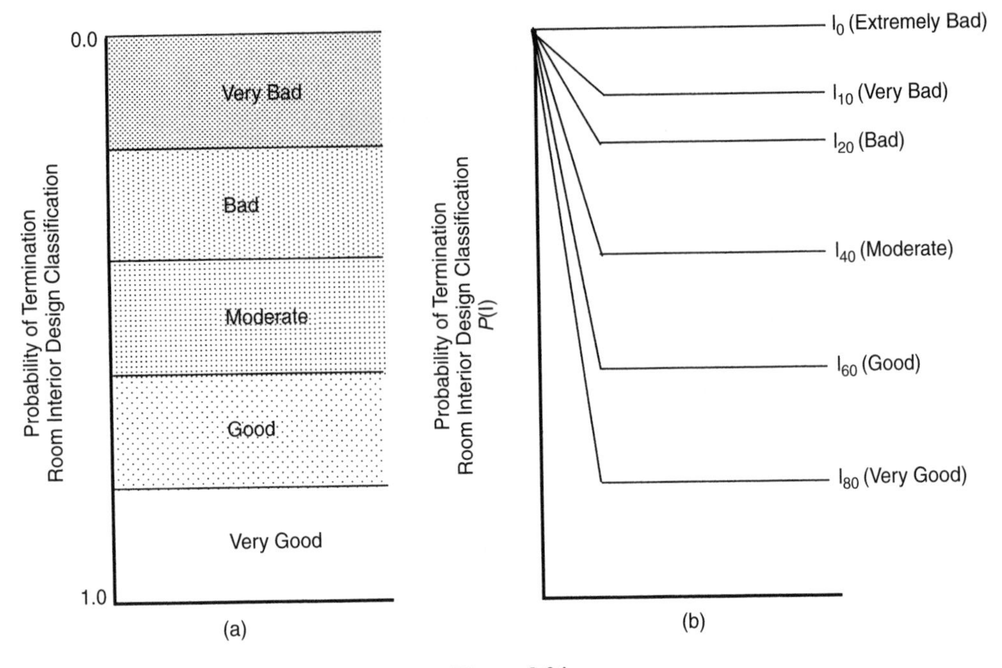

Figure 8.24

The I curve is a graph that represents a room classification (Figure 8.24). We can describe a hazard classification as I_{20}, I_{60}, I_0, etc. The numerical value signifies a probabilistic calibration of the likelihood of FRI. One can select any number from 0 to 100 to represent the I curve for the room.

We suggest the value for I_0 be used for the special case of flammable liquids and gases. One may use a value of I_{10} to represent a regular, *very bad* classification. Otherwise, the numerical values of Figure 8.24 represent an upper bound for the room classifications. However, one should not inhibit communication by rigid adherence to these default values. One may shade the classifications (e.g., I_{25}, I_{70}, or I_{90}) to express values that seem more appropriate for the conditions.

This chapter notes that a room interior design is the basis for the design fire, and different interior designs produce different fire growth characteristics. Some rooms are difficult to drive to FRI whereas others reach that condition easily. For discussion purposes, one could assume that all rooms will reach FRI for all conditions (I_0 always). This conservative approach would produce an L curve for suppression intervention only. However, there is diminished value in this approach because an L curve compares and describes. Because the I curve is a descriptor of the interior design, one may create solutions for difficult problems by integrating effective fuel control into a risk management program. The I curve helps one to develop a sensitivity to fire behavior and describe to others the influence of interior design on building performance.

Example 8.11

While conducting a building evaluation, four possible rooms of origin were identified. Interior design categories were to be selected and I curve values chosen to represent the fire growth hazard potential for the room. The four rooms were described as follows.

Room 1 has contents arranged so that fire spread between fuel packages is difficult. Fuel materials are of moderate combustibility. The room has a ceiling height and size in which FRI could occur, although not easily. A moderate fire growth potential was selected for this room.

Room 2 is small in size and filled with enough combustible contents that, given EB in any one fuel package, FRI would very likely occur. A very bad fire growth potential was selected for this room.

Room 3 is of moderate size with operations involving exposed flammable liquids. An ignition would result in a certainty of FRI. This room has an extremely bad fire growth potential.

Room 4 is large with somewhat high ceilings. Fuel packages are moderately large in area and in height. Fuel package spacing will allow propagation between fuel packages, although ignition of adjacent fuel packages when only one or two were involved is not a certainty. Material contents are moderately combustible. This condition may be more suited to developing an ST curve on which to base the I curve, but because of time limitations, that analysis was not used. The room was selected as a bad fire growth potential classification.

What are appropriate I curve values for these rooms?

Solution

The standard I curves of Figure 8.25(a) represent these room classifications. After looking at the rooms, the individual decided to modify the I curves for rooms 1 and 4 to reflect conditions more suitably. Figure 8.25(b) shows the adjusted curves.

8.20 Thoughts on design fires

The selection of appropriate, standard design fire characteristics is a critical, important part of a building performance evaluation. The philosophy of these selections needs professional discussion. Without guidance from the profession, the individual must use his or her best judgment to select appropriate conditions with which to test the fire defenses.

One should not think of a design fire as a single fire scenario. Rather, products of combustion are tailored for the specific fire defense (resistance) being evaluated. Consequently, the design fire used to evaluate a sprinkler system is not necessarily the same as that for fire department suppression, even within the same building. Perhaps even the rooms of origin may not be the same, because the objective of a performance analysis is to gain a technical understanding of the building.

Time is an important professional commodity. While each building and room has unique characteristics, much similarity exists for fire analyses. For example, although each office space has distinct features, one can develop a very few selected prototypes to serve as models for evaluation. Dimensions and arrangements fall into a modular pattern. Burning characteristics of contents may be selected for the model rooms. Other professions such as architecture and interior design can provide prototype model rooms and their design fire characteristics. A few model rooms can handle a vast majority of evaluation needs. An individualized design fire is created when conditions become different enough from the models that one cannot understand fire defense performance.

It is possible to normalize design fire characteristics in a manner similar to structural engineering. In structural engineering, most live loads are normalized to an equivalent uniformly distributed value. Even though distributed live loads are rare in buildings, standard occupancies and conditions use normalized, equivalent distributed values. Standard values enable more professional time to be devoted to structural performance design needs.

A delicate balance exists in design professions. If equality existed between design fire characteristics and fire defense capabilities, a small deviation in either the fire or the defense capabilities

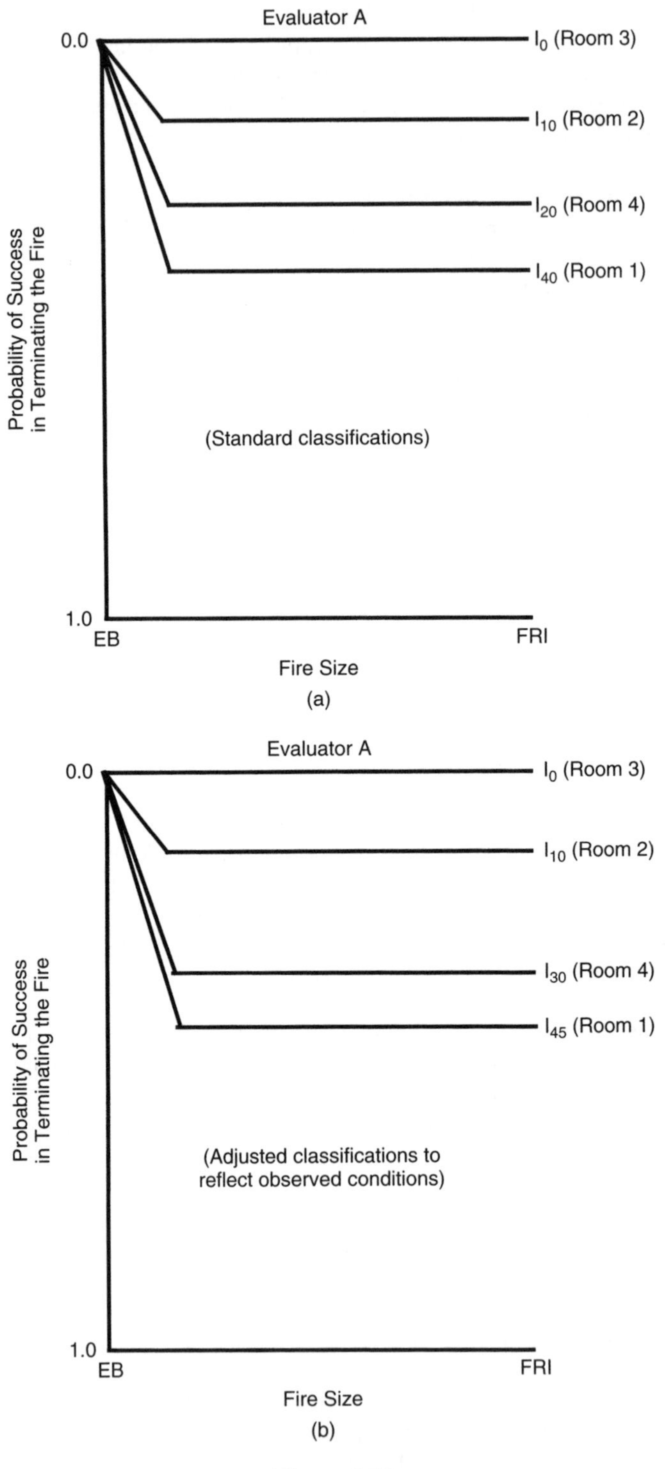
Evaluator A
0.0
I_0 (Room 3)
I_{10} (Room 2)
I_{20} (Room 4)
I_{40} (Room 1)
Probability of Success in Terminating the Fire
(Standard classifications)
1.0
EB
FRI
Fire Size
(a)
Evaluator A
0.0
I_0 (Room 3)
I_{10} (Room 2)
I_{30} (Room 4)
I_{45} (Room 1)
Probability of Success in Terminating the Fire
(Adjusted classifications to reflect observed conditions)
1.0
EB
FRI
Fire Size
(b)

Figure 8.25

would result in failure. Neither fires nor fire defenses are sufficiently understood to allow such precision. If the design fire is unnecessarily large or the fire defense is overdesigned, costs may increase until they seem unfeasible. If a safety index (a measure of the difference between load and resistance) is too small, regular failures and the resulting risks will occur. If it is too large, lack of economy results. Buildings change over time. The structural engineers selected values that were sufficient to allow most building renovations to occur without reevaluation by authorities having jurisdiction. How will fire design characteristics compare with the frequent alterations that take place in buildings and the changes in furnishings and materials that are likely to occur in the future?

Here we do not advocate appropriate values for design or for acceptable risk. Our goal is merely to understand a site-specific building fire performance. After that understanding becomes evident, one can make appropriate decisions and construct risk management programs.

8.21 Summary

The design fire has a major influence on fire defense performance. One part of a design fire (the classification of a room's fire growth potential) establishes a foundation for understanding and explaining fire behavior. The other part of a design fire (describing the products of combustion) identifies the conditions against which the fire defenses are analyzed.

The fire growth potential defines the relative hazard posed by the natural combustion characteristics of the fuels in the space, assuming no extinguishment intervention. Level 1 room classifications provide a rapid assessment of the relative room fire growth potential. We base these classifications on observation and mental estimates founded on fire dynamics and the level 2 evaluation framework. Figure 8.20 provides an organization for the factors that influence the selection. An I curve is a shorthand way of describing the relative hazard for room interior designs with regard to a speed and certainty of fire development.

Reference

1. Alpert, R. L. Ceiling jet flows. *The SFPE Handbook of Fire Protection Engineering*, 3d ed, 2002, Section 2.2.

9 BARRIERS AND MULTIROOM FIRES

9.1 Introduction

Horizontal and vertical barriers serve many functions in a building. They separate spaces to define rooms and provide privacy, security, and noise control. Barriers provide routes for physical, visual, and informational communication with devices such as doors, windows, piping, and electrical conduits. They sometimes provide and hide routes to convey services such as electricity, water, waste disposal, heat, and air through a building. Barriers often support structural loads. Barriers also delay or prevent combustion products moving from one space to the adjacent space.

Because barriers serve so many functions other than impeding smoke and fire, the building's day-to-day operations become dominant in their long-term use, maintenance, and status. These other building functions can have a major influence on a building's fire performance.

Fire endurance ratings have been the basis for organizing prescriptive code requirements for nearly a century and are a major focus of the code and fire community. Although fire endurance ratings have a role in fire safety, they are rarely important in practical building fire performance. This does not imply that flimsy construction is appropriate, but it means that barriers require more comprehensive attention to understand their influence on building performance.

This chapter plus Chapters 10 and 15 relate to three aspects of barrier and structural performance. Although each considers a different aspect of performance, collectively they provide a package that addresses their integration. This chapter focuses on barrier/space modules that produce multiple-room fires. This level 1 evaluation combines observation with simple, rapid analytical techniques. Although these procedures are fast and easy to use, they are very practical and handle the vast majority of fire propagation problems in performance evaluations.

Although this chapter describes a way of thinking that accommodates most practical analyses, fire resistance ratings cannot be dismissed. Chapter 10 discusses ways to translate standard ASTM E 119 and ISO 834 fire test results into field performance tools. One cannot completely divorce barrier performance from structural behavior, and methods discussed in Chapter 15 is a part of the descriptions. These three distinct parts of barrier and structural behavior collectively provide a more complete picture of fire performance.

Building Fire Performance Analysis R. W. Fitzgerald
 ISBN: 0-470-86326-9 (HB)

9.2 Barrier functions

Here are the main roles of barriers in fires:

- Protect people from combustion products while they remain in the building or during building evacuation.
- Protect property or downtime losses from combustion products for the fire duration.
- Affect fire damage and smoke tenability by delaying or preventing movement or by channeling the movement along certain paths.
- Affect the transport of combustion products to detectors.
- Affect sprinkler control or extinguishment effectiveness.
- Affect fire department extinguishment operations.
- Affect fire department search and rescue operations.
- Affect fire propagation time.

Barrier influences may be good or bad. For example, when barriers contain the steam generated from a sprinkler discharge, fire extinguishment is more rapid. On the other hand, barriers can obstruct water discharge to reduce sprinkler effectiveness.

Barriers may help or hinder fire department operations. An architectural plan that involves many turns and decision points will delay finding the fire and laying attack hose lines. A room with few openings will contain steam generation and assist in extinguishment. On the other hand, the room may create backdraft conditions that can increase damage and risks to fire-fighter safety. A room that can be ventilated allows fire fighters to push a fire out of the building. Barriers can block hose-stream application or they can break up the stream into droplets to improve suppression effectiveness. Barriers can help fire fighters establish defense lines to stop or control a fire.

Performance analyses provide a way to understand how architectural layouts and barrier construction affect damage and risk. Although we will remain aware of multiple barrier functions, this chapter will focus on barrier influences involving fire propagation and time durations.

9.3 Concepts for barrier evaluations

A barrier is any surface that will delay or prevent combustion products moving from one space to another. Any barrier that exists is incorporated into the analysis. Therefore, barriers may be

- weak or strong;
- complete or incomplete;
- penetrated or unpenetrated;
- combustible or noncombustible;
- load-bearing or non-load-bearing;
- fire-rated or non-fire-rated.

A door or window, whether open or closed, is a part of the barrier. Any penetrations or openings, whether protected or unprotected from flame and heat, are parts of the barrier. We evaluate all barriers from a field performance viewpoint, whatever their construction, fire resistance rating, or combustibility. In addition to real physical barriers, the zero-strength barrier provides a way to organize the performance analysis in large open spaces. A zero-strength barrier is a fictitious barrier that has no resistance to flame/heat or smoke/gas movement. It zones open spaces to gain a better understanding of time and suppression. Zero-strength barriers are useful to developing risk management programs for large, open buildings.

A performance evaluation moves back and forth between the local behavior and the global perspective of three-dimensional fire propagation. The global time clock starts with established burning (EB) in the room of origin and identifies sequential times for barrier failure and full room involvement (FRI). The local clock normally starts after flashover (Figure 9.1), although an analysis can adapt to situations where a fire can move through a barrier before FRI.

We define three states of barrier performance by ignition potential on the unexposed side. These states are (1) a small, hot-spot ignition can occur on the unexposed side; (2) a large, massive ignition can occur on the unexposed side; or (3) the barrier is successful and no ignition can occur on the unexposed side. These states are considered mutually exclusive and only one state will be present at any time. Thus, if both hot-spot and massive ignitions exist simultaneously, the massive condition will dominate and be considered the only state. Figure 9.2 shows the SVN of these possible outcomes. Figure 9.2 is analogous to a photograph, and the picture (i.e., probabilistic likehood for the events) changes with each increasing increment of time.

We call the small, hot-spot ignition a $\overline{T}$ (pronounced T bar) failure. A $\overline{T}$ (for thermal) failure may result from too much heat transmission through a barrier. It can also result from surface cracking that allows small flames to appear at the unexposed surface of the barrier or from an unprotected penetration. A $\overline{T}$ failure can cause ignition and EB if fuels are present. This EB produces a predictable room fire growth.

A $\overline{D}$ (pronounced D bar) failure occurs when a large opening appears in the barrier. $\overline{D}$ (for durability) failures may result from an open door, a large broken window, partial or full barrier collapse, or a $\overline{T}$ failure that enlarges into a $\overline{D}$ failure. A $\overline{D}$ failure produces a massive influx of fire gases into the adjacent space. This will cause an almost certain FRI within a minute or two, whatever the fuel content in the space.

The failure mode is associated with ignition size and the time to FRI in the adjacent room. When a large influx of heat and fire gases can cause an almost certain FRI very quickly, the barrier has a $\overline{D}$ failure. When the ignition size is relatively small and the room interior design has a role in fire growth, the barrier experiences a $\overline{T}$ failure. A rough rule of thumb suggests that a $\overline{T}$ failure occurs with an opening of less than $100\,\text{in}^2$ ($6.5 \times 10^{-2}\,\text{m}^2$) and an opening of greater than $400\,\text{in}^2$ ($26 \times 10^{-2}\,\text{m}^2$) produces a $\overline{D}$ failure.

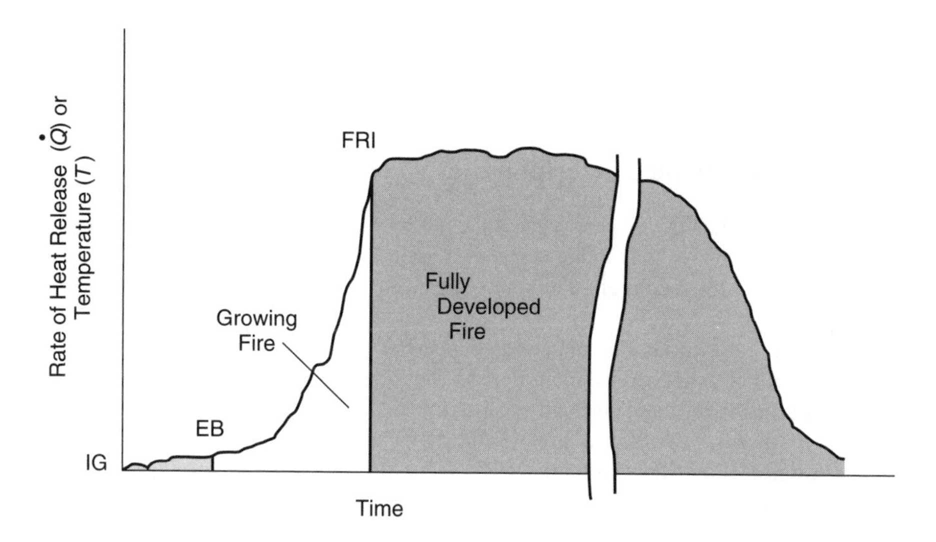

Figure 9.1

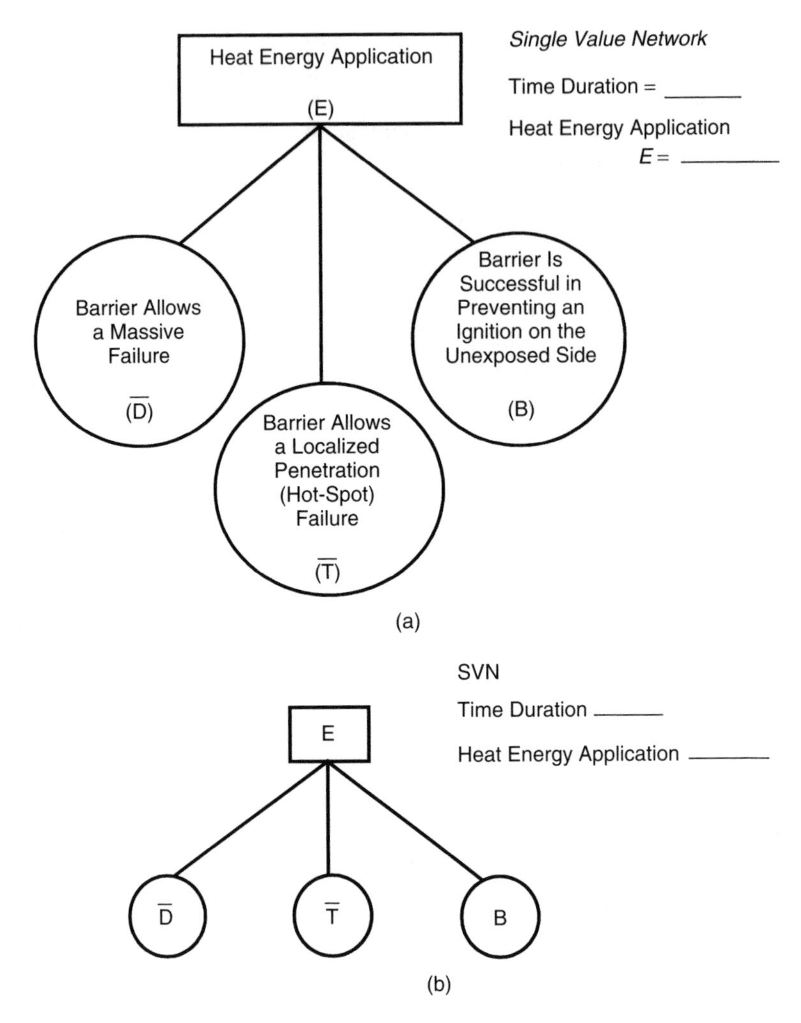

Figure 9.2

9.4 Barrier performance descriptors

All construction materials deteriorate when attacked by elevated fire temperatures. Some assemblies withstand a fire better than others, but eventually all barriers will experience a $\overline{D}$ failure when heat is applied long enough. Let's introduce a conceptual picture of continuous $\overline{T}$ and $\overline{D}$ behavior.

Consider the $\overline{T}$ behavior of an unpenetrated barrier assembly. The early stages of a fully developed fire will not generate sufficient heat to cause a $\overline{T}$ failure. As the fire continues to burn, more heat will transfer into and through the barrier, materials may crack, and the likelihood of failure increases. Figure 9.3 shows single value (SV) performance scales that describe failure estimates at different time durations after FRI. We call the continuous graph that connects the SV failure estimates a $\overline{T}$ curve.

A $\overline{D}$ failure for an unpenetrated barrier results from a partial or full barrier collapse or the disintegration of a relatively large part of the surfacing materials. A $\overline{D}$ failure can also occur

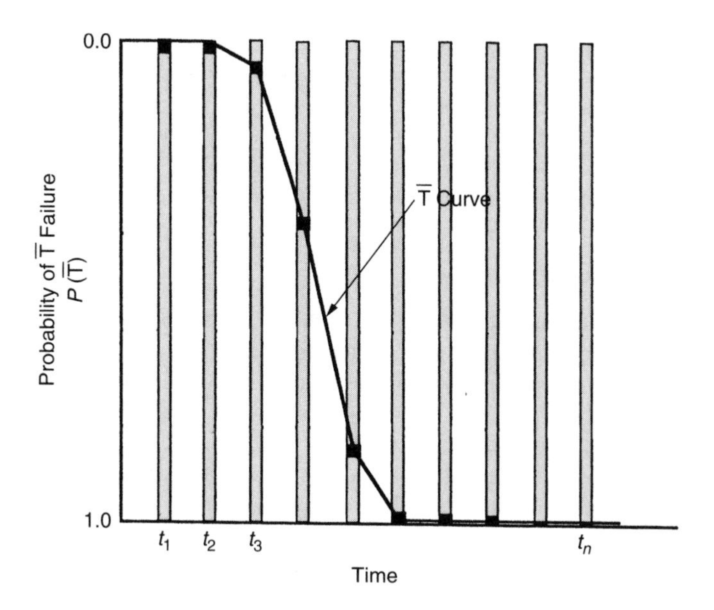

Figure 9.3

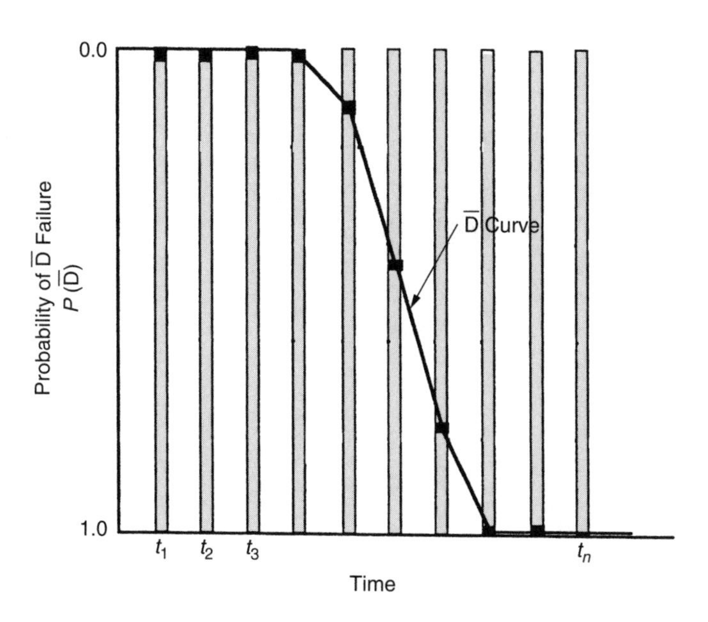

Figure 9.4

when a small $\overline{\text{T}}$ opening enlarges to a $\overline{\text{D}}$ size, as with wood paneling. Normally, but with some important exceptions, a $\overline{\text{D}}$ failure occurs later in the fire than a $\overline{\text{T}}$ failure. Figure 9.4 illustrates SV performance estimates and the continuous $\overline{\text{D}}$ curve.

A fast, consistent way to evaluate barrier field performance involves two stages. First, one develops catalog $\overline{\text{T}}$ and $\overline{\text{D}}$ curves for an unpenetrated barrier assembly. $\overline{\text{T}}$ and $\overline{\text{D}}$ curves are characteristic for any assembly and never change after initial development. One may consider the effort in constructing $\overline{T}$ and $\overline{D}$ curves to be an investment because they are valid for all other buildings. Chapter 10 describes a procedure for translating structural behavior and standard fire tests into catalog $\overline{T}$ and $\overline{D}$ barrier curves.

We obtain field performance curves by multiplying catalog curve coordinates by a barrier effectiveness factor. Chapter 10 uses barrier effectiveness factors to depict field performance and explains a cohesive way to combine the results mathematically.

9.5 The barrier/space module

A network of barrier/space modules describes possible fire propagation paths. Assuming no automatic sprinkler or fire department extinguishment, the task is to establish a time line for fire propagation through these modules. Figure 9.5 shows a network of propagation paths for the building described in Section 4.2.

Although one can look at a barrier in isolation, performance analysis integrates the barrier failure mode and flame/heat movement in the next space. Fortunately, there are ways of sorting out apparent complexities to understand level 1 performance and produce reasonable time line estimates.

A fire analysis starts with a room of origin. After this room becomes fully involved, the fire attacks the subsequent barrier/space module along each path of the network. One studies each

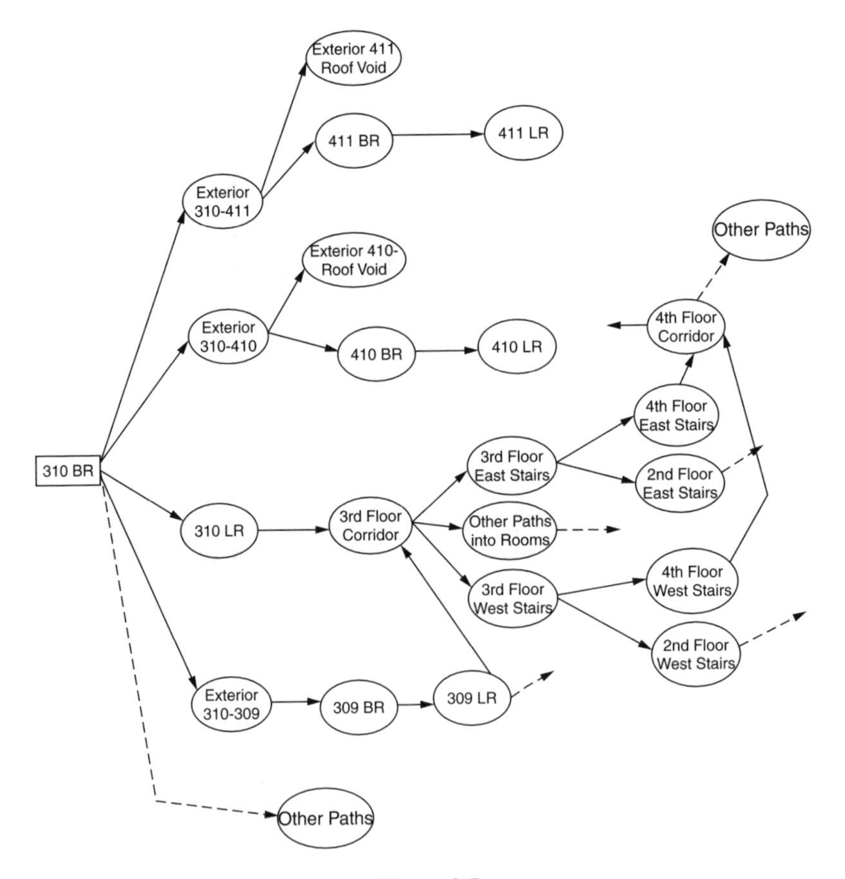

Figure 9.5

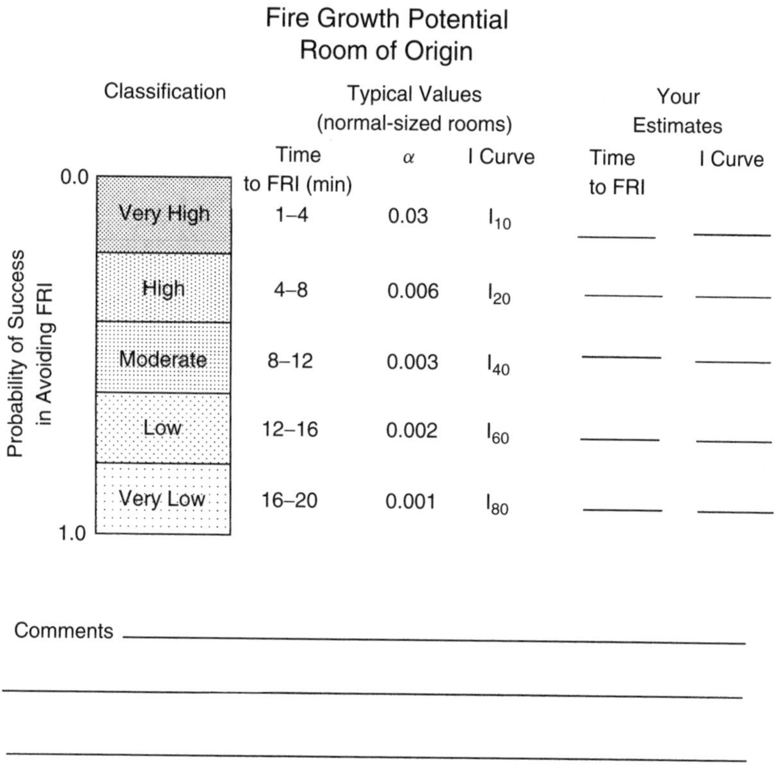

Fire Growth Potential
Room of Origin

Probability of Success in Avoiding FRI	Classification	Typical Values (normal-sized rooms) Time to FRI (min)	α	I Curve	Your Estimates Time to FRI	I Curve
0.0	Very High	1–4	0.03	I_{10}	______	______
	High	4–8	0.006	I_{20}	______	______
	Moderate	8–12	0.003	I_{40}	______	______
	Low	12–16	0.002	I_{60}	______	______
1.0	Very Low	16–20	0.001	I_{80}	______	______

Comments ______________________________

____

____

Figure 9.6

barrier/space module as a sequential unit to estimate the failure mode and associated fire growth in the next space. The process examines each part in isolation and then combines them to describe the site-specific situation being evaluated.

To illustrate the process, Figure 9.6 shows classifications for the room of origin fire growth potential and associated values for I curves and times to FRI. These were discussed in Section 8.14. The typical values provide a sense of proportion for common expectations. We provide space to record an individual's personal estimate for the room. If one uses these typical values, they are deemed to represent the individual's judgment.

Figure 9.7 classifies fire growth potential when a room experiences an ignition subsequently along the path. The representative $\overline{T}$ values describe room fire growth expectations when only a hot-spot failure occurs. We then assume a $\overline{D}$ failure will occur as a separate and independent analysis. Figure 9.7 shows fire growth expectations when a large hole exists in the barrier.

These local independent estimates allow one to manage a methodical, systematic analysis of fire propagation through a cluster of rooms. Examples 9.1 and 9.2 illustrate level 1 evaluations for the building described in Section 4.2.

Example 9.1

Figures 9.8, 9.9, and 9.10 show a floor plan, apartment layout, and elevation for the multistory building described in Chapter 4. Figure 9.5 shows fire propagation paths from apartment 310. Use the form of Figures 9.6 and 9.7 to record estimates for behavior in spaces 310 BR, 310 LR, exterior 310–410, and 311 LR.

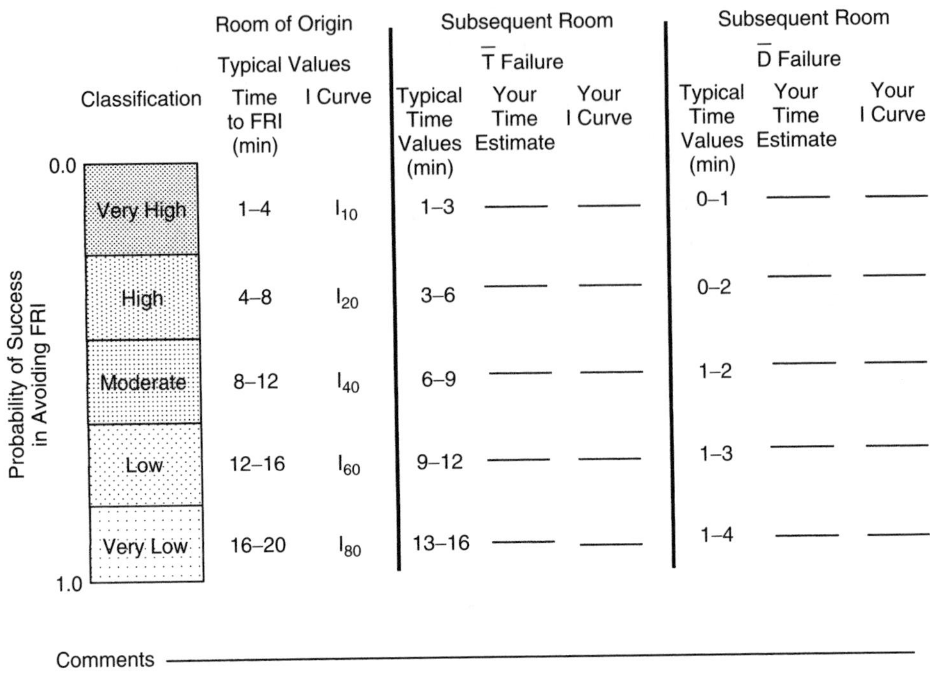

Figure 9.7

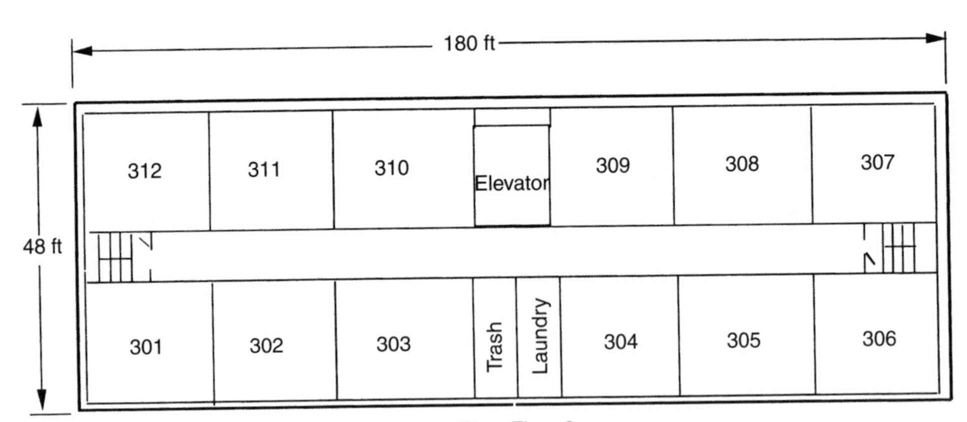

Plan: Floor 3

Figure 9.8

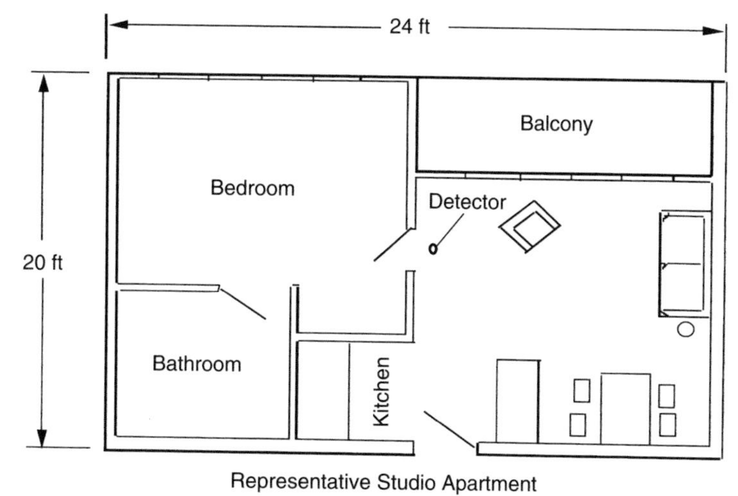

24 ft
20 ft
Balcony
Bedroom
Detector
Bathroom
Kitchen
Representative Studio Apartment

Figure 9.9

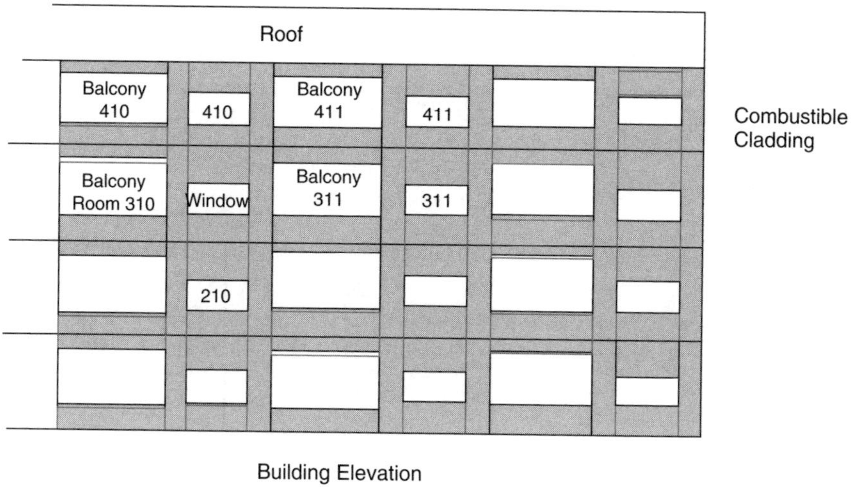

Roof
Balcony
410
410
Balcony
411
411
Combustible
Cladding
Balcony
Room 310
Window
Balcony
311
311
210
Building Elevation

Figure 9.10

Solution

Figure 9.11 shows the estimate for the room of origin, room 310 BR. Figure 9.12 shows the form for room 310 LR, the next room in the scenario. Figure 9.13 shows exterior 310–410, the space outside room 310 BR, and Figure 9.14 shows the estimate for room 311 LR; we will assume the interior design is the same as for apartment 310.

Discussion

This example illustrates a way to record estimates for fire propagation in spaces. A few comments may help with the forms. Neither the times to barrier failure nor the actual failure modes are parts of this evaluation. The room fire growth potential (I curve) and the associated times to FRI

Fire Growth Potential
Room of Origin

Probability of Success in Avoiding FRI	Classification	Typical Values (normal-sized rooms) Time to FRI (min)	α	I Curve	Your Estimates Time to FRI (min)	I Curve
0.0	Very High	1–4	0.03	I_{10}	____	____
	High	4–8	0.006	I_{20}	5	I_{20}
	Moderate	8–12	0.003	I_{40}	____	____
	Low	12–16	0.002	I_{60}	____	____
1.0	Very Low	16–20	0.001	I_{80}	____	____
					____	____

Room 310 BR

polyurethane mattress, stuffed chairs, wood furniture, painted gypsum interior finish

Figure 9.11

Rooms Beyond Room of Origin
Fire Growth Potential

Probability of Success in Avoiding FRI	Classification	Room of Origin Typical Values Time to FRI (min)	I Curve	Subsequent Room $\overline{T}$ Failure Typical Time Values (min)	Your Time Estimate (min)	Your I Curve	Subsequent Room $\overline{D}$ Failure Typical Time Values (min)	Your Time Estimate (min)	Your I Curve
0.0	Very High	1–4	I_{10}	1–3	____	____	0–1	____	____
	High	4–8	I_{20}	3–6	6	I_{30}	0–2	1	I_0
	Moderate	8–12	I_{40}	6–9	____	____	1–2	____	____
	Low	12–16	I_{60}	9–12	____	____	1–3	____	____
1.0	Very Low	16–20	I_{80}	13–16	____	____	1–4	____	____

Room 310 LR

stuffed couch and chair, furniture, curtains and drapes to glass sliding doors leading to balcony, noncombustible interior finish

Figure 9.12

Rooms Beyond Room of Origin
Fire Growth Potential

Classification	Room of Origin Typical Values: Time to FRI (min)	Room of Origin Typical Values: I Curve	Subsequent Room $\overline{T}$ Failure: Typical Time Values (min)	Subsequent Room $\overline{T}$ Failure: Your Time Estimate (min)	Subsequent Room $\overline{T}$ Failure: Your I Curve	Subsequent Room $\overline{D}$ Failure: Typical Time Values (min)	Subsequent Room $\overline{D}$ Failure: Your Time Estimate (min)	Subsequent Room $\overline{D}$ Failure: Your I Curve
Very High	1–4	I_{10}	1–3	___	___	0–1	3	I_5
High	4–8	I_{20}	3–6	___	___	0–2	___	___
Moderate	8–12	I_{40}	6–9	12	I_{45}	1–2	___	___
Low	12–16	I_{60}	9–12	___	___	1–3	___	___
Very Low	16–20	I_{80}	13–16	___	___	1–4	___	___

Probability of Success in Avoiding FRI (0.0 at top, 1.0 at bottom)

Exterior of Building Room 310–410

combustible exterior siding; close contact with window head

$\overline{D}$ failure will produce a large flame front

Figure 9.13

are based on assumed, independent failure mode conditions. Example 9.2 illustrates a way to incorporate time estimates for barrier failure modes.

In Figure 9.11 we record only the values associated with the room of origin estimate. The living room is adjacent to the room of origin in Figure 9.12. Notice that a $\overline{D}$ failure will certainly occur through the open door between the bedroom and the living room. Nevertheless, we estimate living-room fire growth assuming a $\overline{T}$ failure. Then we estimate the fire growth assuming a $\overline{D}$ failure. The framework incorporates these outcomes to give a clear description of the scenarios and to tell a technically credible story.

Figure 9.13 requires some flexibility of definition to describe the analysis realistically. Here the room is the outside space next to and immediately above room 310 BR. Full room involvement becomes a fire size that involves that space and can penetrate barriers into other spaces. Here the potential subsequent spaces are room 410 BR, the space above exterior 410 BR toward the soffit of the overhang, and the outside spaces on each side of the exterior space that defines this room. Zero-strength barriers define the boundaries of the exterior zones.

Values for Figure 9.14 relate to the interior design of room 311 LR. The barrier between room 310 BR and 311 LR is strong and unpenetrated. It will provide a substantial time duration before failure can occur. Whatever the barrier fire endurance, we evaluate this space (room 311 LR) as a room next to one fully involved. The estimate for a $\overline{T}$ failure allows one to focus attention on that aspect alone. Similarly, a $\overline{D}$ failure into room 311 LR allows one to estimate the outcome in isolation. The framework structures the combination of these separate events to obtain a cohesive picture of the building fire.

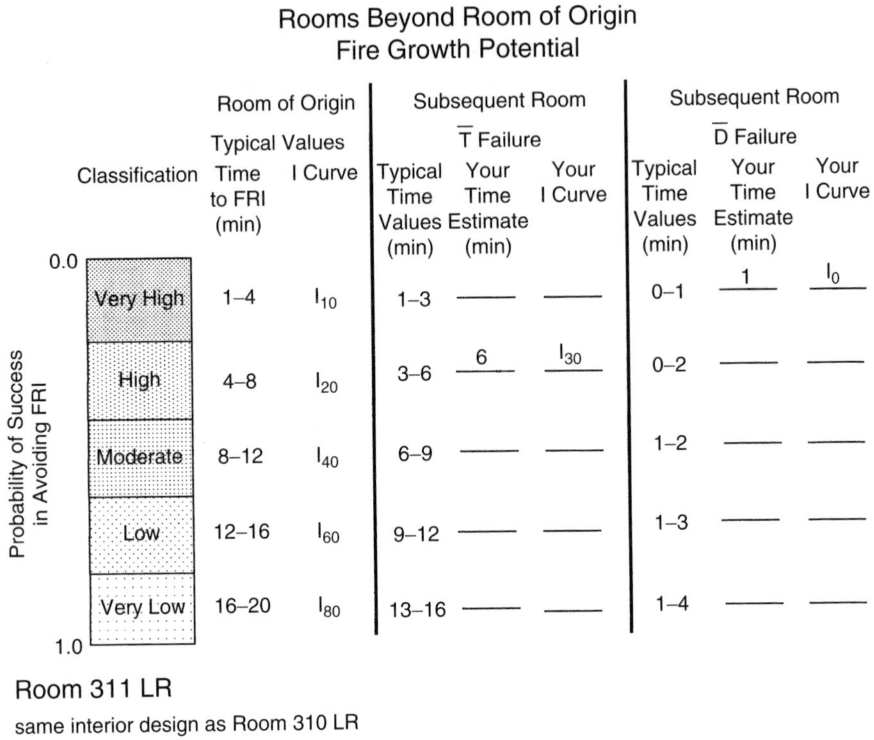

Rooms Beyond Room of Origin
Fire Growth Potential

Classification	Room of Origin Typical Values: Time to FRI (min)	Room of Origin Typical Values: I Curve	Subsequent Room $\overline{T}$ Failure: Typical Time Values (min)	Your Time Estimate (min)	Your I Curve	Subsequent Room $\overline{D}$ Failure: Typical Time Values (min)	Your Time Estimate (min)	Your I Curve
Very High	1–4	I_{10}	1–3	___	___	0–1	1	I_0
High	4–8	I_{20}	3–6	6	I_{30}	0–2	___	___
Moderate	8–12	I_{40}	6–9	___	___	1–2	___	___
Low	12–16	I_{60}	9–12	___	___	1–3	___	___
Very Low	16–20	I_{80}	13–16	___	___	1–4	___	___

Probability of Success in Avoiding FRI: 0.0 (top) to 1.0 (bottom)

Room 311 LR
same interior design as Room 310 LR

Figure 9.14

Example 9.2

Continue the level 1 building analysis of Example 9.1 to develop a time line estimate for the multiple-room fire development.

Solution

Figure 9.5 shows potential fire propagation paths through the barrier/space network. We base level 1 barrier/space performance on observations and judgmental estimates. Although a single spreadsheet can incorporate and order all of the data, we will describe the process separately. Using the building information of Chapter 4, the space fire growth potential classifications are shown in Table 9.1. Table 9.2 shows local time estimates for barrier performance and associated space FRI times. Table 9.3 orders the global time for fire propagation. Figure 9.15 shows a single time line for fire propagation.

Figure 4.7 is a composite of the complete dynamic building/fire system and Figure 9.16 shows a partial completion for this scenario.

Discussion

This example illustrates a level 1 fire propagation time line. We define the building exterior (where fire can propagate externally) as a space in the barrier/space concept. When the window in the exterior wall breaks, the fire can propagate into this space. Full involvement is viewed as a fire that has enough size to cause propagation into the next space (e.g., rooms near room 310

Table 9.1: Room Fine Growth Potential Classification

Room	Classification	Room	Classification	Room	Classification
310 BR	High	Exterior 410–Roof	High	3rd Fl E Stairs	Very low
310 LR	High	410 BR	High	3rd Fl W Stairs	Very low
Exterior 310–411	Low	410 LR	High	4th Fl Corridor	Very low
Exterior 310–410	High	3rd Fl Corridor	Very low	4th Fl E Stairs	Very low
Exterior 310–309	Very low	309 BR	High	4th Fl W Stairs	Very low
Exterior 411–Roof	High	309 LR	High	2nd Fl E Stairs	Very low
411 BR	High	411 LR	High	2nd Fl W Stairs	Very low

Table 9.2: Local Clock Time Estimates for Fire Propagation

From room	To room	Type of barrier failure	Estimated time for barrier failure (min)	Estimated time to next room FRI (min)	Cumulative time (min)
310 BR	–	–	–	4–8	5
310 BR	310 LR	$\overline{D}$	0	1	6
310 LR	3rd Fl Corridor	$\overline{D}$	0	6	12
3rd Fl Corridor	3rd Fl E Stairs	$\overline{D}$	0	1	13
3rd Fl Corridor	3rd Fl W Stairs	$\overline{D}$	0	1	13
3rd Fl E Stairs	4th Fl E Stairs	$\overline{D}$	0	0.5	13.5
3rd Fl W Stairs	4th Fl W Stairs	$\overline{D}$	0	0.5	13.5
4th Fl E Stairs	4th Fl Corridor	$\overline{D}$	0	2	15.5
4th Fl W Stairs	4th Fl Corridor	$\overline{D}$	0	2	15.5
310 BR	Exterior 310–410	$\overline{D}$	1	3	9
310 BR	Exterior 310–411	$\overline{D}$	10	4	19
310 BR	Exterior 310–309	$\overline{D}$	20	9	34
Exterior 410	410 BR	$\overline{D}$	1	4	13.5
410 BR	410 LR	$\overline{D}$	1	3	17.5
Exterior 411	411 BR	$\overline{D}$	1	5	25
411 BR	411 LR	$\overline{D}$	1	3	29
Exterior 309	309 BR	$\overline{D}$	1	5	40
Exterior 410	Roof	$\overline{T}$	5	12	25.5

such as 410, 411, 409, 309, 311). These rooms can become ignited from the external flame front. Also, the roof soffit is another weak barrier that can allow fire propagation into the attic space.

The floor–ceiling assemblies and interior partitions will provide more than ample time for fire department arrival and extinguishment. Even a single layer of 1/2 in non-fire-rated gypsum wallboard provides 30 min in a standard fire test before it reaches the finish rating and the 2 × 4 studs can become ignited. Additional time will be available before the fire spreads to the next space or burns enough of the studs to cause collapse. Here the greater fire-rated gypsum board thickness and fiberglass batt insulation provide even more time.

Table 9.3: Global Clock Fire Propagation

Time from EB (min)	Spaces fully involved with fire
5	BR, apartment 310
6	All of apartment 310 (BR and LR)
9	Exterior wall above BR of apartment 310
12	Third-floor corridor
13	Third-floor stairs (east and west)
13.5	Fourth-floor stairs (east and west)
13.5	BR, apartment 410
15.5	Fourth-floor corridor
17.5	All of apartment 410 (BR and LR)
19	Exterior wall at apartment 411
25	BR, apartment 411
25.5	Roof substantially involved
29	All of apartment 410 (BR and LR)
34	Exterior of apartment 309

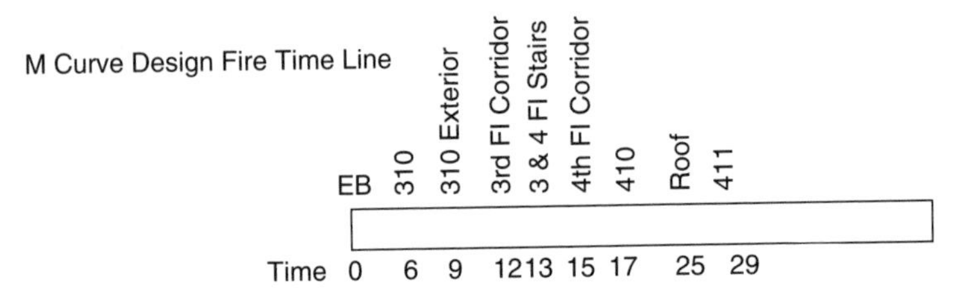

Figure 9.15

	SECTION	COMPONENT	EVENT	2	4	6	8	10	12	14	15	16	18	20	22	24	26	28	30	32
Building Performance	1		Design Fire																	
		a	Size (HRR) (MW)	0.3	1.2	4														
		b	Size (Area or Rooms)			310	310 ext		3rd cor	3-4 strs	4th cor	410				roof		411		
		c																		
		d																		

TIME (minutes)

Figure 9.16

Noise transmission is a major complaint in apartment buildings. Because noise and fire can propagate through the same paths, apartment buildings usually have tighter construction than office and commercial buildings. Interior fire spread between apartment units is not as common, because other building types may not have similar care in construction. Therefore the nonpenetrated barriers should be effective in preventing interior fire propagation for the time frame of interest.

A major decision involves the status of apartment doors that open to the corridor. If these doors are closed, they should provide 20 or 30 min before failure into the next space occurs. On the other hand, if they are open, failure is "instantaneous" with rapid fire spread. Although failure will occur before FRI because of the effluent movement, it is convenient to define barrier failure to occur at FRI when making time line estimates. The lack of door closers increases the likelihood that a door will remain open if an individual leaves the apartment during an emergency.

Whether the door will be open or closed is never "averaged" or selected using statistical data. Instead we would analyze one scenario with the door open and another with the door closed. In this way, we can tell the story: "When the door is open, the following can be expected. And, when the door is closed, the following can be expected." In this way, the owner or occupant can understand clearly the implications of details. For this illustration the scenario assumes the room 310 corridor door to be open and the doors to the other rooms to be closed. A spreadsheet analysis enables one to study a variety of scenarios relatively quickly.

This rapid and convenient level 1 evaluation allows us to "tell the story" of this scenario. For example a fire in the bedroom of apartment 310 will cause flashover within about 5 min and involve the entire apartment within about 7 or 8 min. If the door to apartment 310 is open, the fire will push massive amounts of smoke and fire gases into the corridor, making the third floor untenable within 10 min of the start and spreading up the stairs on each side of the building.

The fire will break room 310's bedroom window at flashover and quickly spread up the side of the building. The window to apartment 410 is likely to break and drive that unit to flashover within about 10–12 min from fire starting in room 310. The fire will continue to spread up the exterior wall and penetrate into the attic within about 15–18 min. The balconies interrupt fuel continuity, making horizontal spread somewhat more difficult. Although horizontal exterior flame spread may be more difficult, available fuel continuity makes it possible but slower than the vertical spread. Whether the corridor door of apartment 410 remains open or closed, the stairwells and corridors of the building will be untenable within about 10–12 min from the fire's start.

9.6 Summary

Barriers greatly affect the fire performance of a building. Buildings have been lost because ineffective barriers allowed fire to move into adjacent spaces before suppression intervention could control the spread. Barrier effectiveness greatly affects the risk and damage level. Even a combustible barrier may buy enough time for the local fire department to respond and extinguish a fire. A few minutes of delay in fire propagation can often make substantial differences to fire department success.

This chapter discusses a process for making level 1 time line estimates for fire propagation. The understanding that develops when making these estimates can be improved with the level 2 discussion in Chapter 10 and the structural behavior described in Chapter 15. Each piece can be studied in isolation and then combined to obtain a complete picture of building performance.

10 BARRIER PERFORMANCE

10.1 Pause for review

Chapter 9 introduced a way of thinking about barriers and building performance. Before we examine barrier performance in greater detail, let us pause to review some important concepts:

1. A barrier is any surface that can delay or prevent fire or smoke moving from one space to another. In addition to physical barriers, we zone open spaces by using a fictitious, zero-strength barrier that has no resistance to heat or smoke.
2. At any instant of time, a barrier will be in one of the following three states:
   - Hot-spot ($\overline{T}$) failure can cause an EB and "routine" fire growth in the next room.
   - Massive ($\overline{D}$) failure allows a large influx of fire gases into the adjacent space that can lead to a rapid and nearly certain FRI.
   - Barrier success (B) does not allow any ignition into the next space.
3. $\overline{T}$ curves, $\overline{D}$ curves, and B curves provide a visual way to describe the dynamic changes for each mode of barrier behavior as the fire grows. We can get a sense of proportion when we show these curves on the same evaluator.
4. A performance analysis looks at sequential space/barrier modules that combine the failure mode with the fire growth of the next space.
5. A time line for multiple-room fires is an essential ingredient for performance analyses. Level 1 time lines are based on observational assessments to recognize construction features and spreadsheets to order and get a sense of proportion for fire propagation in a cluster of rooms. A level 2 analysis provides a better technical understanding of fully developed fires and performance measures. This improves confidence in making judgmental estimates.

10.2 Chapter organization

A performance analysis examines the field performance of barrier/space modules to understand the way a fire can propagate within a cluster of rooms. Although the barrier/space module is the performance unit, this chapter will uncouple this module to examine barrier performance in isolation. The goal is to translate construction knowledge, fire test information, and theory into better performance estimates.

Building Fire Performance Analysis R. W. Fitzgerald
 ISBN: 0-470-86326-9 (HB)

Most individuals think about barriers in terms of the standard fire endurance test. This test is so pervasive in the building industry that we use it as a starting point for performance evaluations. However, translation of standard test results into field performance expectations requires a major reorganization in the way we think about barriers and testing. To start this journey, we briefly review standard test highlights and factors that affect test results. Fuel load is a factor in all equivalent time calculations and will be used as a performance measure.

Here is a brief overview of the performance evaluation process:

- Construct catalog $\overline{\mathrm{T}}$ and $\overline{\mathrm{D}}$ curves based on adaptations of the standard fire test.
- Use fuel load consumption as the measure of fire endurance.
- Obtain field performance $\overline{\mathrm{T}}$ and $\overline{\mathrm{D}}$ curves by multiplying catalog curves by barrier effectiveness factors to adjust for field conditions. Barrier effectiveness factors incorporate applied loading, thermal restraint, construction differences, openings, and automatic barrier protection equipment.
- Adjust curves to maintain mathematical cohesiveness.
- Convert fuel consumption to real-time fire durations.

This chapter organizes a way of thinking about the field performance of barriers. It uses common fire protection practices and established theory as much as possible. However, a gap exists between available test information, theory, and field performance estimates. This gap must be filled with judgment based on accessible knowledge relevant to the application. This chapter describes barrier evaluations in the context of available theory and basic information.

10.3 The standard fire test

The generally accepted basis for specifying fire endurance of barriers is the standard fire resistance test, ASTM E 119 in the United States and ISO 834 in most of the rest of the world. These tests are similar, and their differences are insignificant when constructing catalog curves and evaluating field performance.

The ASTM E 119 test protocol notes that its intent is to compare test behavior, and the results should not be construed as suitable for other conditions or fire exposures. The test measures and describes assembly responses to controlled fire conditions. It does not incorporate all of the factors important for actual fire conditions. Although the standard notes that test results are not intended for acceptance criteria, there is no other practical way of identifying construction assemblies for prescriptive code enforcement. Therefore one should keep in mind that fire endurance test results are a way to satisfy prescriptive code requirements and not an indication of the actual time the assembly will resist a building fire.

Perhaps the biggest misconception in the fire and building code community is that the fire endurance rating describes the time that the assembly should last in a building fire. This incorrect perception is one reason code compliance and building fire performance can be so different. Nevertheless, one can generally assume that a higher-rated assembly may last longer in a building fire than a lower-rated assembly, sometimes. We should realize the actual fire resistance time can be quite different from the rated values.

The ASTM E 119 test establishes fire endurance classifications of floor–ceiling assemblies, roof–ceiling assemblies, walls, partitions, beams, and columns. ASTM E 152 and ASTM E 163 are similar tests for doors and windows. Floor and roof assemblies, beams, columns, walls and partitions, doors, and windows all have their own furnaces and test criteria. Figure 10.1 illustrates a floor furnace. The opening measures about 13 ft by 17 ft (about 4 m × 5 m). A full-size assembly is installed in the furnace, and loads are applied to supporting members.

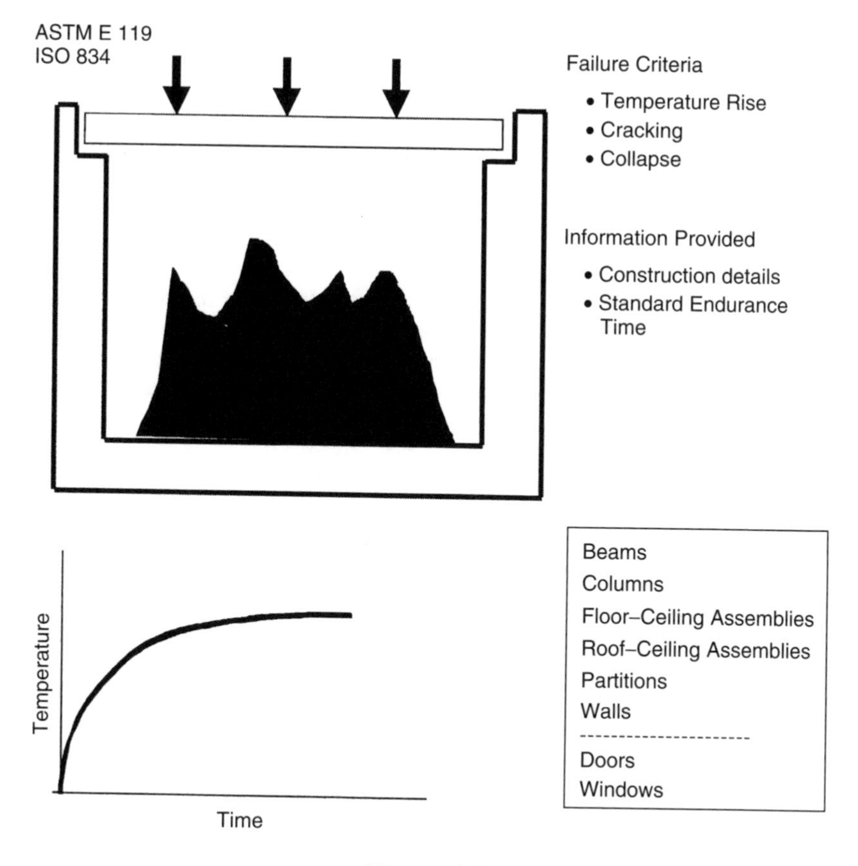

Figure 10.1

A fire is started in the furnace, and the air supply and fuel are adjusted to provide a specified, standard time–temperature relationship on the exposed side of the assembly. The test continues until the sponsor wishes to stop or until failure occurs. Failure is defined as any of the following:

- Temperature on the unexposed side rises to an average of 250°F (139°C), or a hot spot of 325°F (181°C) above ambient.
- The unexposed barrier surface cracks enough for the fire gases to penetrate and ignite combustibles on the unexposed side.
- The assembly cannot support the applied loads.

The first two failure modes are associated with $\overline{\mathrm{T}}$ performance and the third mode relates to $\overline{\mathrm{D}}$ performance.

The time is noted at which failure occurred or the test was stopped. This test exposure time is then rounded down to standard fire endurance classifications of 20 min, 1/2 h, 3/4 h, 1 h, 1 1/2 h, 2 h, 3 h, and 4 h. For example, if an assembly failed at 1:59, it would be given a fire endurance of 1 1/2 h. If it lasted to 2:01, it would receive a rating of 2 hours. A test exposure of 2 h 59 min would also produce a 2 h listing. Therefore a fire endurance rating indicates that the assembly avoided all failure modes in the standard test for at least the rated time duration.

The fire endurance classification (in hours) for restrained and unrestrained test assemblies and the construction details of the assembly are published. Although other information is recorded, neither the failure mode nor information about the assembly behavior during the fire is published. The tests are considered privileged information and additional data is available only from the test sponsor.

The ISO 834 standard is essentially the same as the E 119 test. The differences are in the details, but the philosophy and general practices are similar enough that the basis for developing catalog curves would be the same.

10.4 Standard test discussion

A function of a standard test is to enable results to be repeatable and reproducible among testing facilities. Although the time–temperature relationship is specified, its comparative repeatability is questioned because of differences in thermocouple shielding and the manner in which temperatures are controlled. Some furnaces control temperature by adjusting the airflow and others by adjusting the fuel. Some facilities record the fuel amount, whereas others do not because it is not required by the test standard.

The net heat flux determines how hot the specimen actually becomes. Paulsen [1] found significant differences in heat flux among six European furnaces. Variation in furnace construction, furnace controls, and flame emissivity can cause substantial differences in the heat flux to the specimen. The ability of different furnaces to reproduce results for an assembly is suspect. This could provide an opportunity for manufacturers, if they were so inclined, to select the test facility that gives the best results for their materials and construction. Prescriptive code approval is a motivation to obtain desired fire endurance classifications.

In addition to furnace differences, the thermal restraint provided during testing contributes to the fire endurance. The thermal coefficient of expansion causes construction materials to expand when subjected to temperature rises. When assembly elongation is axially unrestrained, structural elements can expand freely as the temperature increases. If the test furnace restricts the ends from expanding, axial forces are induced. The difference in fire endurance can be substantial because these axial forces provide a prestressing action. The change in fire endurance can vary from a reduction in endurance time for a few assemblies, to a negligible change for others, to a substantial increase for many. Most assemblies will exhibit a better load-carrying capacity and longer fire endurance time when ends are axially restrained against movement. This increase can be as much as 100% or more. A listing of both restrained and unrestrained fire endurance ratings provides some guidance on behavior.

The standard test is supposed to be representative of the details and dimensions used in construction. Physical-analytical functions are very difficult to scale, and the specimen is prepared as well as possible to conform to construction and the test furnace. However, axial, flexural, and shear stresses and their combination, size scaling, flexural and lateral restraint, lateral and local buckling cannot be incorporated simultaneously for a specimen having a 17 ft fixed length. Construction details and workmanship can influence the correspondence between laboratory and field performance in fire situations.

Heat absorption or transmission through assemblies can differ substantially. During a test, the fuel/air mixture is adjusted to maintain the standard time–temperature criteria near the exposed assembly surface. This means that significantly more fuel is needed for some assemblies to maintain the same time–temperature relationship. In other constructions the fuel mixture is reduced substantially. A natural fire is not able to adjust its fuel/air mixture as easily as the test furnace.

A building fire produces positive pressures whereas a standard fire test uses a slightly negative pressure to prevent most smoke and hot gases from infiltrating into the laboratory. This allows cooler laboratory air to leak into the furnace, affecting fire resistance. The plenum temperature of floor–ceiling and roof–ceiling assemblies having suspended ceilings influences fire endurance. When cooler air from the laboratory is drawn into the furnace, one can expect better fire resistance.

The purpose of test standardization is to have comparative results among different laboratories. Although the test protocol remains constant, a fire endurance time is not necessarily the same at all laboratories. Although fire endurance test information contributes to our understanding, one must interpret results carefully when considering performance. Because test time exposures are intended to compare assemblies, they should not be interpreted literally for an actual building fire. Other relevant construction conditions may contribute to an early collapse or to an extended endurance under natural fire exposures. When the goal is to understand and describe performance, one must combine fire test information with structural behavior, knowledge of materials at elevated temperatures, compartment fire dynamics, and judgment.

10.5 Fuel load

Traditionally, the combustible content in a space is described as the fire load or fuel load. Several theories associate fuel load with fire endurance ratings and equivalent standard test time durations. A review of fire load calculation is useful before discussing equivalent test theories.

The fuel load is a measure of the calorific content of the room. It is calculated in terms of equivalent pounds of cellulosic-based fuel per square foot (meter) of area. The floor area is the most common way of normalizing the load, although the bounding surface area may also be used. Examples 10.1 and 10.2 illustrate fuel load calculations in a space.

Example 10.1

Calculate the fuel load for the room shown in Figure 10.2. The weights for the cellulosic and plastic materials are shown in the figure. Express the fuel load as pounds per square foot (psf) of floor area.

Solution

Fuel loadings are expressed in terms of equivalent cellulosic materials. The heat energy content is around 8000 Btu/lb for wood, paper, and other cellulose materials. Values range from 9000 Btu/lb to 20 000 Btu/lb for plastics. A simplification assumes that all petroleum-based materials have a heat content of 16 000 Btu/lb. Therefore 1 lb of petroleum-based material is assumed to have the equivalent neat energy content of 2 lb of cellulosic material.

A completed form for this room is shown in Figure 10.3. Fuel loading is converted to lb/ft^2 (psf) by dividing fuel weights by the floor area of 600 ft^2.

1. Building fuels

- *Structural fuels*: none
- *Service fuels*: none
- *Nonstructural fuels*: $\frac{1}{2}$ in wood interior finish

If we assume that wood weighs 36 lb/ft^3, a slice of 1 ft^2 area and 1 in thickness weighs about 3 lb. The total weight of wood in the room is computed as follows:

$$\begin{aligned}
&\text{Walls } 2 \times 20\,\text{ft} \times 11\,\text{ft} && = 440\,\text{ft}^2 \\
&\text{Walls } 2 \times 30\,\text{ft} \times 11\,\text{ft} && = 660\,\text{ft}^2 \\
&\text{Total} && = 1100\,\text{ft}^2 \\
&\text{Total weight} = (110\,\text{ft}^2)(1.5\,\text{lb/ft}^2) && = 1650\,\text{lb}
\end{aligned}$$

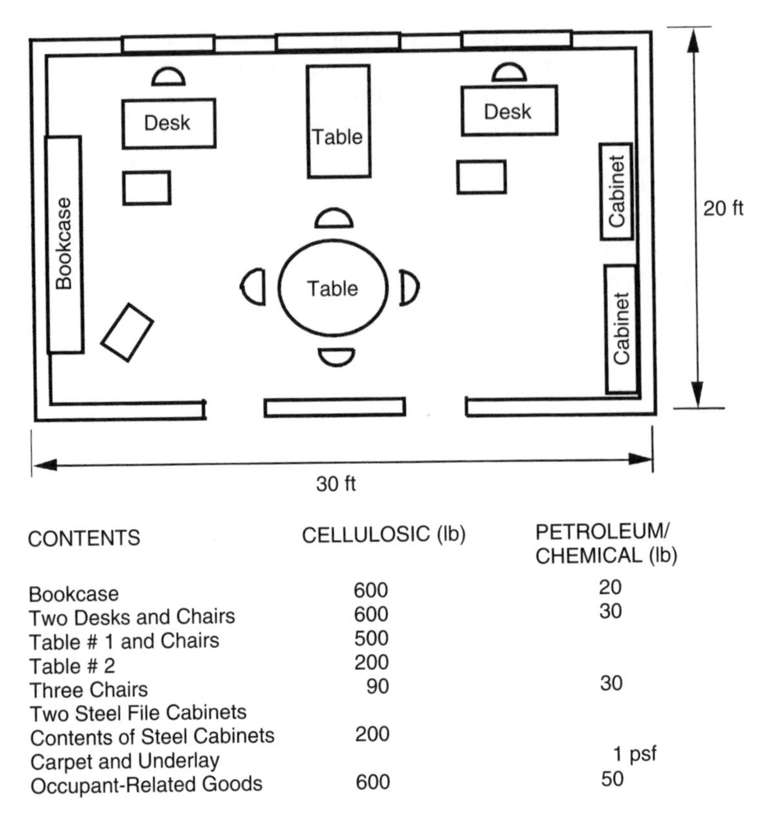

CONTENTS	CELLULOSIC (lb)	PETROLEUM/ CHEMICAL (lb)
Bookcase	600	20
Two Desks and Chairs	600	30
Table # 1 and Chairs	500	
Table # 2	200	
Three Chairs	90	30
Two Steel File Cabinets		
Contents of Steel Cabinets	200	
Carpet and Underlay		1 psf
Occupant-Related Goods	600	50

Figure 10.2

Window areas and door areas were not included in the calculations. The accuracy of the correlation does not warrant the time-consuming sophistication of deducting those areas unless the values are large. To compensate slightly, the door and window trim is not included.

The fuel for the interior finish is computed as

$$\text{Fuel load (interior finish)} = \frac{1650\,\text{lb}}{20\,\text{ft} \times 30\,\text{ft}} = 2.8\,\text{psf}.$$

2. *Contents fuels*

Furniture	Cellulosic (lb)	Petroleum/chemical (lb)
Bookcase	600	
Two desks and chair	600	20
Table and chairs	500	
Two tables #2	200	
Three chairs	90	30
Total	1390	50

ESTIMATING THE WEIGHT OF FUEL

BUILDING FUELS	CELLULOSIC	PETROLEUM/ CHEMICAL
Structural Fuels	—	—
Service Fuels	—	—
Nonstructural Fuels		
Non-Load-Bearing	3.0	
Interior Finish and Trim		
CONTENTS FUELS		
Furnishings	2.3	0.1
Furniture		1.0
Decorations		
Other		
Occupant-Related Goods (ORGs)	1.0	0.1
SUBTOTAL	6.3	1.2
Conversion to Wood	1	2
Wood Equivalent	6.3	2.4
FUEL LOAD	8.5 psf	

Analyst ________ Space ________

Building ________ Date ________

Figure 10.3

The file cabinet contents were not included, because the combustibles were enclosed by a metal cabinet.

$$\text{Cellulosic fuel loading} = 1390\,\text{lb}/600\,\text{ft}^2 = 2.3\,\text{psf};$$

$$\text{Petroleum/chemical fuel loading} = 50\,\text{lb}/600\,\text{ft}^2 = 0.08\,\text{psf}.$$

Decorations

$$\text{Carpet and underlay} = 1.0\,\text{psf(petroleum/chemical)};$$

Occupant-related goods (ORG)

$$\text{Cellulosic fuel loading} = 600\,\text{lb}/600\,\text{ft}^2 = 1.0\,\text{psf};$$

$$\text{Petroleum/chemical fuel loading} = 50\,\text{lb}/600\,\text{ft}^2 = 0.08\,\text{psf}.$$

The subtotal for cellulosics is 6.1 psf. The subtotal for petroleum/chemical is 1.2 psf. Multiplying this by the wood equivalency factor of 2 gives 2.4 psf. The total fuel loading is

$$\text{Fuel load} = 6.1 + 2.4 = 8.5\,\text{psf of floor area } (41.5\,\text{kg/m}^2).$$

Example 10.2

Express the fuel loading for the room shown in Figure 10.2 as psf of bounding surface area.

Solution

The equivalent fuel loading of 8.5 psf (41.5 kg/m^2) of floor area in Example 10.1 can be converted to total equivalent pounds by multiplying by the floor area:

$$\text{Total weight of (cellulosic) combustibles} = 8.5\,\text{lb/ft}^2 \times 30\,\text{ft} \times 20\,\text{ft} = 5100\,\text{lb}.$$

The room surface area is as follows:

Walls	$2 \times 30 \times 11$	=	660 ft^2
	$2 \times 20 \times 11$	=	440 ft^2
Floor and ceiling	$2 \times 20 \times 30$	=	1200 ft^2
Total surface area		=	2300 ft^2

$$\text{Fuel load} = 5100\,\text{lb}/2300\,\text{ft}^2 = 2.2\,\text{psf of surface area}(10.7\,\text{kg/m}^2)$$

10.6 Ingberg theory

The relationship between the standard fire test, natural fires, and building performance has been the subject of questions and controversy for nearly a century. Simon Ingberg was the first to provide a glimpse of the picture that relates these parts. Around the time of World War I, the ASTM E 119 test was being debated. Knowledgeable individuals in the fire community recognized that some construction assemblies behaved better in fires than others, and the distinctions could be roughly ordered with E 119 test results. The question of the link between natural fires and the standard fire test remained.

In the 1920s Ingberg defined fire severity as an area under the time–temperature curve above a base temperature. He proposed an "equal area" theory which states that the fire severity in different fires is equal when the areas under their time temperature curves are equal. Figure 10.4 illustrates conditions where fires are defined to have equal severity. Since the standard time–temperature curve represents one specific fire, all fires can be related to this standard test fire by the equal area theory.

Ingberg [2] reported the results of 10 burnout tests in two different rooms. The test facility was built in 1922 and the reported tests were increased to 16 by 1929 [3]. Some of these burnouts took as much as 13 and 20 h of actual time. The fire loads were noted, and fire severity values using the equal area theory were calculated. Using data from Ingberg [3], Figure 10.5 shows the test results that related fuel loads with standard test times. Two coordinates were plotted for each test fire, one using a base temperature of 150°C and one using a base temperature of 300°C. The best fit of these test results established the relationship between the fire load and the fire endurance test times. The relationship is tabulated in Table 10.1. Although fuel load studies were conducted during the intervening years, the first comprehensive study was published in 1942 [4].

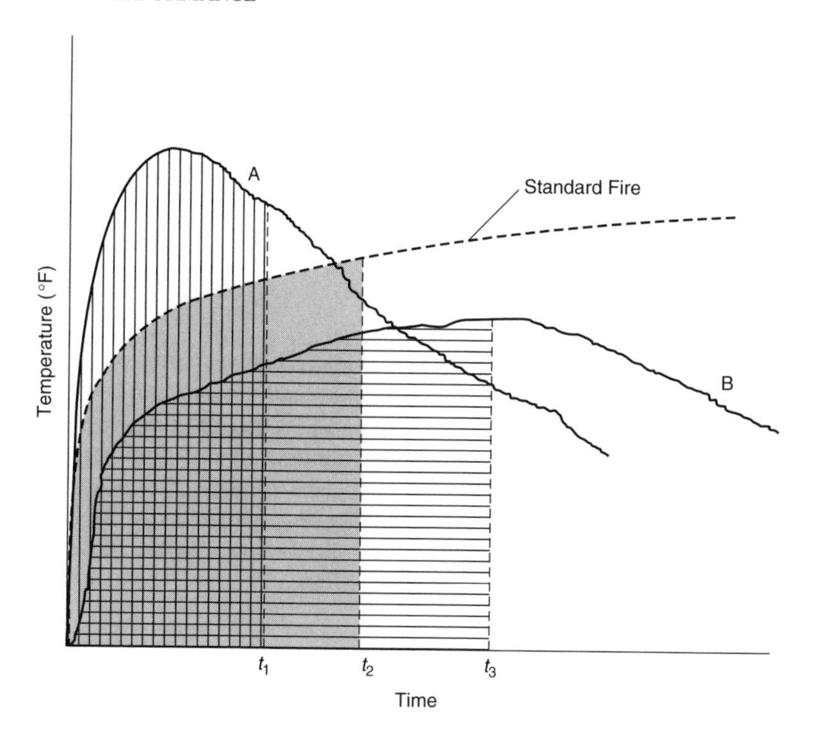

Figure 10.4

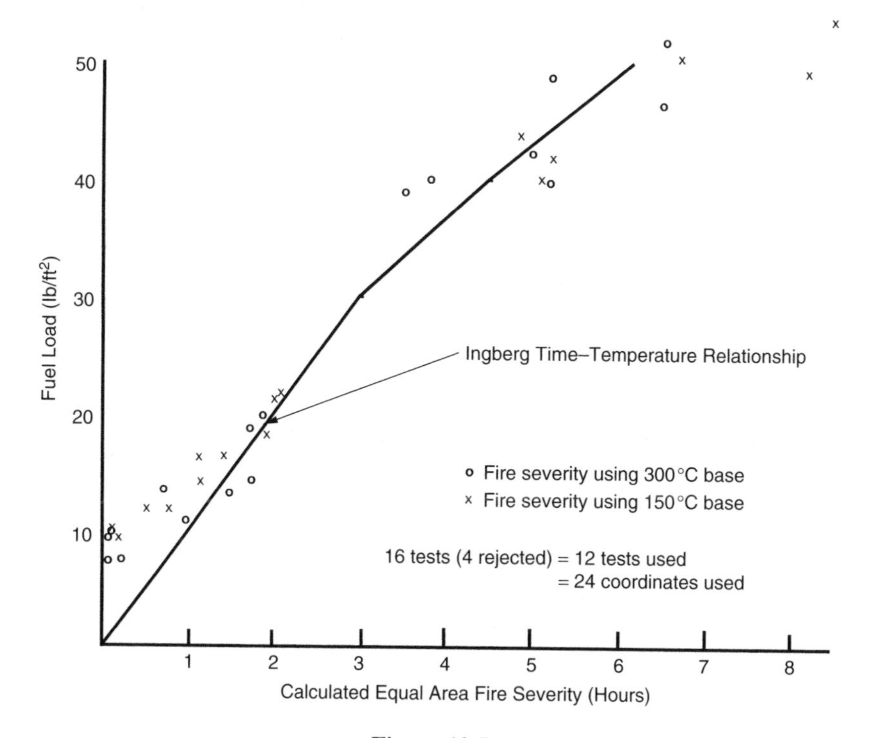

Figure 10.5

Table 10.1

Fuel load (psf)	1	5	10	20	30	40
Test time	6 min	30 min	1 h	2 h	3 h	4.5 h
10^3Btu/ft^2	8	40	80	160	240	320
kg/m^2	4.88	24	49	98	146	195
MJ/m^2	90.753	454	908	1815	2723	3630

10.7 Understanding fire severity better

The Ingberg definition of fire severity and the relationship between fuel load and standard test times has been used for many years by individuals and organizations. Although the concepts are simple to use, there are inaccuracies associated with the theory and its application. Ingberg's original paper noted that radiation energy was very important but it was not included. Ingberg noted that this theory was the best that could be identified at the time.

During the 1960s and 1970s an international cooperative effort was undertaken to understand the postflashover fire and its effect on structural elements. The program integrated experiment, theory, and calculations to increase our understanding of room fire behavior. Some of the factors that influence the time–temperature behavior of postflashover fires are:

VENTILATION

Some flow of air is needed to cause FRI and continued burning. Therefore at least one barrier in the room must have openings that allow the inflow of air and the outflow of fire gases. The ventilation of the room is expressed as $A\sqrt{h}$, where A is the total area of all openings in the room that allow air or fire gases to flow and h is the weighted average height of the openings from the floor. The ventilation ratio of a room is calculated as $A\sqrt{h}/A_T$, where A_T is the total bounding surface area of the room.

FUEL

The quantity, material type, surface area, and arrangement of the fuels influence burning characteristics and the time–temperature relationship for a fire. Because the heat energy of a fully developed room fire is transferred to all the bounding surfaces, expressing the fuel load in terms of room bounding surfaces, rather than floor area, represents the physics better. Nevertheless, conventional practice continues to prefer floor area.

FUEL–VENTILATION RELATIONSHIP

Ventilation has an important effect on the time–temperature relationships for natural fires. The burning characteristics are often viewed in terms of fuel-controlled burning and ventilation controlled burning:

- *Fuel-controlled burning*: when a large quantity of ventilation is available, fuel-controlled burning takes place. The rate of heat release of the fire is limited by the fuel surface available for combustion. Fuel-controlled burning occurs in spaces having large ventilation openings that supply large amounts of fresh air. Temperatures are higher than for ventilation-controlled burning and much heat energy is lost through the openings.
- *Ventilation-controlled burning*: when the ventilation area is restricted, reducing the supply of fresh air, the fire's heat release rate is limited by the air supply available for combustion. This

ventilation-controlled burning causes fuels to burn more slowly, more inefficiently, at lower temperature, and longer.

LOCATION OF VENTILATION OPENINGS

The location of ventilation openings seems to have an important influence on the burning duration. The fire "breathes" by drawing in fresh air and expelling heat and fire gases. When an opening is in a short wall of a long, narrow room, the fuels do not burn uniformly. They seem to burn in phases that move the burning sequentially to the end of the room. This produces longer-duration fires than may be anticipated by a theory which assumes a well-mixed reaction that consumes fuels at a relatively uniform rate. Thus, long, narrow spaces with openings in one end may exhibit a greater fire severity than configurations that have square shapes.

The demarcation between fuel-controlled burning and ventilation-controlled burning depends on the fuel load and the ventilation ratio. Ventilation-controlled burning is more common in building fires and usually, but not always, produces the most severe conditions for a structure. Therefore it is usually assumed that ventilation-controlled burning exists in the fully developed fire. Although higher fire gas temperatures will occur in fuel-controlled burning, the duration of burning is shorter because much of the heat energy is transferred out of the room by the air/fire gas exchange. This reduces the heat energy application against the barriers. Figure 10.6 illustrates temperature relationships for different ventilation conditions.

BOUNDING SURFACE THERMAL PROPERTIES

The insulative qualities of the room have an influence on the time–temperature relationship of the natural fire:

- *Well-insulated (thermally thick) rooms*: the barriers do not lose much heat to the outside, and the room acts more like an oven. Therefore fires in well-insulated rooms burn hotter and more heat energy is applied to the barriers.

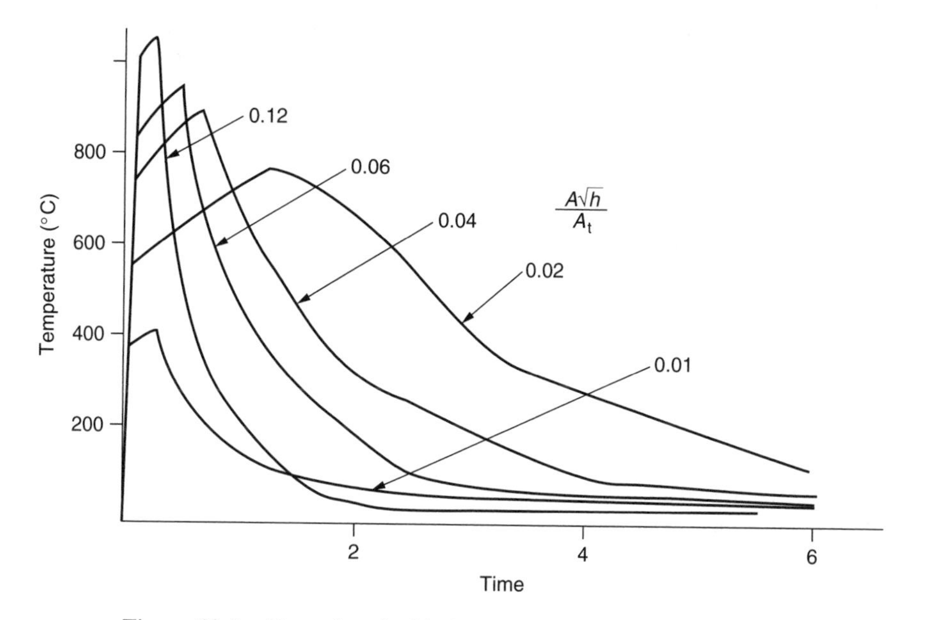

Figure 10.6 (Reproduced with the permission of Lund University)

- *Poorly insulated (thermally thin) rooms*: the barriers transmit more heat to the outside. This heat transfer reduces the heat energy in the room and fires burn somewhat cooler.

The demarcation between thermally thick and thermally thin barriers is difficult to define specifically.

TEMPERATURE

Radiation emitted from a hot surface is proportional to the fourth power of the absolute temperature (Stefan–Boltzmann law). Because materials are particularly sensitive to the radiated heat from a fire, higher temperatures cause a more rapid deterioration in construction assemblies than lower temperatures.

10.8 Other theories

Modern research and computational techniques have expanded knowledge about fire severity, and several theories have been developed to relate fuel loads and room construction to the standard fire test. The most common theories are described below.

INGBERG THEORY

The Ingberg theory was described in Section 10.6. The equivalent time in the standard test, τ_e, is shown in Table 10.1 or it can be calculated from

$$\tau_e = 0.0205L, \tag{10.1}$$

where L is the fuel load in kg/m^2 of floor area ($L < 146\,\text{kg/m}^2$ or 30 psf) and τ_e is in hours.

The Ingberg theory is the simplest way to calculate the required fire endurance time. The equivalent test time is linearly proportional only to the fuel load. The theory neglects the influence of ventilation, thermal properties of the bounding surfaces, and room size. Because the increased radiant energy of high temperatures causes greater material degradation than the standard test fire, some high-intensity fire conditions can make the correlation unconservative.

LAW CORRELATION

Margaret Law [5] analyzed a large number of fire tests involving a variety of room sizes, ventilation ratios, and fuel loads. This correlation is the basis for calculating an equivalent time for a fire test as

$$\tau_e = 0.022\frac{A_F L}{\sqrt{A_V(A_t - A_F - A_V)}}, \tag{10.2}$$

where

A_F = floor area (m^2),
A_V = ventilation area (m^2),
A_t = compartment bounding surface area (m^2).

This equation is an improvement over Ingberg's because it includes ventilation and is based on a relatively large number of tests. However, the compartment thermal properties are not included in the correlation.

PETTERSSON EQUATION

Ove Pettersson [6] expanded on Law's concept to include ventilation height and area as well as a factor relating thermal properties of the bounding surfaces. The equivalent time that an assembly should withstand a fire endurance test is

$$\tau = 0.31CL\frac{A_F}{\sqrt{A_t A_V h_V^{1/2}}} \tag{10.3}$$

where

h_V = average height of ventilation openings (m),
C = bounding surface thermal properties.

The term C incorporates thermal properties of the bounding surfaces as follows:

- $C = 0.09$ when $0 < \sqrt{k\rho c} \leq 720\,\mathrm{J\,m^{-2}s^{-1/2}K^{-1}}$;
- $C = 0.07$ when $720 < \sqrt{k\rho c} \leq 2550\,\mathrm{J\,m^{-2}s^{-1/2}K^{-1}}$;
- $C = 0.05$ when $2550 > \sqrt{k\rho c} \leq 720\,\mathrm{J\,m^{-2}s^{-1/2}K^{-1}}$.

Equation (10.3) incorporates thermal properties of the barriers as well as the height of the ventilation openings to obtain an equivalent standard fire test time.

NORMALIZED HEAT LOAD

Harmathy [7] developed an equivalent barrier fire exposure that relates standard test results with the destructive potential of a fire. Example 10.2 described a room fuel load (equivalent heat content) in terms of the bounding surface area. This normalized (spread) the heat content equally over the entire surface area. Rather than assuming the heat content to be uniform over all surfaces, the *normalized heat load (NHL)* normalizes the total incident heat flux absorbed by the bounding surfaces. The thermal property that is used to "weight" the surfaces is the thermal absorptivity, $\sqrt{k\rho c}$. The NHL then relates natural compartment fires with test furnace fires.

The NHL is defined as

$$H' = \frac{1}{\sqrt{k\rho c}}\int_0^\tau q\,\mathrm{d}t. \tag{10.4}$$

For a room with a natural fire, H' is the total heat energy released by the fire divided by the bounding surface of the room thermal inertia of the boundary:

$$H' = \frac{M_f \Delta H_c}{A_t\sqrt{k\rho c}}. \tag{10.5}$$

The rate of heat release is influenced by the amount and characteristics of the fuel, the ventilation, and the combustion efficiency. Some of the heat is released through toe openings and some is unburned. When these factors are included, the value for H' becomes

$$H' = \frac{(11.0\delta)M_f \times 10^6}{A_t\sqrt{k\rho c} + \sqrt{\phi M_f}}, \tag{10.6}$$

where

M_f = mass of fuel (kg),
δ = combustion efficiency,
ϕ = ventilation factor.

Equation (10.5) is a more conservative representation for the NHL, whereas equation (10.6) provides values closer to expected fire behavior.

When the compartment fire is compared with a test furnace, the equivalent test time can be calculated. This correlation will vary with the test furnace. For the National Research Council of Canada test furnace, the equivalent time in the test furnace may be expressed as

$$\tau_e = 0.11 + 0.16 \times 10^{-6} H + 0.13 \times 10^{-9} H. \tag{10.7}$$

The normalized heat load is a carefully formulated and theoretically sound way to relate compartment fires with standard test results. The correlation can vary with different test furnaces, although the discrepancy may not be significant. The theory relates to thermally thick barriers which absorb all of the heat energy and a small temperature rise on the unexposed face. It does not apply to thermally thin barriers in which the heat is transmitted to the adjacent room and the temperature rises substantially on the unexposed barrier surface.

10.9 Discussion

The methods cited above approximate the standard test time that is equivalent to fuel load burnout. All of the equivalency methods relate fuel loads to an equivalent time in the standard fire test. The terminal point is thermal failure. Surface cracking and collapse failure modes are assumed not to occur. The most important feature is that these theories relate only to equivalencies with the standard fire test. They do not attempt to relate the fire test to building performance. Although none give sufficient information for a building performance analysis, they do provide useful information for making evaluation judgments.

Newton [8] compared these four theories and the COMPF2 computer program with statistical data for office loadings. The mean and +1 and +2 standard deviations for fuel loads and the mean, −1, and +1 standard deviations for ventilation were selected from the Culver [9] statistical data. The analyses involved square rooms with 3 m ceilings that varied in area from 6.25 m^2 to 400 m^2. The floor and ceiling were reinforced concrete, and six combinations of concrete block or gypsum wall constructions were used. The goal was to gain an understanding of real-time relationships and comparative sensitivity of these methods with regard to statistical and practical building conditions.

Spreadsheets enable complex equations to be solved relatively easily to gain an insight into comparisons involving theories applied to statistically related conditions. The "real-time" fire duration was calculated using step functions with stoichiometric wood burning with the associated $A_W\sqrt{h}$ ventilation. Table 10.2 illustrates the type of results that were obtained.

The general observations from this limited study of 90 cases that used statistical data were as follows:

- The ventilation area is the most influential factor in fire severity. Small openings cause long-duration fires that produce the greatest severity. Large openings cause much faster combustion and heat losses through the openings. This creates a high-intensity, short-duration fire of lower equivalent test duration.
- The fuel load has the anticipated direct relationship on fire severity in that more fuel creates a longer fire.
- The material properties for thermally thick barriers (here gypsum and concrete) in the Pettersson and NHL analyses have much less influence on fire severity than does ventilation. The differences between these materials were relatively insignificant.
- Some calculation methods can be affected by room size. The Ingberg theory is only related to fuel load, and independent of all other factors. The NHL pattern seems to give the most

Table 10.2: Typical Results

Fuel load (kg/m^2)	$\frac{A_W\sqrt{h}}{A_t}$	Real time (min)	τ_{NHL}(min)	τ_{Law}(min)	$\tau_{Pettersson}$(min)	$\tau_{Ingberg}$(min)
36	0.01	152	81	136	106	44
36	0.07	22	13	54	40	44
36	0.15	10	27	38	28	44
57	0.01	242	133	216	168	70
57	0.07	35	62	85	64	70
57	0.15	16	36	61	44	70
79	0.01	335	193	299	233	97
79	0.07	48	80	118	89	97
79	0.15	23	44	84	60	97

Room construction: size 5 m × 5 m; floors reinforced concrete; walls wood stud and gypsum.

consistent and logical results for all variables. Fire severity increases directly with room size in the Petterson and Law procedures. Larger rooms produce larger equivalent test times. However, these analyses did not investigate applicability limits of any of the methods. In all fire calculations the range of applicability is always important and rarely identified.

- The NHL method gives the best indication of equivalent test time. Results are consistent and logical. The NHL method has the strongest theoretical base and the heat energy application includes not only the fuel load but also other factors relevant to compartment fires.
- The NHL method is relatively simple, but cumbersome to use without a computer program to order the calculations. However, because catalog curves are a one-time assessment, this time investment is worthwhile. NHL results involve only heat transmission for thermally thick barriers. Cracking or falling away of materials and the barrier's structural integrity are not a part of the analysis. Its primary value is to provide information to construct catalog $\overline{\text{T}}$ curves for some assemblies. Studying trends will help in estimating $\overline{\text{T}}$ curves for untested assemblies.
- None of the methods provides enough information to construct catalog $\overline{\text{D}}$ curves. However, heat transfer calculations combined with a structural analysis (Chapter 15) can provide information to guide catalog $\overline{\text{D}}$ curve construction.

10.10 Awareness pause

The fire endurance of barriers is complicated. Before we continue discussing barriers, it may be useful to pause to reflect on some of the information that has been presented up to now. Here are the major concepts:

- The vast majority of barrier failures are due to openings and weaknesses that are not addressed by the standard fire test and the associated prescriptive fire endurance classification. The most important understanding of barrier performance is gained by a level 1 evaluation described in Chapter 9.
- The fire endurance times prescribed in building codes have practical value because they make it difficult to have flimsy construction in large buildings. The fire resistance time durations should not be taken literally. They enable one to make general comparisons about expected behavior between different construction assemblies. Although prescribed fire resistance requirements do result in more stable construction during a fire, the performance for assemblies does not necessarily correspond to the test rating. Fortunately, as long as the barriers have structural

stability and reasonable construction substance, their fire resistance doesn't seem to be a major factor in systems performance.
- No published investigations of the relationship between building performance and standard fire test results have been discovered. Equivalent theories relate fuel loadings to the standard test. These calculations give a sense of proportion for the factors that influence the destructive potential of barriers and help to make more confident estimates. However, any relationship between test time and field performance is unknown.
- Equivalency theories that relate fuel loads, ventilation ratios, and thermal properties to the fire test are based on heat transmission through the barrier. They do not address material cracking or structural collapse. Fortunately, the structural collapse potential may be estimated as a separate analysis.
- Building performance is influenced greatly by the barrier failure mode. One must understand a barrier's tendency for hot-spot ($\overline{T}$) and massive ($\overline{D}$) failures and their associated time frames. A performance analysis requires a continuum of assembly behavior up to complete collapse. From a practical viewpoint, the fuel load and other conditions rarely cause failure in an unpenetrated barrier. Nevertheless, the amount of fuel is a part of the continuum and is the measure for barrier performance.
- The purpose of a barrier analysis is to understand the fire propagation performance of a building. A level 1 analysis gives a picture of how a cluster of rooms will behave and important features to examine more carefully. A level 2 analysis is necessary for situations where barrier performance is important, because it gives a better insight into performance.

10.11 Estimating barrier performance

Organizing the way of thinking about barriers is the most difficult task in understanding barrier performance. Before we describe a process for evaluating level 2 barrier field performance, we will review some relevant aspects.

PRESCRIPTIVE CODES AND STANDARD TEST ENDURANCE

The major advantage of the standard test is simplicity in regulatory control. However, it is difficult to overcome our biases and comfort after nearly a century of inferring that standard fire endurance times describe field performance. While fire endurance times of 85 and 95 min can be important with prescriptive codes, this difference is usually insignificant for performance evaluations.

A great deal is unknown about the translation of standard fire testing and fire severity theories to field performance predictions. In addition, code-prescribed fire resistance requirements have a very weak technical justification, although a somewhat stronger subjective justification. The same performance uncertainties and computational inaccuracies exist with prescriptive codes. The acknowledgement of these weaknesses during an evaluation can be counterbalanced by judgment in making field performance estimates.

THE PUZZLE

Estimating barrier performance is somewhat like assembling a jigsaw puzzle. One must combine information from several sources to get the complete picture because an integrated technology has not yet been synthesized.

STANDARD TEST KNOWLEDGE

Over the years, individuals who manage and conduct fire tests have acquired a good insight into material and assembly behavior. Valuable information is obtained during a test, but not published because of proprietary interests. Test results and acquired knowledge of material behavior are an important piece of the puzzle. Since relatively few individuals possess this knowledge, one must use available literature and try to acquire additional information throughout a professional career.

INTEGRATION THEORIES

Several theories, such as those of Section 10.8, relate equivalent test time with fuel load, room construction, and natural fires. These theories provide additional information for constructing barrier performance curves. The calculations link natural fire time relationships to get an equivalent time for the standard test. This piece of the puzzle does not incorporate barrier behavior, except perhaps heat transfer relationships. While these theories provide useful information, the onset of a standard test failure mode is not as significant as the way in which the barrier behaves up to and beyond that point.

FIRE INVESTIGATIONS

No published documentation seems to exist to correlate the standard fire test with field performance. A small amount of experiential information seems to indicate that larger barrier sizes and thermal movements of the supporting structure cause earlier failures than may otherwise be expected. One must rely on logical judgments based on available knowledge and information.

CALCULATION METHODS

Calculation methods for structural behavior and heat transfer through barriers are relatively well advanced. Confidence in making performance estimates is improved when calculated information provides knowledge of behavior for a range of fire and construction conditions.

CATALOG FAILURE MODE DIVISIONS

The separation of performance into $\overline{\mathrm{T}}$ and $\overline{\mathrm{D}}$ failure modes allows one to examine each in isolation. A $\overline{\mathrm{T}}$ performance considers only heat transfer through an assembly and cracks that allow ignition into the adjacent space. A $\overline{\mathrm{D}}$ performance examines only partial or complete barrier collapse that allows a massive influx of fire gases into the next space.

BARRIER EFFECTIVENESS

Catalog $\overline{\mathrm{T}}$ and $\overline{\mathrm{D}}$ curves provide independent descriptions of expected fire test behavior for a thermally unrestrained assembly that does not support superimposed live loads. Field performance may be determined by multiplying the catalog curves by a barrier effectiveness factor that incorporates a series of multipliers which shift the catalog values to account for applied loads, thermal restraint, construction influences, openings, and automatic protection equipment. Some conditions will indicate that structural movement can affect material cracking or that thermal penetration can affect deflections. These conditions are incorporated in the multiplier for construction influences.

COHESIVENESS

Each field $\overline{\mathrm{T}}$ and $\overline{\mathrm{D}}$ curve describes an independent failure mode. Mathematically, this produces a 200% total failure for each barrier, which is not possible. The field values must be adjusted so that $P(\overline{\mathrm{T}}) + P(\overline{\mathrm{D}}) + P(\mathrm{B}) = 1.0$.

HEAT ENERGY APPLICATION

The design fire assumes that all heat energy generated will be applied to the barriers. However, some hot gases will leave the space through openings without affecting the barriers. One can adjust the total heat generation to identify the fraction that is applied to the barriers. Catalog curves are based on unlimited heat energy application. Values for $\overline{\text{T}}$ and $\overline{\text{D}}$ can be identified from the curve for any given time interval or level of heat energy application. Normally, the fuel load and fire conditions in a space produce a situation where applied heat energy will be less than the catalog barrier capability. The analysis adjusts the heat energy to reflect these conditions.

JUDGMENT

Although much is known about aspects of barrier behavior, the information is neither sufficient nor well organized. Major gaps appear between what is known and what is needed to evaluate performance. Judgment is the primary intellectual activity that assembles the parts of the jigsaw puzzle to portray barrier performance. Judgment is the only resource to fill the knowledge gaps in translating available information into field performance descriptions.

10.12 Catalog barrier curves

A catalog curve provides a starting point for rapid estimation of field performance. Catalog curves are adjusted by correction multipliers to reflect field conditions. The barrier effectiveness factor integrates correction multipliers to establish field performance $\overline{\text{T}}$ and $\overline{\text{D}}$ curves.

Catalog curves are constructed by envisioning the continuum of barrier behavior during a test fire. The $\overline{\text{T}}$ catalog curve assumes that the barrier will not collapse or deform excessively. Surface cracking and excessive thermal transmission are the functions of interest. The $\overline{\text{D}}$ catalog curve describes the likelihood of a full or partial collapse. For unlimited heat application, all barriers eventually experience a $\overline{\text{D}}$ failure. Because we must understand the continuum of barrier capability, the $\overline{\text{T}}$ and $\overline{\text{D}}$ curves assume unlimited fuel and heat energy application.

The $\overline{\text{T}}$ and $\overline{\text{D}}$ curves are a one-time construction assembly evaluation. Figure 10.7 illustrates hypothetical curves. The vertical axis plots barrier performance because the evaluator can describe $P(\overline{\text{T}}\text{ failure})$, $P(\overline{\text{D}}\text{ failure})$, and $P(\text{B success})$. The horizontal axis plots the heat energy application in terms of the mass of fuel consumed. Any convenient unit may be used. We recommend psf (kg/m^2) of floor area. Additional lines are provided to show scales for adjusted heat energy and real time.

The barrier assembly is viewed as a composite whole without openings or penetrations. We impose two important conditions: (1) the barrier has no lateral restraint to thermal expansion, and (2) the barrier does not support any superimposed live loads. We estimate the behavior that would occur in a standard fire test under these conditions. The estimates combine knowledge of fire testing with structural and material behavior at elevated temperatures, experimental data, failure analyses, and computer models. The full spectrum of available relevant information is combined with judgment to estimate $\overline{\text{T}}$ and $\overline{\text{D}}$ performance.

We face a major dilemma in selecting units for the abscissa. Of the theories identified in Section 10.8, Ingberg and NHL seem to be more suited for consideration. The Ingberg theory has an advantage because the relationship shown in Table 10.1 is invariant with ventilation, barrier thermal properties, and room size. It is more conservative than NHL for small ventilation ratios and less conservative for large ratios. Its major disadvantages relate to these same factors. Fire severity depends on ventilation, thermal properties, and size. The Ingberg theory is clearly inaccurate. NHL is more appropriate because of its careful formulation that incorporates sound

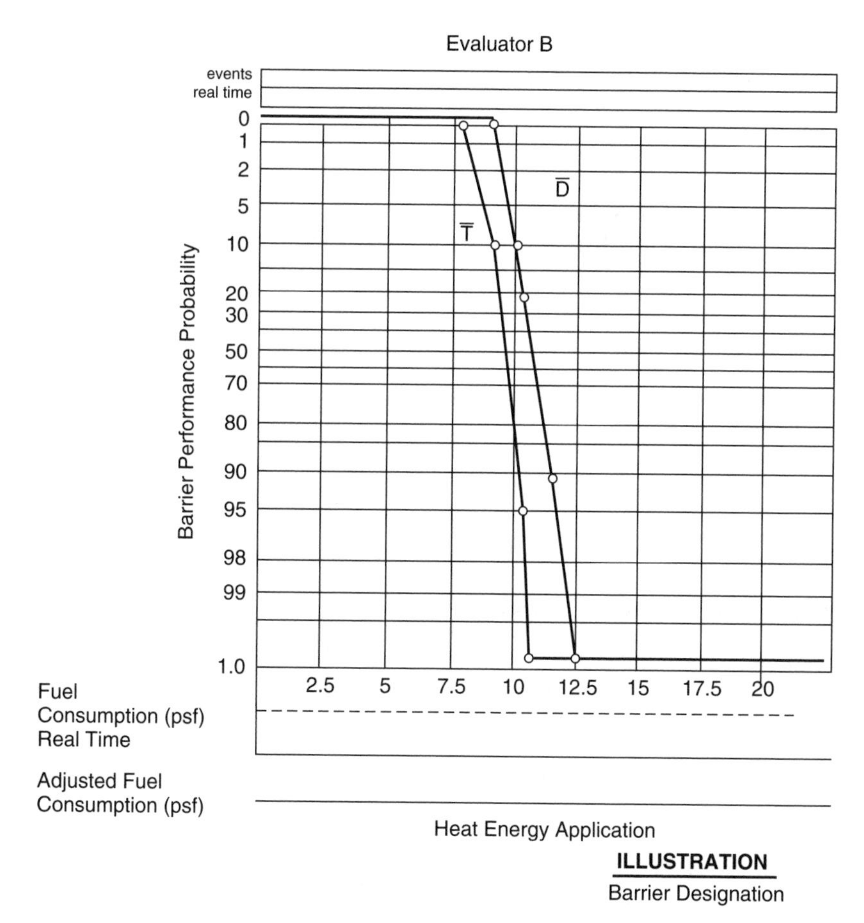

Figure 10.7

theory and more careful experimental study. Unfortunately, this variability makes it more difficult to maintain a consistency for comparisons. Therefore we recommend a two-phase approach.

With great discomfort, the Ingberg theory with the values of Table 10.1 is recommended to establish catalog curves. There are two practical reasons for this recommendation. The first is relative ease and consistency in making comparisons relating to physical construction (resistance) behavior. Theoretical weakness on the loading can be balanced somewhat by closer examination of the resistance side. The second recognizes that problems of barrier destruction before suppression intervention are rare. If an initial performance analysis indicates that a barrier may be approaching potential failure, an NHL analysis will augment information to give better guidance. The Ingberg theory will give an initial screening to recognize potential problems. However, regardless of what theory is used and what the numbers may show, performance estimates are primarily guided by judgment. Construction details that relate to strength are important, and rooms with large ventilation openings require careful scrutiny. Fortunately, weak barriers are relatively easy to recognize for conditions involving well ventilated rooms that produce high temperatures or long-duration fires.

Each coordinate of a $\overline{T}$ curve is evaluated by answering the question, Given X value of heat energy application (fuel consumption), what is the probability that a hot-spot ignition can occur on the unexposed side of this barrier? Each coordinate of a $\overline{\mathrm{D}}$ curve is evaluated by answering the question, Given X value of heat energy application (fuel consumption), what is the probability that a massive ignition can occur on the unexposed side of this barrier?

Example 10.3

In Figure 10.7 read the values for $P(\overline{\mathrm{T}})$ and $P(\overline{\mathrm{D}})$ when the fire has consumed the following amounts of fuel: 7.5 psf, 9 psf, 10 psf, 10.5 psf, 11.5 psf, 13.5 psf, 15 psf.

Solution

The following values for $P(\overline{\mathrm{T}})$ and $P(\overline{\mathrm{D}})$ were obtained from Figure 10.7.

Fuel mass consumption	7.5	9	10	10.5	11.5	13.5	15
$P(\overline{\mathrm{T}})$	0.0	0.1	0.8	0.95	1.0	1.0	1.0
$P(\overline{\mathrm{D}})$	0.0	0.0	0.1	0.2	0.9	1.0	1.0

10.13 Mathematical cohesiveness

The probability for each $\overline{\mathrm{T}}$ and $\overline{\mathrm{D}}$ curve ranges from 0% to 100%. It is not possible for total failure to reach 200% because the system logic must be cohesive, and the mathematics of probability cannot be violated. This apparent inconsistency is relatively easy to address.

Figure 9.2 indicates that

$$P(\overline{\mathrm{T}}) + P(\overline{\mathrm{D}}) + P(\mathrm{B}) = 1.0 \tag{10.8}$$

is the universe at every instant of time and is always valid. When $P(\overline{\mathrm{T}}) + P(\overline{\mathrm{D}}) = 1.0$, $P(\mathrm{B}) = 0.0$ at every time thereafter.

Barriers subjected to a fire's heat energy often first exhibit a $\overline{\mathrm{T}}$ failure. As the heat energy continues to attack the barrier, the probability of a $\overline{\mathrm{T}}$ failure increases. These $\overline{\mathrm{T}}$ failures can increase in size until a $\overline{\mathrm{D}}$ failure develops. If barriers were tested with an unlimited source of fuel, eventually all barriers would experience a $\overline{\mathrm{D}}$ failure. Therefore when $P(\overline{\mathrm{T}}) + P(\overline{\mathrm{D}}) > 1.0$ the value for $P(\overline{\mathrm{D}})$ is unaffected whereas the value for $P(\overline{\mathrm{T}})$ is adjusted (reduced) so that

$$P(\overline{\mathrm{T}}) + P(\overline{\mathrm{D}}) = 1.0 \tag{10.9}$$

for all heat energy or time values after $P(B) = 0$.

This requirement is necessary to maintain cohesiveness in the system. Some barriers are constructed in a way that collapse will occur before the hot-spot failure emerges. This logic is also valid when a $\overline{\mathrm{D}}$ failure precedes a $\overline{\mathrm{T}}$ failure.

Example 10.4

Assume that the catalog barrier curves of Figure 10.7 represent field conditions that were exactly the same as the catalog curve conditions. Determine the mass of fuel consumed when the barrier success, $P(\mathrm{B})$, reaches zero. Calculate values of $P(\overline{\mathrm{T}})$, $P(\overline{\mathrm{D}})$, and $P(\mathrm{B})$ for fuel consumption values of Example 10.3 and construct the barrier success (B) curve and the $\overline{\mathrm{T}}$ and $\overline{\mathrm{D}}$ curves that represent a cohesive condition.

Solution

Equation (10.8) is valid until $P(B) = 0$. Inspection of the $\overline{\mathrm{T}}$ and $\overline{\mathrm{D}}$ curves reveals that the lowest heat energy application at which $P(\overline{\mathrm{T}}) + P(\overline{\mathrm{D}}) = 1.0$ is about 10.1 lb/ft^2. At this value, $P(\overline{\mathrm{T}}) = 0.85$ and $P(\overline{\mathrm{D}}) = 0.15$ and $P(B) = 0$ for the first time. The adjusted values for $P(\overline{\mathrm{T}})$ and $P(\overline{\mathrm{D}})$ are shown in Table 10.3. Figure 10.8 shows the adjusted performance curves.

Table 10.3: Adjusted Catalog Curves

Heat energy (10^3Btu/ft^2)	$\overline{\mathrm{T}}$ value (from $\overline{\mathrm{T}}$ curve)	$\overline{\mathrm{D}}$ value (from $\overline{\mathrm{D}}$ curve)	B value (calculated)	$\overline{\mathrm{T}}$ adjusted value (calculated)	$\overline{\mathrm{D}}$ value
7.5	0.0	0.0	1.0	0.0	0.0
9	0.1	0.0	0.90	0.10	0.0
10	0.8	0.1	0.1	0.8	0.1
10.1	0.85	0.15	0.0	0.85	0.15
10.5	0.95	0.2	0.0	0.80	0.2
11	1.0	0.4	0.0	0.10	0.6
12.5	1.0	1.0	0.0	0.0	1.0

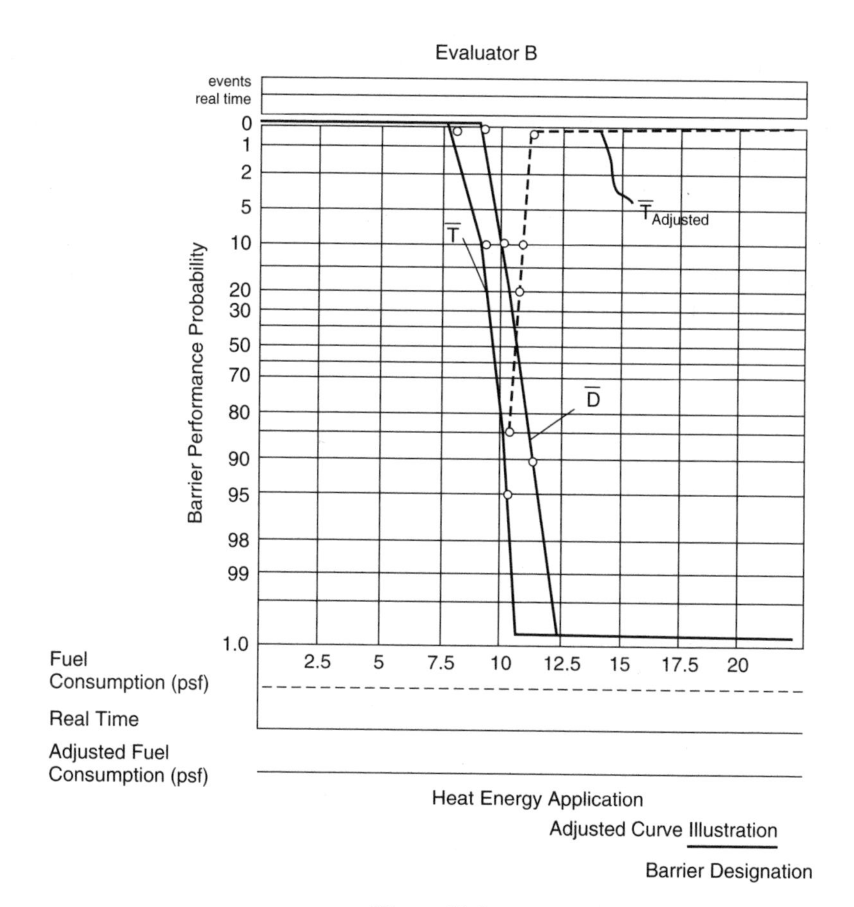

Figure 10.8

10.14 Material behavior in fires

All building materials are affected adversely by the heat from a fire. Some materials are affected at lower temperatures or at shorter time exposures than others. When making judgments concerning the performance of structural materials in a fire, one should be aware of the physical properties that are most sensitive to elevated temperatures. We review the influence of fire on a few of the more common construction materials to help in estimating barrier performance.

STRUCTURAL STEEL

Structural steel is a dominant material for providing strength in modern building construction. High strength and stiffness, ductility, the ability to form a variety of shapes and dimensions, and ease and speed of erection are among the qualities that steel possesses. Structural steel may be used as the skeleton framework for buildings or as the reinforcement for reinforced concrete, prestressed concrete, or other materials. However, steel, like all materials, is affected adversely by the elevated temperatures that are produced by fires.

Structural steel is noncombustible and does not contribute fuel to a fire. Nevertheless, one should not develop a false sense of security of its durability in a fire, because structural steel also loses strength and stiffness when subjected to elevated temperatures. The yield strength is the most significant parameter for load-bearing capacity. The capability to support loads is reduced substantially as the steel temperature rises, causing a reduction in the yield strength. The stiffness, as reflected by the modulus of elasticity, E, is also reduced substantially with an increase in temperature. These two properties interact with the applied loads to affect deformations which can have an important influence on structural performance. Figure 10.9 shows the influence of temperature on yield stress and modulus of elasticity.

The stress limitation is the temperature within the steel and not the ambient temperature of the fire gases. Steel has a high thermal conductivity. Therefore a steel member can transfer heat away from a localized source relatively quickly. This property, in conjunction with its thermal capacity, enables steel to move the heat to other parts of the structure relatively easily. The connectivity

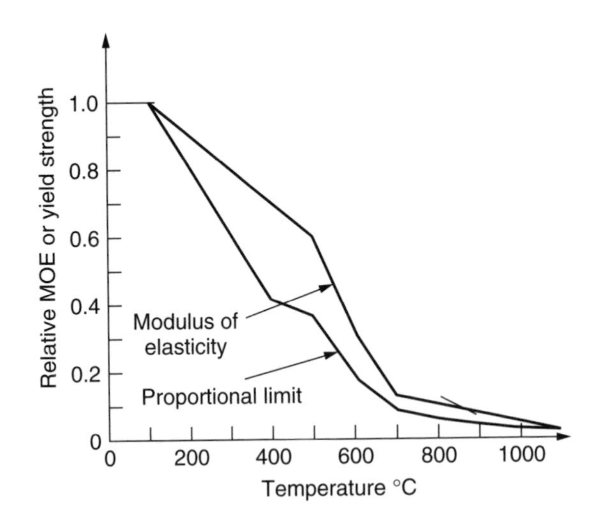

Figure 10.9 (Reproduced by permission of CEN)

and heat transfer interfaces with other parts of the assembly influence the temperature that the steel member feels.

The mass of the steel also has an influence on the fire performance. Heavy sections have a greater resistance to the effects of high temperatures than do lighter sections. This is related to the magnitude of the applied stress and the type of member. Thin, lightweight members in compression, such as in truss or bar joist construction, are more likely to fail earlier than heavy columns.

Structural steel also expands when heated. This property can affect the load-carrying capacity for certain types of construction. For example, when a structural member is heated and its expansion is restrained, compressive stresses, quite similar to prestressing, are induced. These axial stresses, combined with the flexural stresses, can modify the performance of the member. In some cases the tested fire endurance of a thermally restrained member will increase. In other cases, particularly in compressive members, a more rapid failure is likely to occur.

A second influence of heat-induced expansion relates to the structural geometry. When a member can expand and move supporting columns out of alignment, the column load becomes eccentrically applied. This can cause a premature failure because of the additional stresses generated by the eccentric compressive loading. Structural expansion can also move bearing walls to cause premature collapse when accommodation for movement is not incorporated into the design.

CONCRETE

Concrete is a mixture of cement, sand, aggregate, and water. When exposed to heat, concrete loses water and also strength. The type of aggregate has a major influence on the performance of concrete. The two main types of normal weight concrete aggregates are siliceous and calcareous. Siliceous aggregates have an igneous base and calcareous aggregates are limestone. Siliceous aggregates have a lower fire endurance than calcareous aggregates.

Because of the difference in fire performance for different aggregates, the strength at elevated temperatures is compared on a percentage basis of the normal temperature strength. Figure 10.10

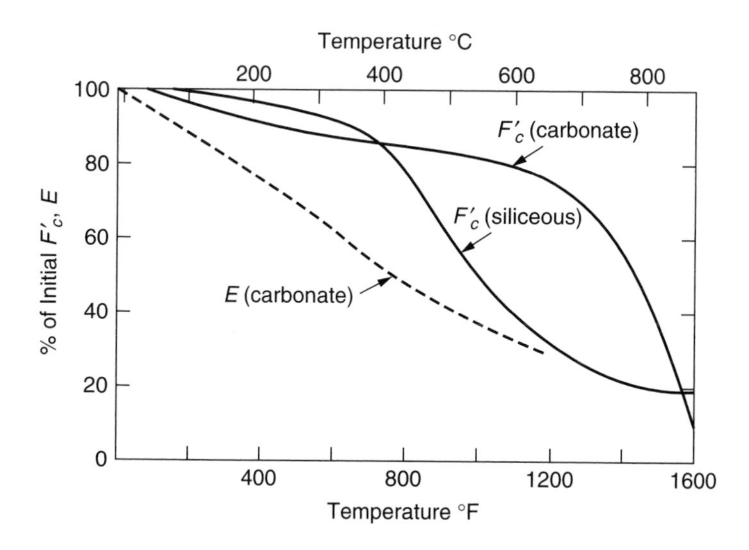

Figure 10.10 (Reproduced with permission of Portland Cement Association)

shows the effect of elevated temperature on the ultimate strength, f'_c, and on the modulus of elasticity, E, of concrete test cylinders.

Normal weight concrete that uses siliceous or calcareous aggregates weigh about 150 lb/ft^3. It is possible to make concrete using lightweight aggregates such as vermiculite, expanded shale, and pumice. Concrete using these lightweight aggregates weighs from 90 to 130 lb/ft^3. Lightweight concrete exhibits a higher fire endurance than normal weight concrete.

As with other construction materials, heat causes concrete members to expand. Reinforced concrete will increase in load-carrying capacity when the ends are thermally restrained by the construction. However, when thermal restraint is low, as at the exterior bays of buildings, the expansion may cause cracks to open and supporting members to deform excessively.

Concrete has a high compressive strength but very low tensile strength. Because flexural loads cause compressive and tensile forces simultaneously in a member, the tensile loads must be carried by structural steel reinforcing bars These reinforcing bars are often called re bars for short. Reinforced concrete is a composite that takes advantage of each material's strengths by locating them in positions of maximum efficiency to produce a structural member. In beam design, the reinforcing bars are placed in positions of tensile stresses. Figure 10.11(a) shows the position of the steel reinforcing for a simply supported beam or slab, and Figure 10.11(b) shows the positions of the tensile reinforcement for a continuous concrete beam.

Concrete structures often give a sense of fire security. However, concrete is affected adversely by elevated temperatures. The sense of security may be appropriate when the structure is constructed in a manner that compensates for the deleterious effects of the loss in strength. But the structure may collapse when construction continuity is not present. The methods of construction have a substantial influence on performance of reinforced concrete.

The moisture content of the concrete influences its fire resistance. Much heat energy from a fire is needed to vaporize the absorbed moisture and capillary moisture in concrete. The water is driven to the unexposed surfaces of the concrete. This process absorbs much of the heat energy

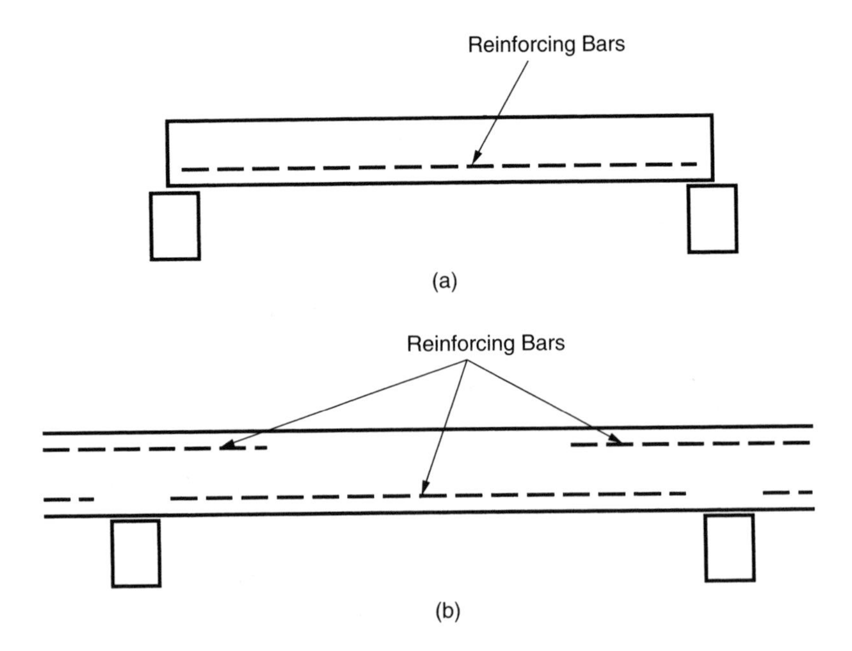

Figure 10.11

of the fire and maintains the temperature of the unexposed side below the failure temperature of a standard fire endurance test criterion for longer periods of time. The voids caused by the evaporation of water contribute to shrinkage and a decrease in concrete strength.

Concrete has a low thermal conductivity and a low thermal capacity, although both properties are influenced significantly by the type of aggregate. These properties enable concrete to be an effective means for protecting steel. Several years ago, encasing structural steel members in concrete was a common way of protection. In reinforced concrete construction, the concrete cover between the fire and the structural steel reinforcing bars provides the fire protection.

A factor that reduces the fire resistance of concrete is its tendency to spall. Spalling is the disintegration or falling away of part of the concrete surface, exposing either the steel reinforcement or the interior regions of the concrete. In a fire, spalling is caused by an explosive build up of steam from the expanding water that cannot be moved to the unexposed surface. The spalling of the protective concrete cover reduces the resistance of the member substantially because the fire gas heat reduces the strength of the exposed steel. In normal weight concrete, siliceous aggregates spall much more readily than calcareous aggregates. Lightweight aggregates perform much better on fire resistance and on spalling in a fire. Sharp edges spall much more readily than rounded or beveled edges.

From a barrier viewpoint, concrete can crack relatively easily to allow a hot-spot ignition into the adjacent space. This cracking does not necessarily lead to an imminent massive failure or collapse. Reinforced concrete in continuous structures usually has collapse resistance beyond the appearance of first cracking.

CONCRETE MASONRY UNITS

A very common use of plain concrete in buildings is through the manufacture of concrete blocks, often called concrete masonry units (CMUs). These blocks are normally made as solid bricks or of larger hollow blocks. CMUs can be manufactured in a variety of sizes and shapes. The void spaces of the hollow blocks may be left unfilled or they may be filled with sand or another inert material to increase the fire endurance.

PRESTRESSED CONCRETE

Precast structural members, and sometimes cast-in-place members, may be prestressed. That is, the steel reinforcing wires are stressed in tension before the concrete is loaded. After the tension of the wires is released, the concrete is initially under compressive stress before any additional loads are applied. This prestressing is designed in a manner that increases the economy of the structural member.

Prestressed concrete normally has a higher compressive strength than most nonprestressed reinforced concrete structures. Also, the prestressing wires are of a much higher strength than normal reinforced concrete structures. Although it has a greatly increased tensile strength, this prestressing wire is very vulnerable to elevated temperatures and must be protected from the heat of a fire.

The thermal and structural properties of the concrete are essentially the same for both types of construction. However, the fire behavior may be quite different. Prestressed concrete structures are usually precast as statically determinate elements. Therefore the opportunity for stress redistribution that is present in monolithic structures is often absent in precast structures. Also, when applied loads are not sufficient, the natural camber caused by the prestressing operation may cause the member to bow and explode upward before collapse.

WOOD

Wood may be used as a main support member as joists, beams, girders, or columns. Wood also may be used as a sheathing on floors, roofs, or walls to serve as a surface for loading or other materials. The function of the material determines its role in a building evaluation. When the materials are used in a structural load-carrying function, the evaluation becomes the structural analysis in Chapter 15. When the materials function as a separation between spaces, they are analyzed as barriers. Sometimes, as in a floor system, the construction serves both functions. This causes no analytical difficulties as the barrier is evaluated for its ability to delay or prevent an ignition into the adjacent space and the frame is evaluated for its ability to support the applied loadings. Time is an important part of the evaluation.

Exposed wood construction augments the fuel of the fire. Also, wood is able to maintain its integrity for a period of time before failure occurs. The analysis evaluates the time duration before the failure criteria are reached. Sometimes the few minutes that wood provides to delay fire propagation into an adjacent space may be enough to enable the fire fighters to extinguish the fire before it grows to a size where defensive action is needed.

When wood burns, it forms a char. This char acts as an insulation that delays pyrolysis and slows but does not stop the continued deterioration of the member. The rate of combustion varies with depth of char and depends on the species and orientation of the wood. The rate of char varies from about 1/25 in/min for soft species of wood to about 1/50 in/min for the harder species. An average value of 1/40 in/min is often used as a general approximation. Thicker members therefore provide substantially more time before failure occurs.

GYPSUM

Gypsum products, such as plaster or gypsum board (often called sheet rock, plasterboard, or dry wall) perform well in a fire. The gypsum has a substantial amount of chemically combined water. During a fire the water slowly evaporates and the gypsum is calcined and disintegrates into a soft powder. This heating action moves slowly through the material and provides time before the fire resistance is compromised.

Gypsum board consists of a plaster core sandwiched between sheets of paper. There are many types of wallboard ranging in thicknesses from 1/4 in to 1 in. A regular gypsum board is the most common type. For better fire resistance, type X gypsum board is often specified because glass fibers in the plaster mixture provide stability by holding the calcined gypsum in place longer during a fire.

GLASS

There are three main forms of glass used in building construction. Ordinary window glass is used as glazing for doors and windows. This type of glass has relatively little resistance to breakage in a fire. Thick plate glass or tempered glass used in curtain walls can withstand high temperatures substantially longer before breaking. To provide greater integrity in a fire after cracking, wired glass is often specified. However, whenever evaluating glass from the point of fire resistance, the entire structural framing and installation becomes important to the analysis. For example, if the glazing does not have adequate gasketing, the fire temperatures will cause a metal frame to deform and any type of glass insert can break readily and very quickly.

A second form of glass is fiberglass-reinforced building products. Products such as prefabricated fiberglass bathroom units and translucent window panels and siding have advantages of economy and aesthetic appeal. The fiberglass does not burn or support combustion. However, the fiberglass acts as a reinforcement for a thermosetting resin. These products are usually about 50% resin, which is very flammable even when fire retardants are incorporated into the composition. Although the fiberglass itself is noncombustible, the entire unit is very combustible. In addition, it produces substantial amounts of smoke during combustion.

A third application of glass in building construction is with fiberglass insulation. Again, the fiberglass does not burn. However, the glass fibers are coated with a resin binder that is often flammable and can propagate flames when they are in an open, unconfined space.

Example 10.5

Figures 10.12 to 10.15 show catalog $\overline{T}$ and $\overline{D}$ curves. Identify which curve best represents the following construction assemblies and discuss the performance descriptions:

- a 2 × 4 wood stud wall with 1/2 in of type X gypsum board on each side;
- an 8 in concrete masonry unit (CMU) wall with the voids unfilled;
- an unprotected 12 in open web joist floor system supporting a 1/2 in corrugated deck and 2 1/2 in of concrete fill;

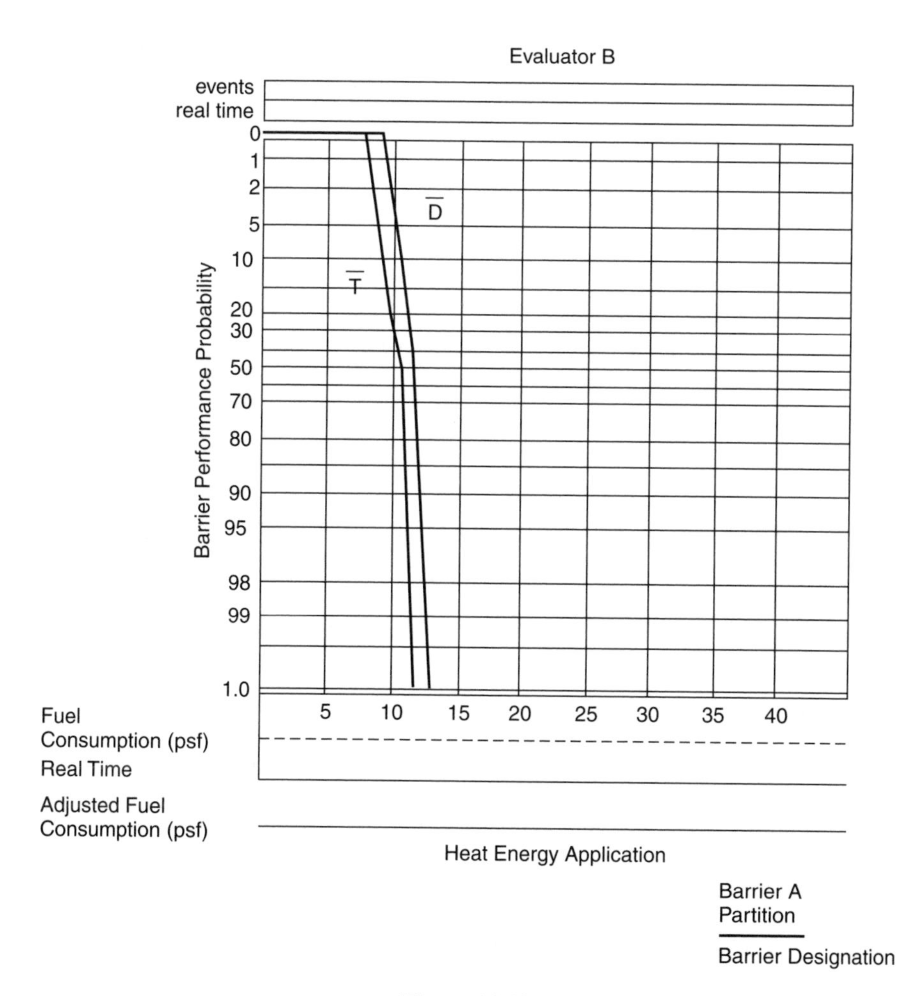

Figure 10.12

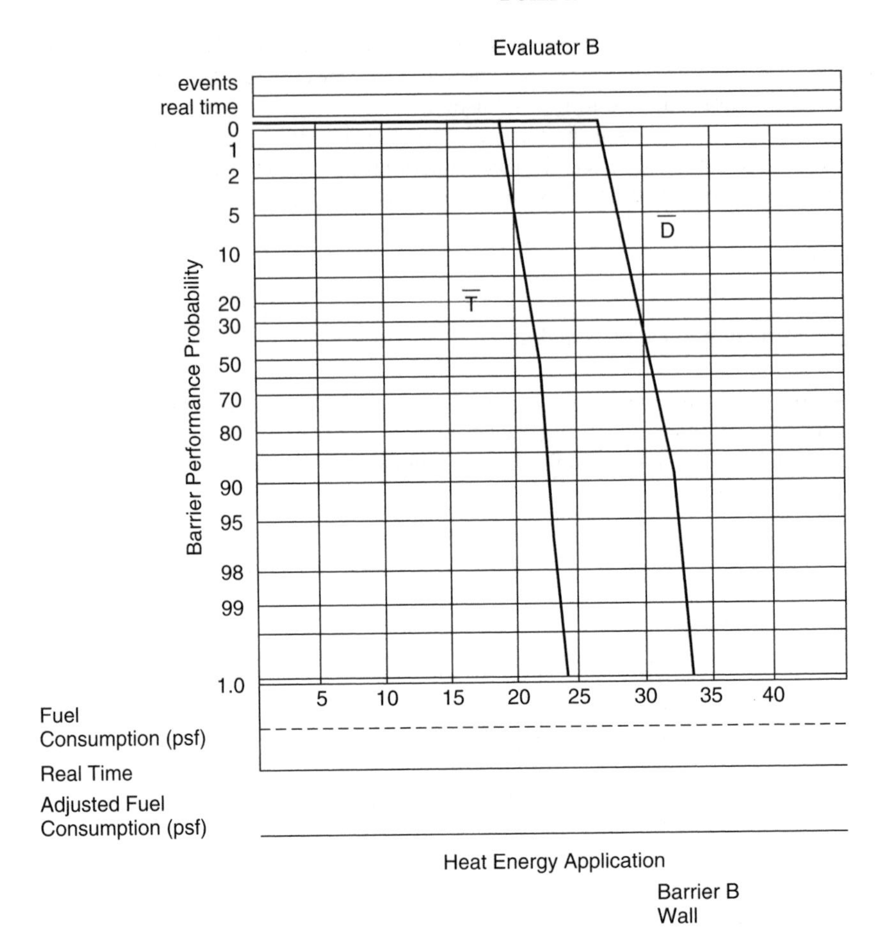

Figure 10.13

- a 12 in open web joist floor system supporting a 1/2 in corrugated deck and 2 1/2 in of concrete fill protected by a membrane ceiling composed of acoustic tiles bearing a classification label.

The Ingberg fuel load relationship (Table 10.1) is shown on these catalog curves to illustrate the differences.

Solution

Figure 10.12 shows a $\overline{T}$ curve that should certainly avoid failure until about 9 psf of fuel is consumed. Then the likelihood of failure increases rapidly until a certainty of failure would occur at a little above 10 psf. Gypsum normally disintegrates or falls off within a few minutes after $\overline{T}$ failure occurs. These characteristics are representative of gypsum board construction.

Figure 10.13 describes the characteristics for a CMU wall. In a fire test, $\overline{T}$ failure usually occurs by cracking between the joints. Even after this type of initial $\overline{T}$ failure, the units should remain stable for a substantial time, thus delaying the $\overline{D}$ failure.

The unprotected open web joists are exposed to high temperatures rather quickly and are unable to transfer the heat before they lose strength and collapse. This $\overline{D}$ failure will occur before the

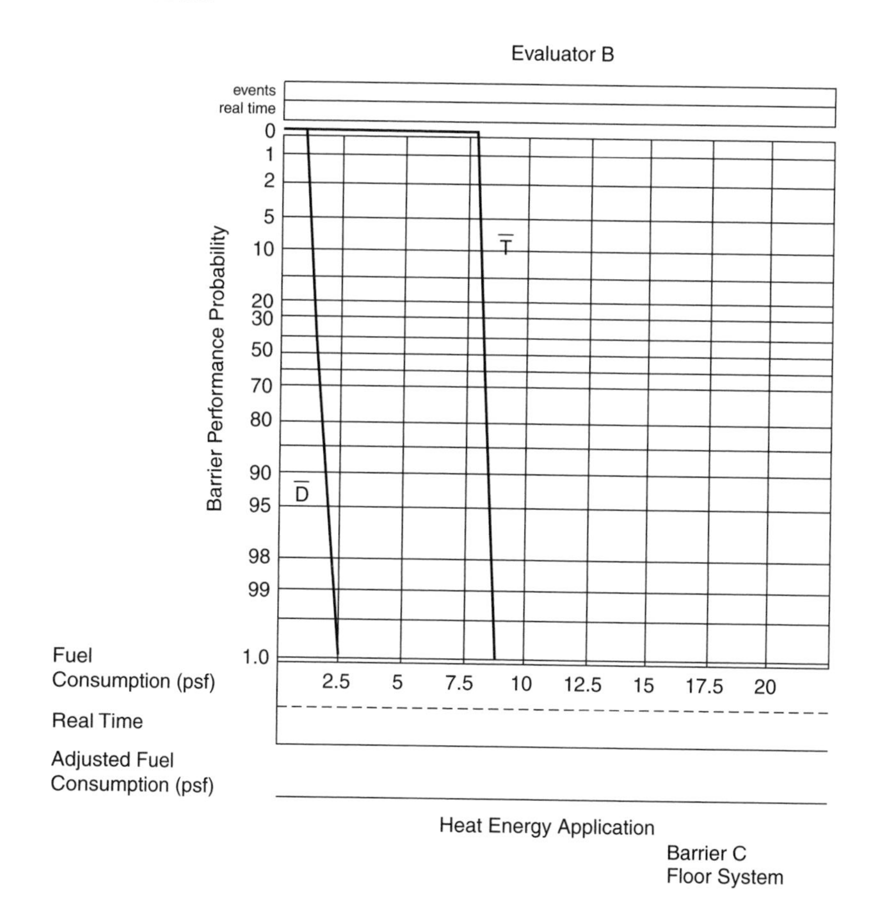

Figure 10.14

heat has had a chance to cause a $\overline{T}$ failure. Thus, Figure 10.14 illustrates expected behavior of unprotected steel joist construction.

A membrane ceiling acts as a barrier to delay elevated temperature increases in the plenum space that weaken the open web joists. Although temperature increases may cause substantial deflections, the stability should remain for the period that the ceiling remains in place. This type of construction benefits by the negative furnace pressure that keeps plenum temperatures lower. Because the $\overline{T}$ and $\overline{D}$ catalog curves envision behavior in a standard fire test, the furnace conditions and ceiling stability are incorporated into the performance shown in Figure 10.15. Barrier effectiveness factors (Section 10.18) recognize differences between test and building fire conditions.

10.15 Heat energy application

The horizontal axis of performance curves expresses heat energy as psf (kg/m^2) of fuel consumed. This measure is easy to observe and estimate in the field, and one can readily recognize when a fuel load will be large enough to challenge a barrier. In addition, because this heat energy measure is a part of all equivalency theories and calculation procedures, one can use any equivalency theory described in Section 10.8 as the basis for constructing catalog curves. In Section 10.12 we recommend using the Ingberg theory as a first approximation. If a performance problem seems to exist, a more careful analysis using NHL may give a better insight for the evaluation.

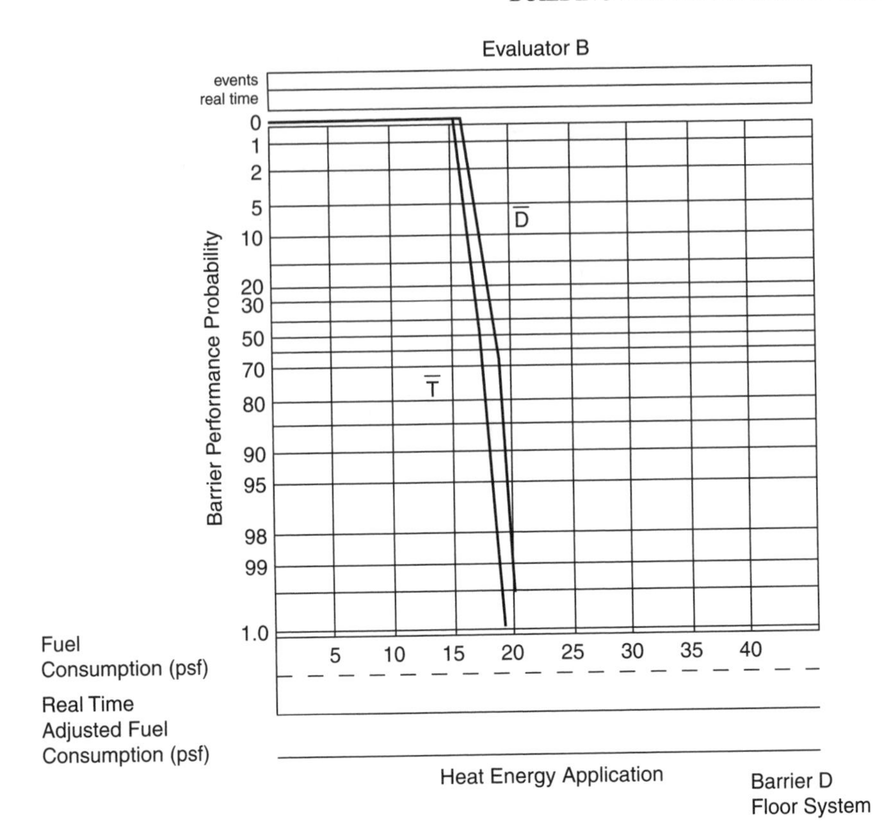

Figure 10.15

The behavior of the catalog $\overline{T}$ and $\overline{D}$ curves is based on heat energy that is applied to the barrier. The design fire identifies all of the heat generated. A conservative approach can assume that all this heat is applied directly to the barriers. However, we know that some of the total heat generated is lost through openings. The design fire can be converted into applied loading with a multiplier. This multiplier is based on observed ventilation and room size conditions and merely shifts fuel load values to reflect the effect of reduced fire severity.

10.16 Real-Time estimates

The fuel consumption time to barrier failure can be adapted to estimate "real" fire propagation times. It is not uncommon that a 10 min delay at a critical moment can give the fire department enough extra time to extinguish the fire manually and prevent extension into additional spaces. Consequently, we attempt to estimate the "actual" time for a barrier to experience $\overline{T}$ and $\overline{D}$ failures.

If one could believe that real time replicates the standard test time, it would be a simple manner to use the correlation described in Table 10.1. That is, for every 1 psf of fuel consumed, 6 min would elapse. However, when heat energy is released faster or slower than the standard fire test, real time to failure will be different than the test time.

A first approximation of real time can use the stoichiometric (i.e., perfect) burning of wood. This can be calculated as

$$m = 5.5A\sqrt{h}, \tag{10.10}$$

where

m = rate of fuel consumption (kg/min),
A = area of openings (m^2),
h = average height of openings (m).

Or in American units,

$$m = 8.86A\sqrt{h}, \qquad (10.10a)$$

where

m = rate of fuel consumption (lb/min),
A = area of openings (ft^2),
h = average height of openings (ft).

Additional lines on the abscissa of barrier performance curves enable one to relate the real time to barrier failure with the fuel loads.

10.17 Constructing catalog curves

The following procedure describes catalog curve construction

1. Identify the assembly and construction details. Gather available relevant information about material behavior at elevated temperatures, failure observations, heat transfer and structural calculations, and test data. Results of standard tests are very helpful. If no standard test data is available, attempt to find an analogy to a similar assembly that has been tested.
2. Using the available information, estimate the test time at which excessive temperature rise on the unexposed surface or material cracking will occur. Extrapolate this time to construct a $\overline{\mathrm{T}}$ curve of judgmental estimates of the likelihood of hot-spot failures.
3. Using the available information, estimate the test time at which a structural collapse or large degradation will occur. Extrapolate this time to construct a $\overline{\mathrm{D}}$ curve of judgmental estimates of the likelihood of a massive failure.
4. Convert these $\overline{\mathrm{T}}$ and $\overline{\mathrm{D}}$ curve estimates for standard test times into an associated fuel load consumption scale using the Ingberg theory. If one uses another theory, defined room conditions must be selected for the calculations.

Example 10.6

Construct catalog $\overline{\mathrm{T}}$ and $\overline{\mathrm{D}}$ curves for a wall built with 2 × 4 wood studs with two layers of 1/2 in gypsum board on each side.

Solution

This assembly is similar to Underwriters Laboratories barrier U301 [10]. The comparison of this unit and assembly U301 is as follows.

	Wall assembly	**UL-tested U301**
Construction	2 × 4@16 in c/c Two 1/2 in gypsum boards each side	2 × 4 fire-stopped @ 16 in c/c Two 5/8 in gypsum boards (UL classification mark)
Loading conditions	No superimposed load	Bearing wall
Expansion constraint	Unknown, but not a factor	
Finish rating	Unknown	66 min
Fire endurance rating	Unknown	2 h

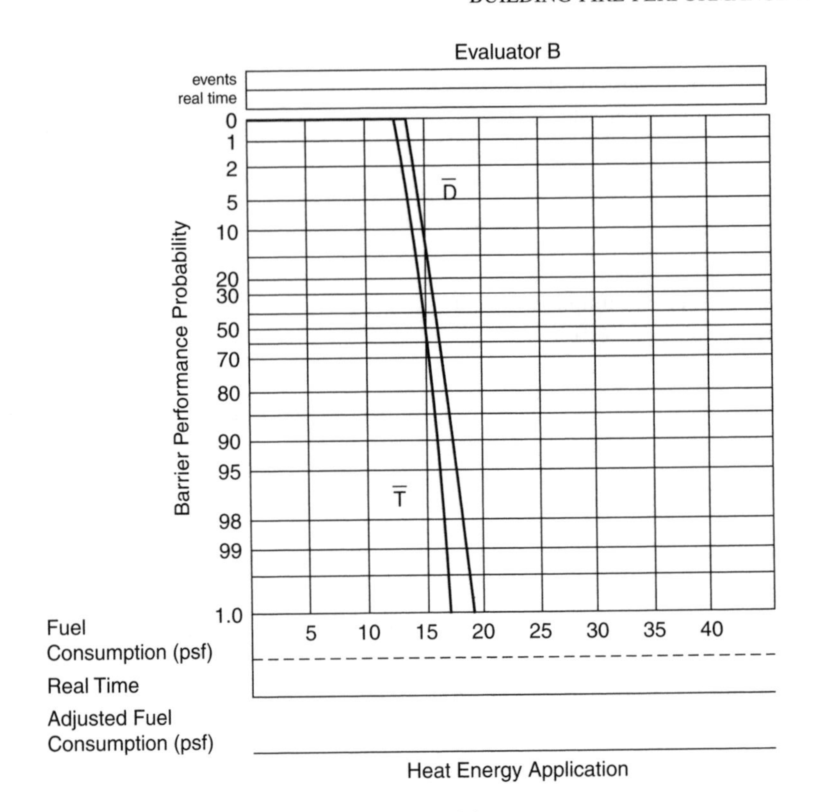

Figure 10.16

Figure 10.16 shows the catalog $\overline{T}$ and $\overline{D}$ curves. The reasoning for the selections is as follows:

- The 2 h fire endurance rating gives a sense that this barrier is substantial and will perform well in a fire. Typically, gypsum fails in a standard test by $\overline{T}$ cracks through the joints a few minutes after the desired rating is achieved. A few minutes thereafter the panels usually disintegrate and fall off, causing a $\overline{D}$ failure.
- Expansion is not an issue in these barriers.
- The finish time describes the time when the temperature in the wood nearest the fire rises to 250°F (average) or 325°F (individual) above ambient. This is close to ignition temperature, although a fire between studs will have relatively little oxygen. Nevertheless, wood drying and charring will reduce strength and the $\overline{D}$ capability somewhat.
- This type of construction is very common, and installation would be similar to that described in the UL *Fire Endurance Index* [10]. The main differences that may be expected might involve nail types, sizes, and spacing and care in fitting and taping. Normal field construction should not cause substantial differences in performance.

Example 10.7

A floor assembly is shown in Figure 10.17. The structural system is composed of W 12 × 14 steel beams spaced 4 ft 0 in c/c supporting a 3 1/2 in concrete slab on cold-formed steel deck. Construct catalog $\overline{T}$ and $\overline{D}$ curves for the following beam conditions:

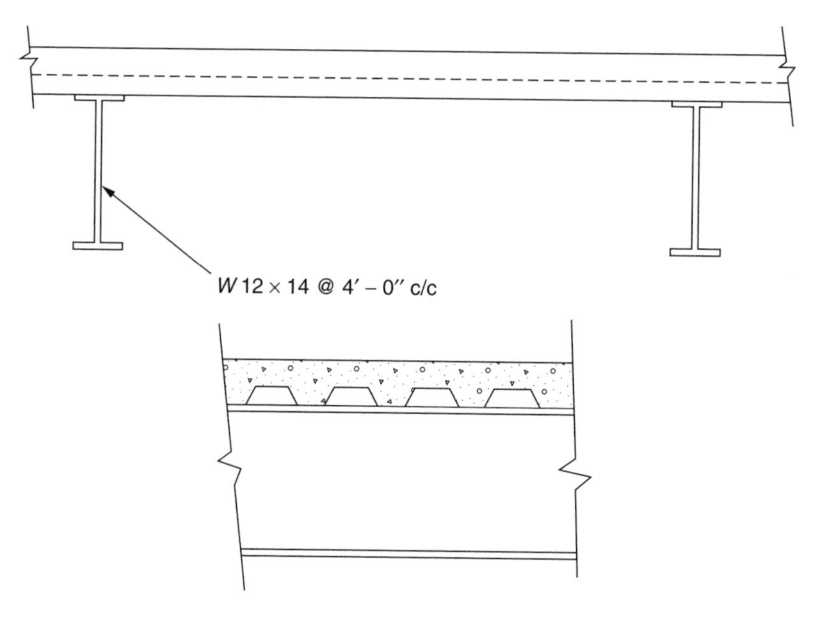

Figure 10.17

(a) unprotected steel beam;
(b) steel beam protected by 1/2 in of spray-applied insulation;
(c) steel beam protected by 1 in spray-applied insulation.

Solution

Because this assembly is not rated, we do not have any published test data. Condition (c) is somewhat similar to UL assembly D907 that uses 1 in of spray-applied insulation, although we do not know if the insulation is the same as that of the UL assembly. The D907 beam has an unrestrained fire endurance of 1 hr. Other beams have ratings of 1 1/2 to 2 h.

These catalog curves will be based on structural calculations that are described in Chapter 15. The structural analysis calculated the strength reduction when the beams were subjected to a fire environment defined by the ASTM standard time–temperature curve. Figure 10.18 shows the ratio of ultimate load to dead load (w_U/w_{DL}) for test time durations. When this ratio becomes 1.0, the assembly will collapse under its own weight. Figure 10.19 shows the catalog $\overline{\mathrm{T}}$ and $\overline{\mathrm{D}}$ curves.

The reasoning for the selections is as follows:

- When the unprotected 17 ft span beam supporting only the construction dead loads is exposed to the standard time–temperature curve, calculations (Chapter 15) indicate a collapse time of about 16 min. This would cause a $\overline{\mathrm{D}}$ failure. The equivalent fuel consumption using the Ingberg theory is are shown on Figure 10.19.
- An average ambient temperature rise on the unexposed side of 250°F (181°C) would cause a $\overline{\mathrm{T}}$ failure. However, a question may arise about concrete slab cracks before this thermal penetration occurs. It is speculated that the temperature reinforcement combined with the cold-formed steel deck will maintain slab integrity for the time until thermal transmission occurs.

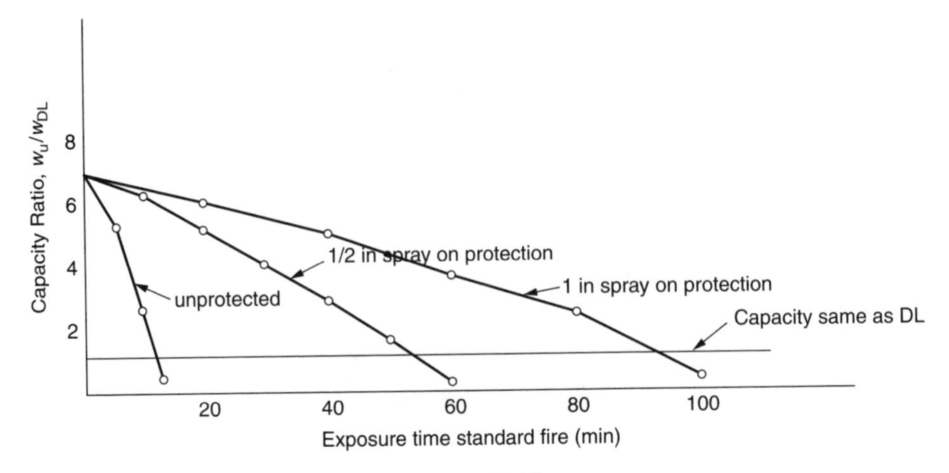

Figure 10.18

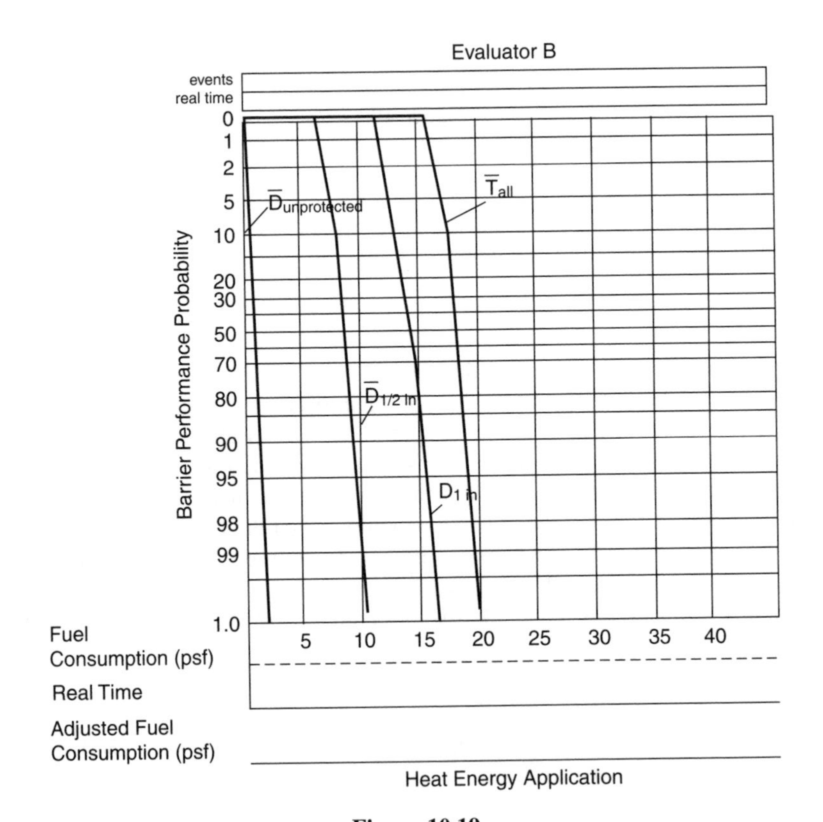

Figure 10.19

- Catalog curves assume that the structural supports remain in place and do not deflect. If cracking is anticipated because of construction support deflections, the reduction in $\overline{\mathrm{T}}$ capability can be incorporated in the barrier effectiveness factor for construction C_C (Section 10.17)
- The $\overline{\mathrm{D}}$ failures will precede a $\overline{\mathrm{T}}$ failure for these assemblies.

10.18 Barrier effectiveness factors

The objective of a barrier evaluation is to understand the way in which the barriers will perform in a building fire. Often this can be as simple as deciding that the construction and field conditions will provide more resistance than needed by the fuel load, and no formal analysis is needed. Detailed analysis is used only when conditions indicate that more information is needed.

The analysis may appear to be very complex, and from a theoretical viewpoint it is. However, the process can be made relatively easily by adjusting catalog $\overline{\mathrm{T}}$ and $\overline{\mathrm{D}}$ performance curves with barrier effectiveness factors that reflect field conditions.

A barrier effectiveness factor is a multiplier that reflects expected change (reduction or increase) in $\overline{\mathrm{T}}$ or $\overline{\mathrm{D}}$ performance due to field conditions. If a barrier is expected to behave the same as the performance curves in a field application, the effectiveness factor is 1.0. However, some construction practices or conditions may make the barrier stronger or weaker than would be indicated by the catalog curves. The effectiveness factor reflects a judgment of the derating or prorating that can be expected.

The influence of the factors is discussed below. To describe the process and give a sense of proportion, value ranges are shown. These values reflect the author's judgment. They are not based on experimentation or group discussions, and the synergism of combining the factors has not been investigated.

INFLUENCE OF APPLIED LOADING

Catalog $\overline{\mathrm{T}}$ and $\overline{\mathrm{D}}$ curves are based on an unloaded (i.e., only dead load) condition. Superimposed loads certainly reduce the $\overline{\mathrm{D}}$ performance and may reduce the $\overline{\mathrm{T}}$ behavior. The magnitude of $\overline{\mathrm{D}}$ reduction can be estimated by structural and heat transfer analysis. The following estimates seem reasonable for routine construction, where C_{L} is the barrier effectiveness factor for applied loading:

$$\text{no load, } C_{\mathrm{L}} = 1.0;$$
$$\overline{\mathrm{D}}, 0.4 < C_{\mathrm{L}} < 1.0;$$
$$\overline{\mathrm{T}}, 0.9 < C_{\mathrm{L}} \leq 1.0.$$

INFLUENCE OF THERMAL RESTRAINT

Catalog $\overline{\mathrm{T}}$ and $\overline{\mathrm{D}}$ curves are based on free movement of the ends when thermal expansion occurs. If the ends are restrained against thermal expansion, they become prestressed. This condition can result in substantial increases in $\overline{\mathrm{D}}$ capabilities and some increase in $\overline{\mathrm{T}}$ resistance. On the other hand, some assemblies may have no change in performance whereas others may experience reduction in resistance. The direction and magnitude depend upon the materials and construction details.

The building construction will indicate the appropriateness of using a thermal constraint effectiveness factor. For example, if a fire occurs in an interior room, the nonfire parts of a large building will provide lateral restraint and it is appropriate to increase the $\overline{\mathrm{D}}$ and perhaps the $\overline{\mathrm{T}}$ behavior. On the other hand, that same construction in an exterior bay is unlikely to have much lateral restraint. Consequently, the C_{L} factor would remain 1.0. Some construction systems, such as prestressed concrete, may experience a reduction in strength when thermal expansion is restricted.

The following estimate ranges seem appropriate for routine construction, where C_{T} is the barrier effectiveness factor for thermal restraint:

$$\text{no thermal restraint, } C_{\mathrm{T}} = 1.0;$$
$$\overline{\mathrm{D}}, 0.7 < C_{\mathrm{T}} < 1.5;$$
$$\overline{\mathrm{T}}, 0.9 < C_{\mathrm{T}} \leq 1.0.$$

INFLUENCE OF CONSTRUCTION

A variety of conditions exist in the standard laboratory test that are not necessarily the same as at the construction site. For example, the laboratory can only test statically determinate systems. Modern construction regularly uses continuous construction that provides increased strength against collapse. Continuous frame construction will increase the $\overline{\mathrm{D}}$ capability, and perhaps the $\overline{\mathrm{T}}$ performance because of decreased deflections.

Another aspect of construction involves scale (i.e., size). The laboratory test involves surface areas of $100\,\mathrm{ft}^2$ (walls, partitions) or $220\,\mathrm{ft}^2$ (floors). Limited information seems to suggest that larger surface areas crack more readily than the laboratory sizes. If this is the case, values of C_C for $\overline{\mathrm{T}}$ will be less than 1.0. It would appear that $\overline{\mathrm{D}}$ values for C_C should also be less than 1.0, where C_C is the barrier effectiveness factor for construction.

Structural deflections can cause cracking of supported elements, such as slabs, ceilings, and panels. C_C incorporates this condition into performance estimates.

Suspended membrane ceilings often provide thermal protection for floor–ceiling assemblies. When they remain in place, they can be effective. Ceiling systems with designed expansion capabilities seem to behave well. However, when expansion is not controlled, deflections in the ceiling support structure cause tiles to fall. Still another concern involves ceiling utility fixtures that allow heat penetration. UL-tested ceiling support structures and fixtures seem to perform well. However, untested fixtures may be suspect when they appear in older buildings or structures in which approved "or equal" substitutes may not have been used.

Estimate ranges are difficult to provide because one feature (e.g., construction continuity with $C_\mathrm{C} > 1.0$) may be offset by a poor suspended ceiling (with $C_\mathrm{C} < 1.0$). Nevertheless, the following estimate ranges seem appropriate (with much less confidence) for routine construction:

no change in condition, $C_\mathrm{C} = 1.0$;
$\overline{\mathrm{D}}$, $0.3 < C_\mathrm{C} < 1.5$;
$\overline{\mathrm{T}}$, $0.8 < C_\mathrm{C} \leq 1.0$.

INFLUENCE OF OPENINGS

Openings and their protectives, if present, have a major effect on fire endurance. Because $\overline{\mathrm{T}}$ and $\overline{\mathrm{D}}$ curves represent field performance, their values can be set at 0.0 when openings are not protected. The following estimate ranges are appropriate, where C_O is the barrier effectiveness factor for openings:

no openings, $C_\mathrm{O} = 1.0$;
$\overline{\mathrm{D}}$, $0.0 < C_\mathrm{O} < 1.0$;
$\overline{\mathrm{T}}$, $0.0 < C_\mathrm{O} \leq 1.0$.

The function of selecting barrier effectiveness factors is to encourage one to look at building construction details as they affect $\overline{\mathrm{T}}$ and $\overline{\mathrm{D}}$ barrier performance. It encodes conditions that cause the barrier to perform differently than the standard fire endurance tests might imply. In effect, the barrier effectiveness factor translates the standard fire test into field performance $\overline{\mathrm{T}}$ and $\overline{\mathrm{D}}$ barrier curves. The goal is to help one understand building performance better and to document expectations. The factor may be expressed functionally as

Barrier effectiveness factor = f(applied loadings, thermal restraint, construction features, and openings)

$$C_\mathrm{e} = C_\mathrm{L} C_\mathrm{T} C_\mathrm{C} C_\mathrm{O} \tag{10.11}$$

10.19 Automatic protection equipment

A small number of barriers are protected by automatic suppression equipment. These barriers fall into two groups. The first involves discharge of water onto physical barriers, such as glass. Exterior sprinklers or water distribution on glass panels are of this type. A second situation occurs when a zero-strength barrier (e.g., escalator opening or open space division) uses a water curtain to provide barrier protection. Water distribution equipment defends barrier openings by the automatic discharge of agents to delay or prevent movement of flame/heat into the adjacent space.

The first situation could use a barrier effectiveness factor to increase the fire endurance of the glass partition. However, the factors would be very large. Reliability of the sprinkler system must also be examined. The second case cannot use a multiplier because the $\overline{\mathrm{T}}$ and $\overline{\mathrm{D}}$ curves both start at zero. Therefore it is suggested that these special cases be evaluated as separate entities. The curves will reflect a judgmental opinion of the effectiveness of the automatic system to prevent $\overline{\mathrm{T}}$ or $\overline{\mathrm{D}}$ failures into the unexposed spaces.

10.20 Field performance $\overline{T}$ and $\overline{D}$ curves

Catalog curves reflect a situation that depicts an estimation of the likely performance of the barrier if it were subjected to a standard test fire with the conditions of no superimposed load and no thermal restraint. Field performance is determined by multiplying the catalog curve coordinates by a barrier effectiveness factor that reflects conditions which may be observed for existing buildings in the field or may be anticipated in the plans for a proposed building. The following examples illustrate the process.

Example 10.8

An individual has inspected a building and the construction is appropriate for the catalog curves of Figure 10.20. Construct field $\overline{\mathrm{T}}$ and $\overline{\mathrm{D}}$ curves to reflect the expected performance of the barrier. The barrier effectiveness was evaluated in the following manner.

Applied loads

The superimposed dead and live loads were estimated as never exceeding 75% of the code requirements. The following adjustment values were selected:

$$\overline{\mathrm{D}}: C_{\mathrm{L}} = 0.6;$$
$$\overline{\mathrm{T}}: C_{\mathrm{L}} = 1.0.$$

Thermal restraint

The axial restraint from other parts of the building will improve $\overline{\mathrm{D}}$ somewhat but will have no effect on $\overline{\mathrm{T}}$. The following adjustment factors account for these conditions:

$$\overline{\mathrm{D}}: C_{\mathrm{T}} = 1.2;$$
$$\overline{\mathrm{T}}: C_{\mathrm{T}} = 1.0.$$

Construction

The floor system was of continuous construction, which improves expected $\overline{\mathrm{D}}$ performance. However, the size and expected deflections will reduce $\overline{\mathrm{T}}$ more significantly and $\overline{\mathrm{D}}$ slightly. The combined effect gives the following estimates:

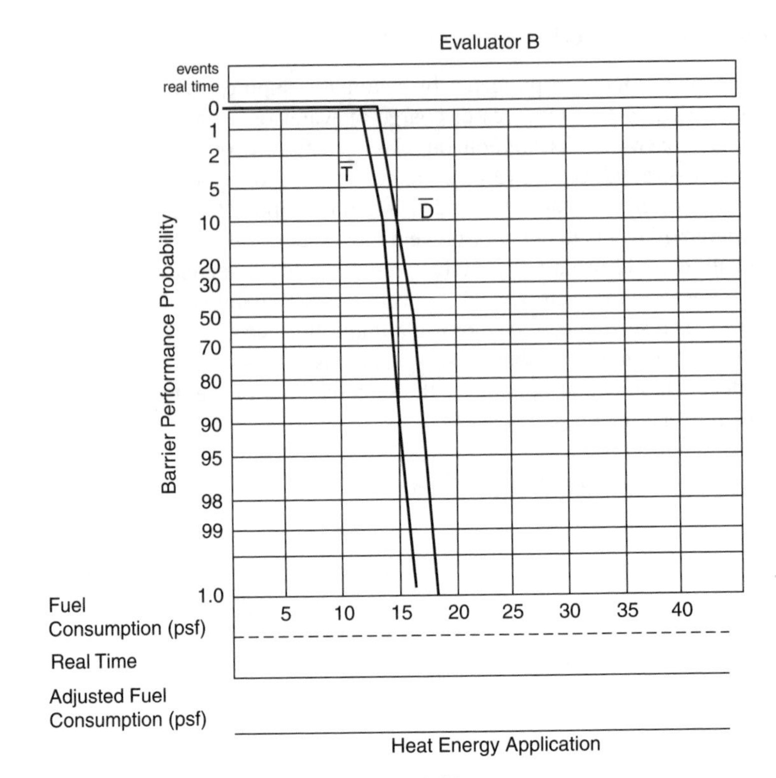

Figure 10.20

$$\overline{\mathrm{D}} : C_{\mathrm{C}} = 1.1;$$
$$\overline{\mathrm{T}} : C_{\mathrm{T}} = 0.7.$$

Openings

There are no through construction openings in the assembly:

$$\overline{\mathrm{D}} : C_{\mathrm{O}} = 1.0;$$
$$\overline{\mathrm{T}} : C_{\mathrm{O}} = 1.0.$$

Solution

The barrier effectiveness factors for the modes of failure are as follows.

For $\overline{D}$

$$\begin{aligned} C_{\mathrm{e}} &= C_{\mathrm{L}} C_{\mathrm{T}} C_{\mathrm{C}} C_{\mathrm{O}} \\ &= (0.6)(1.2)(1.1)(1.0) = 0.79. \end{aligned}$$

For $\overline{T}$

$$\begin{aligned} C_{\mathrm{e}} &= C_{\mathrm{L}} C_{\mathrm{T}} C_{\mathrm{C}} C_{\mathrm{O}} \\ &= (1.0)(1.0)(0.7)(1.0) = 0.7. \end{aligned}$$

Table 10.4

Catalog $\overline{D}$ coordinate (lb/ft^2)	Barrier effectiveness factor	Field performance $\overline{D}$ coordinate (lb/ft^2)
14 @ 0%	0.79	11.1
15 @ 10%	0.79	11.8
16.5 @ 50%	0.79	13
17 @ 90%	0.79	13.4
18 @ 100%	0.79	14.2
Catalog $\overline{T}$ coordinate (lb/ft^2)	**Barrier effectiveness factor**	**Field performance $\overline{T}$ coordinate (lb/ft^2)**
12.5 @ 0%	0.7	8.8
13.5 @ 10%	0.7	9.4
14.5 @ 50%	0.7	10.1
15 @ 90%	0.7	10.5
16 @ 100%	0.7	11.2

The catalog $\overline{T}$ and $\overline{D}$ coordinates are multiplied by the barrier effectiveness factors of 0.70 and 0.79, respectively. Several values are shown in Table 10.4. The field performance curves are shown in Figure 10.21(a) and the curves adjusted for cohesiveness are shown in Figure 10.21(b).

Example 10.9

Assume that the inspector for the building of Example 10.8 discovered several rooms of the same floor construction with through construction electrical cable holes. The only fire propagation protection was the floor outlets. Construct field $\overline{T}$ and $\overline{D}$ curves to reflect the expected performance of these barriers. The barrier effectiveness was evaluated in the following manner.

Applied loads

Same as Example 10.8:

$$\overline{D} : C_L = 0.6;$$
$$\overline{T} : C_L = 1.0.$$

Thermal restraint

Same as Example 10.8:

$$\overline{D} : C_T = 1.2;$$

$$\overline{T} : C_T = 1.0.$$

Construction

Same as Example 10.8:

$$\overline{D} : C_C = 1.1;$$
$$\overline{T} : C_T = 0.7.$$

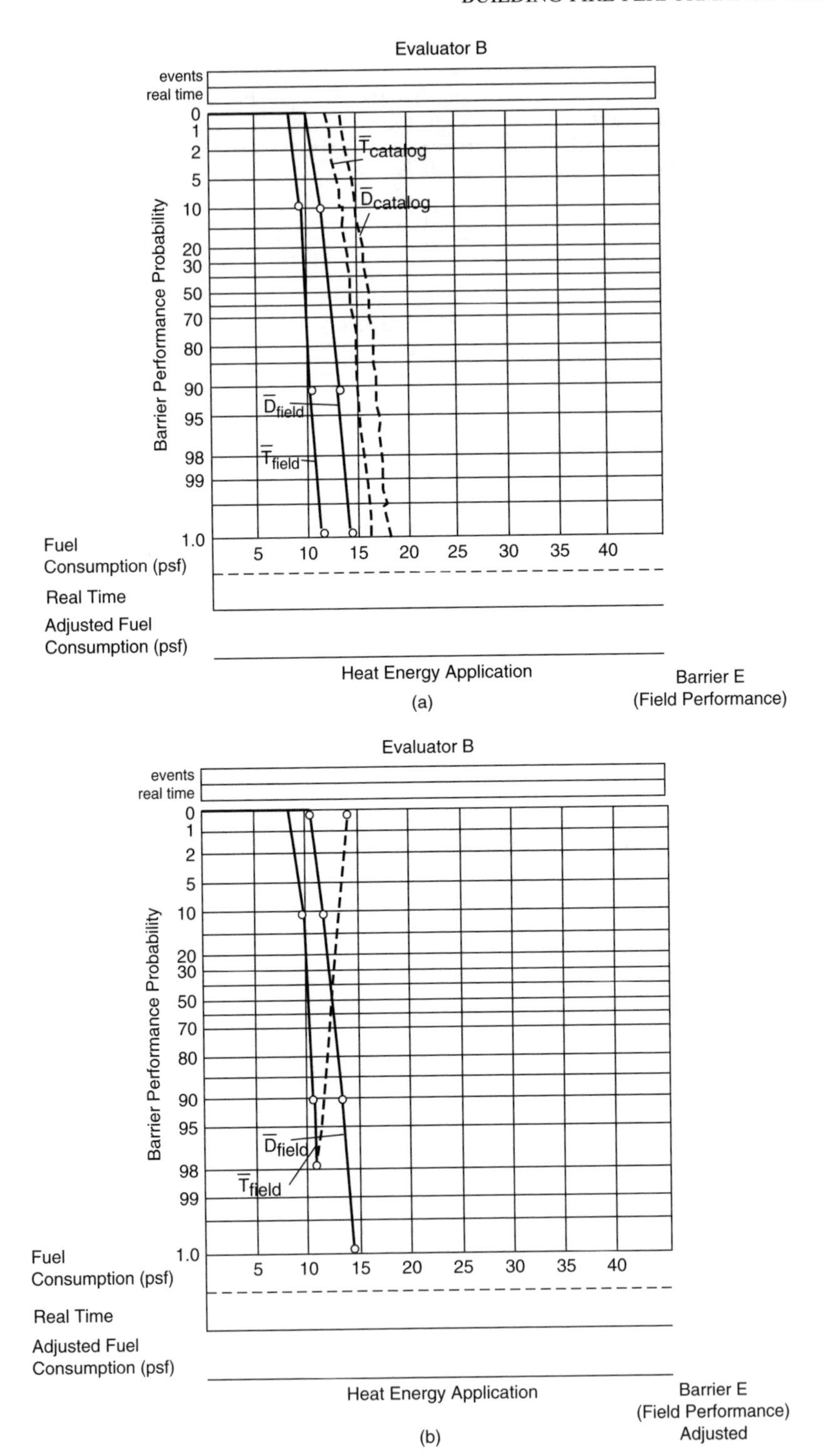
Evaluator B
events
real time
Barrier Performance Probability
0
1
2
5
10
20
30
50
70
80
90
95
98
99
1.0
$\overline{T}_{catalog}$
$\overline{D}_{catalog}$
$\overline{D}_{field}$
$\overline{T}_{field}$
Fuel Consumption (psf)
5 10 15 20 25 30 35 40
Real Time
Adjusted Fuel Consumption (psf)
Heat Energy Application
Barrier E
(Field Performance)
(a)
Evaluator B
events
real time
Barrier Performance Probability
0
1
2
5
10
20
30
50
70
80
90
95
98
99
1.0
$\overline{D}_{field}$
$\overline{T}_{field}$
Fuel Consumption (psf)
5 10 15 20 25 30 35 40
Real Time
Adjusted Fuel Consumption (psf)
Heat Energy Application
Barrier E
(Field Performance)
Adjusted
(b)

Figure 10.21

Openings

The unprotected through construction holes have only minimal resistance to a $\overline{\mathrm{T}}$ failure. The structural capacity is reduced so little that $\overline{\mathrm{D}}$ is estimated as being unaffected. The following adjustment factors were selected:

- $\overline{\mathrm{D}}: C_{\mathrm{O}} = 1.0$;
- $\overline{\mathrm{T}}: C_{\mathrm{O}} = 0.05$.

Solution

The barrier effectiveness factors for the modes of failure are as follows.

For $\overline{D}$

$$\begin{aligned} C_{\mathrm{e}} &= C_{\mathrm{L}}C_{\mathrm{T}}C_{\mathrm{C}}C_{\mathrm{O}} \\ &= (0.6)(1.2)(1.1)(1.0) = 0.79. \end{aligned}$$

For $\overline{T}$

$$\begin{aligned} C_{\mathrm{e}} &= C_{\mathrm{L}}C_{\mathrm{T}}C_{\mathrm{C}}C_{\mathrm{O}} \\ &= (1.0)(1.0)(0.7)(0.05) = 0.035. \end{aligned}$$

The $\overline{\mathrm{T}}$ and $\overline{\mathrm{D}}$ coordinates are multiplied by the barrier effectiveness factors of 0.035 and 0.79, respectively. Several values are shown in Table 10.5. The field performance curves are shown in Figure 10.22(a) and the curves adjusted for cohesiveness are shown in Figure 10.22(b).

Example 10.10

Assume that the fuel loading in the spaces is 8.5 psf. What can one expect for the performance of the barriers in Examples 10.8 and 10.9?

Table 10.5

Catalog $\overline{\mathrm{D}}$ coordinate (lb/ft^2)	Barrier effectiveness factor	Field performance $\overline{\mathrm{D}}$ coordinate (lb/ft^2)
14 @ 0%	0.79	11.1
15 @ 10%	0.79	11.8
16.5 @ 50%	0.79	13
17 @ 90%	0.79	13.4
18 @ 100%	0.79	14.2
Catalog $\overline{\mathrm{T}}$ coordinate (lb/ft^2)	**Barrier effectiveness factor**	**Field performance $\overline{\mathrm{T}}$ coordinate (lb/ft^2)**
12.5 @ 0%	0.035	0.44
13.5 @ 10%	0.035	0.47
14.5 @ 50%	0.035	0.51
15 @ 90%	0.035	0.52
16 @ 100%	0.035	0.56

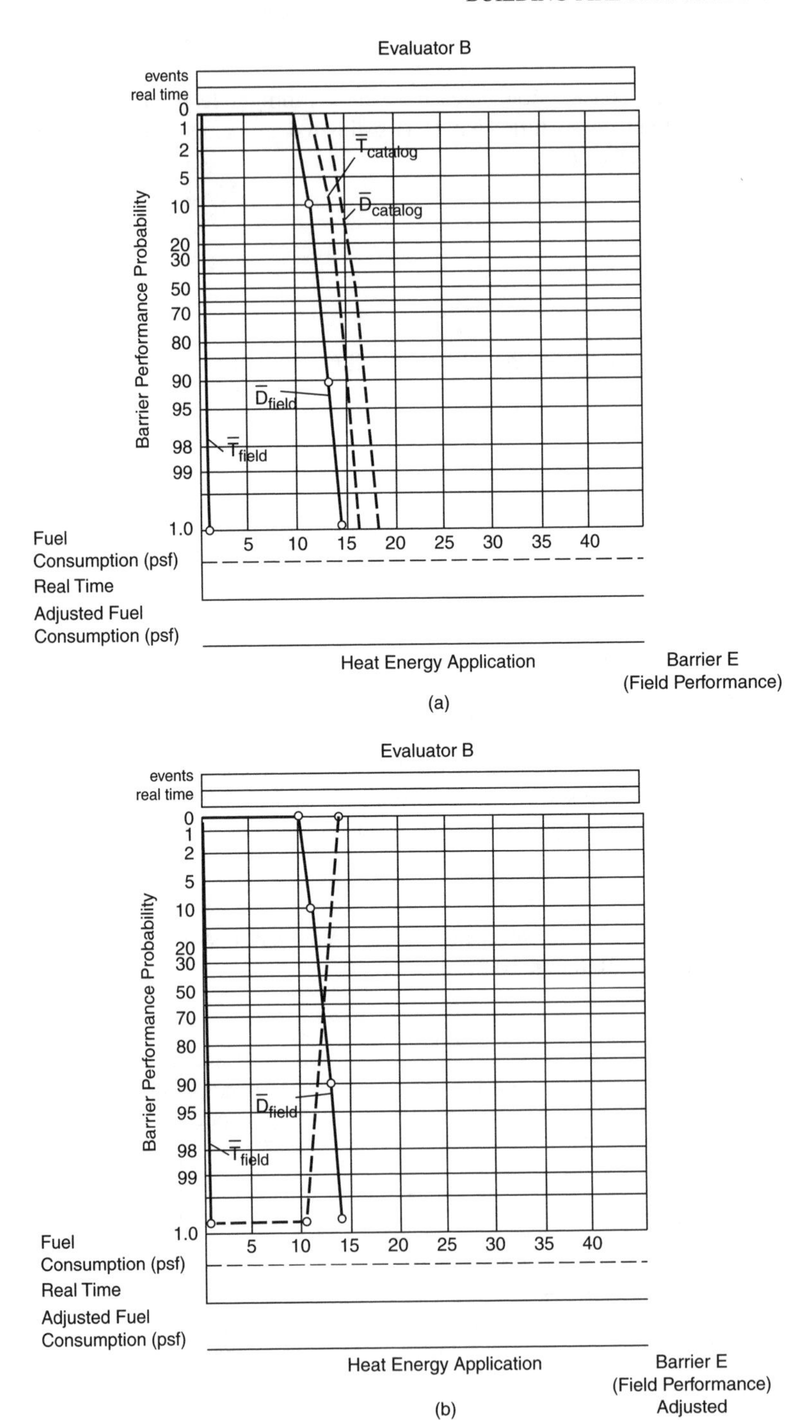
Evaluator B
events
real time
Barrier Performance Probability
0
1
2
5
10
20
30
50
70
80
90
95
98
99
1.0
$\overline{T}_{catalog}$
$\overline{D}_{catalog}$
$\overline{D}_{field}$
$\overline{T}_{field}$
5 10 15 20 25 30 35 40
Fuel Consumption (psf)
Real Time
Adjusted Fuel Consumption (psf)
Heat Energy Application
Barrier E
(Field Performance)
(a)
Evaluator B
events
real time
Barrier Performance Probability
0
1
2
5
10
20
30
50
70
80
90
95
98
99
1.0
$\overline{D}_{field}$
$\overline{T}_{field}$
5 10 15 20 25 30 35 40
Fuel Consumption (psf)
Real Time
Adjusted Fuel Consumption (psf)
Heat Energy Application
Barrier E
(Field Performance)
Adjusted
(b)

Figure 10.22

Solution

If complete burnout occurred, one would expect that the barrier described by the field curves of Figure 10.21(b) would not experience any failure during the fire. The barrier of Figure 10.22(b) would certainly experience a $\overline{\text{T}}$ failure but not a $\overline{\text{D}}$ failure.

10.21 Applied heat energy

Room openings affect the burning time (Section 10.16), the maximum temperature, and the amount of heat energy that is transferred out of the room. Thus, the heat energy that is actually applied to a barrier will be less than the heat that is generated. The fuel load defines the total heat energy that is available to attack barriers. The design fire identifies the cumulative total heat energy released by the fire. This value may be adjusted to reflect heat loss through openings that does not attack the barrier. This correction factor may be determined from one of the equivalent exposure theories if its value justifies the cost of analysis.

10.22 Combustible barriers

The catalog and field performance $\overline{\text{T}}$ and $\overline{\text{D}}$ curves described above assume that the exposed barrier surface is noncombustible. Many barriers found in buildings are combustible or have combustible elements such as wooden doors or panels. These barriers often delay fire propagation long enough to enable the fire service to extinguish the fire. The $\overline{\text{T}}$ and $\overline{\text{D}}$ catalog and field performance curves can reflect these conditions by changing the abscissa to show real time.

Catalog curves for combustible barriers are developed in the same manner as for noncombustible barriers with the exception that the probability values are established on the basis of time. The question to ask in determining coordinates is, Given X value of time duration for the fire, what is the probability that a hot-spot $\overline{\text{T}}$ or a massive $\overline{\text{D}}$ ignition can appear on the unexposed side of this barrier?

The $\overline{\text{T}}$ and $\overline{\text{D}}$ curves for combustible barriers are easier to construct because performance can be calculated with greater confidence.

Example 10.11

A barrier is constructed using 2 × 4 studs with a 1/4 in plywood panel on one side and a 3/4 in panel on the other side, as shown in Figure 10.23. Construct $\overline{\text{T}}$ and $\overline{\text{D}}$ curves for this barrier.

Solution

(a) Consider first that the exposure is on the side with 1/4 in panels. If the plywood burns at 1/40 in/min, then the time to penetrate the panel is about 10 min. At this time the following two actions will occur.

1. The 3/4 in panel will be attacked by the fire. Using the same burn rate, one can expect a burn through about 30 min after the panel begins to burn. The total time for the first $\overline{\text{T}}$ failure would be about 10 min + 30 min, or about 40 min after the start of the test fire. As the panel continues to deteriorate, the $\overline{\text{T}}$ holes will enlarge to produce $\overline{\text{D}}$ failures. Estimate the $\overline{\text{D}}$ failure to be well established about 2 min after the $\overline{\text{T}}$ failure occurs. This would be about 42 min after the start.

2. The studs will ignite and burn. The char will deepen and begin to weaken the load-carrying capacity. Eventually the studs will collapse, causing a $\overline{\text{D}}$ failure. Using the same burn rate on both sides of the stud, the 1 1/2 in thickness will be consumed after about 30 min. Adding the 10 min to penetrate the 1/4 in panel indicates the studs would be consumed about 10 min + 30 min = 40 min

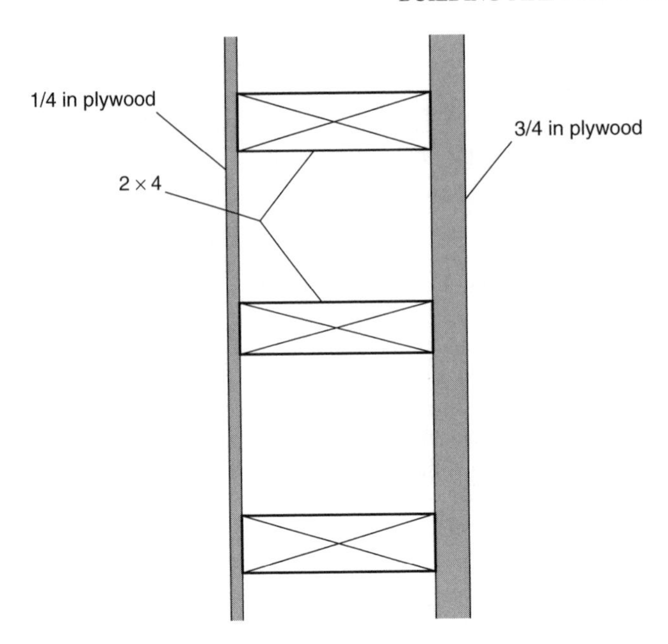

Figure 10.23

after the fire start. Actual collapse would occur somewhat more quickly because the studs would collapse before burn through. The 3/4 in panel will provide some support for the studs, but not much.

The $\overline{\text{D}}$ failure times of 42 min for the 3/4 in panel burn through and perhaps 36 min for stud collapse indicate that $\overline{\text{D}}$ and $\overline{\text{T}}$ would occur at about the same time as shown in Figure 10.24(a).

Note that the catalog $\overline{\text{D}}$ was based on no superimposed load, such as a nonbearing partition. If this were a load-bearing partition, collapse would occur earlier. If the superimposed load were the allowable value for a 2 × 4 Douglas fir no. 2 partition 8 ft high (2200 plf), one could expect collapse after about 10–12 min. The time after the start of the fire is 10 min + 10 min = 20 min. This would indicate an applied load factor of $C_{\text{L}} = 0.3$ for the $\overline{\text{D}}$ curve of Figure 10.24(a).

(b) Consider now that the fire starts on the 3/4 in side. Using the same 1/40 in/min rate, the 3/4 in panel will burn through in 30 min. Continuing to the 1/4 in panel, another 10 min could be expected. This provides about 40 min to initiate a $\overline{\text{T}}$ failure. Thus, the $\overline{\text{T}}$ curves could be expected to be the same, regardless of the location of the fire.

Again using the time for destroying the studs as 30 min, the $\overline{\text{D}}$ time for stud disintegration is 30 min + 30 min = 60 min. Even if the studs collapse slightly sooner, a time of around 50 to 55 min seems reasonable. Thus, $\overline{\text{D}}$ for stud collapse would be longer than the $\overline{\text{T}}$ for penetrating through the two panels (30 min + 10 min ≈ 40–42 min). Figure 10.24(b) shows these catalog curves.

The catalog curves indicate about the same time for a $\overline{\text{T}}$ failure, but the mechanism is very different. This would be evident in the barrier being able to support applied loads for a short time longer.

10.23 Summary

The goal of a performance evaluation is to understand how a building will behave. The simplicity of using test endurance ratings with prescriptive codes is inadequate for performance applications

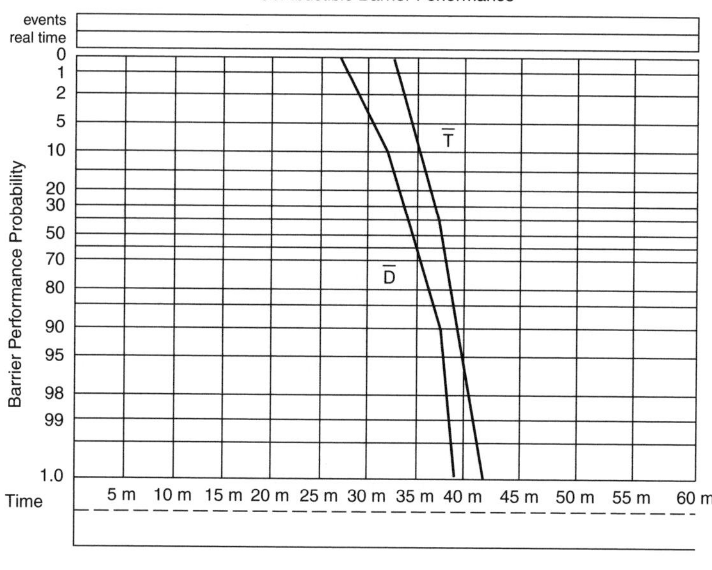

(a)

Wood partition:
Fire on ¼ in side

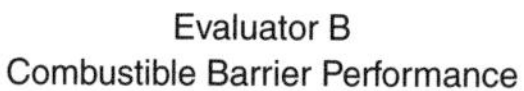

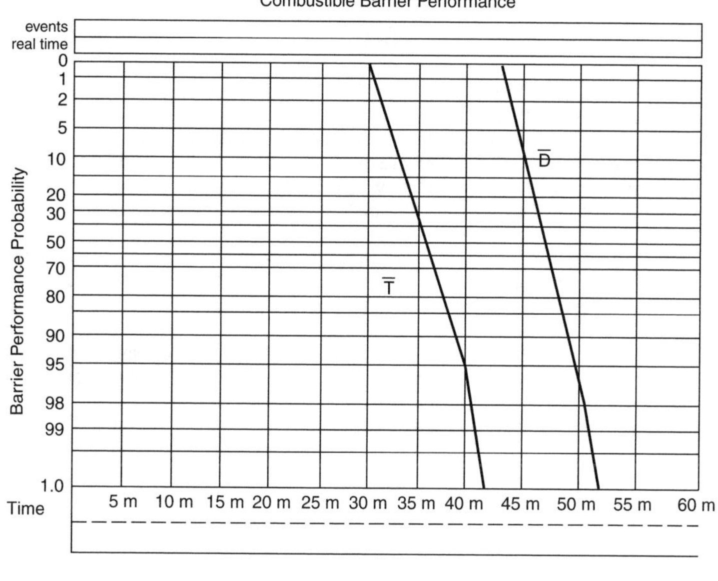

(b)

Wood partition:
Fire starts on ¾ in side

Figure 10.24

because the fire endurance rating is not as important as barrier holes. This chapter notes weaknesses in historical code practices and reliance on fire endurance equivalency. Whether one uses prescriptive methods or performance design, those weaknesses remain. Performance understanding requires one to examine construction practices and normal functional uses of barriers. The process is more rigorous than prescriptive practices, but the process enables one to describe barrier weaknesses more clearly.

Level 1 evaluations (Chapter 9) enable one to recognize likely paths and ease of multiroom fire propagation. Initial recognition of the type and locations of barrier weaknesses gives a good understanding of the complete fire performance. Often a level 1 analysis is satisfactory for performance evaluation needs. When one requires a more sophisticated performance knowledge, the procedures can be augmented in level 2 and level 3 analyses.

In most buildings the number of different barrier materials and constructions are very small. However these barriers incorporate many different penetration types, sizes, locations, and protectives, as well as construction boundary conditions. To simplify the process of barrier evaluation for field conditions, a two-step process is used.

1. Select (standard) catalog $\overline{\mathrm{T}}$ and $\overline{\mathrm{D}}$ curves for the barrier.
2. Adjust the catalog curves with a barrier effectiveness factor, C_e, to reflect the influence of field conditions on performance.

Most of the effort goes into constructing catalog $\overline{\mathrm{T}}$ and $\overline{\mathrm{D}}$ curves to describe the incomplete range of behavior. This involves a blend of theory, material and structural behavior, and judgment. However, this is a one-time investment because the catalog curves can be used for all future applications. The barrier effectiveness factor enables one to produce field performance estimates easily and quickly.

References

1. Paulson, O., *On heat transfer in test furnaces.* Technical University of Denmark, Lyngby, 1979.
2 Ingberg, S. H. Tests of the severity of building fires. *NFPA Quarterly*, Col. 22, No. 1, July 1928.
3. Ingberg, S. H. The severity of fires in buildings. *Architectural Engineering and Business*, May 1929.
4. Fire Resistance Classifications of Building Constructions, Building Materials and Structures Report 92, Superintendent of Documents, Washington, D.C., October 1942, May 1929.
5. Law, M. A relationship between fire grading and building design and contents. JFRO Internal Note No. 374 (1970).
6. Pettersson, O. The connection between a real fire exposure and the heating conditions according to standard fire resistance tests–with special application to steel structures. ECCS, CECM-III-74-2E, Chapter II, 1974.
7. Harmathy, T. Ż. The fire resistance test and its relation in real world fires. *Fire and Materials*, Vol. **5**, No. 3, 1981.
8. Newton, B. A. Analysis of common fire severity methods and barrier performance. MS thesis, WPI, 1994.
9. Culver, C. G. Survey results for fire loads and live loads in office buildings. Technical Report NBS Building Science Series 85, May 1976.
10. *UL Fire Resistance Directory*, Vol. 1. Underwriters Laboratory, Northbrook, Illinois, 2001.

11 DETECTION AND INITIAL ACTIONS

11.1 Introduction

Detection is the event that launches the building's fire response. It is a critical event in manual suppression and in alerting occupants for emergency actions involving life safety. Detection can also play a role in other fire defenses by initiating actions such as closing doors, recalling elevators, changing operations of air-handling systems, or triggering equipment operation. Fire detection is important to the general risk management of a building.

The fire size at detection is the starting condition for transferring actions to another component of the fire defense system. For example, after detection and the subsequent fire department notification, the focus shifts to the manual suppression evaluation. Similarly, detection and alerting the occupants is a package that must precede life safety actions. Time durations, operational effectiveness, and activities that follow detection are central to a performance evaluation.

Detection may involve humans only, equipment only, or both. Equipment evaluations can be relatively methodical and orderly, but human interactions involve greater variability and uncertainty. The level 2 analysis structures the process, and the framework portrays the effect of conditions that affect performance. Level 3 evaluations allow one to examine the sensitivity of the detection and initial actions to possible ranges of variation. Level 1 evaluations provide the initial understanding by organizing the factors into preliminary performance estimates.

Figure 11.1 shows the part of the complete process that relates to detection and initial actions. For simplicity, we show only the design fire and initial actions for fire department notification. Chapter 18 discusses the life safety aspects of detection and initial actions.

PART A: DETECTION

11.2 Instrument detection

Fire detectors may provide spot protection, line detection, or volume or space detection. Spot detection occurs when a detector is located in a fixed position and is able to sense certain products of combustion over a defined floor area. Heat detectors and smoke detectors are two common instruments for spot detection. Spot-type heat and smoke detectors have different sensitivities

Building Fire Performance Analysis R. W. Fitzgerald
 ISBN: 0-470-86326-9 (HB)

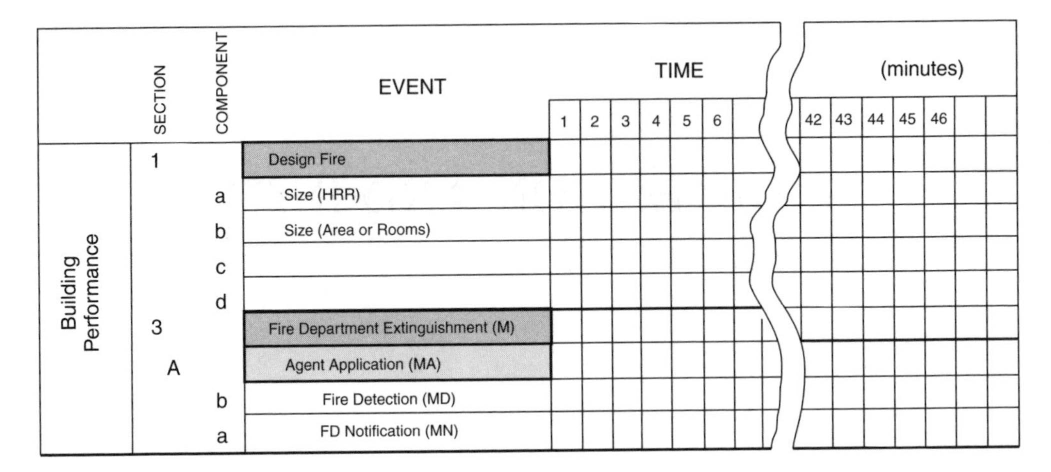

Figure 11.1

to fires in different locations. Heat detectors are much slower to react than smoke detectors in most fire conditions. Other factors that directly affect detection include ceiling height, HVAC operation, heat or airflow obstructions such as walls, sloped ceilings, or irregularities caused by exposed beams and girders.

Line-type detection occurs when the detection instrument has a one-dimensional linear form. For example, a photobeam smoke detector will have a transmitter sending a light beam to a receiver several hundred feet away. When smoke enters the path of the light beam and obscures enough of the light to correlate to the sensitivity setting of the detector, an alarm signal will be initiated. Another type of continuous line detection uses wires in an open circuit held apart by heat-sensitive insulation. When the insulation melts, the wires make contact and initiate an alarm. Line-type heat detection provides continuous coverage along the length of the cable rather than an area protected by a spot-type heat detector. Line-type detection is generally less sensitive (e.g., operates at a higher temperature) than spot-type detectors.

Volume or space detection involves active air sampling that uses fixed perforated tubes in a space. Air is continuously drawn through the tubes to a detection chamber. When smoke-contaminated air reaches the detection chamber, the small smoke particles are sensed, producing an alarm. Aspirating or active air-sampling smoke detectors are the more sensitive detector type when compared to the average spot-type smoke detectors. These detectors are normally used in clean room environments or areas where very expensive equipment is involved. Passive air-sampling smoke detectors include standard duct-type smoke detectors that rely on air entering a sampling tube to reach the detector.

11.3 Detection instruments

Fire detection has become sufficiently sophisticated in recent years that it is often possible to tailor detection system performance to its intended function. The relationship between the type of detector, its sensitivity, and its response to the products of combustion in the environment are all parts of a performance evaluation. We shall identify some detector characteristics to establish a base for performance evaluations.

Detectors may be classified according to the combustion products that they are designed to sense. Instrument classes may be described as follows:

1. Heat detectors
   a. Fixed temperature
   b. Rate of rise
   c. Rate compensation
2. Smoke detectors
   a. Ionization
   b. Photoelectric
   c. Photobeam (light obscuration)
   d. Air sampling detectors
3. Flame detectors

Each of these detectors responds to the presence of specific fire signatures. A fire signature is the measurable or sensible product of combustion that can be detected by an instrument or causes an awareness of the fire. Detectors will respond to the fire signature reaching a predetermined threshold level of magnitude or rate of change.

HEAT DETECTORS

Several types of heat detector are available. A fixed-temperature heat detector actuates when the sensing element reaches a specified thermal value. On the other hand, a rate-of-rise heat detector responds to a rate of temperature change that exceeds a calibrated value. It is also possible to obtain rate compensation or rate anticipation detectors that respond to the temperature of the surrounding air reaching a predetermined level. Combination detectors that can respond to a fixed temperature threshold and a rate of rise also are available.

Generally, a fixed temperature heat detector is the least costly and most reliable. However, it also has the attribute of being the least sensitive in detecting a fire. Consequently, fixed-temperature heat detectors are placed in locations where heat will not dissipate before reaching the detector, thus delaying actuation. They are also placed in locations where a larger fire at detection is within acceptable performance for the building. Rate-of-rise detectors are more sensitive to rapid changes in temperature and usually detect a fire of smaller size. A rate-of-rise detector actuates when the rate of temperature increase exceeds a predetermined level.

SMOKE DETECTORS

Smoke detectors sense airborne products of combustion carried into the sensing chamber by air movement. Smoke detectors may be identified by two types of operation: ionization and photoelectric. An ionization smoke detector has an effective electrical conductance between two charged electrodes in the sensing chamber. The air between these electrodes has been made conductive by ionizing the chamber with a small amount of radioactive material. When smoke particles enter the sensing chamber, the normal current flow due to the ion migration is reduced. The electronics of the detector senses this decrease in current flow and actuation occurs. Ionization detectors are particularly sensitive to invisible airborne products of combustion, such as those generated by a flaming fire.

A spot-type photoelectric smoke detector uses a light beam that propagates through the air within the detector housing. When smoke particles enter the air between the light source and the receiver, they reflect or scatter the light beam. In either case the amount of light reaching the photosensitive receiver changes, and the electronics of the conductor starts the detection response. A photoelectric detector is more sensitive to visible smoke, such as would be generated by a smoldering or smoky fire.

Generally, ionization detectors are more sensitive to flaming fires near the detector. Ion detectors remote from the fire source do not provide as rapid detection as photoelectric detectors. Photoelectric smoke detectors are more appropriate when visible particles are common. They are more stable in locations near kitchens or furnace rooms. The reliability of ionization and photoelectric smoke detectors has increased in recent years. The unwanted (nuisance) alarm rate is still high in homes using ionization spot detectors, but it has declined in recent years. Smoke detectors have the distinct advantage of more rapid fire detection in large open spaces where smoke is less likely to dissipate than heat. The placement of the detection instrument is critical in fire performance.

A photobeam smoke detector is commonly used in line detection for large open spaces. A photobeam detector transmits a light source at one end and has a photosensitive receiver at the other end. The distance between the transmitter and the receiver may vary from several feet to several hundred feet, depending on the type of coverage. When smoke obscures part of the light beam, the receiving device senses less light and that initiates the detection response.

An active air-sampling smoke detector uses a piping network of sampling tubes to draw air from the protected space into a detection chamber. Three types of detection chamber are commonly used. One uses a photoelectric light-scattering system that irradiates the sampled air. Another uses a laser source with a particle-counting procedure. Either of these devices can sense smoke particles at very low concentrations. The third system is a cloud chamber type in which air samples are introduced into a high-humidity chamber within the detection device. The humidity of the air is increased and the pressure is lowered slightly. Moisture in the air condenses on any smoke particles, forming a cloud that is sensed by the photooptics in the detection device.

FLAME DETECTORS

Flame detectors that provide extremely rapid fire detection operate by "seeing" the radiant energy produced by combustion. Flame detectors may operate in the ultraviolet (UV) or the infrared (IR) wavelength range. The receiver of a flame detector receives radiation and filters out all wavelengths other than those for which the detector is designed. Selection depends on the type of hazard being detected. It is important to match the type of detection with the type of fire expected. Because flame detectors detect radiant energy in a line of sight, they must be able to see the fire. Consequently, the space between the potential fire source and the detector must be free of obstructions. Flame detectors require care in design and installation and are used in high-hazard industrial areas where response must be very rapid.

All detectors have a restorable or nonrestorable operating mode. A restorable detector is one where the detector returns to normal operation when the fire signature is removed. A nonrestorable detector is destroyed by the combustion products and must be replaced. For example, a fusible element heat detector operates by melting the eutectic metal. After operation the detector cannot be restored to its original operating condition. On the other hand, a bimetallic snap-disk detector operates by the differential expansion of two different metals on heating. The difference in the coefficient of expansion of the two metals causes a bending that activates the detector. When the heat is removed, the sensing element returns to its original condition without additional restoration efforts. Smoke detectors can be restored to normal operating condition by clearing the detector of smoke and resetting it from the fire alarm control panel.

11.4 Detection operational performance

Detection performance is defined by the fire size at the time of detection. An analysis involves the type of detection instrument, its reliability, its sensitivity to the combustion products, the

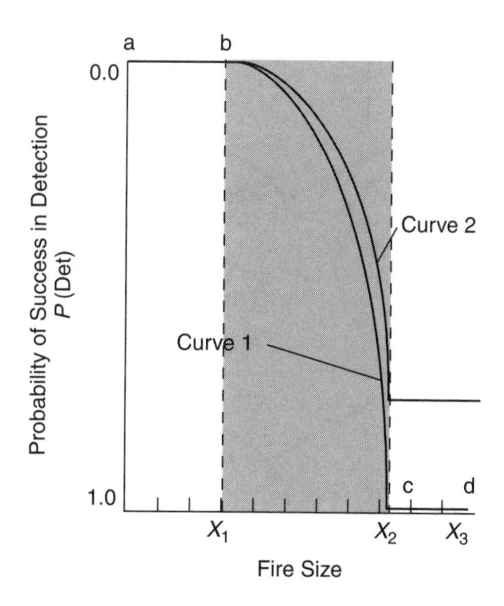

Figure 11.2

distance from the fire, and the obstructions to products of combustion movement. The dynamics of fire growth also influences the fire size at detection.

An understanding of detection performance is obtained by analyzing the likelihood that a detector will actuate at a selected fire size. Figure 11.1 lines 1 a, b, and c describe the design fire with line c available to show the products of combustion specifically associated with the detector being evaluated. The cells of Figure 11.1 line 3Aa describe the dynamic detector performance. Each individual cell is a single value (SV) estimate of the likelihood that a detector will actuate at that fire size. Figure 11.2 shows representative curves for detector performance. Segment a to b indicates a certainty that the detector will not actuate within this range of fire size. Segment c to d shows a certainty that the detector will actuate. Examination of selected fire sizes within the window of uncertainty from b to c provides an opportunity to understand the fire and detection with some clarity.

Each SV estimate uses the relevant information available at the time of the analysis. For example, the type of detector and its location are specified. The conditions of the fire scenario are identified, such as conditions along the path through which the combustion products must travel to reach the detector. The status of open or closed doors is defined.

A single value (SV) cell analysis starts with selecting a time and the associated fire conditions from a column of Figure 11.1. The network of Figure 11.3 structures the evaluation. The first event estimates the probability that enough products of combustion will reach the detector for the fire size selected. This estimate examines the transport of fire signature products from the fire to the detector. The transport involves the distance associated with the building geometry and construction features. Figure 11.4 identifies the factors associated with this estimate. The second event assesses the operating sensitivity of the detector. Figure 11.5 lists the factors that affect this evaluation.

11.5 Detector reliability

The SV network of Section 11.3 assumes complete reliability of the detection instrument. One can differentiate detection system quality with a reliability analysis of Figure 11.6 to guide judgments.

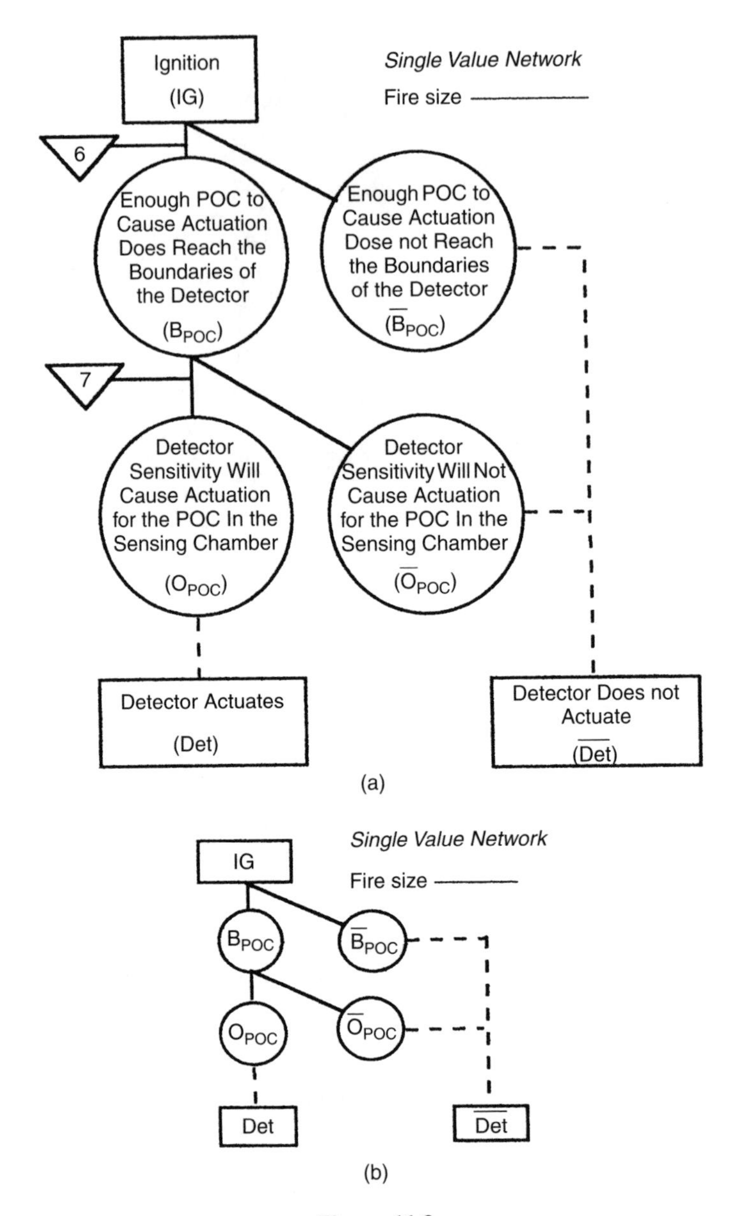

Figure 11.3

The reliability event (rdi) describes manufacturing quality control. It is the instrument reliability when removed from the shipping carton. Detection installation (idi) relates to the quality of the construction control. The instrument reliability (rdi) and the construction control (idi) collectively may be considered as the mission readiness. The final event regarding actuation (adi) represents detection system maintenance.

It is possible to have a completely reliable system at installation if *all* detectors are tested and respond at the time of installation. At that instant, rdi, idi, and adi all have a certainty of success, and the operational event, $P(\text{DI}) = 1.0$. When only a sampling of detectors are tested,

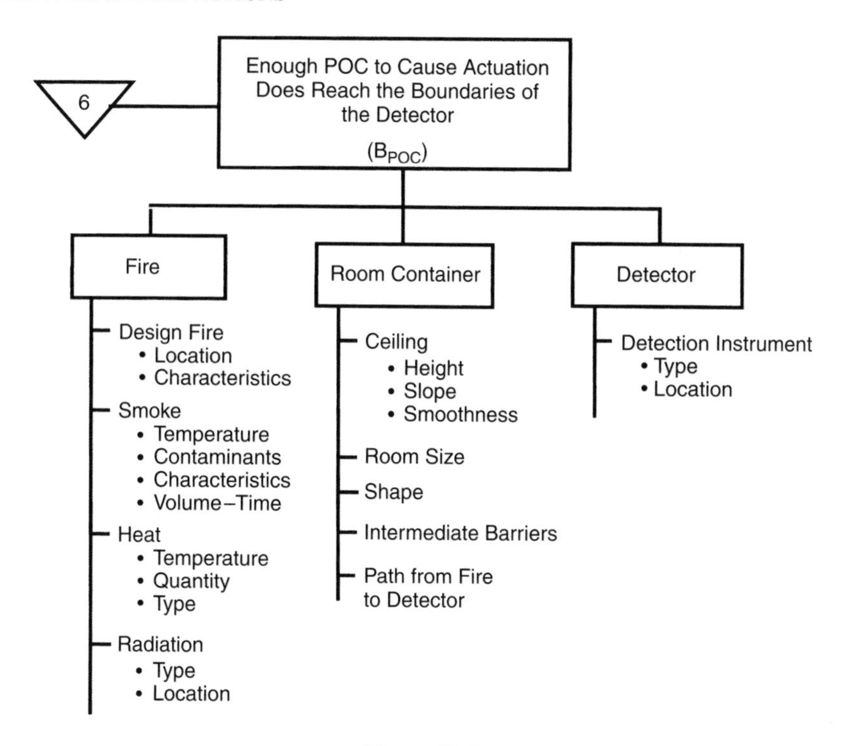

Figure 11.4

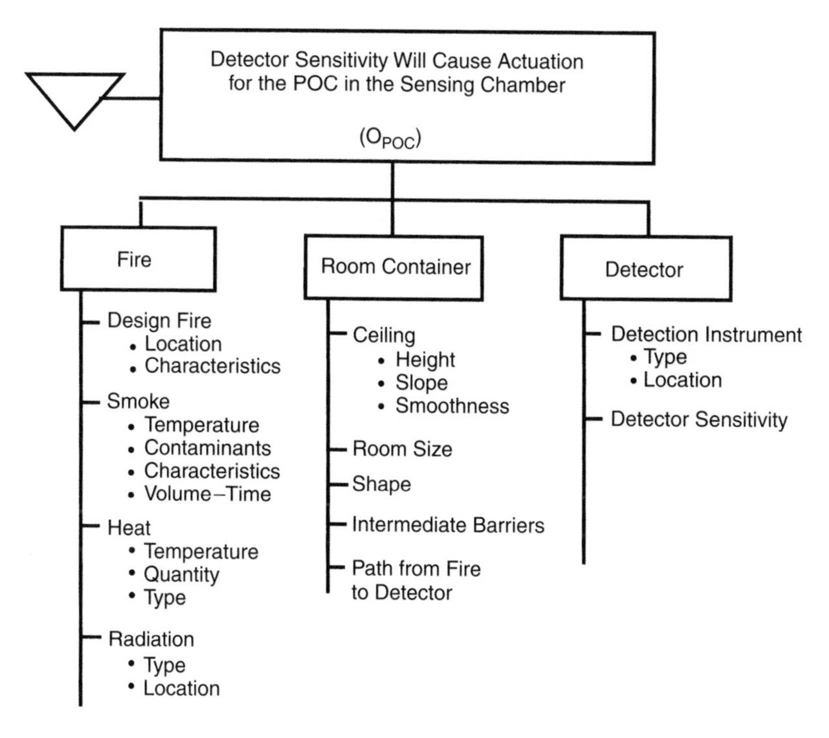

Figure 11.5

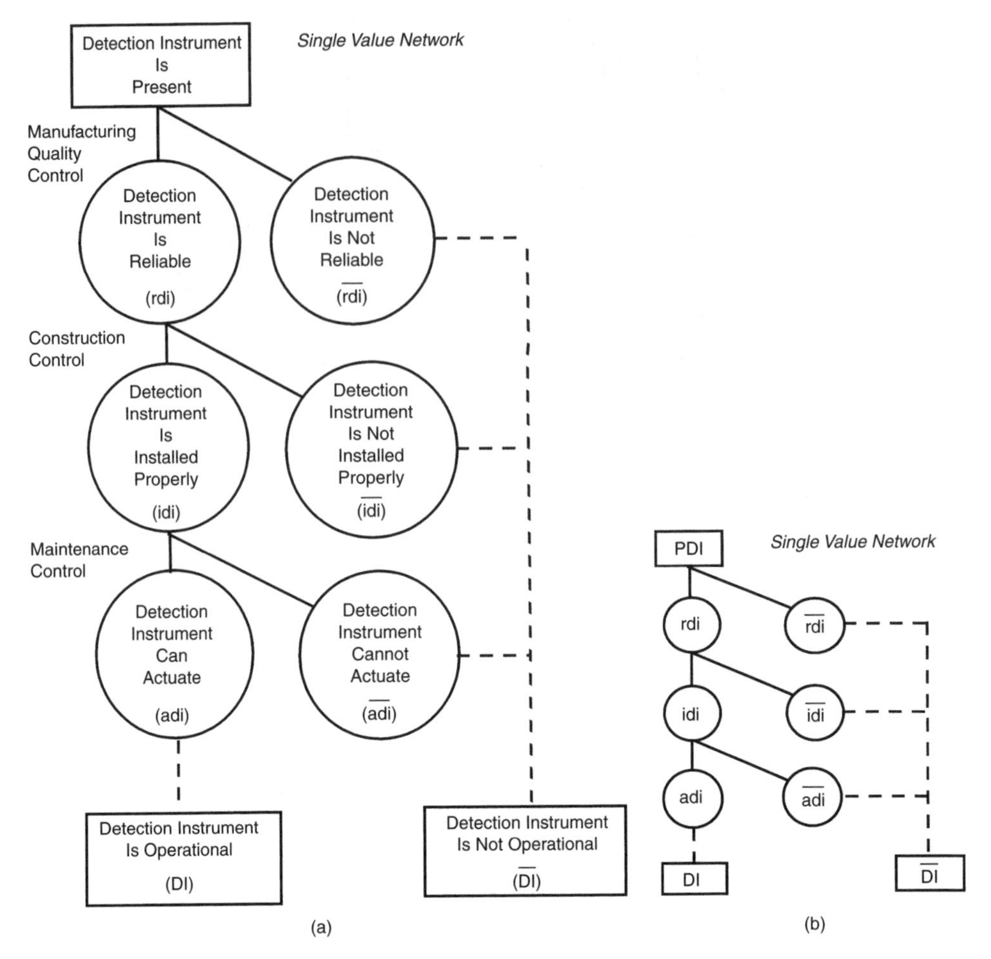

Figure 11.6

the estimate for events rdi and idi will reflect the confidence of the individual in the quality of the detectors and their installation. A value of 1.0 shows the inspector is completely confident that all detectors have been installed properly and will work. Values of less than 1.0 suggest that care and workmanship seem deficient.

Event (adi) describes the long-term maintenance of the system. If the building manager has in place a long-term, effective maintenance plan, the reliability of the detectors is likely to remain high and $P(\text{adi})$ is close to 1.0. On the other hand, if maintenance of detectors is not a priority, the reliability will deteriorate and its operational level will worsen. After a successful installation testing, only event actuation, adi, can change over time.

When one multiplies the probability value for $P(\text{Det})$ of Figure 11.3 by the reliability factor, $P(\text{DI})$, each coordinate changes to reflect the system reliability. Curve 1 in Figure 11.2 shows detector operations when the system has perfect reliability. Curve 2 shows a system with less than perfect reliability. The relative position of these curves reflects reliability.

11.6 Instrument detector evaluations

Examples 11.1 and 11.2 illustrate level 2 evaluations.

Example 11.1

A smoke detection system was evaluated for design fire products and times noted in Figure 11.7. The detection system and the smoke transport paths for different potential fire locations and ventilation conditions for the room were identified. The following estimates were selected:

- The smallest fire size for which the detection system could actuate is 200 kW.
- The detector certainly would actuate for fires larger than 600 kW.
- An intermediate fire of about 400 kW was selected to look more carefully at the details of detector performance.

This analysis provided the following information. The detector was located in a room next to the room of origin with an open door between the rooms. A transom over the door, exposed beams, and partial shielding of the detector created some interference with smoke movement from the fire to the detector. The probability that enough smoke would reach the detector before the fire grew larger than 400 kW was estimated at 40%.

While the specific detector sensitivity was not determined, the individual has some knowledge of this detector type. Based on this knowledge, this detector was estimated to have a 95% probability of actuation if the design fire smoke were to reach its boundaries. This estimate would increase as more smoke is generated and reaches the detector.

Assuming perfect detector reliability, sketch the detection curve.

Solution

The window of uncertainty is bounded by the values of 200 kW and 600 kW. Figure 11.8 shows the SV calculations for a fire size of 400 kW. Figure 11.9 shows the detection curve.

Example 11.2

The building of Example 11.1 has 40 smoke detectors distributed throughout the building. A sample of four detectors was tested and all were operational. The detector of Example 11.1 was one of the four detectors. What reliability value should be assigned for the detection curve?

	SECTION	COMPONENT	EVENT	TIME				(minutes)				
				1	2	3	4	5	6	7	8	
Building Performance	1		Design Fire									
		a	Size (HRR) (kW, MW)	40	140	320	580	900	1.3			
		b	Size (Area or Rooms)									
		c										
		d										
	3		Fire Department Extinguishment (M)									
	A		Agent Application (MA)									
		b	Fire Detection (MD)	0.0	0.0	0.4	1.0	1.0				
		a	FD Notification (MN)									

Figure 11.7

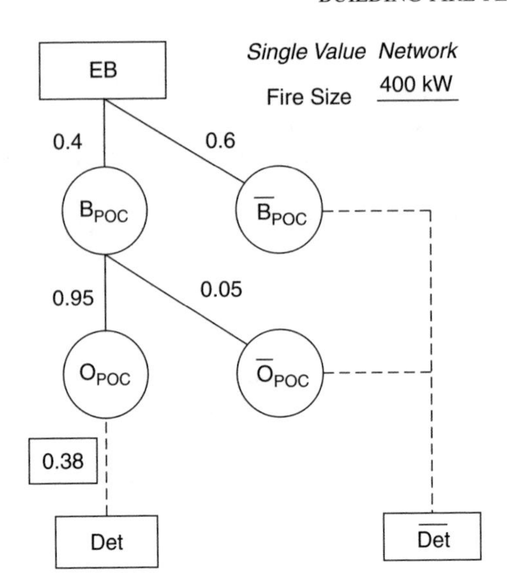

Figure 11.8

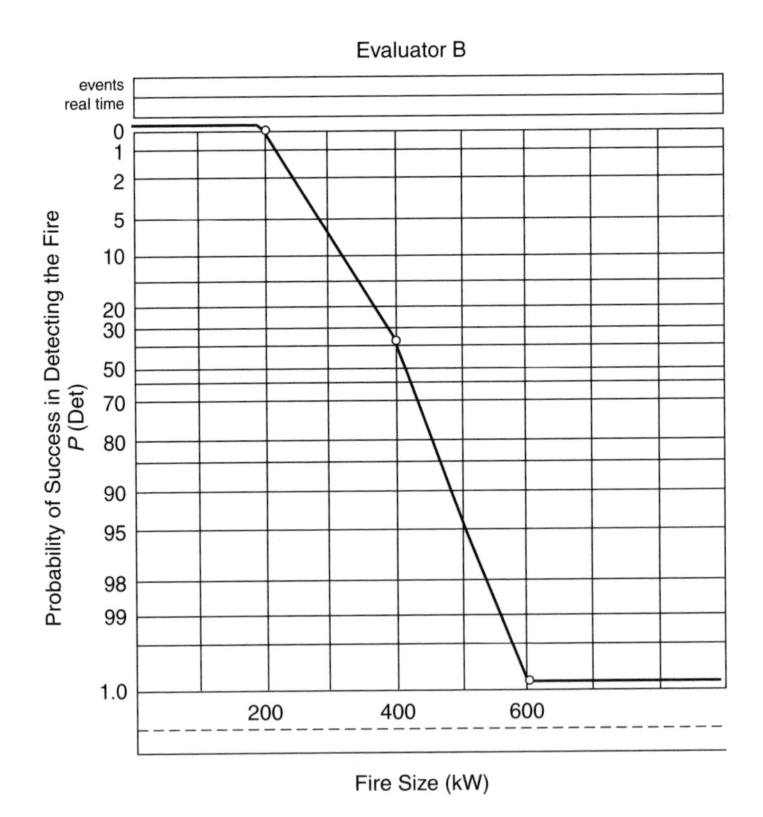

Figure 11.9

Solution

Because the detector of Example 11.1 was specifically tested and found operational, its reliability for the immediate future is 100%. The building environment and the maintenance policies would guide whether this value should be derated for the longer-term future.

Discussion

Although prescriptive codes usually require 100% acceptance testing of systems, sometimes only a small percentage are actually tested. Then the inspector requests a certification that the rest of the system was tested by the contractor. When the acceptance test is conscientiously completed, the initial operational readiness is reliable. Not all existing detection systems have been tested in the past and some existing buildings may have deficient systems.

A question of reliability for untested detectors arises. Let us assume that another detector had never been tested during its life. Therefore it is not known if the detector was operational when it left the factory (quality control, rdi) or if it was installed properly (construction control, idi). A probabilistic value could be expressed for rdi as the manufacturer's statistical quality control, if it were known. However, what value should be assigned to an unknown contractor's construction quality control? Is a 10% sample sufficient to conclude that all detectors were installed properly and assign a reliability of 100%?

This becomes a matter of resource allocation. If a detector is important to the success of subsequent fire defense performance, it should be tested to ensure 100% reliability. If reliable detector operation does not significantly affect subsequent fire defense performance, one may trust in the results, or question the need.

11.7 Human fire detection

Humans can detect a fire by sensing the products of combustion as in feeling heat, smelling smoke, seeing light, or hearing noise. Often a human will sense an abnormal condition, such as an unusual odor or a trace of smoke. However, the cue may not be strong enough to cause the individual to believe that a hostile fire actually exists. It is common for an individual to investigate the cause of the abnormal condition. We define human fire detection as that instant when an individual clearly recognizes or believes that a hostile fire exists. The detection event is completed when that recognition or belief occurs. Subsequent actions by the individual are not a part of the detection process.

It is important to recognize that the analysis of human detection rarely involves a single description for any building. We construct at least two performance curves: one for occupied conditions and one for the unoccupied status. Detection is merely one link of manual suppression and life safety. The analysis must reflect conditions that have relevance to building performance and its risk characterizations. Figure 11.10 shows the SV network to estimate the likelihood of human detection. Figures 11.11 and 11.12 describe the factors that influence the likelihood of human fire detection.

11.8 Combined detection modes

Fire detection can occur by instruments or by humans. In many buildings, both modes are present. When multiple modes are available, each mode of detection is evaluated independently with an associated continuous probability curve. We calculate composite curve coordinates by using the SVN of Figure 11.13. Because these events are independent, their order in Figure 11.13 is unimportant to the calculated result.

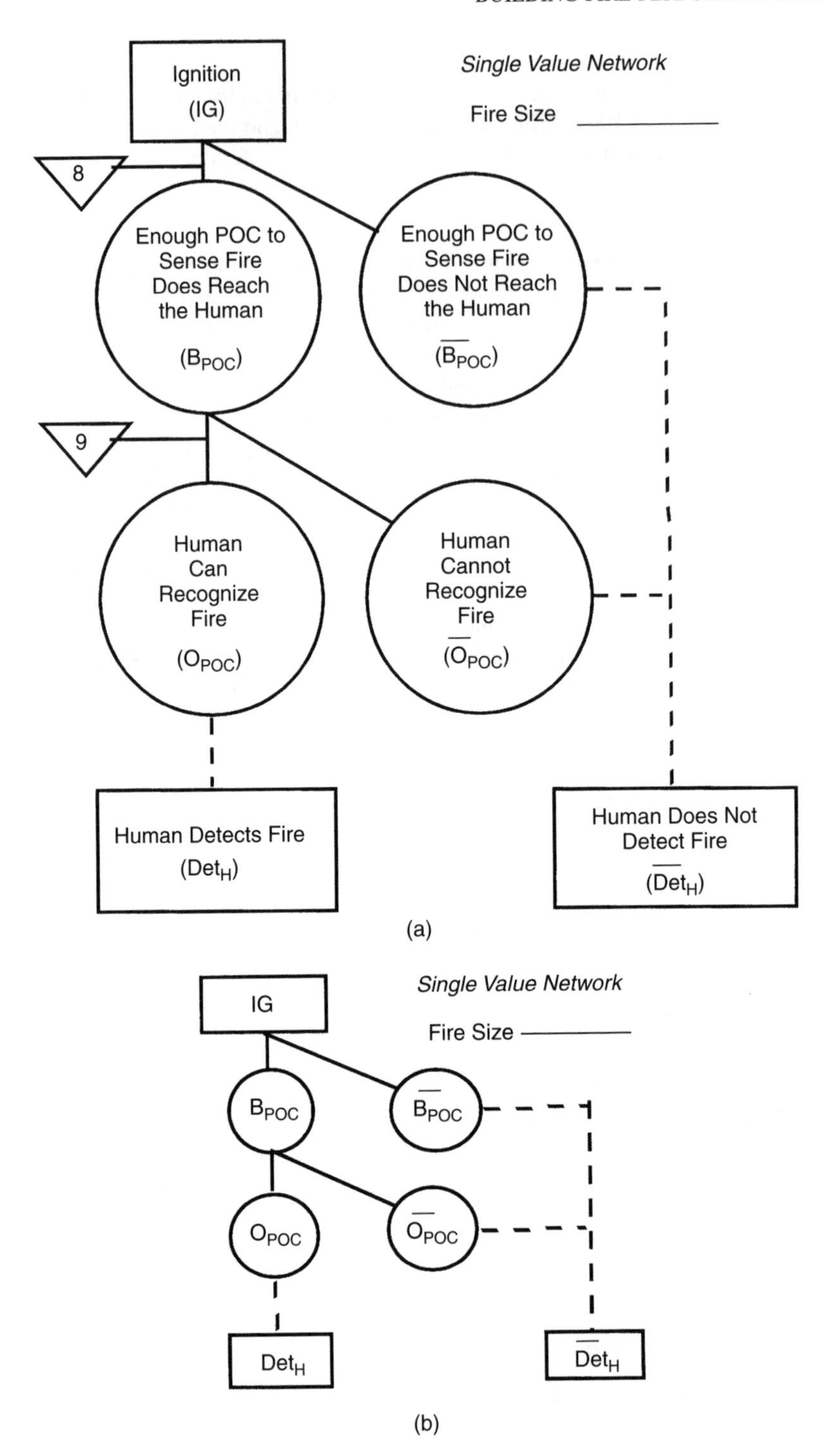
Single Value Network
Fire Size
Ignition
(IG)
8
Enough POC to Sense Fire Does Reach the Human
(B_{POC})
Enough POC to Sense Fire Does Not Reach the Human
($\overline{B_{POC}}$)
9
Human Can Recognize Fire
(O_{POC})
Human Cannot Recognize Fire
($\overline{O_{POC}}$)
Human Detects Fire
(Det_H)
Human Does Not Detect Fire
($\overline{Det_H}$)
(a)
Single Value Network
Fire Size
IG
B_{POC}
$\overline{B_{POC}}$
O_{POC}
$\overline{O_{POC}}$
Det_H
$\overline{Det_H}$
(b)

Figure 11.10

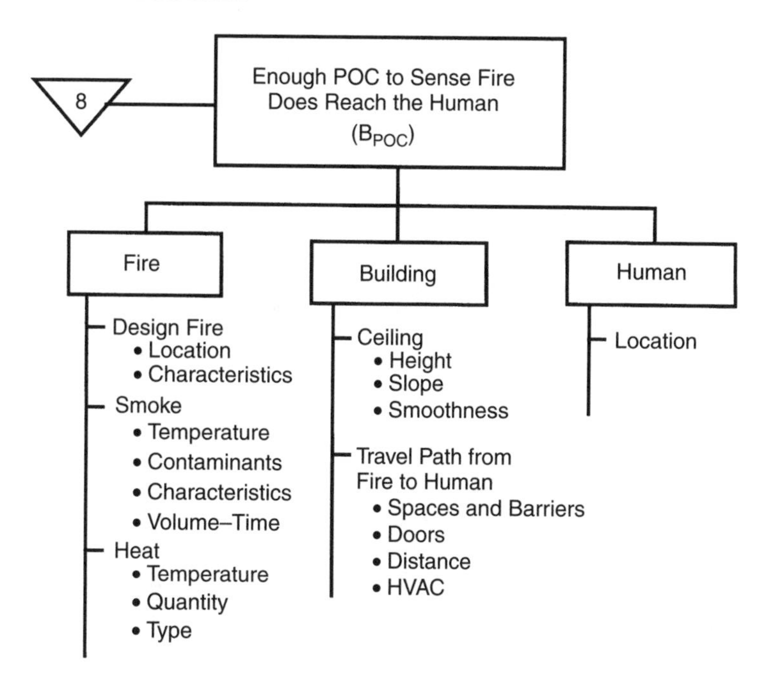

Figure 11.11

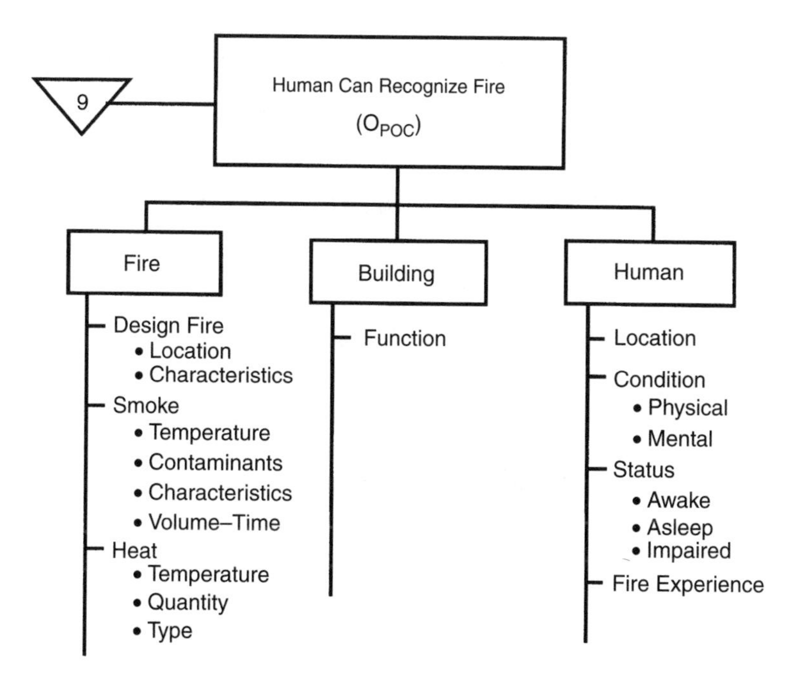

Figure 11.12

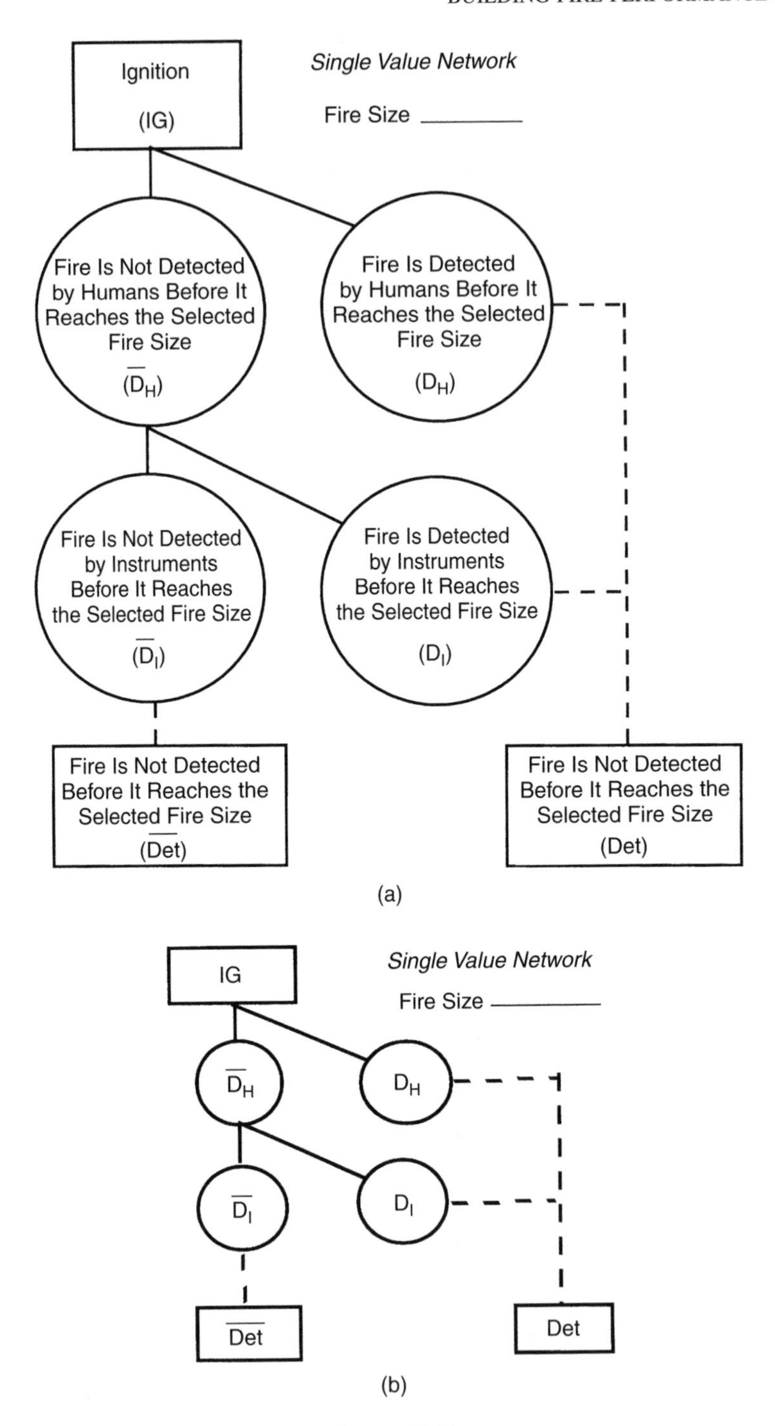

Ignition
(IG)
Single Value Network
Fire Size
Fire Is Not Detected by Humans Before It Reaches the Selected Fire Size
($\overline{D_H}$)
Fire Is Detected by Humans Before It Reaches the Selected Fire Size
(D_H)
Fire Is Not Detected by Instruments Before It Reaches the Selected Fire Size
($\overline{D_I}$)
Fire Is Detected by Instruments Before It Reaches the Selected Fire Size
(D_I)
Fire Is Not Detected Before It Reaches the Selected Fire Size
($\overline{Det}$)
Fire Is Not Detected Before It Reaches the Selected Fire Size
(Det)
(a)
IG
Single Value Network
Fire Size
$\overline{D_H}$
D_H
$\overline{D_I}$
D_I
$\overline{Det}$
Det
(b)

Figure 11.13

Table 11.1

Fire size (kW)	Detector with building occupied	Detector with building unoccupied	Human with building occupied	Human with building unoccupied
Ignition	0.0	0.0	0.0	0.0
EB (20)	0.0	0.0	0.0	0.0
40	0.0	0.0	0.4	0.0
60	0.6	0.6	0.8	0.0
80	0.99	0.99	0.95	0.0
	(reliability limit)			
160	0.99	0.99	0.97	0.0
320	0.99	0.99	0.98	0.0
			(reliability limit)	

When combining separate detection curves, one must use consistent scenarios. Conditions are never averaged or weighted. The intent is to tell a story that the building management and its users clearly understand.

Example 11.3

Human detection and an installed instrument detection system were evaluated for a building. Both occupied and unoccupied conditions were studied to gain an understanding of a range of performance. Table 11.1 shows the results of the SV evaluations. Construct detection curves for each of the four conditions. Then construct a composite curve for the occupied condition and the unoccupied condition.

Solution

Figure 11.14 shows the four curves that describe conditions of Table 11.1. Curves A and B are superimposed because the presence or absence of humans in the building has no influence on the actuation of the detection instruments.

Figure 11.15 shows detection curves for the occupied and unoccupied building conditions. Curve A combines the occupied values for instrument and human detection by using the SVN of Figure 11.13 for each coordinate. The combined success reliability at 80 kW is 0.9995. Curve B combines the unoccupied values of Table 11.1. Because the unoccupied detection is 0.0 for all fire sizes, the curve becomes the same as the unoccupied instrument curve. We use evaluator C to show the combined results.

11.9 Level 3 detection evaluation

The performance measure for detection is the size of fire at the instant of detection. A level 3 evaluation examines the range of fire sizes that can occur as different, but reasonable variations in fire growth, building features, or detector sensitivity can exist. A level 3 evaluation is only justified when its cost is less than the value of the additional information.

Appendix C describes different levels of evaluation using detection as the illustrative model. Figure 11.16 illustrates the concept of level 3 studies. The dashed line describes a level 2 evaluation and the solid lines illustrate an expected variation envelope for building operations. The function of a level 3 analysis is to develop an understanding of the performance sensitivity to expected variations. For example, the thermal lag of different detectors having the same

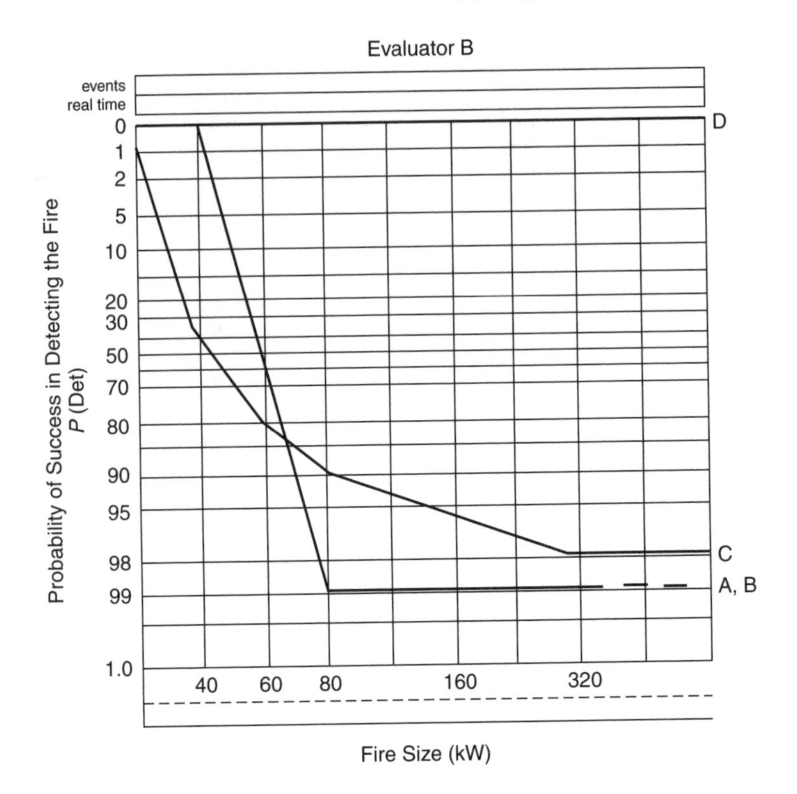

Figure 11.14

temperature listing can vary substantially. A level 3 analysis enables one to understand why the range in possible fire sizes for a 135°F (57.2°C) can range from X to Y. If certain conditions or values of response time index (RTI) are important to the building performance, one can ensure that those conditions are specified and inspected. If a range of performance is acceptable, close attention to those details may not be critical to the building's behavior. How much is enough becomes a judgment call that can be tailored to the needs of the problem.

11.10 Level 1 detection

A level 3 evaluation is rarely necessary. A level 1 evaluation is always done. A level 2 evaluation examines selected details to increase knowledge about building performance. One of the many roles of a level 1 evaluation includes decisions on whether one needs a level 2 investigation, for what components, and which details need attention. All three levels use the same analytical process. Differences relate to the depth of investigation to obtain the understanding needed for the problem being solved.

Although level 1 evaluations are rapid and based on observations and mental estimates, they require attention to details that influence performance. Physical features such as design fire characteristics and location, detector type and location, and building conditions that influence the transport of fire products must be defined before the evaluation.

Detection is defined as either (a) an instrument senses products of combustion and goes into alarm or (b) a human is aware of a fire's existence. The awareness or actions of a human after

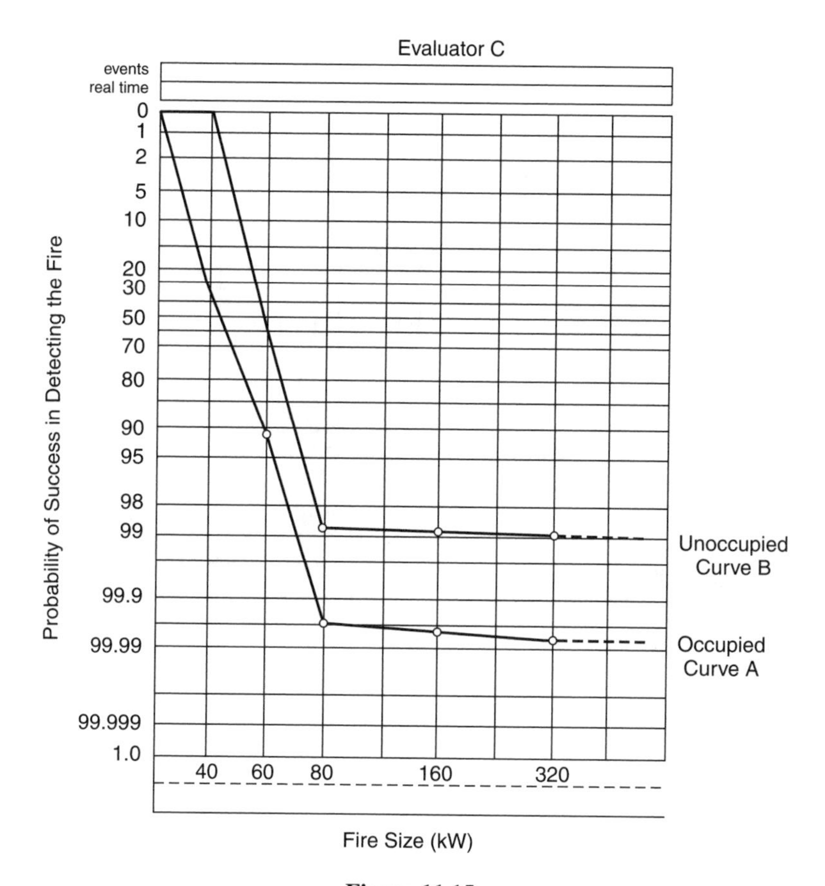

Figure 11.15

a detector goes into alarm are not a part of the detection evaluation. Detectors and humans are evaluated independently. Section 11.8 describes their combination, if it is necessary.

The initial task for evaluating building detection gathers as much information about the system as possible within the time available. The most important factors are the general type of detector (e.g., smoke, heat) and their locations. Locating detectors on small-scale floor plans enables one to recognize the extent of coverage and the building features that influence combustion product transport from the fire to the detector. One augments this information with additional detector characteristics and building features that affect operational effectiveness and reliability. Information is often easier to obtain for proposed new systems because of access to more complete design documentation. Information may be more difficult to obtain for existing buildings because gathering details requires time and accessibility and some details must be surmised from other evidence.

11.11 Instrument detection analysis

The continuous value (CV) detection curve is a graph of the probability estimates across row 3Aa of Figure 11.1. The thought process for estimating the probability of any cell involves answering the following questions:

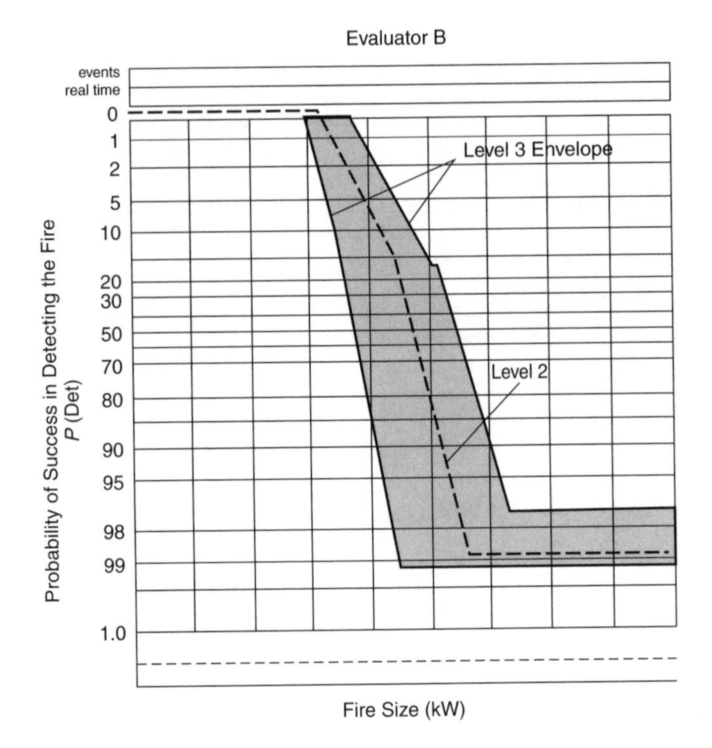

Figure 11.16

- Will enough combustion products be generated and transported to the detector housing to cause actuation of the most sensitive detector of the type being evaluated?
- Does the detector have the sensitivity to actuate for the POC in the environment before the fire grows to a larger size?

The first question relates to the fire size, design fire characteristics, and path from the fire to the detector. The second question addresses detector sensitivity. Figure 11.17 shows these factors.

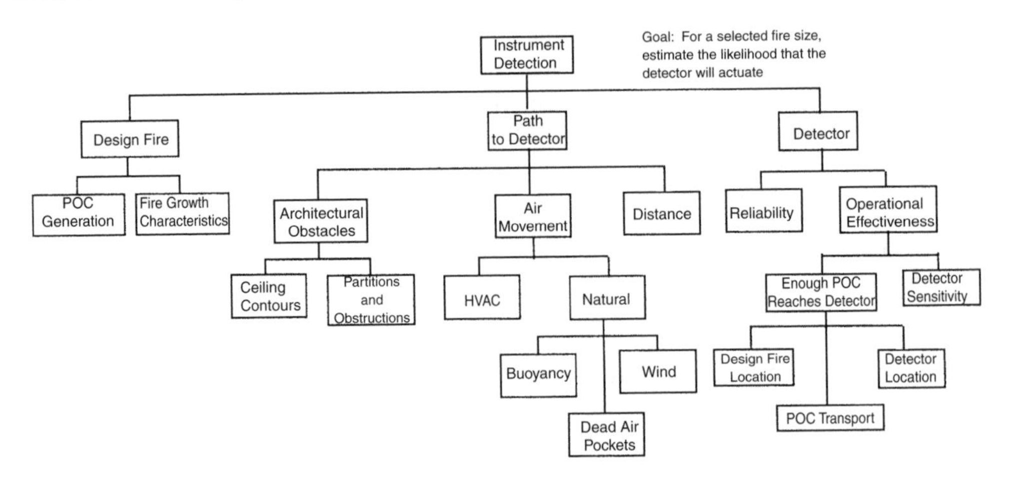

Figure 11.17

Probability estimates represent judgments based on the best information available. They may use verbal descriptors or probabilities to give a sense of proportion for expected behavior.

Reliability is defined as the probability that a detection instrument will actuate when enough products of combustion enter its sensory area. Reliability involves a manufacturing quality control of the instrument, an installation quality control, and maintenance control. Acceptance testing of each detector ensures complete reliability at the time of testing. The reliability of the detection system can deteriorate over time if the system is not maintained adequately. One can derate the performance curve to reflect deficiencies.

Example 11.4

Figure 11.18 shows the studio apartment described in Chapter 4. A fixed-temperature, 135°F (57.2°C) heat detector is on the living-room ceiling near the bedroom. Construct a level 1 detection curve for a fire that starts in the wastebasket beside the couch.

Solution

The performance question is not, What is the likelihood that the detector will eventually actuate? Instead the question is, What is the likelihood that the detector will actuate when the fire is 1 kW, 5 kW, 10 kW, 50 kW, $\cdots$, n kW?

Assume that we take photographs at various stages of a fire which starts in a wastebasket beside the couch, as shown in Figure 11.19(a) to (d). These fire sizes provide SV benchmarks to estimate the likelihood of fire detection.

We estimate there is no chance that the detector will actuate for a fire the size of Figure 11.19(a). We further decide there is a certainty that the detector (assuming perfect reliability for this exercise) will have actuated for the fire in Figure 11.19(d). There may be a small chance that the

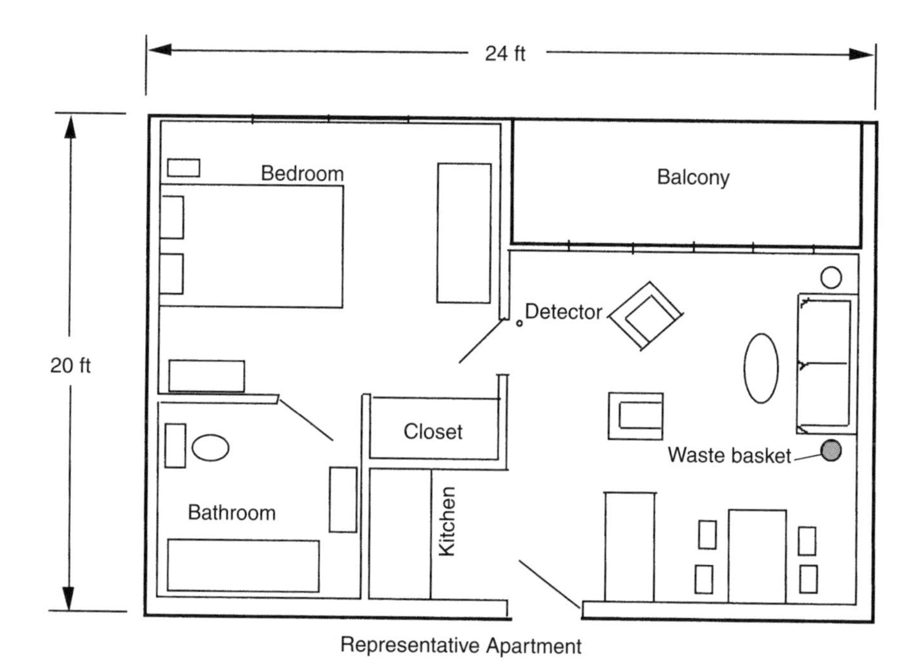

Figure 11.18

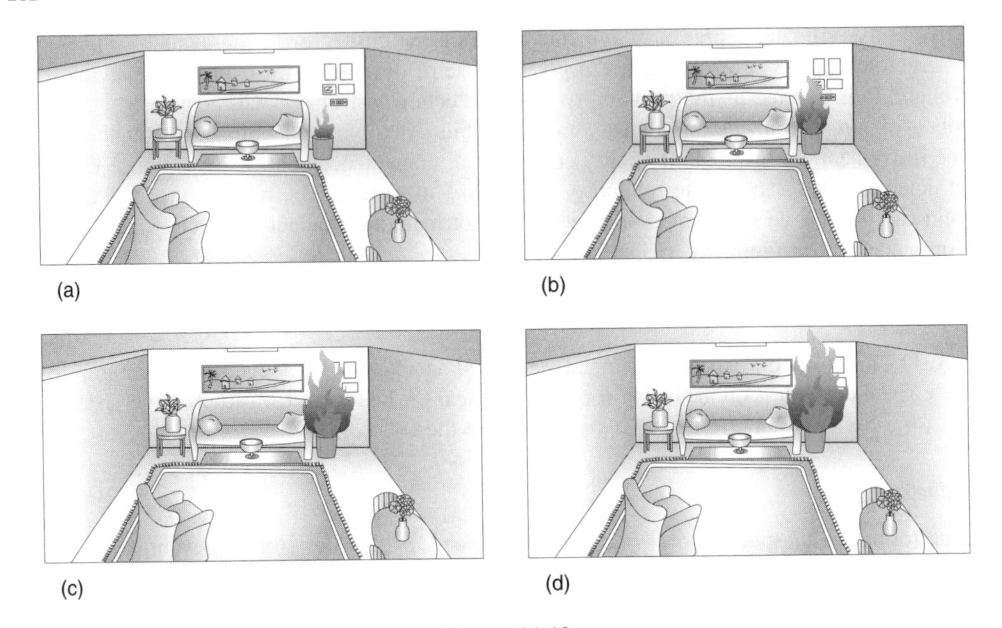

Figure 11.19

detector can actuate when the fire is the size in Figure 11.19(b), and a much better likelihood for a fire the size in Figure 11.19(c). This initial set of observations is similar to analyzing photographs of different fire development stages and predicting the status of the detector at each instant of time. Rapid, continuous mental estimates for this scenario provide the basis for sketching a level 1 detection curve.

While the above description gives an initial indication of performance, communication can be enhanced with verbal descriptors for the heat detector actuation:

- *certainty* that it will not actuate;
- *very small* likelihood of actuation success;
- *small* likelihood of actuation success;
- *moderate* likelihood of actuation success;
- *good* likelihood of actuation success;
- *very good* likelihood of actuation success;
- *certainty* that it will actuate.

Only a brief time is available to make the assessment, and inadequate information exists concerning items such as the fuel detector sensitivity. Based on the available information and knowledge, one can envision performance and estimate a relationship between fire size and detector actuation. Table 11.2 illustrates a thought process for evaluation. For now, do not be concerned about whether you agree or disagree with the values assigned, but focus on the process.

Verbal descriptions of estimations are a useful way to begin thinking about performance evaluations. However, the utility is magnified when relationships are calibrated numerically. A probability measure is the most convenient way to describe the expected outcomes. The evaluation encodes the information that is available at the time of assessment. The value judgment is the individual's personal belief of the performance expectations and is expressed as a probability. Figure 11.20 shows the level 1 detection performance for this scenario. In this illustration,

Table 11.2

Fire size	Detector actuation classification	Reasoning
Fire-free status	Certainly no	No fire
Wastebasket involved (Figure 11.22(a))	Certainly no	Small fire; distance to detector
Couch arm ignited and fire starting to grow along arm and back	Very small	Looks like a size close to a threshold; detector sensitivity unknown
Couch side involved (Figure 11.22(b))	Moderate	Fire size and heat release increasing; detector sensitivity unknown
Flames near ceiling (Figure 11.22(c))	Very good	Fire size; room container small; conditions appear to be near a threshold to certainty
Flames across ceiling (Figure 11.22(d))	Certainly yes	Fire size; path to detector direct and clear

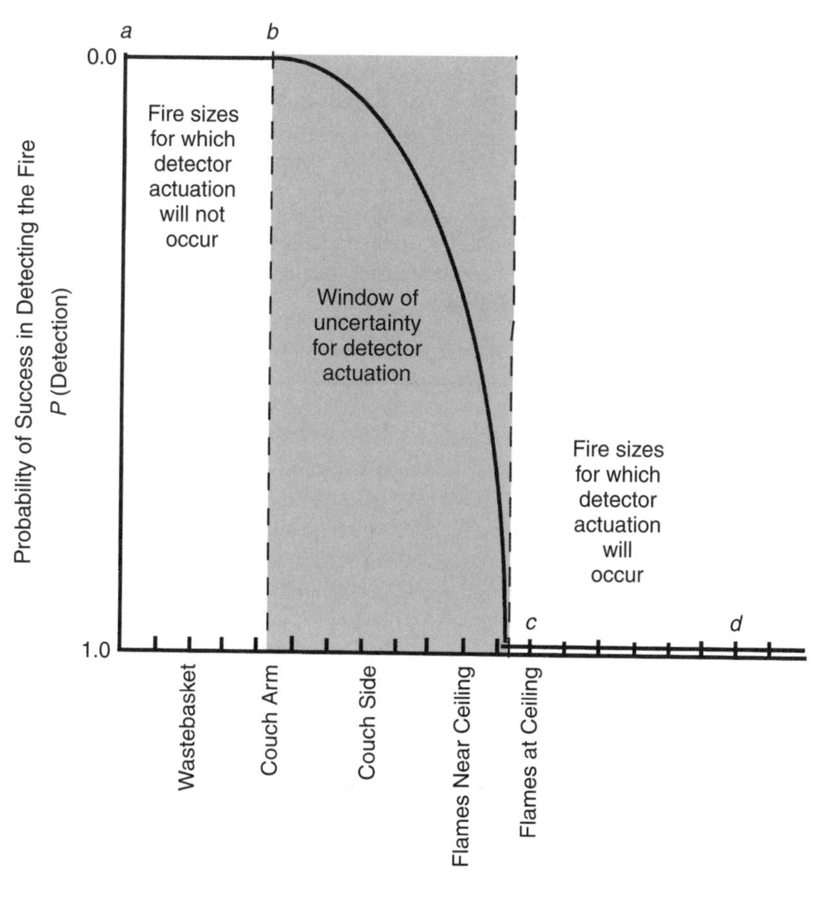

Figure 11.20

intermediate numerical values were not specified. Using a sense of proportion for position on the graph allows one to describe the general behavior and characteristics of the functional behavior.

Example 11.5

Several scenarios have been described for the studio apartment of Figure 11.18. Match the scenario with the appropriate performance curve of Figure 11.21. In Figure 11.21 consider the room to be each separate space, surrounded by boundaries of walls, floors, and ceilings. This apartment has three rooms of living room LR (kitchen, living, dining); bedroom BR; and bath BA. Although the abscissa of Figure 11.21 shows fire sizes and FRI, each relates to the room in which the fire originated. Assume perfect reliability for all detectors. Reliability factors can be incorporated into the curves later.

- *Scenario 1*: fire in BR; door between BR and LR open; transom wall extends 12 in from ceiling to door soffit; 135°F (57.2°C) heat detector on the ceiling above the door at LR, 6 in from wall.
- *Scenario 2*: same as scenario 1 except that the heat detector is on the transom wall above the door at the ceiling intersection.
- *Scenario 3*: same as scenario 1 except that the door between LR and BR is closed.
- *Scenario 4*: same as scenario 1 except that a photoelectric smoke detector replaces the heat detector.
- *Scenario 5*: fire in kitchen; photoelectric smoke detector above door of LR, 6 in from wall.

Solution

The purpose of this exercise is to show that one can convey simple graphical information that differentiates details of expected performance. As one expands techniques of evaluation, documentation of the logic and decision rationale becomes easier to construct. The graphical representations that correspond to the scenarios are as follows:

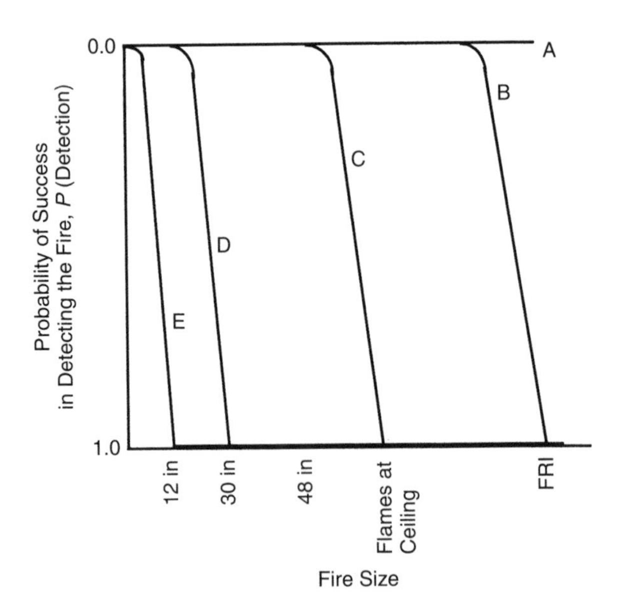

Figure 11.21

- Scenario 1 matches curve C.
- Scenario 2 matches curve B.
- Scenario 3 matches curve A.
- Scenario 4 matches curve D.
- Scenario 5 matches curve E.

The primary audience for the CV performance curves is you. You can describe your visual thinking to depict the dynamic changes in expected performance as time or fire conditions change. The logic for estimates provides the basis for documentation, if needed. If it seems sensible, use your CV curves to express performance to other people.

11.12 Human detection analysis

Evaluation of human detection follows a pattern similar to that of instrument detection. The questions for any fire size are as follows:

- Will enough products of combustion to sense the fire reach the individual?
- Will the individual believe that a hostile fire exists at the design fire selected?

Figure 11.22 shows factors that influence the evaluation of the fire size at human detection.

Typically the uncertainty can be wide because of the broad range of factors that influence fire detection. A part of this is due to the mobility of humans in a building and the fact that an individual may be in different locations when a fire occurs. Also, the training, experience, and status of individuals are quite variable. For example, an individual may be awake or asleep; alert or impaired by the influence of alcohol or drugs; mobile or having restricted movement; or a variety of other conditions. All individuals do not react in the same way to an abnormal condition.

The factors noted above contribute not only to variability in the fire size when detection occurs, but also to differences in reliability. Often one uses specific scenarios when analyzing the role

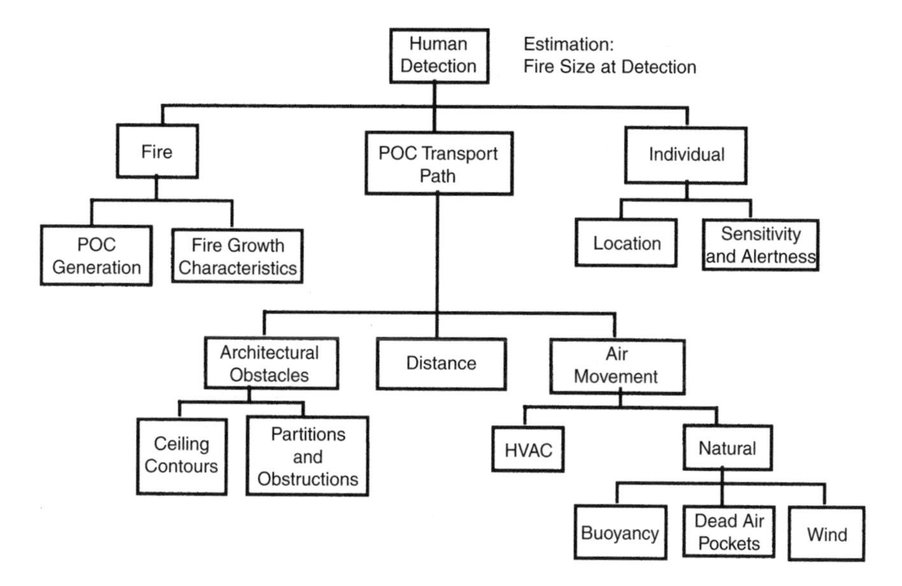

Figure 11.22

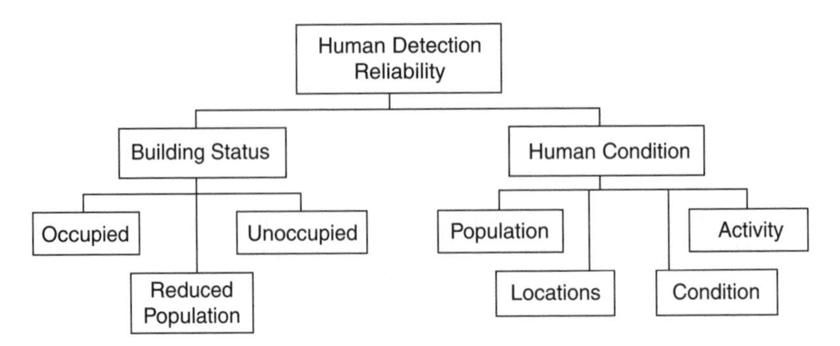

Figure 11.23

of humans in detection, and conditions of Figure 11.23 may apply to many possible scenarios. For example, is the building occupied or unoccupied? Is the individual awake or asleep? Alert or impaired by drugs or alcohol? Where is the individual located for the scenario under study? Level 1 analyses allow a rapid assessment for different conditions to gain an understanding of the sensitivity and reliability of an individual in fire detection. This understanding enables one to tell the story to individuals who may not have considered a range of conditions.

PART B: ALARM AND NOTIFICATION

11.13 Alarm and subsequent actions

After a fire is detected, a variety of activities are set into motion. Most of these can be placed into two groups. One relates to occupant safety and the other relates to fire department notification. Figure 11.24 shows common actions that follow fire detection. This chapter looks at fire department notification. Chapter 18 discusses occupant safety.

Fire department notification (MN) occurs when the message of a fire is received and understood by the local fire service communication center. A performance analysis estimates the time duration from established burning to the instant when the fire service communication center is aware of

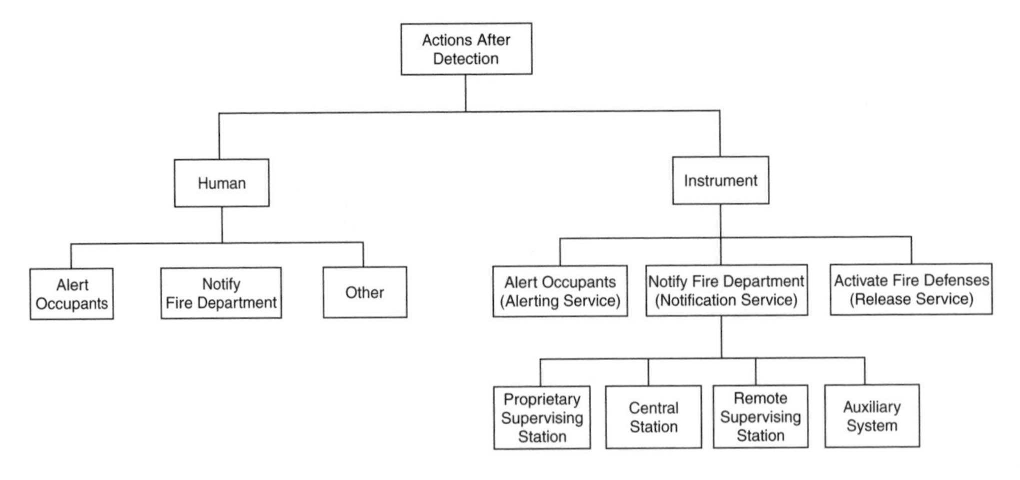

Figure 11.24

the fire. This time duration and its reliability are critical factors in understanding a building's performance regarding fire department operations.

Occupant alerting is the action most commonly associated with fire detector actuation. If the occupant becomes aware of an alarm, several activities may take place. The occupant may decide to leave the room and attempt to evacuate the building. The occupant may search for and alert other occupants, hide, investigate, decide to remain in the room until the emergency is over, notify the local fire department, or take other actions that may seem appropriate for the conditions and situation of the moment. Possible occupant actions after becoming aware of a fire are part of a building performance evaluation.

A third type of action involves building changes. Release services use detection to trigger changes in building operation by recalling elevators, closing doors, changing the air-handling controls, or modifying other active or passive fire defenses. When an alerting service is part of a functional active or passive fire defense, the detection event is incorporated into the evaluation of that defense. Although this type of action is important in building evaluations, it is not a part of the occupant-alerting or fire department notification process. The release service is a part of the associated fire defenses.

Local fire departments are notified of a fire by an individual or some type of automatic transmission. There are four types of automatic transmission equipment associated with notification of the local fire department. One involves a type of alarm transmitted to a supervisory station maintained by the building owner. This type of protection is called a *proprietary supervising station system* and is common in hotels, large offices, apartment buildings having a security staff, larger department stores, and many manufacturing businesses. Two other types of third-party notification involve signal transmission through a *central station* or a *remote supervising station*. The fourth type of notification is an *auxiliary system* that transmits a direct signal to the fire service communication center. The principles of operation and basic functions of each are described below.

11.14 Proprietary supervising station system functions

A proprietary fire alarm system serves a single property or a group of noncontiguous properties under the ownership of a single corporate entity. The alarm and signaling devices connect to a supervising station where trained personnel responsible to the owner of the protected property are in constant attendance. This type of system is widely used in large offices and other commercial and industrial organizations that monitor and act on a variety of emergency services. Besides the fire alarm function, a proprietary supervising station may monitor security and other controls. For example, it may trigger fire suppression systems or control building functions, such as occupant alert, elevator recall or emergency operation, HVAC and smoke control, and emergency communications. The proprietary supervising station monitors trouble signals, guard services, and actual fire alarms.

A variety of requirements are specified for the construction of the facility and the design, installation, testing, and maintenance of equipment. Personnel training and standard operating procedures are important to prompt and reliable actions in an emergency.

Buildings that use a proprietary supervising system are normally large and complex. Floors and subsections are separately zoned to identify locations more specifically. Modern equipment is sophisticated and can identify specific locations with digitally coded information. This type of information can reduce the time to fire detection and enable the local fire department to locate the fire more rapidly. The organization of the detection and fire location systems and standard operating procedures should be an integral part of any careful building evaluation.

The time duration for retransmitting a fire alarm signal normally ranges from 15 to 45 s. Here are some conditions that may result in a longer delay of retransmission to the local fire service communication center:

- The attendant may send a guard to investigate before calling the fire service communication center.
- The attendant may notify the owner of the emergency before transmitting the signal to the local fire service communication center.
- The attendant may attempt to verify that a signal means a fire and not a false alarm by awaiting the actuation of other detectors.

11.15 Central station functions

A central station is a facility operated by a person or company whose business is to furnish a variety of services relating to fire protection systems in protected properties. Normally the central station is remote from the protected building, and a single facility will service many buildings from a geographical region. The operating company is listed by Underwriters Laboratories (UL) or Factory Mutual Research Corporation (FM Global). The local authority having jurisdiction (AHJ) verifies the character of the installation by a UL certificate or an FM Global placard. A minimum of two individuals must staff the station at all times to ensure prompt and continuous attention to all signals. To be listed, a variety of quality and redundancy conditions must be incorporated into the system. The main conceptual difference between a proprietary system and a central station is that the proprietary system is under the ownership and control of a single corporate entity. On the other hand, a central station services many different clients.

A central station takes action when signals are received from the water flow alarm of sprinkler systems, actuation of special hazard suppression systems, manual fire alarm boxes, and fire detectors. If the building has a guard service, the central station also monitors the periodic reporting of the guard. The signals may be of operational supervision, trouble, or actual alarms. When a signal reaches the central station, a prescribed set of procedures take place.

Central station services involve a variety of activities for the subscriber on the premises and remote from the premises. On the premises, the central station company is responsible for installation of the equipment and its testing and maintenance. In addition, the company provides a runner to reset the equipment manually when required. A central station company monitors all signals from the protected property, manages the entire system, keeps records, and retransmits fire alarm signals to the fire service communication center. Highly protected risk (HPR) industrial facilities and commercial facilities having property or operational continuity of high value normally use a central station.

When an alarm actuates, the operators are expected to retransmit the alarm without delay to the fire service communication center. After providing notification to the local fire service communication center, the subscriber is notified by the fastest available means.

One may expect retransmission of an alarm to a fire department to be completed within 15–45 s. One can estimate this event more accurately by studying records of past alarm retransmissions.

Although the record of listed central station facilities is good, here are some possible causes of delay:

- contacting the owner before retransmitting the alarm to the fire service communication center;
- poor operator training resulting in "missed" or incorrect responses to an alarm;
- leaving a system in a test mode after testing is complete.

11.16 Remote station functions

A remote station fire alarm system transmits the signal from a building fire alarm system to a remote location, usually a fire service communication center, which should take appropriate action. The remote station facility normally receives signals from properties of several owners. In this context it is similar to a central station. However, there are several important distinctions between the two types of system.

Often a local fire department is unwilling to receive routine supervisory or trouble signals from buildings because of personnel or operational constraints. However, the local fire department may be willing to allow all alarm signals to be received at a location acceptable to the local AHJ. This supervising facility is not required to be listed by UL or FM Global. Nor is the facility required to provide the redundancy of a central station. However, there are specific requirements to comply with the standards for remote supervising station fire alarm systems.

Trained personnel are required to be present at all times, recognize the type of signal received, and take prescribed actions. An alarm signal must be retransmitted directly to the fire service communication center whereas trouble and supervisory signals may be handled differently.

The retransmission of an alarm to the local fire service communication center should take 15–60 s. Although the reliability of remote stations can be excellent, it is not always as high as that of central stations. Here are some factors that can influence the reliability and speed of alarm retransmission:

- The attendant may notify the owner of the emergency before transmitting the signal to the local fire service communication center.
- The attendant may be asleep when the alarm is transmitted; this will delay retransmission.
- Poor operator training may result in "missed" or incorrect responses to an alarm.
- A system may be left in a test mode after testing is complete.

11.17 Auxiliary fire alarm system functions

An auxiliary fire alarm system transmits a fire alarm actuation directly to the local fire service communication center. This system is the fastest and most reliable means of fire department notification. However, it can be installed only if the community has a public fire alarm reporting system and the AHJ grants permission for the connection and accepts all the equipment. If the community does not have a public fire reporting system, the owner cannot install an auxiliary fire alarm system.

A public fire alarm reporting system is most commonly associated with the street box for reporting fires by the public. These publicly accessible boxes are frequently located at or near the main entrance of schools, hospitals, and places of public assembly. The AHJ may designate locations for additional boxes. The public system may be used if the distribution cables and wires and the power supply and circuitry are available for signal transmission. It is possible also to use coded wire or radio reporting systems or telephone reporting systems. The major issues for these systems are that the community has an approved public fire alarm reporting system available and uses a direct connection circuit.

Auxiliary fire alarm systems must be installed and maintained by the owner of the building. Normally, only fire alarms are used in auxiliary systems, although other public emergency calls may be incorporated if they do not interfere with the transmission and receipt of fire alarms. When other types of supervisory or trouble signals are included in the system, a remote station system is recommended.

The time from detector actuation to the community fire service communication center is instantaneous. The reliability is extremely high, although the following factors may reduce the reliability if they are present:

- poor maintenance of cable plant;
- weather-related conditions that adversely affect the cable plant;
- damage to auxiliary boxes due to traffic accidents.

11.18 Occupant actions

Occupant actions often play an important role in building performance during a fire. Among the many potential activities are the following:

- Discover the fire.
- Initiate manual fire suppression during the early stages of fire growth.
- Notify the local fire department.
- Alert other occupants to the existence of the fire.
- Begin egress of the building.

After receiving a cue from a fire, other actions often taken by occupants may include

- investigate;
- hide;
- attempt to rescue others;
- leave the building.

Many of these actions may be sequential. For example, an individual may discover a fire. Subsequent actions may be to alert others, notify the fire department, fight the fire, or take another potential action. Human behavior and fire involve many activities and decisions. Human actions and decisions depend on the situation and other factors such as personal experiences with fire, perception of threat, education, and the actions of others.

The performance of many buildings depends on actions taken by humans at certain moments in the process. Event uncertainty is great, both in time duration and outcome success. Sometimes the human factor is timely and valuable. In other cases, human actions or inactions are a major factor in fire losses. An evaluation incorporates a value judgment for the performance expectations of the human activity link in the process, given the conditions and information that are available for the site.

Performance evaluations and documentation are needed when human responses are an identifiable link in a fire defense component. The only considerations at this time are the timeliness and likelihood for an individual to notify the fire department of a fire. The uncertainties of human behavior make evaluation very difficult. Consequently, the window of uncertainty is wide and the reliability is often low.

11.19 Human notification of the fire department

After detection and human awareness of the fire, many actions may take place. Considering only fire department notification, the next step is to decide to call the fire department. After deciding to notify the fire department, two additional events must occur. The first is actually sending the message. This event considers details such as travel distance and ability to locate a telephone, physical and mental condition of the individual, and the ability to send the message. The second is the accurate receipt of the message by the fire department. Delays can occur because of language differences or because the excitement may cause an individual inadvertently to give an incorrect address. This is a common occurrence.

Notification involves decision making of individuals who are in a stressful situation. This evaluation becomes one of the more uncertain assessments in the process. Nevertheless, the exercise of making these estimations provides an insight into potential weaknesses in the building's fire defenses and clues to potential life safety problems.

11.20 Fire department notification analysis

The notification process provides a link to fire suppression. Whether one considers human notification or a fire detection and alarm system notification, the events in the process are the same. Figure 11.25 shows the SVN diagram to guide time duration and probability estimates for the notification process. Figures 11.26 to 11.28 identify factors that influence notification events.

The time duration between fire detection and fire department notification is linked. It is generally more useful to start by developing a time line for the process. Then one may examine the reliability and express performance in probabilistic terms.

Figure 11.29 shows a network diagram to identify signaling system reliability. As with the detection system, one may establish a 100% reliability at the time of installation by testing the complete system. After establishing this initial reliability, one need only incorporate a derating function to reflect the building's maintenance policies.

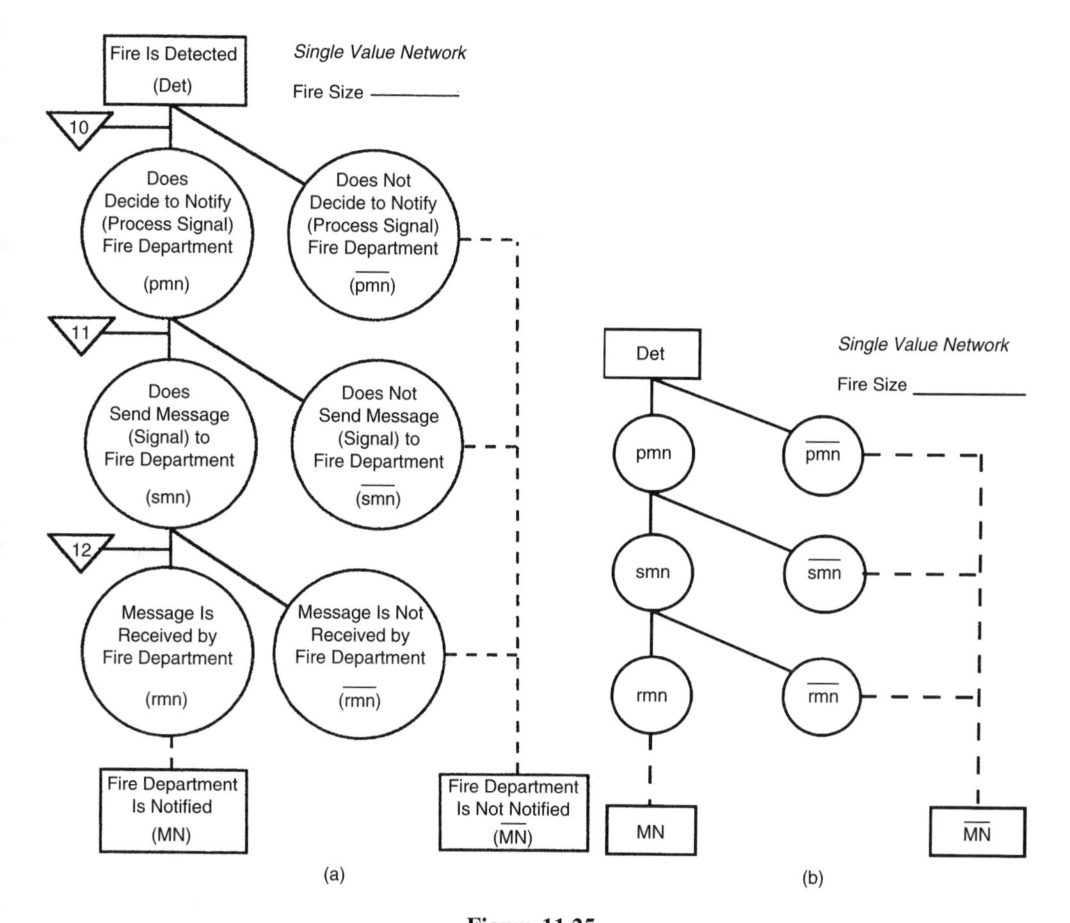

Figure 11.25

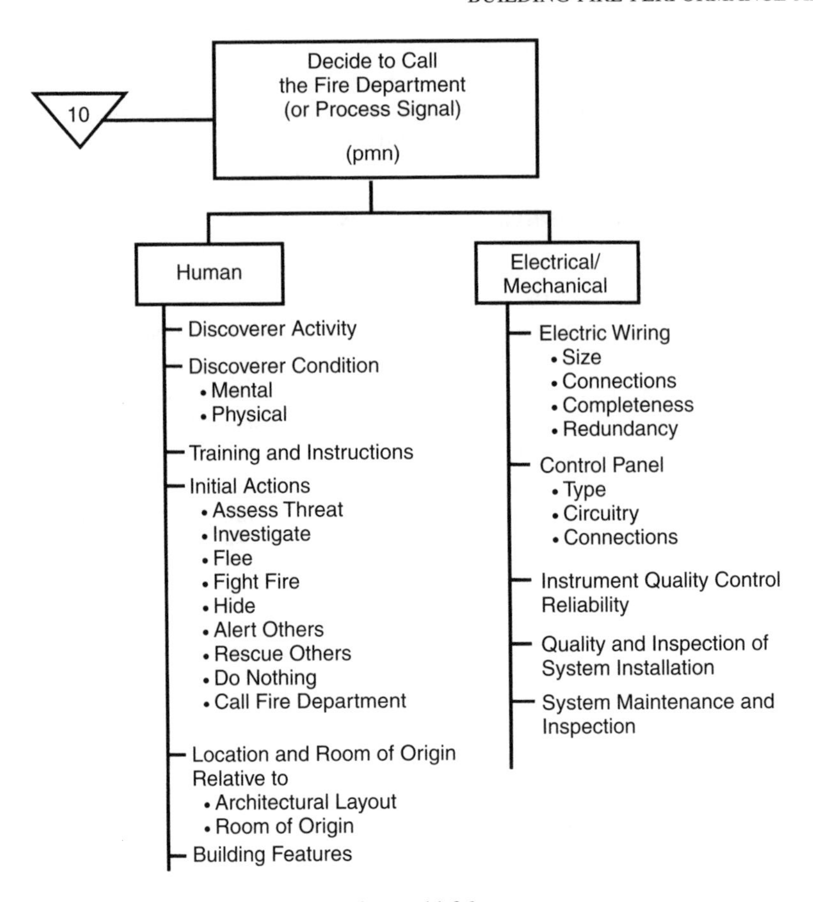

Figure 11.26

Example 11.6

A small office building was evaluated for the time duration and likelihood of fire department notification after detection. Two states were considered. One with the building fully occupied and the other for periods during which the building is unoccupied. Fire department notification is based on a human link because no remote alarm capabilities exist. An analysis of the building, the prefire instructions and training, and the normal activities of the occupants leads to the following evaluation.

Given detection in a fully occupied building, the assessments for fire department notification are shown in Table 11.3. The probabilistic values are based on a SVN analysis. The probability values in the last row were calculated from the SVN of Figure 11.25.

Table 11.3

Time after detection (min)	0	1	2	3	4	5	6	7	8	9	10
Decide to notify FD $P(\text{pmn})$	0.0	0.0	0.1	0.4	0.8	0.9	0.96		0.98		0.99
Send message $P(\text{smn})$	0.0	0.0	0.0	0.1	0.6	0.7	0.9		0.98		0.98
Message received $P(\text{rmn})$	0.0	0.0	0.0	0.0	0.2	0.6	0.8		0.95		0.97
FD notified $P(\text{MN})$	0.0	0.0	0.0	0.0	0.1	0.38	0.69		0.89		0.94

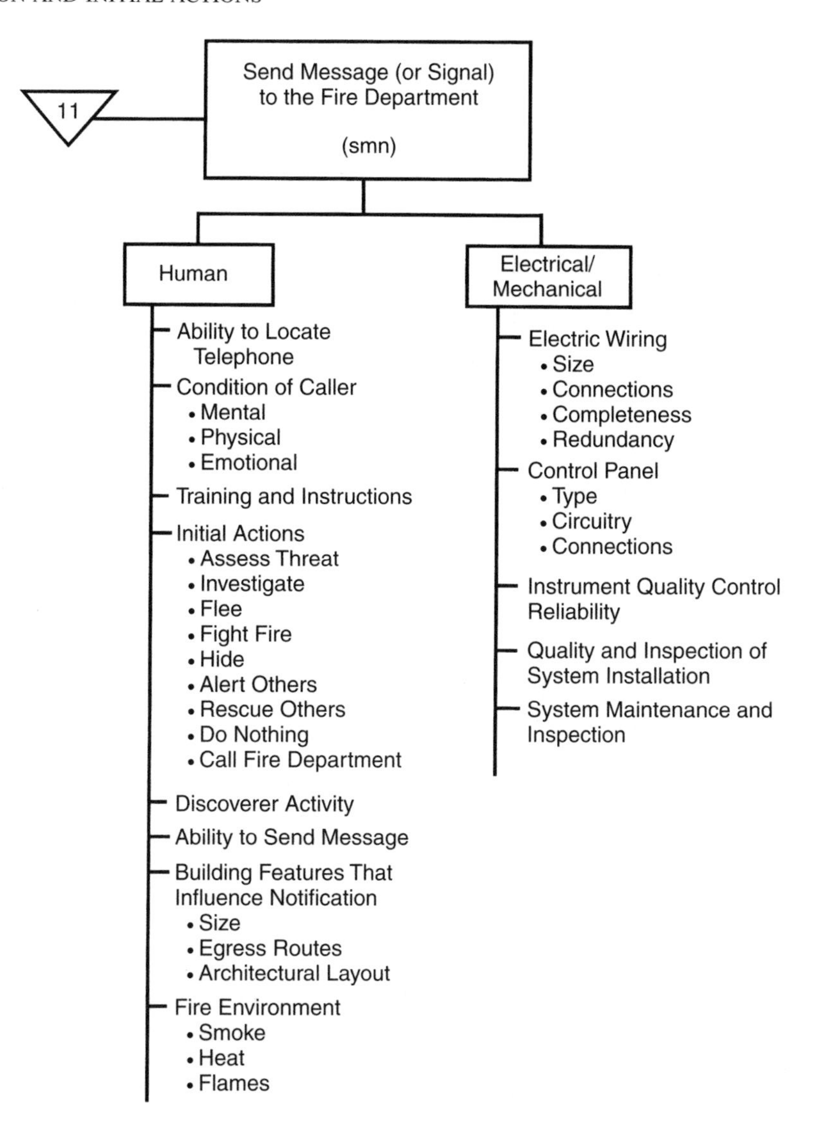

Figure 11.27

When the building is unoccupied, instrument detection will occur, but notification will not occur except by the chance actions of a passing citizen. This chance occurrence is not considered in a building analysis, so the building's success probability of fire department notification is zero.

Solution

Figure 11.30 shows notification curves for these two conditions. Although hypothetical values are given here to describe the process, an evaluation would document the value judgments.

The problem statement and Figure 11.30 describe time durations given detection. The time lag for detection is not included. Figure 11.31 shows the segments of Figure 4.7 that relate to this analysis. Figure 4.7 always shows "global" time. Values for MN had been established using

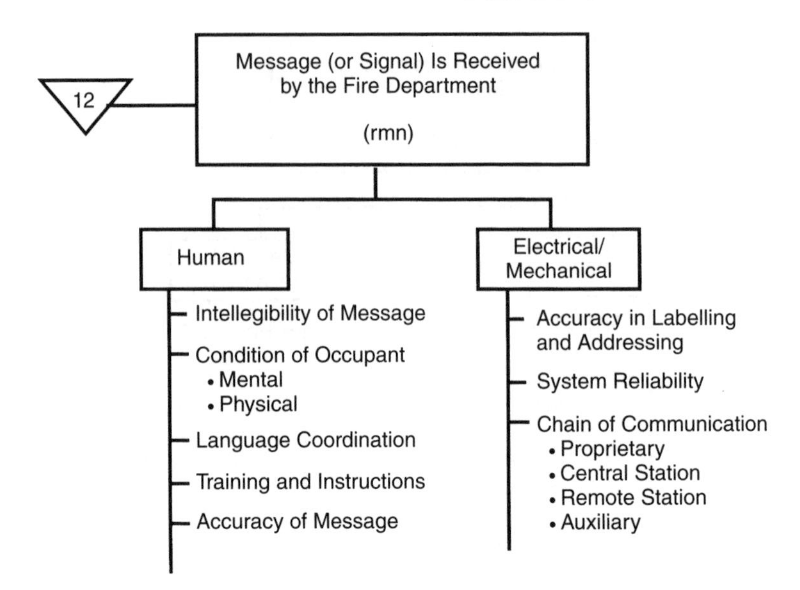

Figure 11.28

a "local" time, but were adjusted to global time for the spreadsheet. The intermediate values for pmn, smn, and rmn are not shown, although these additional rows could be included in the spreadsheet.

Figure 11.32 shows the notification curves for a detection time lag of 4 min. Each cell of a vertical column is a performance curve coordinate. They may be the result of subcalculations, as these values for MN, or an independent evaluation. Each horizontal row of cells defines a CV performance curve.

11.21 Level 3 analysis

A level 3 analysis develops a deeper insight into performance by examining variability of details. For example, perhaps a specialist wants to study the detection and alarm system in isolation. The objective may be to portray the range of detection performance for different fire locations or growth characteristics. Alternatively, it may look at the influence of different actions after detection. This may include a range of fire conditions, detection times and probability values, or occupant states for different scenarios. The analytical framework portrays comparisons with an envelope of performance curves. The understanding that emerges becomes the basis for better communication. Level 3 evaluations allow one to explore in isolation a range of relevant factors or questions. The detailed local knowledge may be incorporated into the global framework to recognize the impact on building performance.

11.22 Level 1 analysis

A level 1 evaluation is a rich and important source for gathering, processing, and understanding performance information. Although one need not express probabilistic values explicitly, the thought process remains the same for all analytical levels. As we cultivate observational skills and

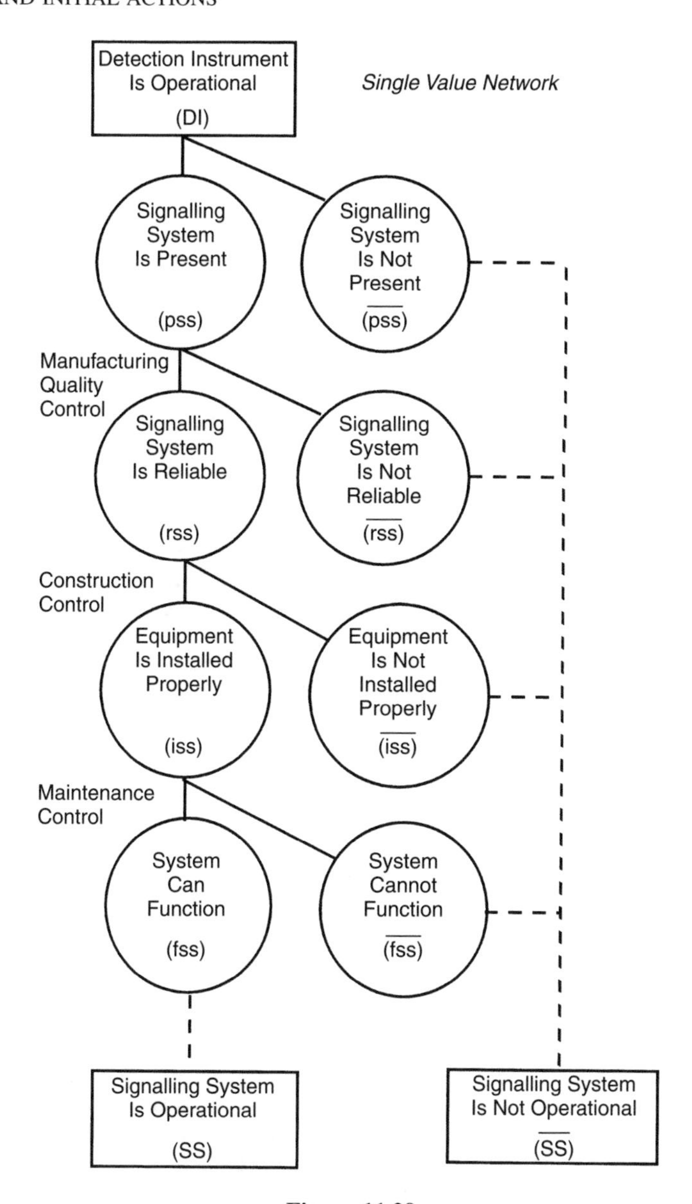

Figure 11.29

estimates, building performance becomes more evident. Level 1 notification evaluations provide information for the M curve and give an insight into life safety risk characterizations.

11.23 Summary

Detection, alerting of occupants, and fire department notification are performance events that relate to other fire defenses. Details are important. Detection time is influenced by detector placement

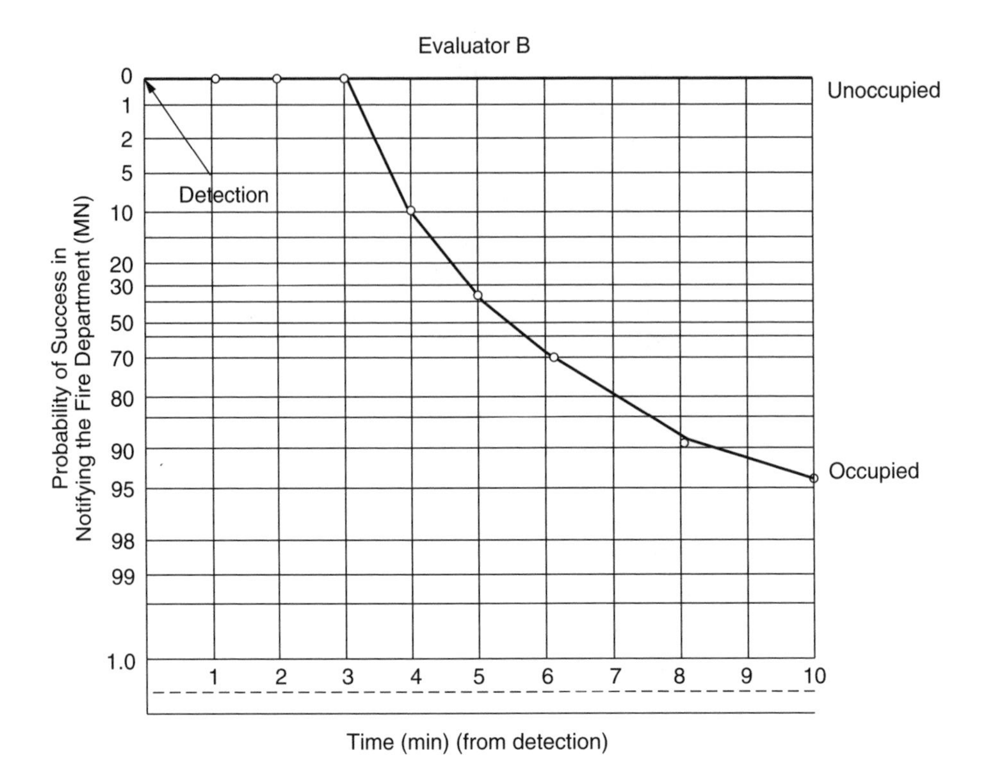

Figure 11.30

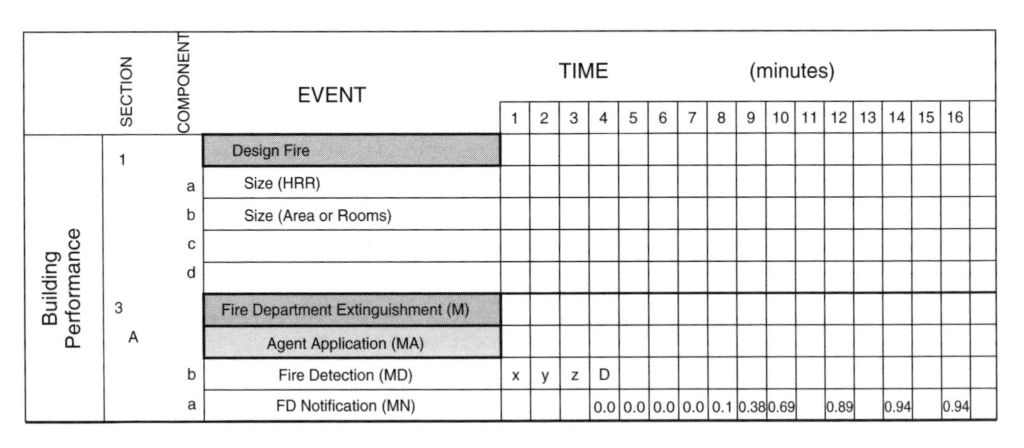

Figure 11.31

and sensitivity, fire location, and combustion product transport, itself influenced by obstructions to flow, ceiling height, and distance. In addition, factors such as interconnectivity, reliability, and power supply are incorporated into the evaluation. Notification examines the process of processing and transmitting information to the fire communication center. Humans can play

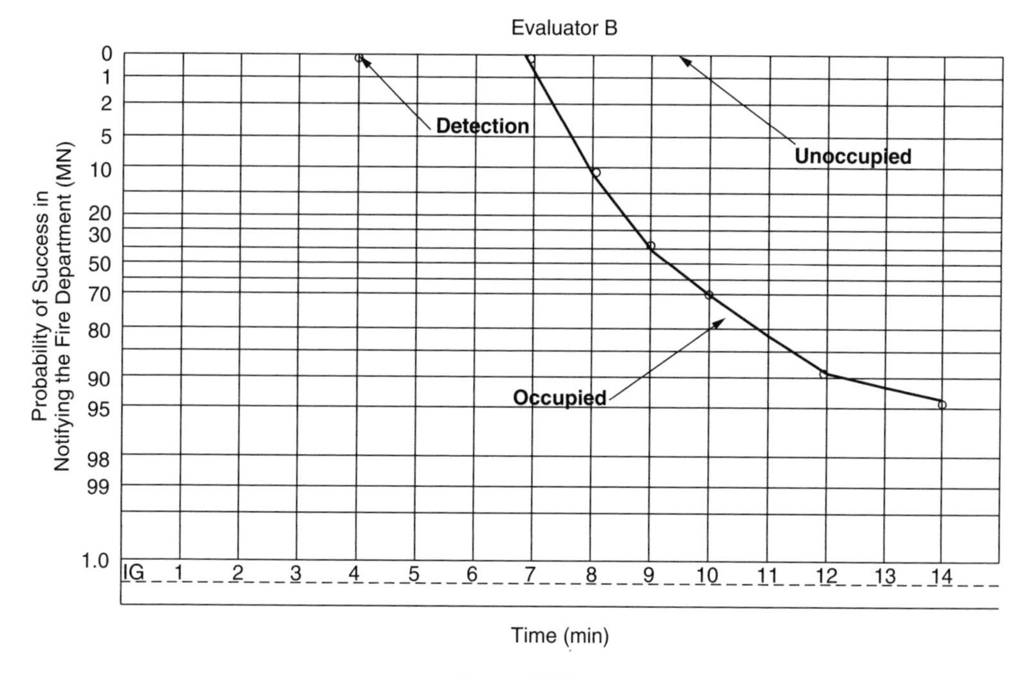

Figure 11.32

a variety of roles in the process. Performance analyses that use probabilistic value judgments give an insight into the strengths and weaknesses of the building detection and notification systems.

12 FIRE SERVICE MANUAL SUPPRESSION

12.1 Introduction

Fire department extinguishment is an active fire defense that can have a major role in a building's fire performance. For example, fire damage in nonsprinklered buildings relies exclusively on barrier effectiveness and fire department extinguishment. Although the fire service has evolved over the years with resourcefulness and ingenuity in fighting fires, the building itself plays a major role in the success or failure of manual fire fighting. The world has more experience in manual fire suppression and less quantitative data with which to evaluate a building's influence on fire fighting than any other part of the entire fire safety system.

A building fire can be a very complicated situation. A fire ground commander must manage the available forces to search for and rescue occupants, locate and extinguish the fire, and protect important building assets. These tasks must be accomplished within a brief time period for a situation in which information is incomplete, often inaccurate, and dynamically changing throughout the incident.

Building and site conditions may help or hinder the manual fire combat. Sometimes the building and fire department interaction results in a successful operation in which people are not injured and relatively little damage occurs. At other times, life loss and major damage to the building and neighboring structures can occur. The building design and construction have a major influence on the success or failure of fire ground operations.

A fire department manual extinguishment (M curve) analysis evaluates building and site conditions for fire suppression. An M curve analysis does not evaluate local fire departments in isolation. Rather, its function is to understand and identify the influence of the building's site, architectural, and fire defense features on manual suppression effectiveness.

This chapter first gives an overview of M curve procedures and describes fire department operations. Then, the level 2 framework structures a way of thinking for analyzing building performance augmented by local fire department operations. Finally, guidance is provided to blend events to order level 1 performances and details.

12.2 Overview of the M curve analysis

A building analysis using fire department extinguishment (M curve) measures performance by estimating the fire damage. The fire damage is the floor area or number of rooms destroyed by

Building Fire Performance Analysis R. W. Fitzgerald
 ISBN: 0-470-86326-9 (HB)

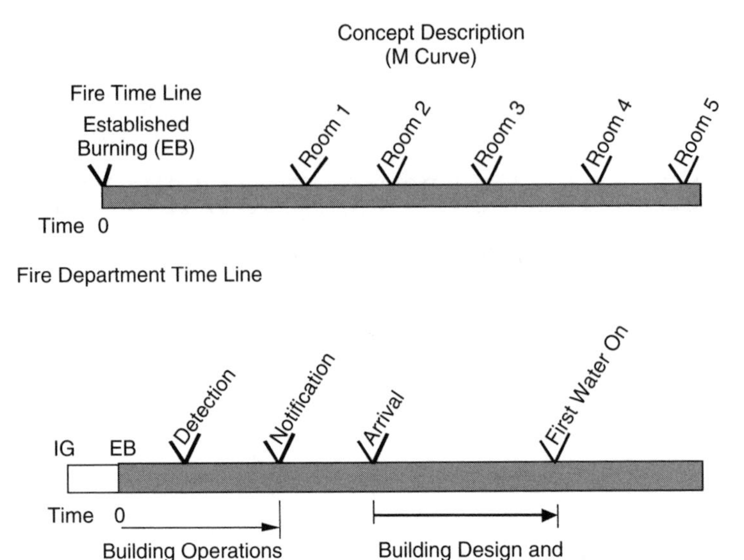

Figure 12.1

flame and heat. Making performance estimates develops a good understanding of the influence of building features on the ability to mount a fire attack.

An M curve evaluation integrates two distinct parts of the building/fire system. The first is a time line that identifies fire propagation through the building barrier/space modules. A second time line tracks fire suppression activities and estimates time durations for each critical event. Their comparison describes the expected fire size at first water application. Figure 12.1 illustrates these time lines. In the fire propagation time line, R_n indicates the sequential room involvement over the time duration as described in Chapter 9. Uneven intervals show that barrier effectiveness or room conditions may cause different propagation times in the barrier/space modules.

The time line for first water application shows the following events:

- the fire size at detection;
- the time duration between detection (MD) and fire department notification (MN);
- the time duration between notification (MN) and the arrival (MR) of the first-in apparatus and the other responding companies;
- the time duration to establish a water supply, find the fire, stretch attack lines, and discharge water on the fire (MA).

When one has a sense of the fire size at first water application, the likelihood of fire department extinguishment without propagation to additional rooms is estimated. Understanding the building features that influence fire-fighting and local fire department capabilities forms the basis for the estimates. Additional knowledge is obtained from answers to the following questions:

- Is enough water available to extinguish the fire?
- Will the water discharge be continuous until the fire is extinguished?
- Can the fire department extinguish the fire before it extends to a larger area?

When effective barriers confine a fire to a single average-sized room or even a small cluster of rooms (e.g., an apartment), extinguishment can be easily achieved. However, when barrier openings allow greater interior fire spread or when exterior propagation involves additional floors, extinguishment becomes difficult. We gain an insight into the importance of building features and their locations by making probabilistic estimates for the events that define the functional process.

M curves define a distinct scenario such as an occupied or unoccupied building and day or night conditions. Each scenario tells a story that will give a different insight into the building's capabilities to work with the local fire department resources. Generally, one uses at least two M curves to gain a good understanding of the building performance.

12.3 First steps

The first step in an M curve analysis is to select a room of origin. Section 7.5 discusses criteria for this selection. Then we construct a time line for fire propagation through a cluster of rooms as described in Chapter 9.

Detection and notification are an integral part of an M curve analysis even though these events are the management responsibility of the building and its occupants. The fire size at detection and the time duration from detection to fire department notification define the segment from established burning (EB) to fire department notification (MN).

The time duration between EB and MN of Figure 12.1 can be established in one of two ways. The first is to base the time durations on actual building conditions. This has an advantage of assessing a building's existing status. A second approach is more useful for identifying new design performance specifications or retrofit changes to an existing detection and notification system. This approach assigns an arbitrary time duration to MN and estimates the fire size after response (MR) and initial water application (MA). Because the fire size at first water application is influenced so greatly by notification, one may identify acceptable damage levels and design the detection and notification system accordingly. This enables one to describe damage in terms like this: If notification occurs within 2 min of EB, the expected damage is X. If MN is delayed to 12 min from EB, the estimated damage will be Y.

Returning to the time lines of Figure 12.1, assume the fire propagation time line and time durations for detection (MD) and notification (MN) have been established. Now attention shifts to the fire department/building system. Time estimates are made for fire department arrival (MR) and the first application of water to the fire (MA). This analysis gives a sense of expected fire size and conditions. This knowledge enables one to estimate the ease or difficulty of fire control and extinguishment without fire extension to larger sizes.

Time is the common element that relates fire growth and fire intervention. Figure 5.1, Section 1 and Sections 3A and B provide a more detailed look at the events of Figure 12.1. Each cell of Figure 5.1 relates a fire size (Section 1) with the following two questions (Sections 3A and B):

- What is the probability that water will be applied at the fire size being studied?
- Given that water is applied at that fire size, what is the probability that the fire can be extinguished before it grows to a larger size?

These events are sequential and conditional. Collectively they represent an *interval of time*, rather than an instant of time. Using the time line analysis and answering these questions is an indispensable approach to performance evaluation. The understanding that unfolds in an M curve analysis gives a clear insight into the building architecture and site features that influence fire performance.

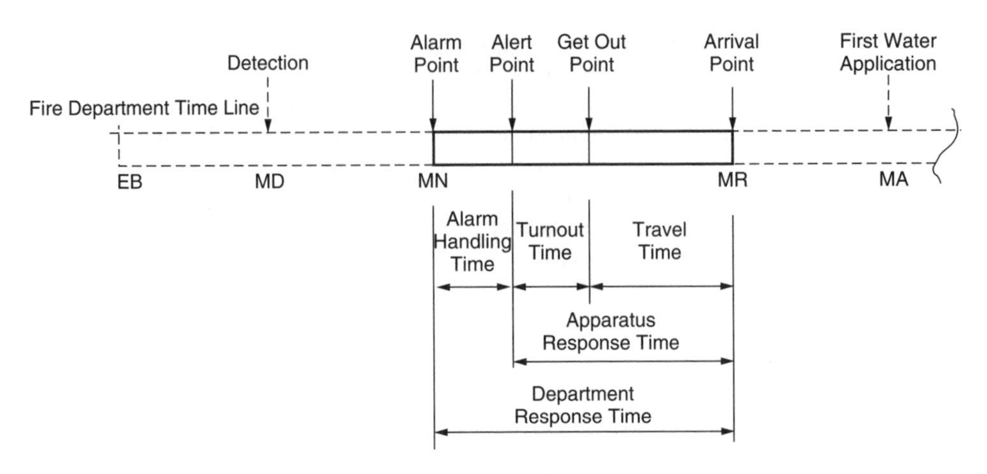

Figure 12.2

12.4 Fire department response

The response time duration is a combination of several distinct operations between fire department notification (MN) and arrival at the site (MR). The standard operating procedures (SOP) of any fire department are important to identifying a response time for a building. These SOP as well as fire station locations, the number and type of apparatus responding, and conditions that affect travel time all influence the response time. The operations that make up the response time are

- alarm-processing time;
- turnout time;
- travel time.

Each of these activities is under the management control of the local community and may vary from one community to another, even when the communities are near to each other. One can establish an alarm-handling time and turnout time, and then use those estimates for every building in the community. The travel time is easy to determine for any site location. One may estimate durations for alarm-handling time, turnout time, and travel time from measurement, statistical data collected from audiotapes, calculation, or other information sources. Figure 12.2 shows a schematic of the process.

12.5 Types of fire department

There are three groups of fire fighters that make up community fire services. They may be described as full-time, paid-on-call, and volunteer fire fighters. Full time fire fighters are responsible for continuous staffing of fire stations plus responding to fire and nonfire emergencies, training, and inspections. Fire stations are generally staffed 24 hours a day with full-time fire fighters. The number of personnel can vary greatly among communities.

Volunteer and paid-on-call fire fighters have employment outside the fire service and respond to emergency calls. They participate in regular training programs to build or maintain their fire-fighting skills. Paid-on-call fire fighters earn wages for the time that they respond to emergency calls. Volunteer fire fighters do not receive any compensation for response to calls. Paid-on-call and volunteer fire fighters normally experience a narrower variety of incidents than full-time fire fighters.

Communities provide experience a narrower fire services that are appropriate for the local fire problem. A building evaluation examines the response time and the availability of fire-fighting forces. Suburban and rural communities generally use volunteer and paid-on-call fire fighters. During daytime hours when many volunteer members may be at work in other communities, limited staffing may attend a fire. On the other hand, considerable staffing may be available at night or on weekends.

A variety of mix 'n match combinations can exist in local fire departments. Some communities have only full-time fire fighters. Others have only volunteer or paid-on-call fire fighters. Many have a nucleus of full-time fire fighters augmented by volunteer or paid-on-call fire fighters during those emergencies that are beyond the capability of the on-duty staff. Some communities have only a full-time chief and part-time fire fighters.

The organization of the community fire department influences the type and response speed and the number of personnel who can attend the emergency. Standard operating procedures will vary enormously from community to community, and the M curve analysis must consider these procedures.

12.6 Fire companies

A variety of equipment is available for a fire department to do its work. The core units for most building fires include engines, ladder trucks, rescue vehicles, and the staffing to accomplish their functions. Specific customs, equipment, procedures, and words may vary between geographic regions and communities. The goal here is to provide a basic familiarization and vocabulary of fire fighting.

The function of an engine company is to extinguish the fire. The vehicle (apparatus) used to deliver the water is usually called an engine or pumper. An engine company establishes a water supply at the building site from underground piping system and hydrants, by drafting from an available body of surface water or cistern, or by drafting from water brought in by tanker trucks. Although an engine carries a water storage of 500 gallons or more on the vehicle, a reliable or sustained extinguishing water supply is external to the vehicle. The water on the vehicle provides rapid delivery to a small fire or a temporary initial supply until a more reliable external water supply has been established. The engine company lays supply lines to transport water from the water source (e.g., hydrants, ponds, tank reservoirs) to the pumper. Attack lines are stretched from the pumper to the fire. Alternatively, attack lines may feed the fire department connection (FDC) and additional attack lines may start from the standpipe connection and move toward the fire. Figure 12.3 shows common arrangements.

Water application to a fire requires people to do the necessary tasks. A workable engine company staffing is commonly viewed as an officer and three fire fighters. The actual staffing in a community may be greater or less than this number. The staffing may vary from an officer and one fire fighter to an officer and five fire fighters. Personnel availability, local politics and protocol, and the local SOP are the usual bases for engine company staffing. No matter the actual number of personnel present, the work remains the same. Insufficient staffing means that the water application will take much longer or standards of safety will be reduced.

A ladder company does everything on the fire ground except apply water to the fire (and sometimes it does that too). A ladder company may provide assistance with forcible entry, search and rescue, ventilation, and property protection by moving it and covering the contents with salvage covers. Ladder companies also carry ladders for placing against the building to allow occupants to escape or the engine company to enter. A ladder company has the important capability to deliver an elevated master stream in a defensive fire-fighting operation. An officer and three fire fighters usually form an efficient team, although the actual staffing will vary considerably

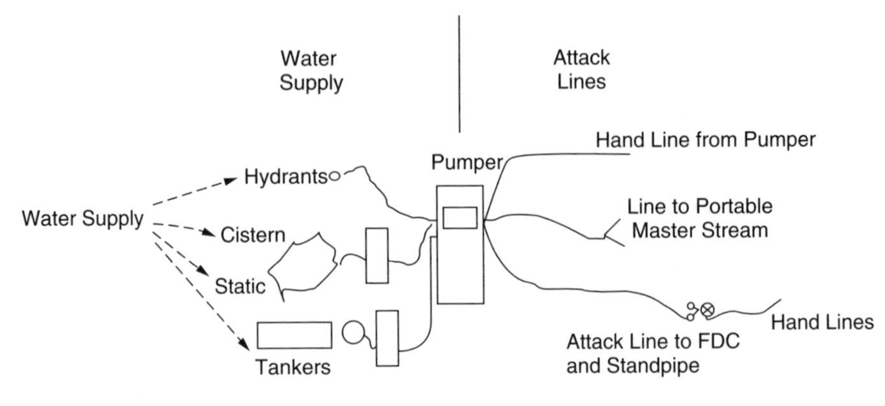

Figure 12.3

among different communities. If a ladder company is not available, engine company personnel must do those duties at cost of time and efficiency.

A quint is a combination vehicle that has been gaining popularity in recent years. A quint combines the usual engine capabilities of pumper, booster (water) tank, and hose with an aerial device (usually 55–75 ft) and ladders. Several quints provide a flexibility for each arriving vehicle to assume a different sequential role in the fire-fighting process to adapt on-site needs more easily.

A rescue company or rescue squad provides a combination of engine company and ladder company functions. Rescue companies are often viewed as elite, fast response units that can conduct an initial search and rescue or fight the fire. The actions depend on the needs at the specific site at the time of arrival. Often the rescue company will have an officer and five fire fighters, although the number may vary. Because of the function and nature of a rescue company, this number does not usually change substantially.

Besides engines (pumpers), ladder trucks (trucks), and rescue vehicles, many fire departments have a variety of special-purpose vehicles for specialized emergencies. For example, local departments may have special vehicles to deal with hazardous materials, water rescues, additional hose supply, or forest fires. Tank trucks to transport water are common in many rural communities. The equipment reflects local community needs and resources.

Within any individual community, the number of vehicles, their storage locations, the staffing, and the number of vehicles assigned to a first alarm or to multiple alarms will vary substantially. Menke [1] noted that the number of personnel at a first alarm may vary from 8 to 37. Some communities initially send a single engine to the scene. A common first response is two engines, possibly one ladder truck or quint, and a chief. Sometimes a rescue company augments this response.

The role and perception of fire departments have been changing significantly. A generation ago, fire departments were perceived to have only fire-fighting functions. Today communities recognize that the fire department is the local emergency management agency. Many fire departments provide emergency medical services of varying types to a community or region. In addition, the local fire department is the first service considered for emergencies such as collapse, rescue, terrorism, and flooding. The range of emergency services is broad and increasing in scope and complexity.

A building fire evaluation requires an understanding of the number and type of vehicles that attend first and multiple alarm responses, and the expected staffing. Often staffing can vary during different parts of the day, particularly in volunteer and paid-on-call departments. A study of the local fire department SOP and the community resources is an essential part of an M curve evaluation.

12.7 Alarm handling

Notification (MN) is the instant at which the community fire service alarm center is aware of the fire. This is not the instant when an individual transmits the message. Nor is it when a device or third party becomes aware of the fire. For example, a telephone call to 911 or a fire detection transmitted to a central station or remote station does not complete notification. The alarm point occurs only at the instant when the *fire service* alarm center becomes aware of the emergency. Notification (MN) completion defines the beginning of the fire department response (MR) event.

We define the *alert point* as the instant when the fire emergency is transmitted to the responding fire companies. In full-time fire departments, the alert point becomes the instant when a bell, horn, or other device is sounded in responding fire stations. In some communities the alert point is the instant when the responding fire fighters receive the first sound of a radio or telephone message calling them to the fire emergency. In other communities the alert point may be the sounding of a siren at a central location.

The *alarm-processing time* is the time duration between notification and the alert point. Each community has its own protocol for identifying the responding companies and alerting them. The alarm-processing time can vary between communities. In some locations the alarm-processing time may be as little as 20 or 30 s. In other communities, as much as 2 min could elapse before responding companies are alerted. A time duration of about 1 min is common.

Communities that combine fire and police dispatching may be less sensitive to the differing needs of the fire service regarding alarm handling than are dispatch centers dedicated only to fire department operations. During a fire, fire ground messages may be relayed to the dispatch center and back to the fire ground. Dispatchers who are sensitive to the differences between police and fire conditions and have the appropriate vocabulary to communicate effectively are essential when sharing communication among different public safety departments.

Collectively, the *alarm-handling time* includes all the activities between notification (MN) received and dispatching the responding companies. An M curve evaluation involves an analysis of the SOP of the community alarm system. The communication center is usually a vital link in the management of the fire service and in the safety of the building being evaluated.

12.8 Turnout procedures

The *get-out point* is the instant when the wheels of the responding apparatus cross the threshold of the fire station. The *turnout time* is the duration between the alert point and the get-out point. Turnout time can vary between communities depending on the type of fire department and the departmental SOP.

When fire fighters live in the station, turnout time is often 20–40 s during the day and 30–60 s at night. In paid-on-call and volunteer companies the turnout time depends on the SOP. These departments use two common procedures. In one method, responding fire fighters are alerted of the fire emergency and a fire fighter will drive the engine from the station directly to the fire scene. Fire fighters individually drive their own vehicles to the fire ground, some often arriving before the first apparatus. When the department has a driver on duty or when the driver lives and works near the fire station, the turnout time may be short. When the driver must travel to the station, the turnout time is longer. A second common procedure has responding fire fighters travel to and assemble at the fire station. After enough fire fighters have responded to staff the company, the apparatus leaves the station. This procedure may involve a greater turnout time. Some volunteer companies have members living in the stations to reduce turnout time.

A complete M curve analysis requires the analysis of the departmental SOP for turnout and response to the fire scene. This involves the first engine and additional fire apparatus that will respond. Some communities have only one engine. Others have several engines and other types of emergency vehicles. Automatic aid and mutual aid agreements are common among neighboring communities in some jurisdictions.

It is important to understand the local community resources and procedures. Time durations for the first company arrival as well as a complete response can vary significantly. Community and mutual aid time and resources may be influenced by other emergencies such as brush fires, electrical storms, wind storms and blizzards, and other natural and man-made emergencies that produce a high frequency of emergency calls. The fire grows exponentially while individuals move linearly, and a few minutes can dramatically change the suppression tactics and damage for a building.

12.9 Travel time

The *arrival point* is the instant at which the wheels of the fire vehicle stop at the building site; *travel time* is the time duration between the get-out point and the arrival point. Different pieces of apparatus will have different travel times because of station locations and different speeds of travel. Because an effective fire attack may require several fire companies, one identifies travel times for all the responding vehicles.

The travel time is among the easiest estimates. One knows the distance between the station and the site, as well as the route difficulty, terrain, and natural impediments to travel. For example, limited access highways, bridges, and railroad crossings often limit movement along particular routes. An M curve analysis identifies this information.

Most pieces of fire equipment are large, heavy, and cumbersome to drive. Corners, hills, and unimproved roads slow travel speeds. Traffic congestion may slow travel time if gridlock occurs. Ordinary automobile congestion does not usually reduce travel time much, because private vehicles usually clear the way for emergency vehicles. Weather can influence the travel time. Fog-reduced visibility, ice, and heavy snow can reduce speeds of emergency vehicles. Nevertheless, one may construct reasonable travel time estimates.

The only additional time duration to be incorporated into the travel time involves site access. The *respond to site* event is completed when the wheels of the engine stop and fire ground operations begin. There are situations when the building may be visible, but site access problems, such as insufficient turning radii, speed bumps, difficult terrain, and narrow, incomplete, or blocked streets delay the vehicle placement at the site. Site access is another part of the response analysis.

12.10 Setup time

The *agent application* point (MA) is the instant at which the first water is applied to the fire from the attack lines. The *setup time* is the duration between the arrival point and the agent application.

The duration for this setup period will vary tremendously depending on the site conditions, the building and fire conditions, the staffing and types of equipment responding, and their associated activities. The development of time line events associated with the setup time provides a major part of the performance understanding for manual suppression. The evaluation process provides a good insight into the building design features that influence this performance and the extent of damage that is likely to occur.

Because of their importance in understanding building performance, we will describe some fire ground operations to provide a background for evaluations. In addition, we will discuss architectural and site factors to create an understanding of building evaluation for fire suppression.

When a fire department learns of a fire emergency and the responding forces are traveling to the site, they never know exactly what they will encounter when they arrive. However, when they do arrive and begin to take action, a generalized process unfolds that can portray the activities which must take place. Because of the potential for many contingencies and conditions, one must recognize that the fire ground commander adapts decisions to meet the perceived needs of the situation. Although the organization and activities may be related to any type or size of structure, the mental image on which the process is modeled is an architecturally designed commercial, business, or residential building.

Over the next few sections we shall try to convey a sense of fire ground activities for a working fire. The goal is to provide an organized, descriptive framework of operations so that one can recognize the general process. This understanding is important to be able to recognize the implications of site and building design decisions on fire fighting.

The process of fighting a fire involves accomplishing a number of discrete activities. One should not infer that this procedure is step-by-step. The activities are done when they need to be done as determined by the conditions perceived at the time. Some activities must be completed sequentially. Others may be undertaken simultaneously. In many ways, the process is similar to that of a construction project. Staffing, equipment, and existing conditions influence time durations to complete discrete activities. We have borrowed some concepts of construction management to organize building evaluations for fire control and extinguishment.

12.11 Fire ground operations

The first arrival is usually an engine company. This first-in company places itself in what appears at the time to be the most appropriate position to carry out initial activities. This involves balancing several potentially conflicting needs, including gathering information about the location and size of the fire, learning if there are any potential life safety threats, establishing a water supply, and stretching attack lines. Shortly after the first company arrives, a chief and other companies arrive.

The fire ground commander, also called the incident commander, is in charge of the fire scene. Initially it is the officer of the first arriving company. However, this position is transferred when a more senior officer arrives. The fire ground commander who is in responsible charge of the scene and all operations is placed in a situation of making rapid decisions with incomplete, often inaccurate, and constantly changing information. In addition, these decisions must be made during periods of maximum distraction and emotional stress within very short periods of time. The process becomes decision making under uncertainty in a very pure form.

Experience, knowledge, and training are important to a fire ground commander's educational background. Timely, astute decisions can make the difference between a good stop and a fire that got away. The other important element to the success or failure in limiting the fire size is the influence of the building site and design features. The collection, storage, and rapid information retrieval of those features during preincident operational planning can greatly affect fire suppression and safety.

An initial *size-up* becomes the basis for the fire ground commander's decisions. A size-up is a mental model of the entire system that makes up the fire ground conditions at the time of arrival. A size-up begins on route and becomes more defined on arrival at the site when better information becomes available.

Fire ground operations use available community resources, manpower, and equipment to rescue endangered individuals and to reduce the destructive damage of a hostile building fire. The size-up provides the information on which to base decisions about the strategy and tactics to achieve that goal. An ideal size-up would include the type of information described below. Obviously, time constraints of required early action preclude collection of all the information initially. However, when complete information is not available, the fire ground commander makes decisions based on the available knowledge at the time.

LIFE SAFETY

How many occupants are in the building? Where might they be located? Are they capable of self-evacuation or guided evacuation? How certain is the information?

BUILDING CONSTRUCTION INFORMATION

In buildings of this type, do barriers generally contain a fire to one or few rooms, or do they allow the fire to propagate easily? What is the collapse potential? Can the building construction be used to advantage in controlling the fire and limiting its propagation? Does this building present a safety hazard to fire fighters?

FIRE

Where is the fire? What is its size? What is burning, and what is the fuel loading? Are hazardous materials present? Are they involved in the existing fire? Can they become involved if the fire extends?

WATER SUPPLY

How many hydrants are available and where are they located? How much fire flow is available? Are static water supplies available for drafting? Must tankers supply water?

FIRE-FIGHTING RESOURCES

How many fire companies and fire fighters are available? What other types of apparatus are present or available? Where is the equipment now placed? Should it be relocated? If additional equipment responds, where should it be placed? Are personnel needed for a primary search and rescue or are they available for fire fighting? How many hose lines can we lay and where should we position them? Are fire fighters available for relief, particularly in hot weather and difficult fire-fighting conditions? What fire flow can be delivered internally and what is available externally?

FIRE ATTACK

Where are the building access locations? Are there any natural or man-made obstacles to approaching the building from different sides? Where are the stairwells? Are standpipes available and where are they located? Are they wet or dry? How do we gain access to the floor of origin? What fire attack routes are available? How clearly are they recognizable? Will doors have to be forced to reach the fire location? How many? What size and how many hose lines can we stretch with the staffing available? How long will it take to stretch the hose lines? Must we position hose lines at critical locations to defend people or valuable property from fire extension? Will these positions help or hinder fire attack or control? Where and how can we ventilate?

ENVIRONMENTAL CONDITIONS

What is the time of day? What are the weather conditions (e.g., extremely hot or cold, storms, precipitation)? What site conditions will influence movement (e.g., mud or snow)? How will the wind conditions affect fire fighting? What are the smoke, heat, and visibility conditions within the building? What are the heat and visibility conditions near the fire?

EXPOSURES

Are there any other external buildings exposed to the fire? Should we protect those exposures before or after attacking the existing fire? Are there internal exposures that should be protected?

BUILDING SERVICES INFORMATION

What fuels are present, where are they stored, and where is the fuel shutoff located? What is the electrical system and where is the shutoff? Is emergency generation available? Do we have emergency lighting? Where? Does the HVAC system continue operation, shut off, or shift to an emergency mode of operation at detection? Can its operation be changed to help fire fighting? Where are the controls and how is it done? Are elevators in operation or is there a recall at detection? Is there a fireman's key for emergency use? What other building services information is available that would influence life safety or fire fighting?

RESOURCE AUGMENTATION

Does standpipe water need to be augmented from the building's fire department connection? Does the building have a sprinkler system at the fire location? Would connection to the sprinkler siamese enable the system to operate? Are additional alarms or mutual aid needed?

Better decisions can be made with more available information. However, the fire will not wait. As it continues to burn, the conditions change. The fire companies will take initial actions according to experience, preincident building information, preconceived operational planning, and immediate needs before a size-up is completed. The fire ground commander will modify a plan to reflect conditions that are evident or that change during the fire.

Communications and the chain of command are essential features in fire ground operations. New information about conditions inside the building must be transmitted to the command area. Communication is crucial to update status reports, deploy resources, and to learn the outcomes of any actions.

Life safety is a primary concern for a fire ground commander, but the focus of this chapter is on evaluating a building for manual suppression. Manual suppression involves five coordinated operations:

- find the fire;
- establish a water supply;
- stretch attack lines and initiate agent application to the fire;
- prevent extension of the fire to other spaces;
- extinguish the fire.

Depending on the size-up, the fire ground commander may employ an interior offensive attack to extinguish the fire aggressively. Alternatively, the commander may select an exterior or interior defensive procedure to prevent fire extension to threatened exposures outside or within the building. Building and fire conditions may change over time, requiring the fire ground commander to alter tactics from an offensive mode to a defensive mode. An M curve evaluation provides an understanding of the building and anticipated suppression operations to estimate the likely extent of fire damage.

12.12 Finding the fire

Although locating the fire may seem a trivial problem, the process is not as easy as it may appear. Small buildings do not usually pose much difficulty. However, finding the fire in large or complicated buildings can be time-consuming and difficult. Dense smoke obscures visibility, and a fire is often located by feeling the heat rather than by seeing the flames. Consequently, accurate information given to the fire officer can reduce the time to agent application in a complicated building. Accurate information on the fire location is needed to place attack hose lines effectively and to anticipate potential life safety or property protection concerns.

An officer can learn of the fire's location from an occupant, security person, or a building annunciator. An annunciator, technically called a *visible alarm signal appliance*, is a unit having a panel that identifies the location of the fire. If a building has an annunciator, the unit should be located for easy recognition. Often the visible display panel may be near the entrance or in the lobby. At other times the annunciator may be located so unobtrusively that it cannot be found easily.

After finding the annunciator, the display should give clues to the location of the fire. These clues should be specific, direct, and unambiguous. For example, if the first detector actuated on the south wing of the seventh floor, an indicating light should identify that location. Unfortunately, some annunciators are not that specific. Additionally, the location may be difficult to read or ambiguous. Therefore a part of an M curve evaluation is to study the location and quality of the annunciator and the likelihood of its providing a clear, unambiguous fire location.

If the fire officer is not directed to the fire by an occupant or by an annunciator, a reconnaissance search must be initiated. Clues, such as the lowest smoky floor level, are a start. An air-handling system that does not automatically shut down at detection and spreads smoke throughout the building can thwart smoke clues. A flaming fire visible to the outside gives a strong indication, if that side of the building provides sight access. In any event, the fire officer uses the clues available to scout out the fire location. Recognize that visibility may be poor and locked doors may delay access to some building locations.

Locating the fire can sometimes be difficult and time-consuming. The ease and expected time duration for locating the fire are parts of the M curve analysis.

12.13 Establishing supply water

The establishment of supply water for attack lines is another part of the evaluation process. Engines and quints will carry water in their tanks. Often the 500 or 750 gallons is sufficient to provide enough water to extinguish a smaller fire or to start a fire attack. However, to provide enough water for a larger fire or building, three principal sources of water supply are

- underground public or private water distribution systems with hydrants;
- a body of water such as a lake, pond, stream, or cistern from which water can be drafted;
- tanker trucks (tenders) that shuttle water and deposit it into portable tanks at the fire scene for attack pumper use.

The most common water supply in communities with underground piping is through the hydrants. Supply hose lines move water from the hydrants to the pumpers. The pump provides the desired pressure and delivers water to the building supply lines, standpipes, or attack lines. Underground water distribution pipe sizes, distances of the hydrants from the building, the number of hydrants available, the distance to the pumpers, and available staffing are part of a supply evaluation.

If underground piping and hydrants are not available, water must be drafted from an available pond or stream to the pumper. Sometimes dry hydrants with piping extending to a pond may be used when the pond does not have ready access from the road. Establishing this type of water supply can be very time-consuming and its reliability is uncertain during periods of drought or deep freezing. If these water sources are not near the building, alternate sources must be considered.

Tankers provide the third supply source. Tank trucks relay water to a temporary reservoir that must be set up. The pumper drafts from this reservoir to provide water for supply lines. An analysis evaluates tanker discharge scheduling, staffing needs, and reliability during different times of the day and seasons of the year.

Whatever the source of fire-fighting water, supply hose lines must move the water to the attack pumper. Staffing, terrain, and weather conditions are important to the successful establishment of supply water. The laying of supply hose lines from the source to the pumper can involve a variety of possible evolutions, depending on the situation. The most common of these variations is the use of relay pumpers to boost the water quantity or pressure. But, whatever the specific evolution, laying large-diameter hose lines requires a lot of manpower. Obstacles such as hills, fences, long distances, and weather conditions can have a major influence on the time to establish supply water.

Site conditions are very important in an M curve evaluation. They influence the location of hydrants and water supply. The site also affects apparatus placement, building readability for fire-fighting features and devices, and the ability to reach parts of the building to initiate a fire attack.

12.14 Initial fire suppression considerations

Although stretching attack lines and applying extinguishing agent cannot be completed until the fire is located, these activities often accompany reconnaissance operations used to locate the fire. Some activities must be sequential and others may be done simultaneously.

An initial major consideration is what type of attack to use. The fire ground commander must select whether to use a defensive mode or an offensive mode. This may or may not be independent of whether to mount an exterior or interior attack. The size-up becomes the basis for the initial decision. As more information is acquired and the success or failure of initial operations becomes apparent, the fire ground command strategy may change to fit the needs of the situation.

If the fire ground commander believes the fire can be confined and extinguished without further fire extension, an offensive attack may be used. An offensive mode of fire fighting involves laying attack hose lines and aggressively applying agent to the fire. When this can be done, the fire is often extinguished relatively quickly.

There are two mechanisms used to extinguish fires small enough to be put out by direct attack. One is to apply water into a confined room in which the fire that has not yet reached full room involvement. The water converts to steam, prevents air from continuing the combustion process, and smothers the fire. To use this form of extinguishment, the fire fighters must have the modern protective clothing, often called bunker gear. This clothing protects against burns, and the self-contained breathing apparatus (SCBA) enables a fire fighter to breathe in this atmosphere.

A faster and more effective way of extinguishment "pushes the fire out of the building." When a room is ventilated to the outside and water is discharged, the expanding steam and fire gases drive the fire out of the building. Among the many benefits of ventilating a fire at the proper time is that visibility is improved and the hot combustion products are pushed out of the building.

Envisioning the fire being "pushed" is very useful in building evaluations. To some extent, the fire department can control the direction in which the fire will move. If people or valuable property (e.g., company financial records or data storage) are threatened, the direction in which the hose lines are placed and the building openings through which the fire is pushed become important. Among the factors involved in the building's architecture are considerations of possible fire extension to other building parts during extinguishing operations. The building architecture provides fire attack routes and contributes to the fire movement.

An aggressive fire attack is usually an interior attack and normally results in a relatively rapid knockdown. However, the fire ground commander faces a variety of factors that enter into the attack decision. The threat of fire extension to exposures is among the more important considerations. An exposure is an adjacent building or a part of the building currently on fire that poses a threat for ignition and fire extension.

When exposures are threatened, the fire ground commander must decide if the best strategy is to aggressively fight the fire or to protect the exposures from fire extension and to allow the fire to burn itself out. A third approach protects exposures and also applies water to the burning fire, but not in an aggressive attack. These latter strategies are described as defensive modes of fire fighting.

Among the many determinants for this decision are the number of hose lines that can be employed and their effectiveness; the availability of external appliances, such as deck guns, ladder pipes, and multiple 2 1/2 in attack lines; the water supply; the building construction and potential for collapse; and the effectiveness of building construction in preventing extension of the fire to the exposures.

12.15 Critical fire size

The *critical fire size* is the fire size at which the fire ground commander will use a defensive mode of fire fighting rather than an offensive mode. Every building has a critical fire size. A complicated building may have several critical fire sizes, depending on the fire location and the building construction. Inherent in a building analysis is the recognition that different critical fire sizes can exist in different parts of a building.

The critical fire size depends on the building design and the local community resources and capabilities. Fires that reach magnitudes larger than the critical fire size before manual suppression can become effective have the potential to grow to enormous proportions. Fires smaller than the critical size at the time of first water application have a better likelihood of extinguishment before the fire can spread.

When enough rooms are fully involved and the fire department cannot extinguish the fire with an aggressive fire attack, the critical fire size has been reached or exceeded. Even if the fire ground commander changes to a defensive mode, or has started in a defensive mode, the building or a large segment of the building will be lost.

The critical fire size is unique for areas in a building. Different spaces or room clusters within the same building may have different critical fire sizes. The evaluation concept recognizes that an offensive attack can extinguish fires smaller than the critical fire size within a short time duration. Because a critical size fire is the demarcation between offensive and defensive fire fighting, fires larger than the critical size will result in much greater damage to substantial parts of the building and potentially greater losses from halted production.

12.16 Interior fire attack

The general components of an interior fire attack involve the factors described below. A building analysis would note building design features that affect the local fire department's ability to extinguish the design fire.

- Gain building access.
- Find the fire.
- Gain floor access with the necessary equipment.
- Advance attack hose lines to encounter the room or rooms of involvement and complete initial agent application.
- Estimate fire control success of the initial attack.
- Evaluate need and success of additional attack hose lines to contain or extinguish the fire.
- Check for fire extension, particularly in adjacent combustible concealed spaces.
- Evaluate extinguishment and containment success for different stages of the design fire.

Entry points spaced around the building provide building access. Most often these entry points are entrance doors, and occasionally they are access panels or windows. The first arriving officer must learn the type and severity of the emergency. This is often done by going into the main entrance or another entrance. Rarely is there a problem with access during normal hours of building operation. However, what is the situation when an emergency fire alarm actuates, nothing is showing, and the building is completely locked? If a key to the building is not available, the officer must make a decision to use forcible entry or to wait for a responsible individual to open the building. If a fire is burning, the delay will allow substantial fire growth to occur before the fire attack begins. If the alarm is a malfunction and forcible entry is used, the owner must repair the damage to the building. One element of a building evaluation identifies procedures for entry of fire officers during an alarm when the building is closed. If no special provision is made (e.g., secure locations for building keys), does the owner have realistic expectations of possible outcomes in an alarm response?

Building access is the first step in the process. The next phase is to find the fire. Is the annunciator panel easy to find when no one is present to show its location? Will the panel show the location of the actuation clearly, distinctly, and unambiguously? If a fire officer must scout the fire by reconnaissance, can the building be lighted or must flashlights be used? Building security is often a major interest. Are guard dogs on the premises? Must several locked doors be forced to gain needed access? Preincident planning can obtain much of this information.

This discussion attempts to create an awareness of factors that influence water application time durations. An architectural design usually focuses on normal building operations. Fire can occur during occupied or unoccupied conditions. One must construct enough M curve scenarios to understand the major situations. Each building has its own distinct characteristics that must be incorporated into an evaluation.

After the general fire location is known, attack hose lines must be stretched. Constructing a time line forces one to understand the building features that influence the process. For example, must hose lines be stretched from the attack pumper to the fire or are standpipes available to connect attack lines closer to the fire? Stairwells serve as "beachheads" for launching a fire attack more safely. Standpipes in stairwells may be wet or dry. The time and water quantity to charge a dry standpipe or to increase the pressure in a wet standpipe are part of an analysis.

Building obstacles, environmental conditions, and available staffing to accomplish the necessary tasks influence stretching attack hose lines. Identifying fire attack routes is a first step. One may envision a direct route from the entry point to the fire, but this may not be the actual route selected. Stretching hose lines is hard physical work that takes time to accomplish. Fire fighters act on the best information available. Their decisions initially may be based on inaccurate or misleading information that causes inefficient route selection. Retracing a route to start again is time-consuming and difficult, particularly after a hose line has been charged. Therefore accuracy and completeness of information to clearly locate the fire, along with a sense of building circulation patterns, become important ingredients in an analysis.

ARCHITECTURAL OBSTACLES

After a fire attack route has been selected, one identifies the architectural obstacles along the path from the entry point to the fire. An architectural obstacle is a building feature that requires a time duration to move a hose line through or beyond it. Architectural obstacles may be classified as

- doors and locks;
- corridors;
- corners (turns);
- stairs;
- large rooms (decisions);
- clutter and contents impediments;
- hazards.

The type of door, its hardware, wall construction around the door, and the locked or unlocked status affect the time to lay a hose line. The barrier must be crossed. If keys are not available for locked doors, forcible entry will occur. Forcible entry requires appropriate tools, staffing, techniques, and time.

The length of corridors is important because of the time it takes to traverse the distance and because it is easier to pull the first few feet of hose than the last few feet. The time to stretch a hose line along a corridor is not a linear function.

Corners are important in moving charged and uncharged hose lines. Hose connections can bind at corners, and stretching a charged hose line around corners is a major task. When a hose is bound at a corner, a fire fighter must return and dislodge the connection or ease the line. A fire fighter may be needed at corners to help to advance the line. If that level of suppression staffing is not available, anticipate delays in laying the line. Often three corners (turns) are considered a rule of thumb for a single company to maneuver a hose without substantial delays.

Stairs offer similar problems. In addition to 180° turns, the physical effort to carry the heavy hoses and other equipment up stairs causes fatigue. Standpipes are normally installed in stairwells of buildings to shorten the time needed to lay an internal attack line to a fire.

Large rooms having several doors leading to potential attack routes can cause delay in traversing the distances and in requiring decisions on the proper door to use. A mistake in selecting the proper door may not be readily apparent, and retracing steps takes time.

Clutter and contents can impede the laying of hoses if it becomes necessary to thread the hose around the contents. Again, connections can get hung up in going around corners and charged lines are very difficult to move.

Finally, hazards can cause an individual to pause to consider personal mortality and safety. Electrical hazards are generally recognized. However, when one encounters a "laboratory" sign, biological, chemical, and nuclear hazards come to mind. The room may contain a computer laboratory. Alternatively, it may contain chemical or biological toxins. Caution is usually exercised, and this can influence the fire's outcome and fire suppression. Unfamiliar buildings may have holes in floors. In nonvisibility situations these conditions and disorientation can pose significant hazards. Movement can be very slow when safety concerns arise. Prefire information is very helpful in these situations.

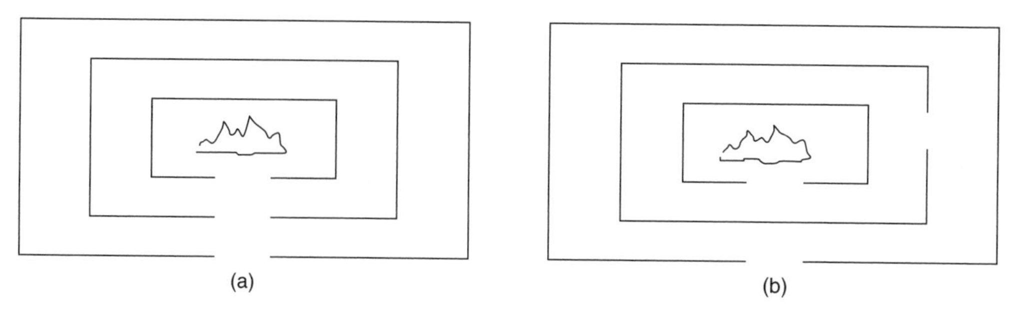

Figure 12.4

Elevators may reduce time to agent application by moving equipment with which to fight the fire. The recall of elevators and the control of their subsequent use by the fire department can greatly reduce the time to agent application in tall buildings.

The architectural layout does make a difference. Figure 12.4 illustrates architectural circulation conditions that affect the time to lay hoses.

ENVIRONMENTAL CONDITIONS

Environmental conditions affect the time to stretch a hose line. Inside a building these conditions may involve visibility, heat, fatigue, or safety. Outside the building, extreme hot or cold temperatures, precipitation, and site conditions like mud or deep snow affect time to complete the necessary tasks.

Visibility is the most common environmental problem inside a burning building. Smoke near a fire is often so dense that one cannot see the fire or building features. Movement by touch and go requires longer time durations to traverse the distances or to find the fire. Ventilation practices of opening windows or roofs to the outside will raise the smoke level to a position where fire-fighting operations are easier.

The building looks very different under normal operating conditions and lighting than it does during a fire. Environmental conditions have a significant influence on fire ground operations. Dense smoke can reduce visibility to nearly zero, even if lighting is available. This significantly affects the time to accomplish the tasks needed to attack a fire. The general fire location to effectively place attack lines is a "macro" fire location. A "micro" fire location relates to fire fighters applying water to the seat of the fire. Smoke conditions are often so dense that a nozzleman cannot see the fire. To apply water in smoke, the fire fighters may need to listen for the fire or to feel the heat to direct hose streams. Thermal imaging cameras help to locate the seat of a fire or hidden fires. Ventilation is important in helping to locate a fire in a smoky environment.

Not only can the fire's heat affect an individual, but the work needed to accomplish the tasks causes accelerated fatigue. Heat from the fire and heat retained inside protective clothing increases discomfort and fatigue and affects performance. Fatigue can occur because the operations are physically and emotionally demanding, motion is restricted, and the effort over a short time duration is vigorous. All these factors increase time durations for agent application.

STAFFING

Staffing has a great influence on the time to agent application and the type of offensive or defensive mode that can be mounted. Considering the tasks related to achieving agent application with a single hose line, the operation is a team function. Personnel available to do the tasks influence the time duration to complete an operation. Insufficient personnel need more time to apply water. Adequate personnel will reduce the time duration. During the operations, the fire is growing independently and exponentially. If the fire has more time to grow, it may exceed the capability of the fire department to extinguish it with an aggressive interior attack. When this occurs, the department is forced into a defensive mode, and a longer fire duration and greater damage will ensue.

The numbers of hose lines and people influence the tactics of fire suppression. As a rule of thumb, one engine company stretches one hose line and can extinguish one room. Two engine companies with two hose lines can extinguish two fully involved rooms. We do not wish to discuss room sizes and extinguishing capabilities of hose lines, but this gives some sense of proportion for evaluating fire suppression. It also provides a sense of the advantage in stretching several hose lines at a fire. Several hose lines enable a fire ground commander to select the most appropriate attack lines to use for the situation when conditions are understood better. Multiple hose lines can defend barriers and prevent extension to adjacent rooms while the primary attack line is extinguishing a fully involved room. This reduces damage in some building fire situations.

CONSTRAINTS

Constraints to interior fire-fighting operations can be important. One is the capacity of SCBA air tanks. Normally 20 min of use is a reasonable estimate. If the fire is not extinguished within the time limitation provided by the tanks, replacement crews may be needed. Again, staffing available at the site becomes an important function in an evaluation.

A second constraint relates to actual hose distances. Normally an engine company will carry two lengths of hose to the standpipe or supply hose connection. When the actual distance exceeds 100 ft, additional lengths must be obtained. This takes additional time, which increases the time for agent application. The stretch from 40 ft to 60 ft is very different than the stretch from 90 ft to 110 ft. Consequently, fire attack path distances from standpipe connections and the fire location are important.

In summary, the building and site features have a major influence on the mode of fire extinguishment and its effectiveness. The community resources of water supply, number and type of apparatus and equipment, and available staffing are also important. The M curve analysis integrates these factors to describe the likelihood that the building/fire department system will extinguish a fire of $1, 2, \ldots, n$ rooms. An M curve evaluation should ensure that no one will be surprised at the outcome of a hostile building fire. In particular, the influence of building features and their locations in the process becomes clearly understood.

12.17 Callery method

Time durations for laying attack lines are difficult to estimate. Even individuals who have fire ground experience normally think in terms of task completion rather than time durations. Callery [2] developed procedures to evaluate building fire attack routes. Unfortunately, the time durations were based on estimates, and the development of practical tools must await resources to obtain field data involving architectural obstacles and staffing for the simulation program. When the decision-support structure is developed, buildings can be evaluated for manual fire attack consistently and easily.

We can easily envision and measure the time to walk along a building fire attack route during conditions of no fire, clear visibility, no way-finding difficulties, and an unobstructed, continuous path. This hypothetical friction-free condition establishes a minimum time to traverse the fire attack route. A variety of building, staffing, and environmental conditions require more time to stretch a hose line along that path. These building conditions create a friction to the unobstructed movement. Callery's attack difficulty index is the ratio of the time to stretch an attack line to the friction-free time. This "building friction" can be expressed as an equivalent distance analogous to friction-related equivalent lengths in sprinkler piping design.

The Callery method breaks a building attack route into architectural segments, such as doors, corridors, stairs, turns, and way-finding decisions. Equivalent distances for these architectural segments are obtained through a series of multipliers that reflect such conditions as status of doors, and environmental conditions. Some of these multipliers are nonlinear to reflect factors such as the last ten feet of hose being harder to stretch than the first ten feet. Two other factors are hose size and whether it is uncharged, charged, or has water flowing. Staffing may have little effect when runs are short, but can have significant impact with long distances. Movement affected by environmental conditions is incorporated.

Although reliable values are not yet ready for publication, a framework for estimating time durations for a specific building has been developed. Completion requires some experimental data for computer simulations and field testing to validate its accuracy. When completed, this tool will enable buildings to be evaluated for local fire department suppression with greater confidence.

12.18 Concepts in fire extinguishment

The final event in an M curve analysis involves the completion of manual extinguishment. This may be expressed as blackout, an old fire-fighting term, or by other terms, such as knockdown and tampdown. Whatever term is used, fire extinguishment (except for hot spots) is the event being evaluated.

Eventually the fire will go out and the fire department will have contributed to this outcome. The success of this event is not that the fire will eventually go out, but will the fire department extinguish the specified fire before it extends to a larger size? In other words, given that the agent application (MA) component is successful, what is the probability that the building/fire department system will prevent the fire from spreading to additional spaces or larger areas?

The building is one major determinant of manual suppression success. The number of hose lines and their positioning is another. The question is one of fire extension. A fire is *under control* when the fire ground commander is confident that the fire will not extend to additional spaces. The fire may not be out, but its future course of action is understood and, in time, it can be extinguished without further extension.

The major considerations of extinguishment without fire extension are

- the effectiveness of barriers in preventing fire extension;
- the perception of structural collapse;
- the number of hose lines that can be placed effectively;
- sufficient water quantity and continuity;
- the size of rooms;
- the number of rooms already involved;
- the ability to ventilate fire-involved spaces when necessary;
- the potential for fire extension into concealed spaces.

Each consideration has an influence on the others. The evaluation integrates their effects in the probability assessment that manual suppression will extinguish the fire before it extends to additional spaces. The estimate is a judgmental decision whether the fire will be controlled within the area under consideration or whether it will extend to additional spaces.

12.19 M curve construction procedure

The framework for an M curve evaluation is based on the physical performance of the building in preventing fire extension and on functional fire-fighting operations. Before describing M curve construction, let us pause to review the process for manual suppression evaluation. This review provides a step-by-step procedure to evaluate a single coordinate of the M curve.

The goal of an M curve analysis is to help one understand the building's fire performance for the manual suppression fire defenses. An evaluation is based on specific, selected fire scenarios that do not "average" different conditions. The goal is to understand the fire performance through telling a story using plausible, reasonable scenarios that will enhance communication between all parties.

Step 1: Select a room of origin and the fire scenario that is a good representation for the building The room of origin will be a representative room or a room in which ineffective barriers would allow easy fire spread to other rooms. Alternatively, a room of origin may be selected to evaluate conditions where fire-control or fire-fighting difficulties are evident. Most buildings require only one or a small cluster of rooms to understand the M curve performance well. Complicated or large buildings may require a few rooms of origin to give a more comprehensive understanding of the performance expectations.

Step 2: Construct a time line for fire growth This is the M curve design fire. It provides FRI time durations for the room of origin and all sequential fire propagation rooms. This means that barrier effectiveness (Chapter 9) is an important constituent in the designated fire. The connectivity and fire extension procedure described in Chapter 9 is a convenient technique on which to base design fire time lines. The influence of doors, windows, and other barrier penetrations is a major consideration in the scenario development.

Step 3: Select a time delay from established burning to fire department notification This time delay for notification may be based on expected conditions in the building, or it may reflect "what if" conditions for which the individual may wish to describe expected outcomes. When different time delays are expected for occupied and unoccupied conditions, two different M curves will result. Other scenarios may be selected if one needs additional information for special conditions.

Step 4: Construct a time line for fire department attack and initial agent application Given the time duration from established burning to fire department notification, the critical events for this time line include

- time duration for fire department response;
- time duration for first water application of the initial fire attack.

The response time involves alarm-handling time, turnout time, travel time, and site access time. These values can vary between communities, but they are relatively easy to determine. The time from arrival to applying first water in the fire attack is more difficult to assess. However, this analysis provides the insight into the building and site features that influence fire fighting. The major activities are gaining building access, establishing a water supply, finding the fire, stretching hose along the attack route to the fire, and discharging agent. The process is sensitive to staffing, and building features can influence activities and the time to completion.

Step 5: Compare the time lines for fire size and agent application This comparison gives a first indication of the fire size at first water application and a sense of proportion for the scope of the extinguishment operation. There are many uncertainties that will be evident in the time line construction. A network analysis addresses these uncertainties.

Step 6: Select a fire size to evaluate The fire size is the abscissa value for the M curve coordinate being evaluated. One makes a separate M coordinate evaluation for each fire size of interest. Frequently the room of origin (or a part of it) is the first evaluation. Then sequentially larger numbers of rooms are evaluated until the outcome for additional rooms is recognized clearly. Evaluation needs and the level of understanding will guide how many and which rooms to include.

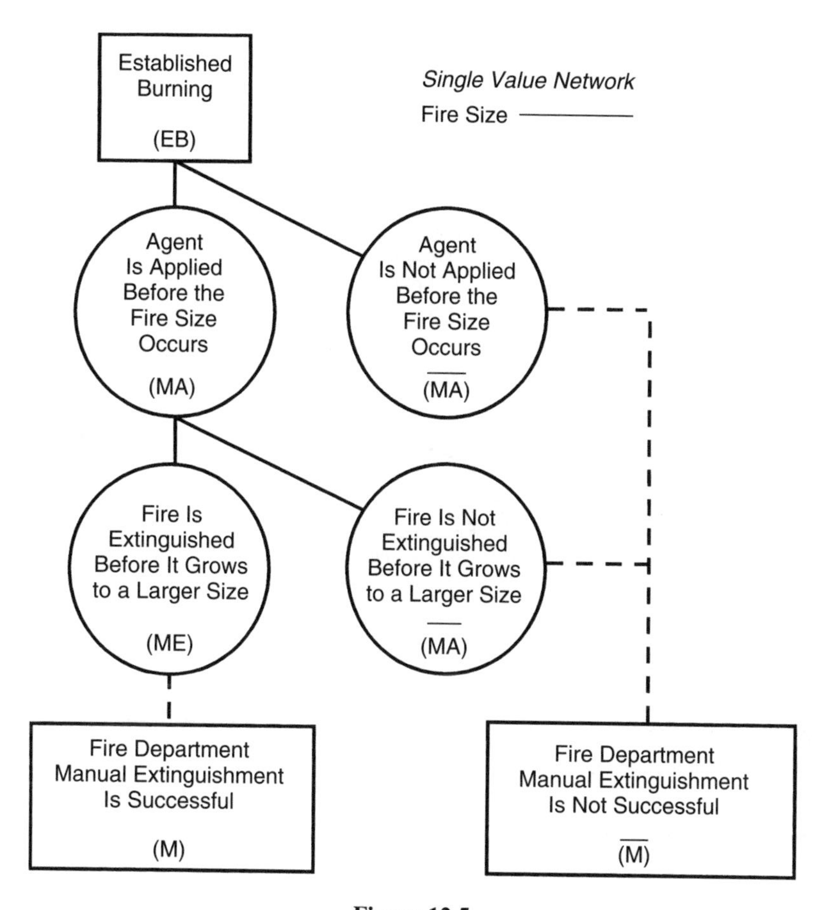

Figure 12.5

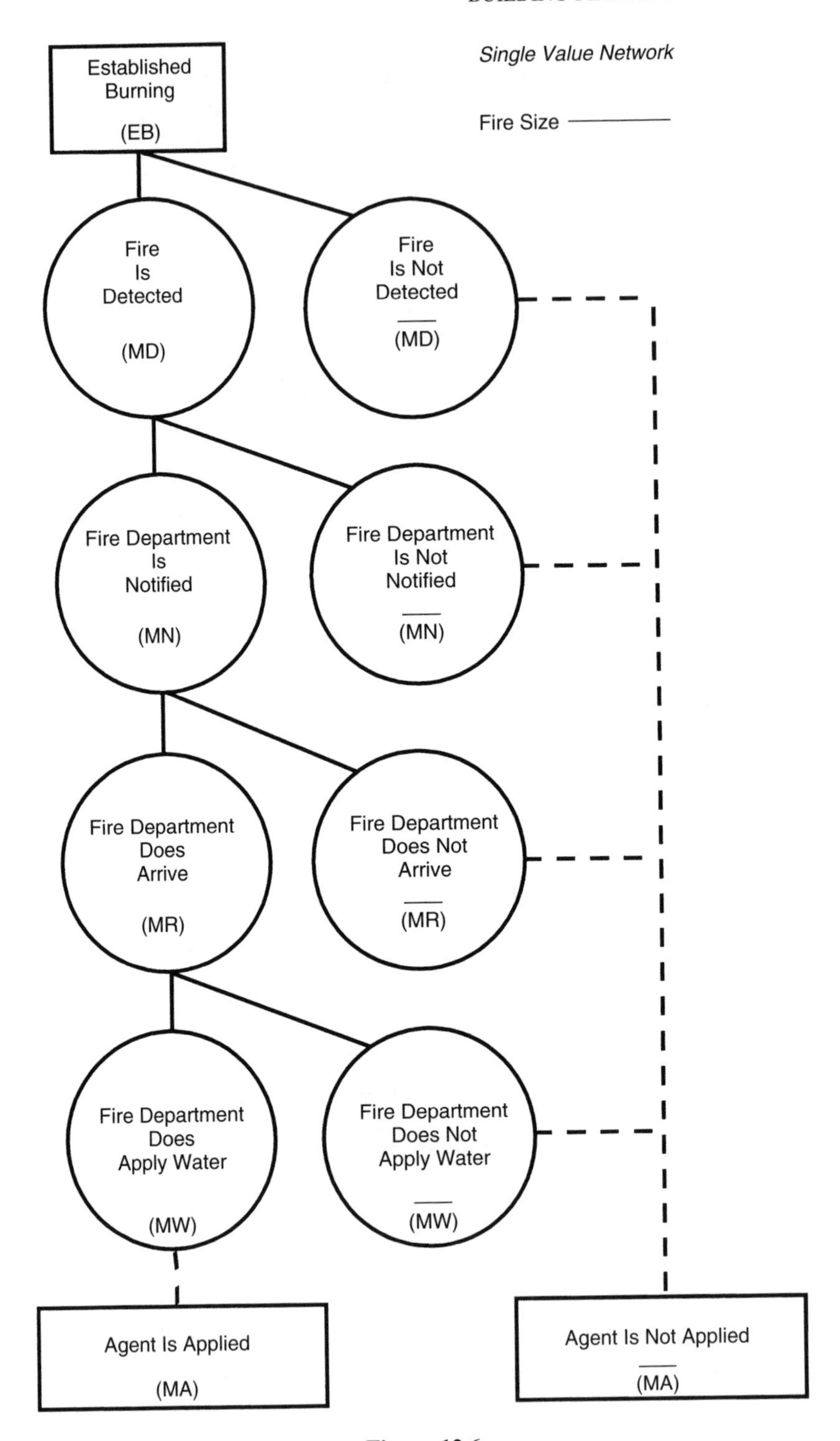
Single Value Network
Fire Size
Established Burning
(EB)
Fire Is Detected
(MD)
Fire Is Not Detected
$(\overline{MD})$
Fire Department Is Notified
(MN)
Fire Department Is Not Notified
$(\overline{MN})$
Fire Department Does Arrive
(MR)
Fire Department Does Not Arrive
$(\overline{MR})$
Fire Department Does Apply Water
(MW)
Fire Department Does Not Apply Water
$(\overline{MW})$
Agent Is Applied
(MA)
Agent Is Not Applied
$(\overline{MA})$

Figure 12.6

Step 7: Using the understanding that unfolds during time line construction, make the value judgments for the SVN diagrams This process enables the building/fire department synergism to be expressed in terms of performance. One can clearly recognize variables inherent to the analysis and uncertainties that are present. Probabilistic values for the events represent their significance.

Step 8: Construct the M curves and document the basis for the building performance The M curve provides a visual representation of the dynamic relationship between the fire propagation and fire suppression performance. This visual descriptor communicates performance expectations, and associated documentation can provide the basis for the estimates.

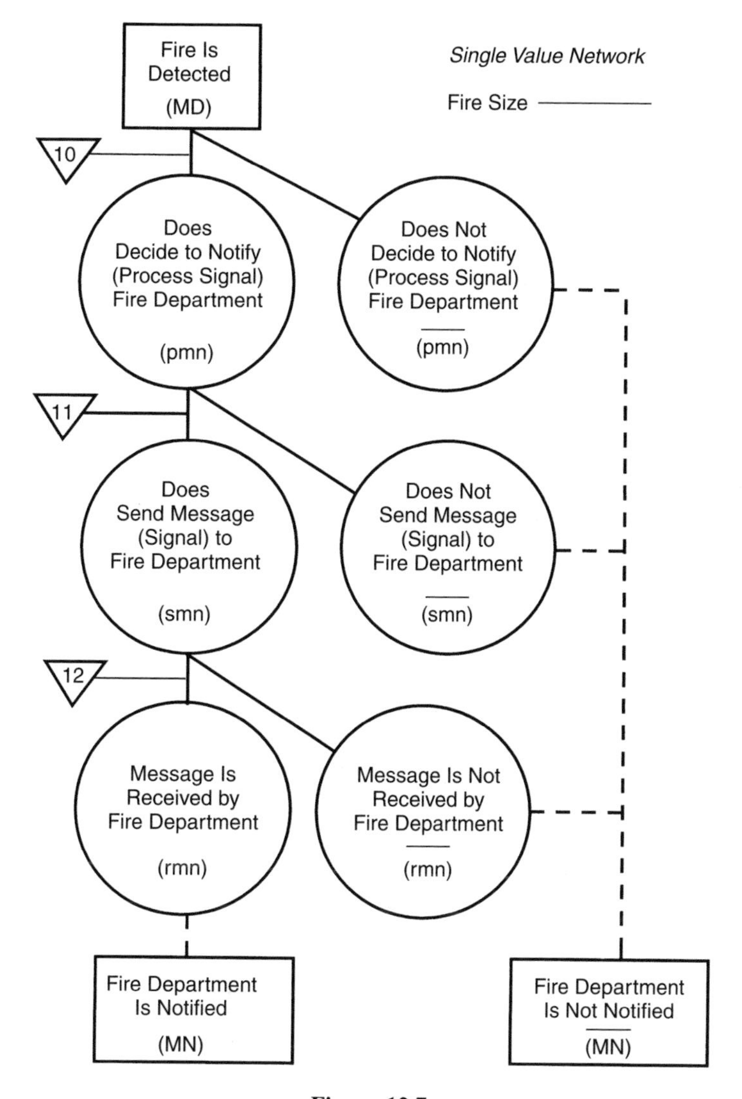

Figure 12.7

12.20 Level 2 analysis

A level 2 M curve is straightforward. One develops time lines and examines only enough M coordinates to establish a reasonable performance understanding. The building evaluation establishes the documentation rationale. Figures 12.5 to 12.9 group the manual suppression activities into functional network packages. Figures 12.11 to 12.16 show the factors that influence event evaluations. Figure 12.10 shows the hierarchical structure. Detection and notification were described

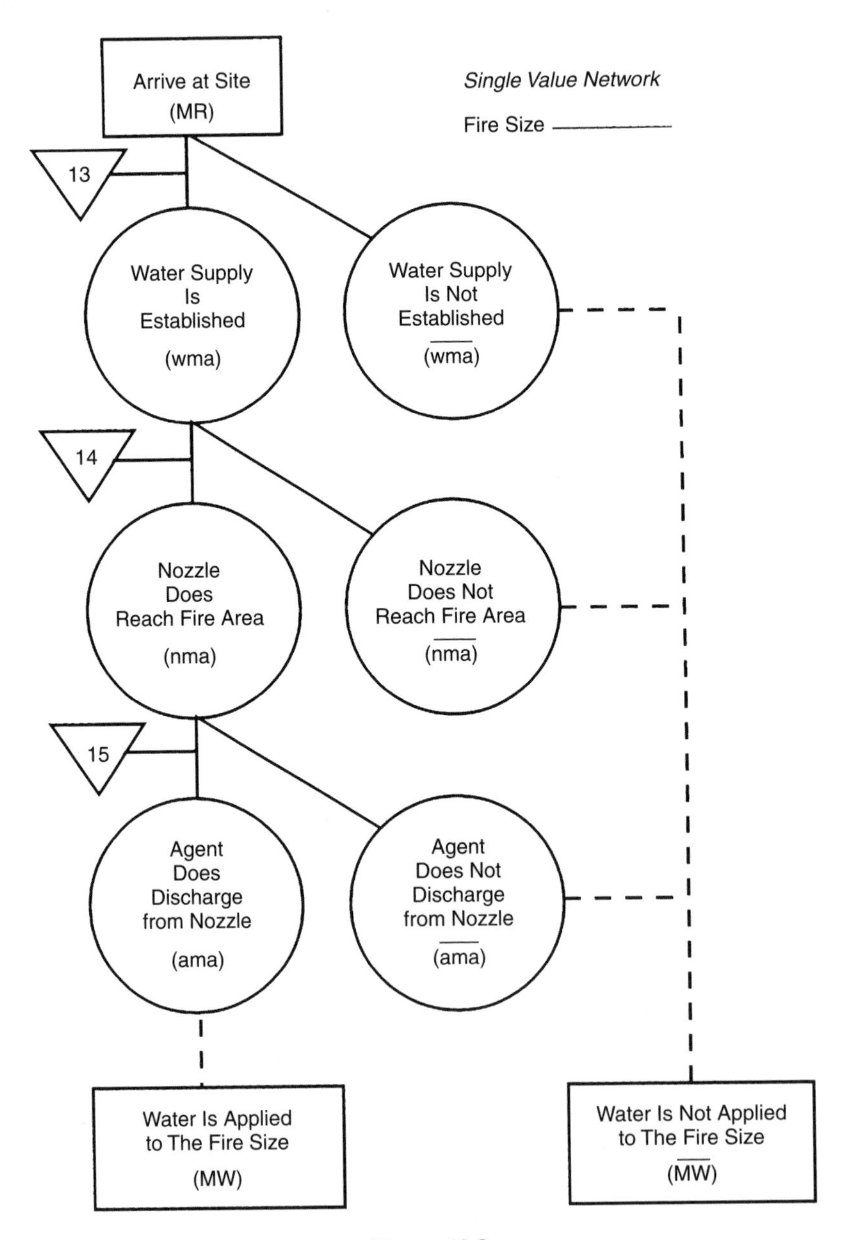

Figure 12.8

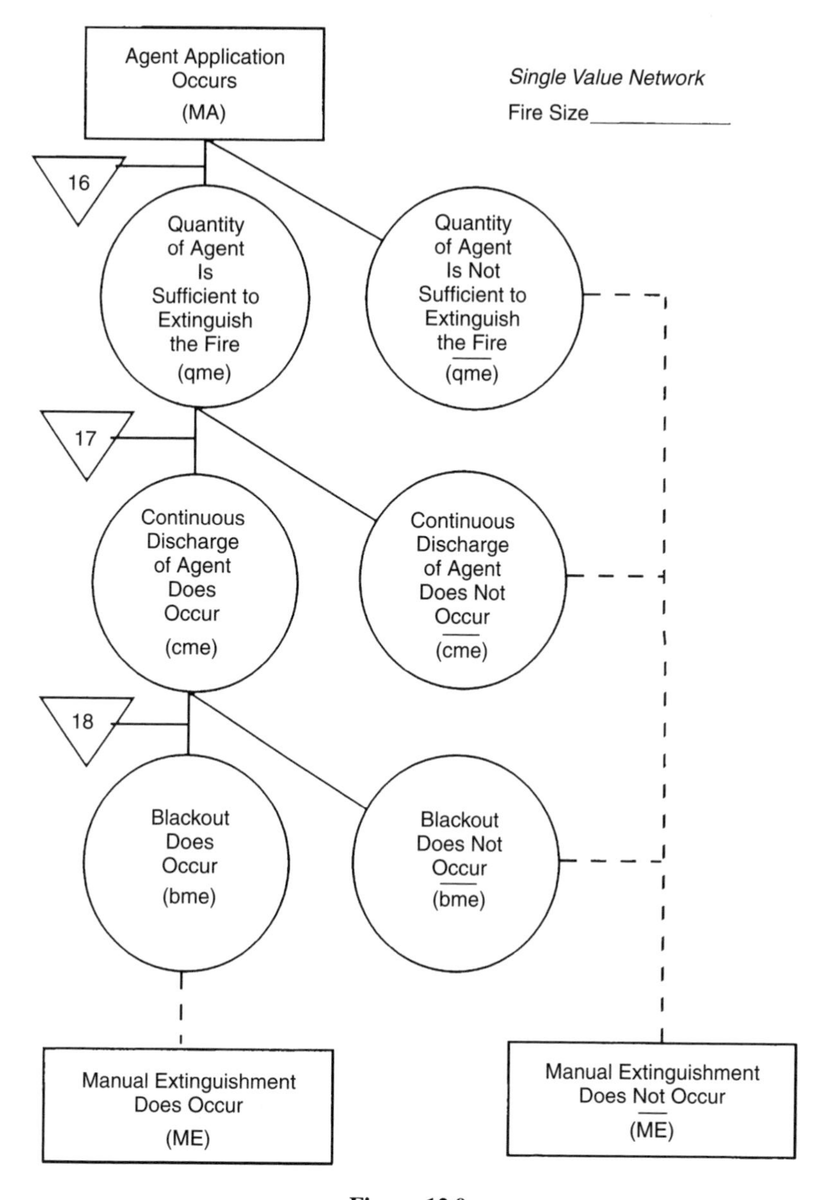

Figure 12.9

in Chapter 11. The following example illustrates calculation procedures after probabilistic values have been selected.

Example 12.1

The evaluation scenario involves a moderately large (4000 ft^2) room occupied with alert, awake individuals pursuing their routine work. A fire area of about 800 ft^2 was selected as a zone size to evaluate the M curve. After constructing the time lines and examining critical events,

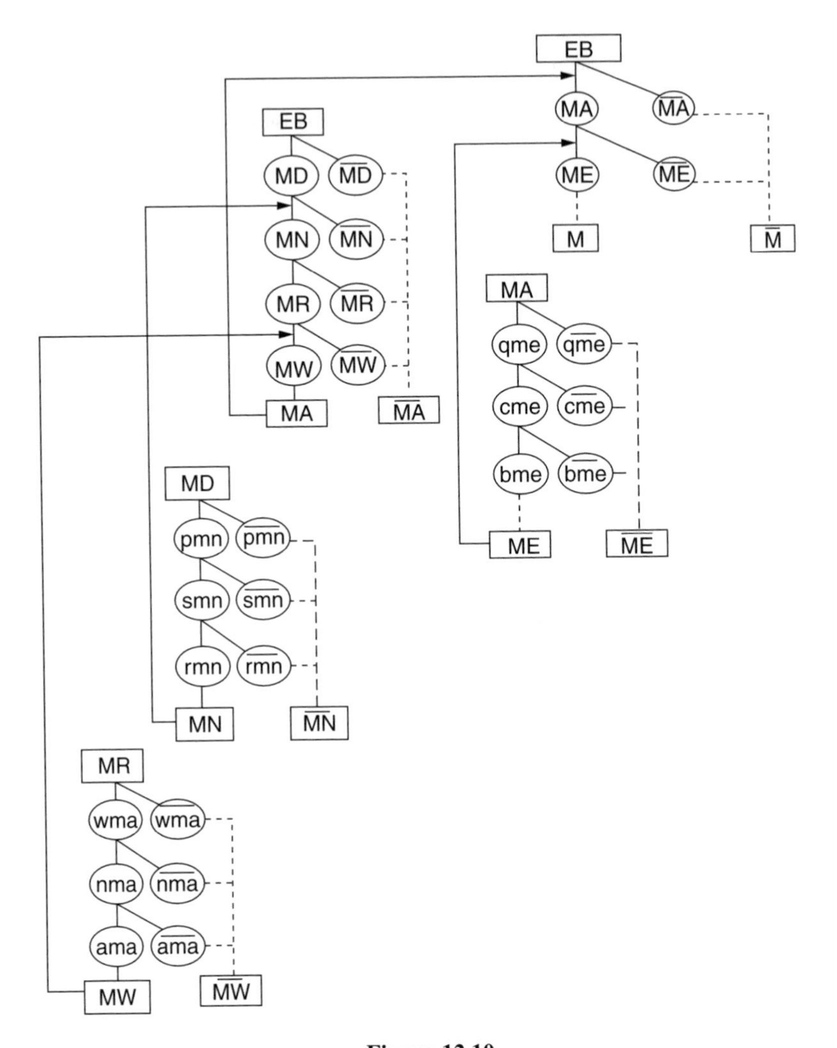

Figure 12.10

the following SV probabilities were estimated for the 800 ft^2 fire size occurring about 12 min after EB:

- Probability that the fire will be detected is $P(\text{MD}) = 1.0$.
- Given detection, the probability that an occupant will decide to notify the fire department is $P(\text{pmn}) = 0.999$.
- Given the decision to notify the fire department, the probability that the message will be sent is $P(\text{smn}) = 0.99$.
- Given the initial contact with the fire department communication center, the probability that the message will be received correctly by the dispatcher is $P(\text{rmn}) = 0.999$.
- Given that the dispatcher received the message correctly, the probability that the fire department will respond to the site is $P(\text{MR}) = 0.99$.

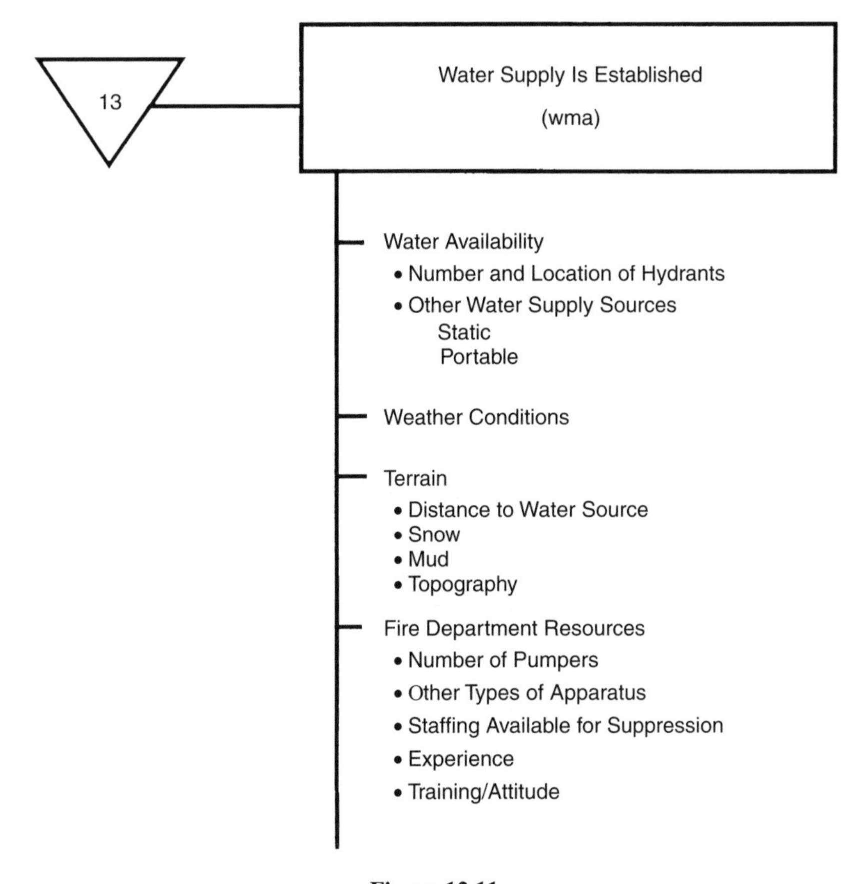

Figure 12.11

- Given that the fire department responds to the site, the probability that the fire department will establish a water supply for the supply lines is $P(\text{wmw}) = 0.95$.
- Given that the fire department responds to the site, the probability that the fire department will find the fire, begin an interior fire attack, and have the hose lines in position to apply water is $P(\text{nmw}) = 0.6$.
- Given that the hose lines are in position and water is requested, the probability of agent discharge is $P(\text{amw}) = 0.995$.
- Given that initial agent application occurs, the probability that sufficient water is available to extinguish a fire of 800 ft^2 is $P(\text{qme}) = 1.0$.
- Given that the water is sufficient, the probability that water can be applied continuously until the fire is extinguished is $P(\text{cme}) = 1.0$.
- Given that the quantity and continuity of water is sufficient to extinguish a fire of 800 ft^2, the probability that the fire department will extinguish the fire before it grows to a larger size is $P(\text{bme}) = 0.4$.

Calculate the SV probability that the fire department will extinguish the fire of 800 ft^2 without it spreading to a larger size.

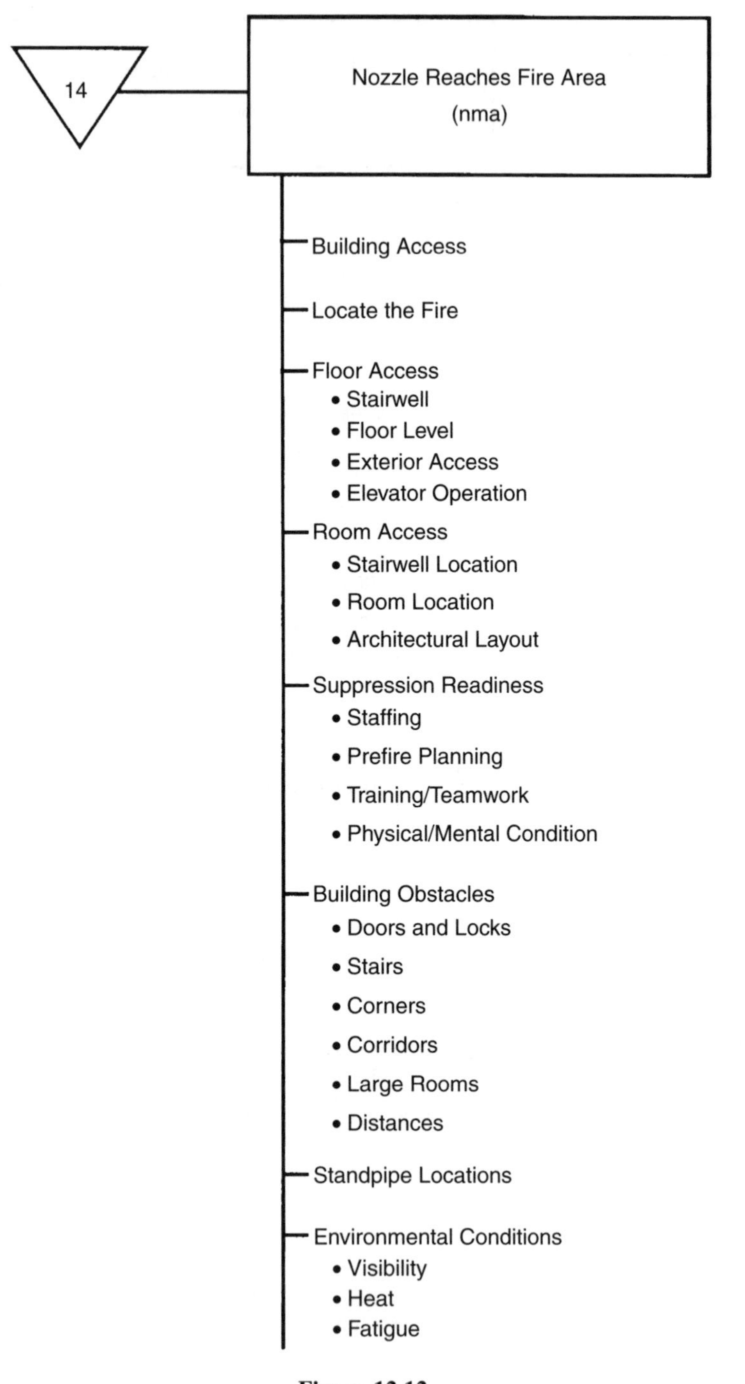
14
Nozzle Reaches Fire Area
(nma)
Building Access
Locate the Fire
Floor Access
• Stairwell
• Floor Level
• Exterior Access
• Elevator Operation
Room Access
• Stairwell Location
• Room Location
• Architectural Layout
Suppression Readiness
• Staffing
• Prefire Planning
• Training/Teamwork
• Physical/Mental Condition
Building Obstacles
• Doors and Locks
• Stairs
• Corners
• Corridors
• Large Rooms
• Distances
Standpipe Locations
Environmental Conditions
• Visibility
• Heat
• Fatigue

Figure 12.12

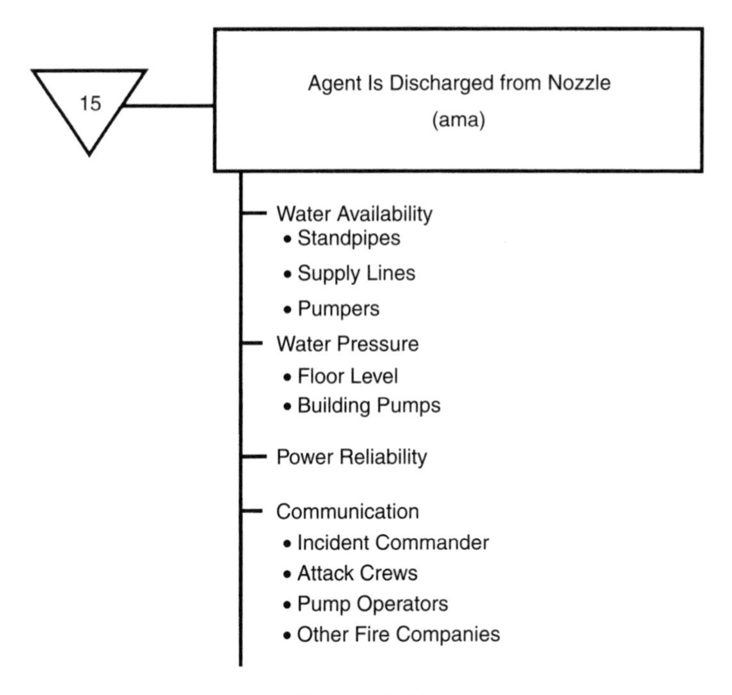

Figure 12.13

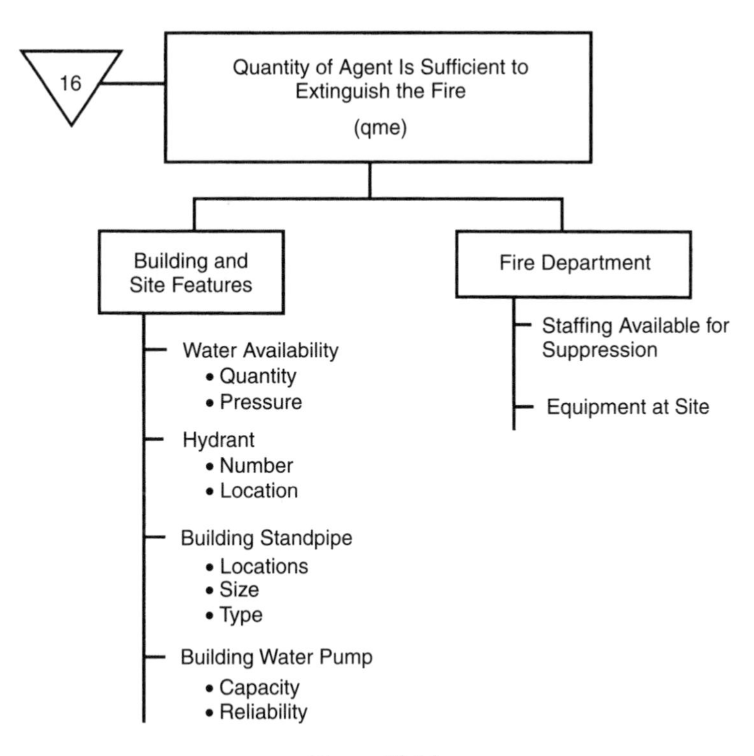

Figure 12.14

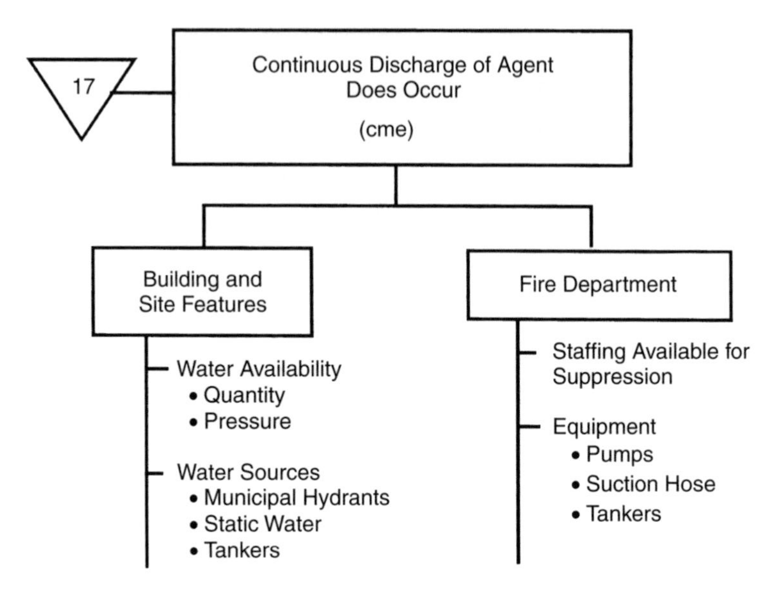

Figure 12.15

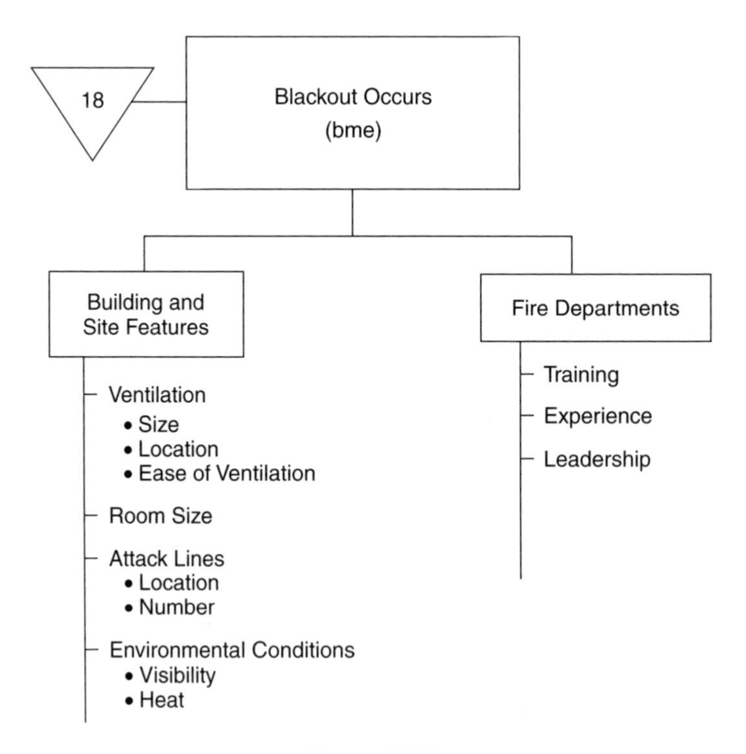

Figure 12.16

Solution

The probability of notification given detection is $P(\mathrm{MN}) = 0.989$, as shown in Figure 12.17. The probability that water application will occur at about 12 min (fire reaches about 800 ft^2) is $P(\mathrm{MW}) = 0.567$. This value is determined from Figure 12.18 with the values for wma, nma, and ama.

The probability that agent application (MA) will occur before the fire has reached 800 ft^2 is calculated from Figure 12.19 as $P(\mathrm{MA}) = 0.555$. Given agent application, the probability that

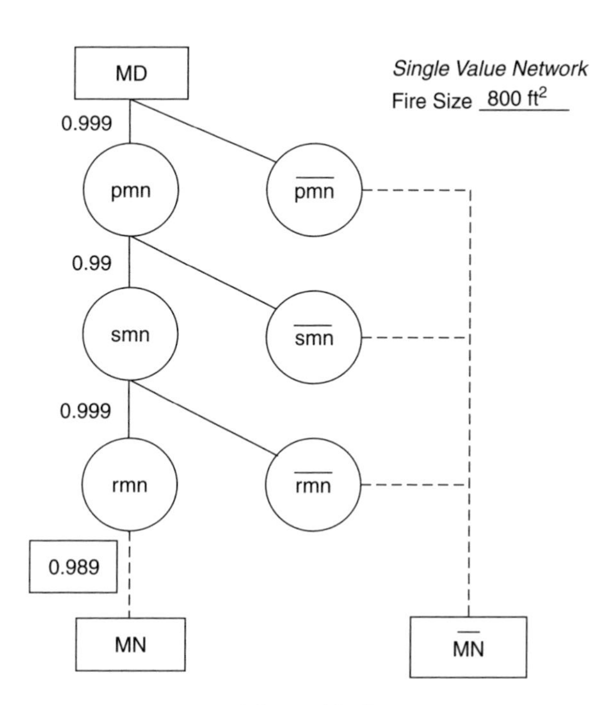

Figure 12.17

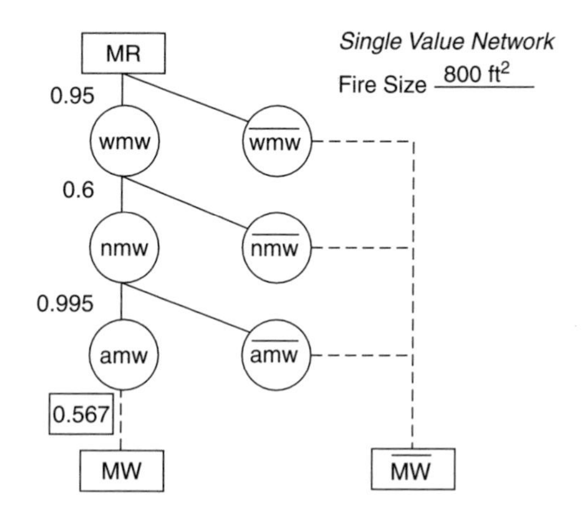

Figure 12.18

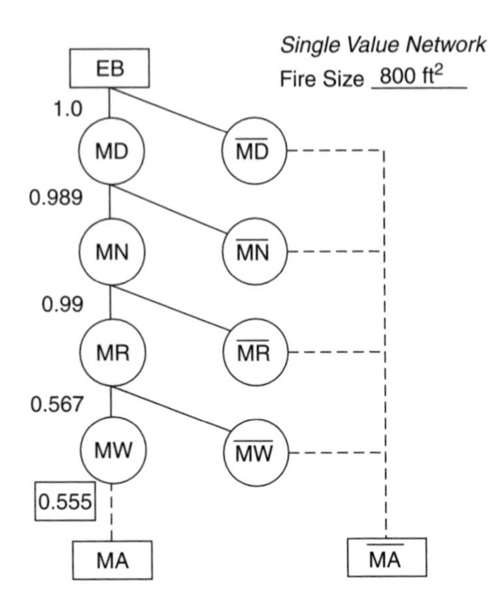

Figure 12.19

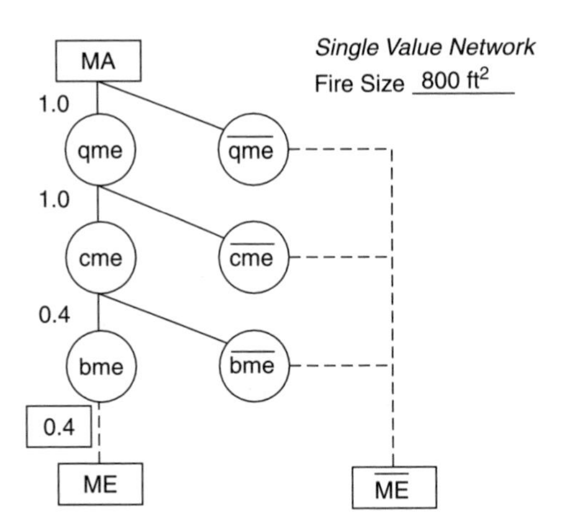

Figure 12.20

the fire department will extinguish the fire before it extends beyond 800 ft^2 is $P(\text{ME}) = 0.4$, as shown in Figure 12.20. Incorporating these values for $P(\text{MA})$ and $P(\text{ME})$ into Figure 12.21 yields a probability of $P(\text{M}) = 0.222$ that the fire will be extinguished without extension beyond 800 ft^2.

Discussion

The SVN of Figure 12.21 shows that the local fire department has a chance of only about 1 in 5 ($P(\text{M}) = 0.22$) of extinguishing a fire within the 800 ft^2 zone. The trend of the numbers for this situation would lead one to expect that the fire could be controlled before destruction of substantially larger areas. One reason for the good performance is the timely and reliable notification of the fire department. Many buildings do not have this notification quality.

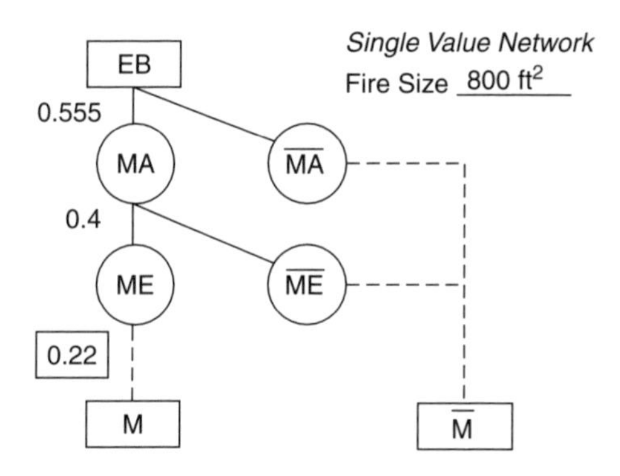

Figure 12.21

Major problems with manual fire suppression evidently occur after response to the site. More specifically, the problem centers around a difficulty in applying water. The value of $P(\text{MW}) = 0.567$ identifies one source of the problem. Its exact nature would have been recognized during the time line evaluations and the translation of building or fire department inadequacies into probabilistic estimates. The problem could be due to a fast fire development, delays in establishing attack water, or difficulties in stretching attack lines. Although specifics were not indicated in the problem statement, the documentation would show the exact nature of the difficulty.

The very low likelihood of extinguishment without extension after water application suggests a second problem. This would seem to be due to a lack of barriers that forces the fire department to push the fire into other areas not yet involved. The total area of damage would be larger than the 800 ft^2 that was selected for SV analysis.

A sense of the situation could be described better if another larger area were analyzed, perhaps about 1500 ft^2. Certainly water does not seem to be a problem, and the fire department is on the scene. The analysis would answer questions about the ease of fire extension (fuel availability) and the ability to lay enough hose lines to control fire movement and extinguish it.

This level 2 analysis identifies the general weaknesses of the building/fire department system. Here it is the ability to stretch hose lines into position and the lack of barriers to prevent extension of the fire to additional spaces after extinguishment begins. Those are building features. An 800 ft^2 fire is relatively large (nearly the size of a small house), and smoke and heat would be spreading through the large room. Multiple hose lines having coordinated application would be necessary for effective extinguishment. The type and number of fire companies and staffing would be examined to ensure that the local fire department can provide these resources.

12.21 M curve construction

The goal of an M curve analysis is to understand the building's ability to work with the local fire department in limiting fire damage. The SV networks and factors shown in Figures 12.5 to 12.16 direct the analytical thought process. The SV numbers reflect an assessment of the effectiveness of each step. Documentation can amplify the reasoning associated with performance weaknesses.

Example 12.1 and its discussion were based on a level 2 analysis for the dynamic study of a building fire. Figure 12.22 shows those parts of the complete system (Figure 5.1) that relate to the M curve. The 12 min column shows estimates from Example 12.1.

In the composite array (Figure 5.1), the vertical columns contain SV events. The horizontal rows identify performance curves. Thus, row 3 shows the M curve performance change over

	SECTION	COMPONENT	EVENT	TIME (minutes)													
				1	2	3	4	5	6	7	8	9	10	11	12	13	14
Building Performance	1		Design Fire														
		a	Size (HRR)												800 ft^2		
		b	Size (Area or Rooms)														
		c															
		d															
	3		Fire Department Extinguishment (M)												0.22		
	A		Agent Application (MA)												0.555		
		a	Fire Detection (MD)												1.0		
		b	FD Notification (MN)												0.989		
		c	FD Arrival on Scene (MR)												0.99		
		d	First Water Application (MW)												0.567		
	B		Fire Extinguishment (ME)												0.4		
		a	Quantity of Water Sufficient (qme)												1.0		
		b	Continuous Water Discharge (cme)												1.0		
		c	Blackout Occurs (bme)												0.4		

Figure 12.22

the time. Row 3Ab shows notification curve coordinates over time, and row 3A describes the likelihood of agent application, MA, as time moves along.

Level 1 evaluations, described later in this chapter, enable one to develop a sensitivity to the building performance. Curves are sketched from observation and mental estimates to give a sense of proportion of performance. A level 1 evaluation helps an inspector decide what level 2 fire size analysis will provide the best information about the building. After one level 2 analysis is completed, performance curves can usually be sketched with greater confidence. The inspector can decide how much analysis is enough to make informed decisions about performance, risk, and risk management.

12.22 Building fire brigades

Occupant extinguishment is not a part of an M curve evaluation. Chapter 20 discusses the role of occupants. However, some buildings have an identified, on-scene building fire brigade. Normally, only industrial and institutional occupancies have building fire brigades, and not all organizations have them. When a building fire brigade does exist, the fire suppression capabilities are evaluated using the same process as for the local fire department extinguishment. The amount and type of equipment, the organization and command structure, regularity of training, and fire-fighting experiences influence the effectiveness of a building fire brigade. In evaluations one must be careful to develop an objective analysis of building fire brigade capabilities. If a building fire brigade exists, its M curve is evaluated separately from the local fire department M curve.

The building fire brigade may fulfill two functions. One is that a building fire brigade may put out the fire while it is still small. The second function changes the fire growth and reduces the

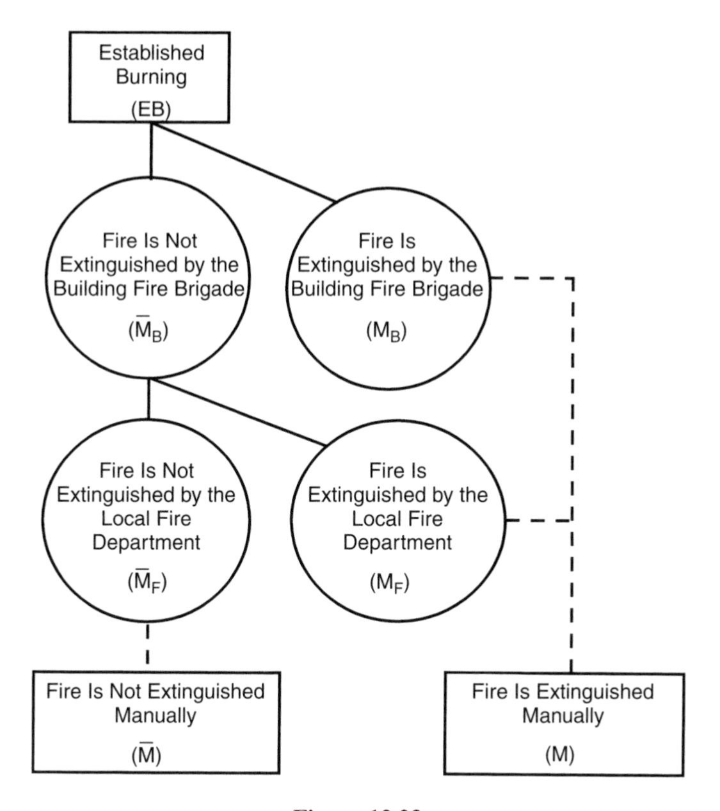

Figure 12.23

size that will be encountered by the local fire department. The local fire department becomes a mutual aid resource. The SVN of Figure 12.23 shows integration of the two organizations.

12.23 Level 3 analysis

A level 3 analysis provides a sensitivity to the impact of variations in parameters that affect fire defenses. For example, if the local community reduces engine company staffing by one person, what is the effect on the M curve? How will a fire growth increase of 10% affect the M curve? How will the M curve change if the building installs an auxiliary fire alarm system? A central station system? What is the effect of installing smoke curtains on fire damage? Completion of a level 2 analysis provides the information that targets specific questions for more in-depth examinations.

12.24 Level 1 analysis

A level 1 analysis uses visual observations and mental estimates to blend available information for a rapid understanding and description of performance. It provides a sense of the building behavior and the fire damage. The M curve thought process uses the factors and relationships of this chapter. The coalescing of knowledge, building information, and local fire department data enables one to sketch M curve performance.

Figure 12.24 shows the parts of the process. The general building features, room of origin, and scenarios are chosen. The process then becomes one of envisioning the unfolding of scenarios and the influence of factors that affect the time lines and outcomes. The objective is to understand as much as possible about the building performance within a short time. The procedures methodically dissect the building/fire system and estimate level 1 behavior for the major components. Then we put the components together to describe performance. The factors that are important to performance are identified and documented. The relative continuous curve descriptions help to scale performance and time sequencing.

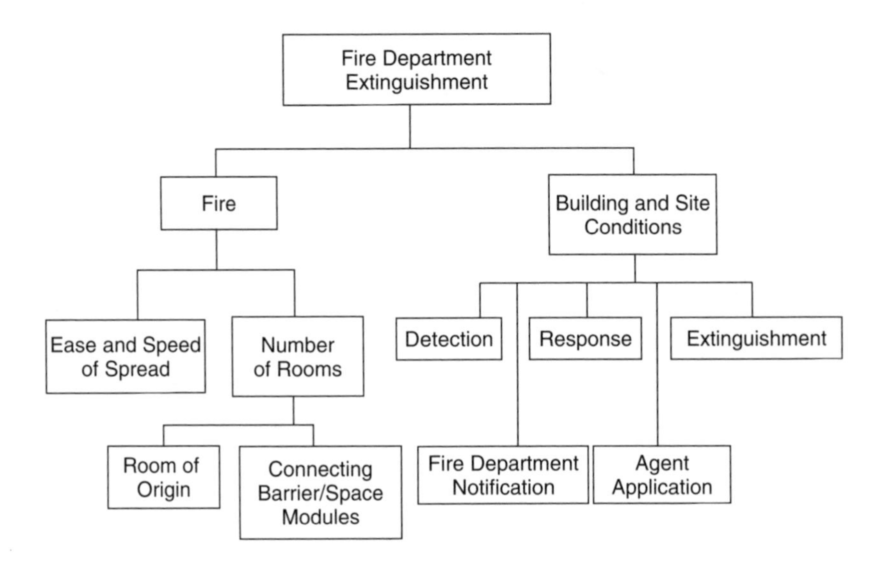

Figure 12.24

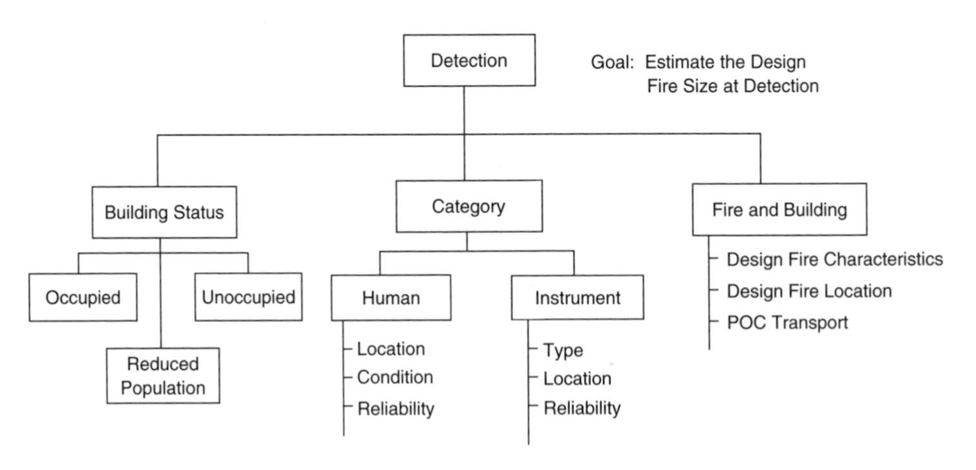

Figure 12.25

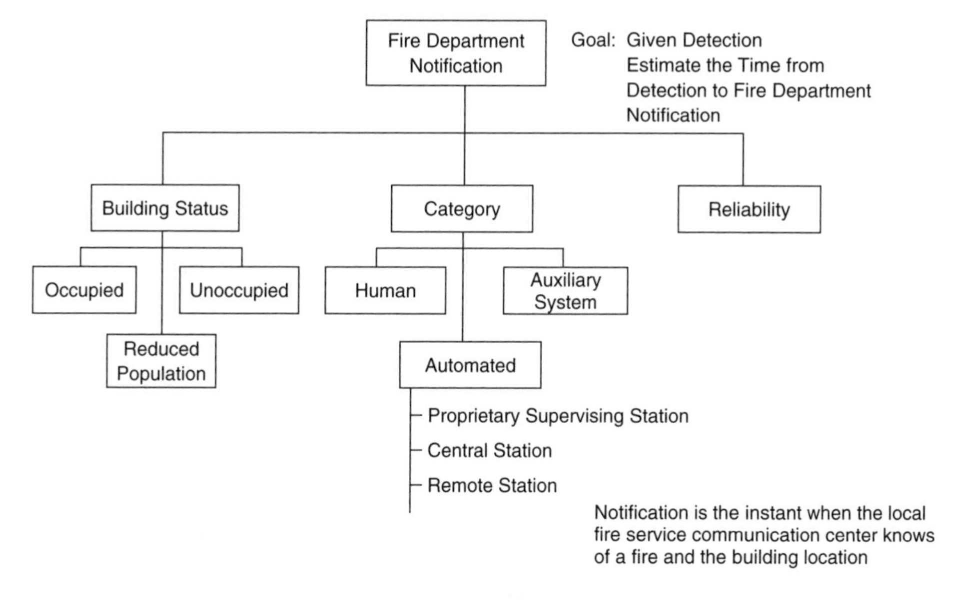

Figure 12.26

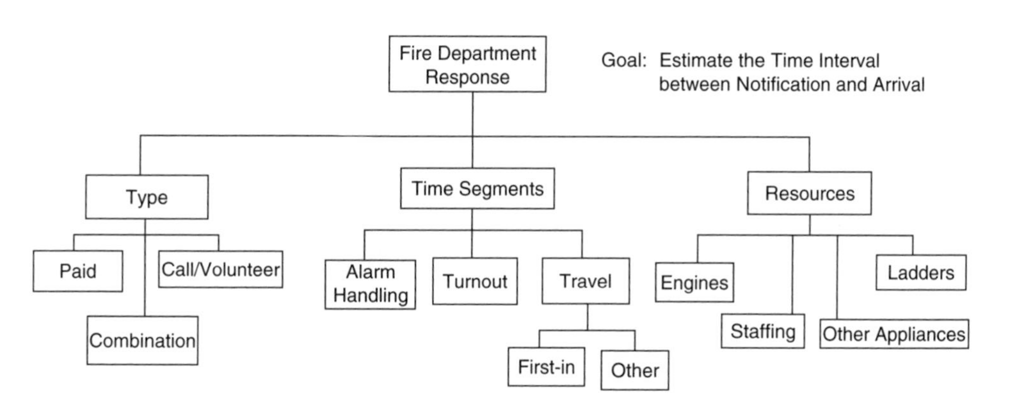

Figure 12.27

Figures 12.25 to 12.28 show the major factors that one considers in a level 1 evaluation. The synergistic integration of these components with the supplemental factors shown in Figures 12.29 to 12.32 enables one to get a sense of proportion for M curve performance and the factors that influence that performance.

12.25 Summary

On the fire ground, the fire department is responsible for

- search and rescue of occupants;
- protection of exposures within and outside the fire building;

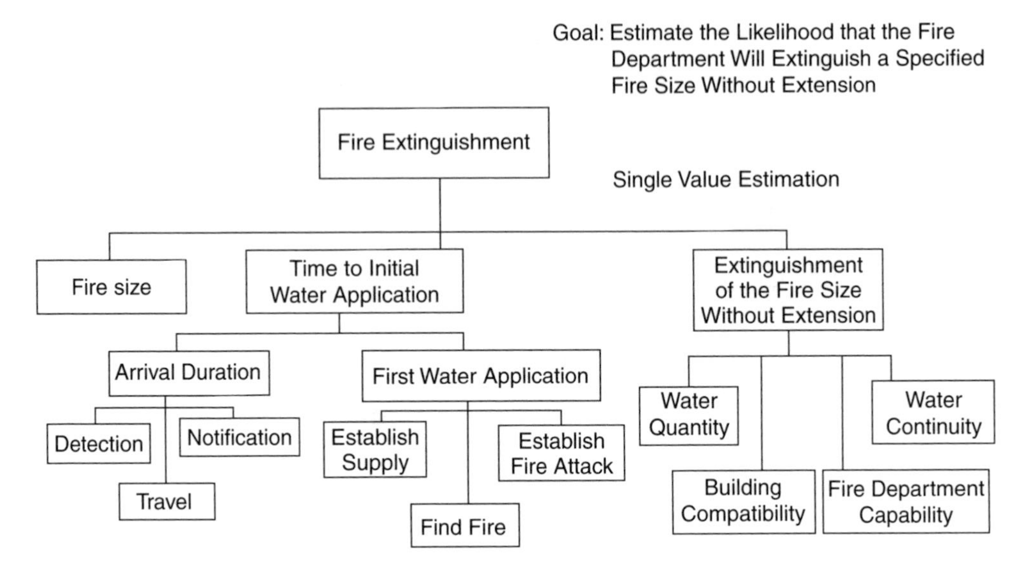

Figure 12.28

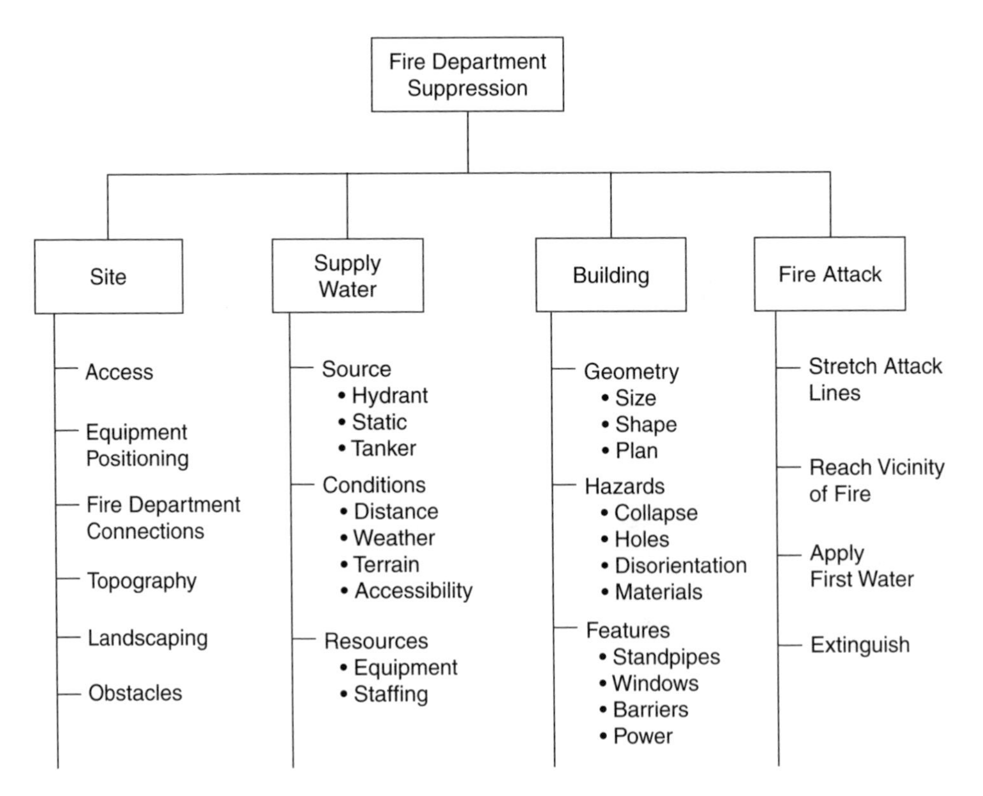

Figure 12.29

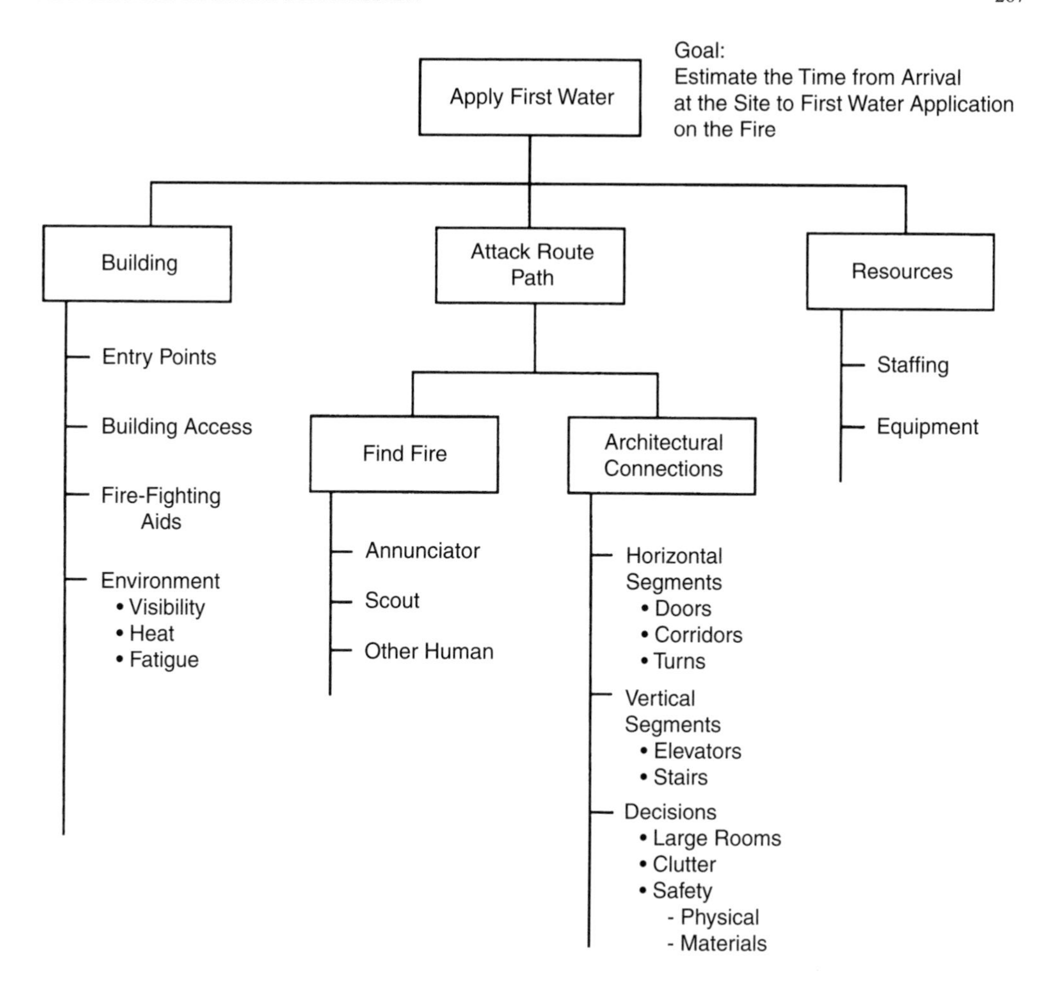

Figure 12.30

- protection of property;
- removal of smoke;
- extinguishing the fire.

The fire ground commander must allocate available on-site resources to do the best job possible in a situation where information is often inadequate or wrong, and conditions change as time progresses. This chapter considers only fire suppression.

Local fire department suppression is the primary line of defense for nonsprinklered buildings. Some building designs help fire department manual extinguishment to be very successful. Other building designs give the local fire department little chance of extinguishing a fire with an offensive internal attack. Some designs give little chance for extinguishment before complete destruction. It is important to be able to differentiate between these situations and document the building and site design features that influence the outcomes.

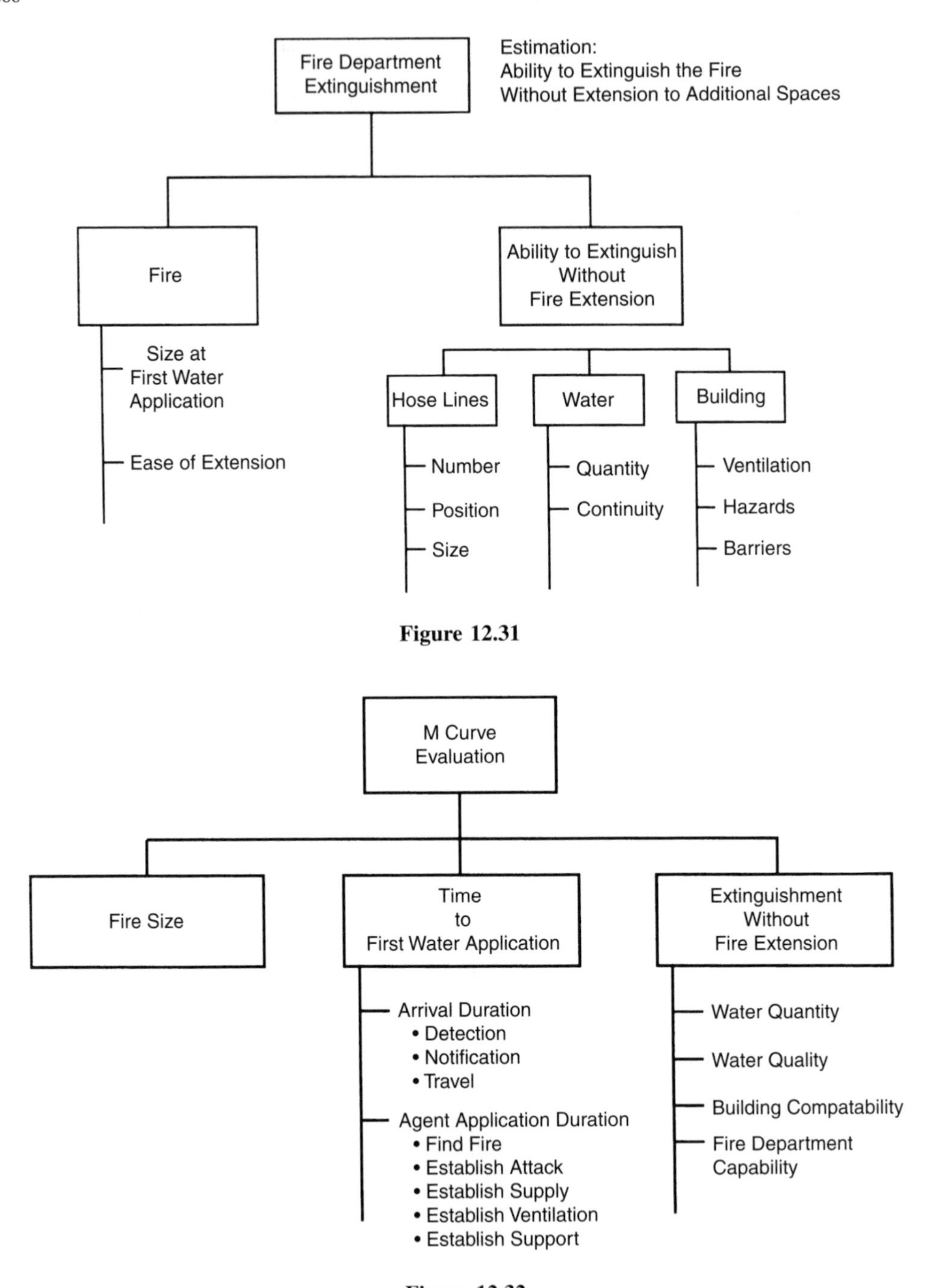

Figure 12.31

Figure 12.32

References

1. Menke, W. K. Predicting effectiveness of manual suppression. MS thesis, WPI, 1994.
2. Callery, J. F. Building evaluation for manual suppression. MS thesis, WPI, 1996.

13 AUTOMATIC SPRINKLER SUPPRESSION

13.1 Introduction

Although concepts of sprinklers and similar early devices were conceived and constructed in the early nineteenth century, the first practical and extensively used sprinkler system was developed by Henry Parmelee in 1878. The first automobile was built at about the same time in 1886, and "mass production" of 13 cars of common design first occurred in 1896. Since their beginning, the technology of sprinkler systems has at least paralleled, and has probably exceeded, the evolutionary technology of the automobile during essentially the same time period. New developments and improvements in sprinklers and sprinkler system technology continue to appear at a rapid pace.

Over the years, the automatic sprinkler system has had an outstanding record of success in controlling or extinguishing unwanted fires. Nevertheless, there have been cases in which a building protected by an automatic sprinkler system has sustained major losses. Questions arise as to what factors produce success and what flaws induce failure, and how one can compare the quality of different systems. When a sprinkler system is the primary fire defense, one examines its installation and operation details to determine expected performance.

This chapter briefly describes sprinkler system functional performance to organize sprinkler system (A curve) evaluations. An A curve describes the likelihood that 1, 2, 3, 4, ..., n sprinklers will control the design fire. Alternatively, the A curve indicates the likelihood that flame/heat damage will be controlled within successively increasing floor areas. An evaluation concentrates on the effect of automatic sprinkler system features on reliability and operational effectiveness.

13.2 Sprinkler extinguishment

A large number and variety of sprinklers are available for installation in ceilings or on sidewalls. Each type will fuse (i.e., open) at a specified heat condition and deliver a particular water droplet size, density, and distribution pattern onto the fire. A sprinkler can be designed for a design fire and its room environment. To envision conventional sprinkler operation, let's describe a possible scenario.

Assume that you are able to observe the interior of a relatively large room with a number of sprinklers in the ceiling. Ignition and established burning occur in a fuel package and a fire

Building Fire Performance Analysis R. W. Fitzgerald
 ISBN: 0-470-86326-9 (HB)

plume rises and begins to develop a hot layer at the ceiling. As the fire continues to burn, more heat and higher temperatures accumulate at the ceiling.

Now consider a sprinkler such as that of Figure 13.1. The sprinkler frame (a) is connected to a pipe. The water in the nozzle (b) is held back by a cap (c), which in turn is kept in position by levers (d). A fusible link (e) maintains the levers in position. The link has a heat-actuating element (f) that softens when heated. When the temperature reaches a predetermined level, the link separates, causing the levers to disengage and the cap to be blown off by the water pressure in the nozzle. Water flows from the nozzle and strikes the deflector (g), causing a pattern of water distribution to the floor below.

Return to the fire in the fuel package described earlier. As the fire grows, the hot-layer temperature at the ceiling continues to rise. When a link absorbs enough heat energy, it fuses, releasing the cap from the sprinkler. Only sprinklers near the fire will fuse, allowing the water

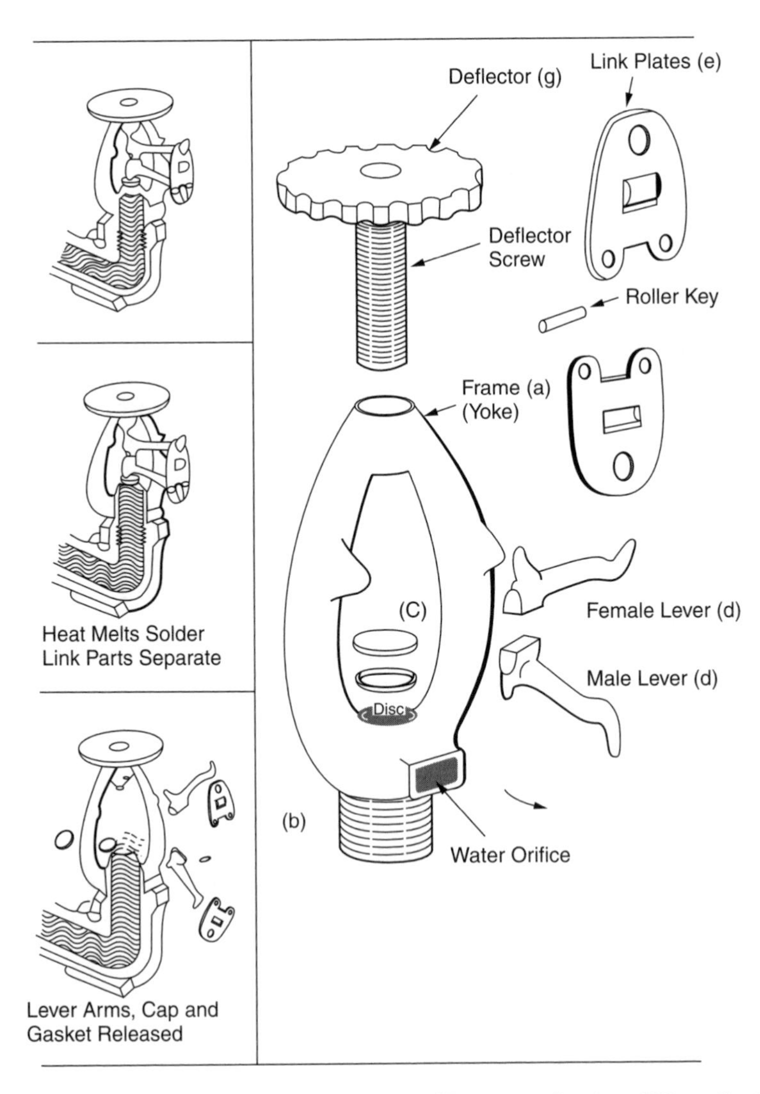

Figure 13.1 (Reproduced by permission of Insurance Services Offices, Inc.)

to impinge on the deflector and discharge a pattern of water droplets onto the fire. Many fires are extinguished with only one sprinkler operation, thus reducing the fire and water damage substantially. Although additional sprinklers are sometimes needed, most fires are controlled by the operation of only a few sprinklers. Occasionally the fire conditions and sprinkler capabilities are mismatched and the fire may grow larger and faster, causing more sprinklers to fuse. The water supply now becomes important because when too many sprinklers open, water pressure and quantity is reduced, allowing the fire to grow beyond the capability of the sprinkler system.

Here are some common characteristics of typical sprinkler system operations:

- Each sprinkler is independent. Only sprinklers fused by the heat transfer from the fire will operate.
- Sprinklers normally do not fuse quickly in a fire. Depending on the ceiling height and type of fusible link, one would normally expect the sprinkler to fuse as the fire grows to 350–500 kW (about 4–5 ft in height). We might note that the common fire size at sprinkler actuation is between the enclosure point and the ceiling point of the realms of fire growth described in Section 8.5. It is possible to install fast-response sprinklers to fuse when the fire is smaller.
- Most commonly, only one or a few sprinklers will open during a fire because of the speed and effectiveness of suppression.

After the sprinklers fuse, water continues to flow until the sprinkler system is shut down. Water damage in unoccupied buildings can be reduced drastically by the installation of water flow alarms connected to the local fire department or to a supervisory service.

13.3 The sprinkler system

The sprinklers described above are only one component of a complete sprinkler system. The function of the other parts of the system are to supply enough water at sufficient pressure and duration to control or extinguish the fire. This is accomplished by the water supply, the water distribution system, pumps and other operational devices, where needed. The automatic sprinkler system may be organized into five major components:

- The water supply and distribution system that brings the water to the building.
- The control system of valves, pumps, and other devices that links the external water supply system to the internal building sprinkler piping system.
- The building's piping distribution of risers, mains, and branches.
- Sprinklers that discharge water to the fire.
- A monitoring system for trouble and alerting alarms.

Figure 13.2 shows a sprinkler system schematic. The water supply reaches the building through a feed main. A water control valve is needed to shut off water for system maintenance. Water control valves must be open for water to flow to a fused sprinkler. The water flows vertically through a riser to a feed main to a cross main to the branch lines, to which the sprinklers are connected. A variety of pumps and other devices may be installed to maintain or adjust pressure.

13.4 Types of sprinkler system

This simple description of sprinklers and the sprinkler system provides a basic organization from which variations can be introduced to meet specialized design requirements. Building needs and

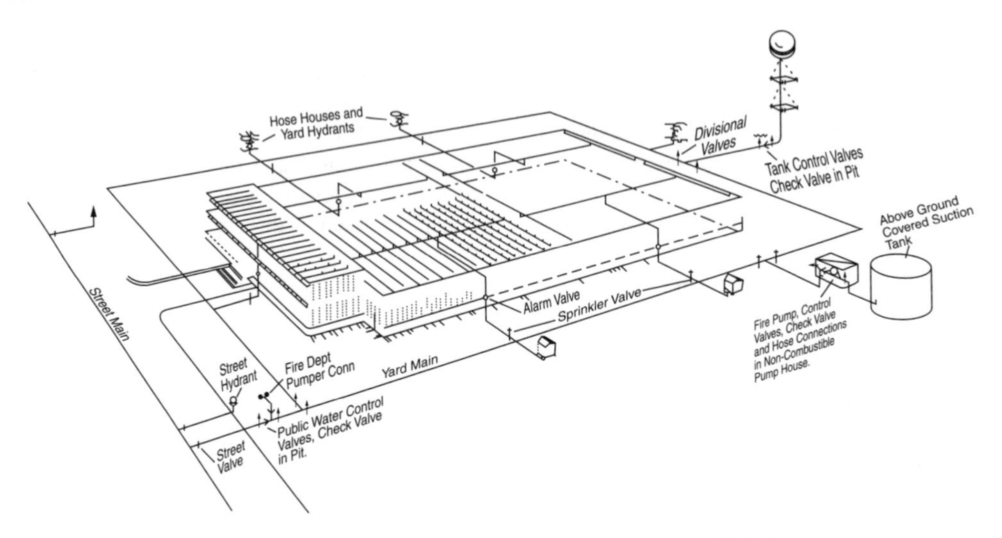

Figure 13.2 (Reproduced by permission of FM Global)

designs can vary substantially. The sprinkler system must be able to adjust to the functional needs of the building operation and still perform the role for which it was intended.

Building sprinkler systems are designed to address specific hazards that will exist under varying environmental conditions. Therefore a sprinkler system for an apartment or office building will have different requirements from that of an unheated warehouse or an industrial operation in which a large, fast fire could occur.

Sprinkler systems may be classified as *wet pipe, dry pipe, deluge*, and *preaction.* The appropriateness of the installation features, as well as evaluation techniques for reliability and design effectiveness of a specific installation are site specific. The goal here is to provide a basic understanding of sprinkler systems to structure A curve evaluations.

WET-PIPE SYSTEM

The piping for a wet-pipe sprinkler system is filled with water under pressure. When a sprinkler fuses, water can flow from the orifice immediately. This is the most common type of sprinkler system in buildings that do not have freezing potential or other special needs.

DRY-PIPE SYSTEM

A dry-pipe system is used in spaces that are subject to freezing or in spaces where a greater water flow control is desired. The piping contains air or nitrogen under pressure, rather than water. Compressors maintain the necessary pressure. The water supply is held back at a dry-pipe water control valve located in a heated space. When a sprinkler fuses, the release of air causes a pressure drop which automatically opens the dry pipe valve and releases water into the system. However, water movement can be delayed until the air in the pipes is removed. Quick-opening devices, called accelerators or exhausters, can be installed to quickly expel air and gases in the pipes, allowing water to reach the open sprinklers more quickly.

DELUGE SYSTEM

A deluge system can apply water over a relatively large area by supplying many open sprinklers simultaneously. This type of system is used for special types of hazard where the immediate application of large quantities of water is needed to control a potentially rapidly expanding fire. Flammable liquid storage facilities, aircraft hangars, and extra hazard industrial operations might install a deluge system.

A deluge system is usually actuated by a heat detection system. When a heat detector actuates, water is released by a deluge valve and fills the dry-piping system to operate. A variety of special control devices, such as water control valves, releasing mechanisms, and supervisory equipment, are available for installation.

PREACTION SYSTEM

A preaction system is used where room contents are very sensitive to the discharge of water. Library stacks or valuable exposed files illustrate concerns that owners may have for an inadvertent discharge of water. A preaction system is a dry-pipe system with an additional control linked to a sensitive fire detection system. If the sprinkler fuses but the detector does not sense a fire, then no water will flow. On the other hand, if the detector senses a fire, the water control valve is opened to permit initial water flow through the dry pipes. However, if the sprinkler does not fuse, then no water will discharge. A preaction system will activate and discharge water only when the sprinkler fuses and the detection device senses a fire.

SPECIALIZED DESIGNS

Other designs may be devised by combining sprinkler system components to meet specialized needs. For example, rather than using a conventional automatic sprinkler system on a ship, an automated system may be used. This design uses a conventional sprinkler system with a manual opening of the water control valve. Creative automatic suppression designs can meet distinctive needs for specialized functions or hazards.

13.5 Agent application

We define the reliability of a sprinkler system as the likelihood that water will discharge when a sprinkler fuses. This event is called agent application (AA). The question will enough water be delivered at an appropriate pressure and with an appropriate distribution pattern is addressed by the system's operational effectiveness (AC). Agent application is evaluated once for the complete system and operational effectiveness is examined for each space or zone of coverage.

A quantity of water under pressure is delivered to the building from an external water source through the underground piping system on the property. At a building, the water that eventually enters the sprinkler system piping network is controlled by a group of valves, pumps, and other devices. These water control system components are the primary influence for the system reliability.

Valves act to control the water flow from external supply pipes to the internal sprinkler pipes. Valves are used in all systems and must be open to allow water to move into the sprinkler piping network. Water control valves are sometimes closed because of inattention or ineffective supervision. Many installations have several sectional control valves. All of the valves, public and private, that control the feed of water in a zone must be open. During routine maintenance and daily building operations, the fire protection needs may be neglected. This neglect is a common cause of closed sprinkler valves.

Sprinkler water supplies are usually separated from the building's potable water. However, when this is not feasible, public health authorities have concerns about the possibility of contaminated sprinkler water flowing back into the potable supply. This can occur when sprinkler water pressures are greater than the domestic pressures. One-way check valves often address the problem satisfactorily. Reduced pressure backflow preventers are sometimes used. Improperly specified or designed backflow preventers can reduce pressure or prevent water from reaching the sprinkler, thus reducing the reliability of the system. Backflow prevention devices are controversial with fire protection engineers and public health authorities. Fire protection professionals feel that backflow preventers reduce sprinkler reliability and water flow whereas health officials feel they are always necessary to avoid waterborne diseases.

Quick-opening devices (QODs), such as accelerators and exhausters for dry-pipe systems, improve the speed with which water can be applied to a fire. They are not as reliable as wet-pipe systems that do not need such equipment. Detection in preaction systems adds another device which can malfunction and which needs periodic maintenance. All devices that affect the operation of a system to deliver water to a fused sprinkler contribute to agent application reliability.

Another potential problem that arises more frequently with dry-pipe systems than with wet-pipe systems is blockage of the pipes. Pipe scale and debris can prevent water from flowing through the piping system to the open sprinkler.

Pumps perform several functions such as pushing water through the piping system, increasing water pressure, and maintaining high static pressures for operation or water delivery. They can influence agent application (AA) and design effectiveness (AC) components.

Fire pumps must function under conditions more severe than the routine operational conditions for other kinds of pump. Fire pumps operate only under fire conditions or test conditions, and with infrequent use they are not maintained like other operational pumps in constant use. Fire conditions may interrupt electricity or fuel delivery for pump operation. If pumps are an important part of the sprinkler system, the reliability of the pumps, controllers, drivers, accessories, and power for emergency conditions is a part of an evaluation.

In summary, if a sprinkler fuses and calls for water, a number of system components must function to deliver the water to the location of the fire. A reliability analysis includes proper system design combined with a regular inspection and maintenance program.

13.6 Framework for analysis: agent application

Agent application (AA) expresses a probability that water will flow from the nozzle when the sprinkler system calls for it. Its evaluation considers two components:

- All supply valves are open (vaa).
- Water does reach the sprinkler (waa).

Figure 13.3 shows an SVN diagram of these events. These events are independent and the position or evaluation sequence is not important. Figures 13.4 and 13.5 show the factors that influence performance.

During an evaluation, the probability of agent application (i.e., the reliability) is normally determined only once for a building or sprinkler zone, regardless of the number of building spaces.

Example 13.1

A wet-pipe sprinkler system is evaluated as a part of a complete building performance analysis. In evaluating the agent application (AA) component, company attitude showed genuine concern for fire safety in general and toward maintenance of all fire protection systems in particular. Long-term maintenance records indicate regular and documented attention. Valves are supervised, and examination of company practices suggests that after an impairment, valves have an extremely

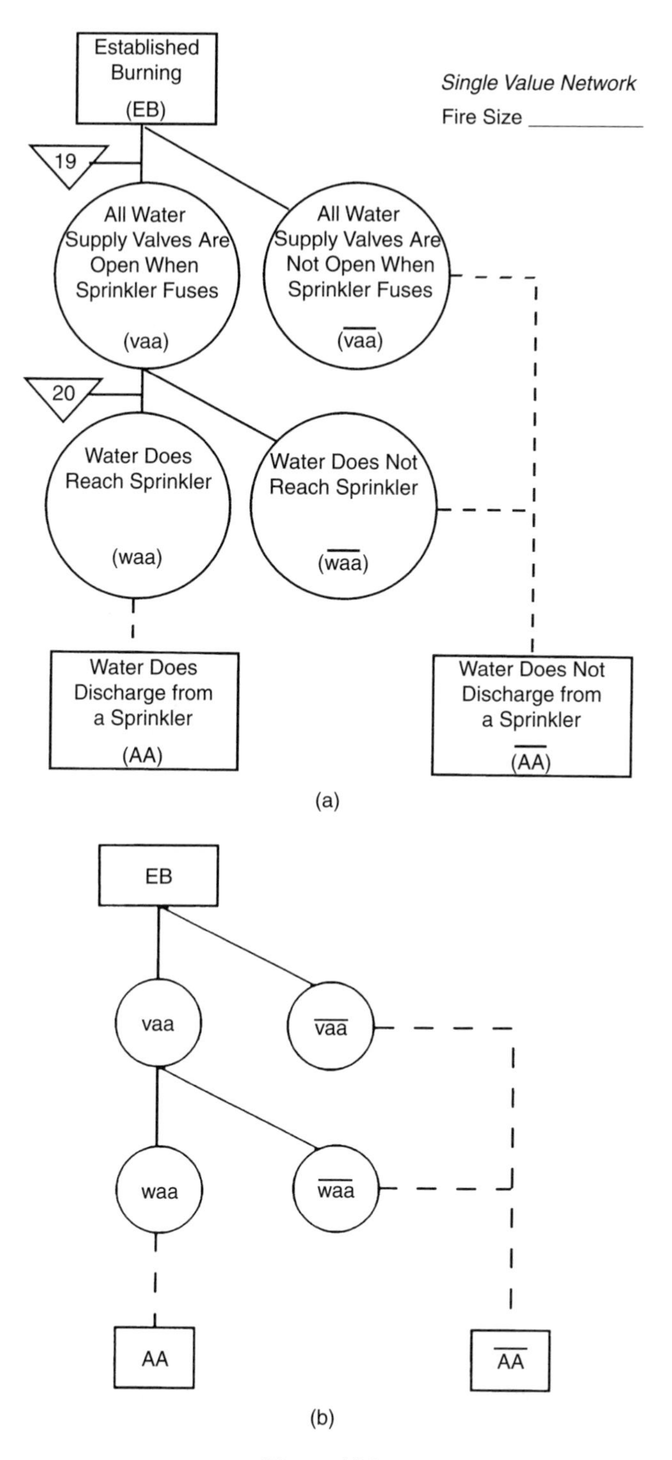
Established
Burning
(EB)
Single Value Network
Fire Size
19
All Water
Supply Valves Are
Open When
Sprinkler Fuses
(vaa)
All Water
Supply Valves Are
Not Open When
Sprinkler Fuses
($\overline{vaa}$)
20
Water Does
Reach Sprinkler
(waa)
Water Does Not
Reach Sprinkler
($\overline{waa}$)
Water Does
Discharge from
a Sprinkler
(AA)
Water Does Not
Discharge from
a Sprinkler
($\overline{AA}$)
(a)
EB
vaa
$\overline{vaa}$
waa
$\overline{waa}$
AA
$\overline{AA}$
(b)

Figure 13.3

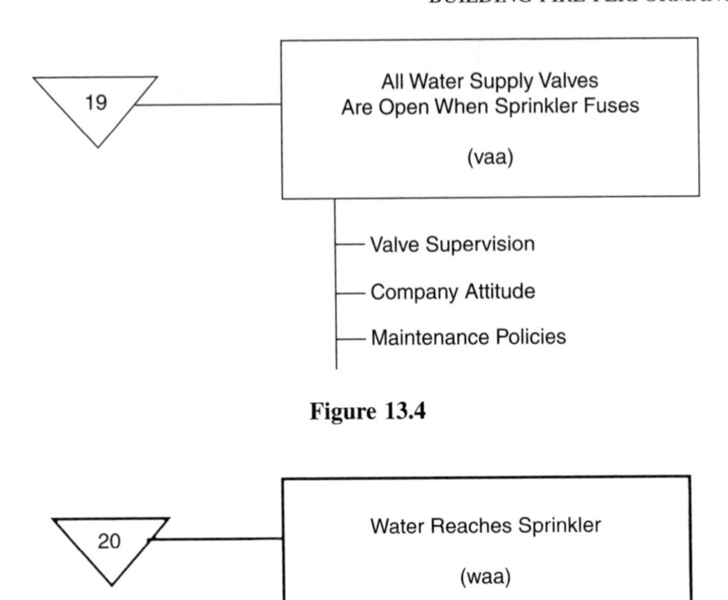

Figure 13.4

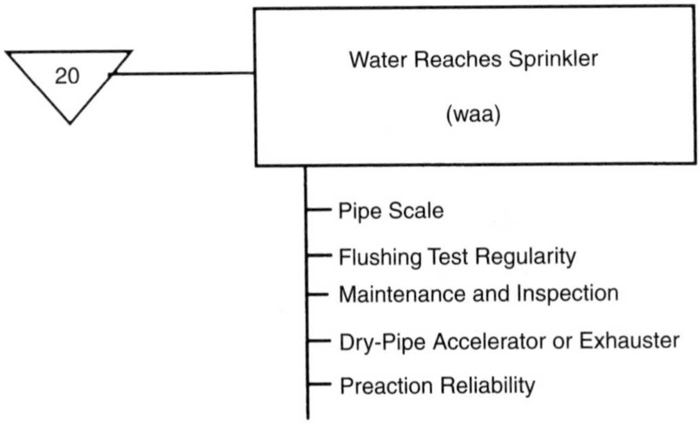

20
Water Reaches Sprinkler
(waa)
Pipe Scale
Flushing Test Regularity
Maintenance and Inspection
Dry-Pipe Accelerator or Exhauster
Preaction Reliability

Figure 13.5

high likelihood of being reopened promptly. The probability that all valves will be open when needed during a fire was assigned as 0.995.

This system has regular maintenance and shows no sign of restricted flow in the pipes. Given a fire and sprinkler fusing, the individual estimates that there is almost a certainty that water will flow in the pipes and assigned it a probability of 0.999.

Determine the probability of success for the agent application component of the A curve. Designate this Building A.

Solution

The values are shown in the SVN of Figure 13.6 and $P(\text{AA})$ is calculated as 0.994.

Example 13.2

During the A curve analysis for a manufacturing plant, it was noted that the building had a dry-pipe sprinkler system. Although control valves were open at the time of inspection, the company seems indifferent to maintenance and valve supervision. No automatic valve supervision was installed. There is rapid turnover in employment by security and maintenance personnel, with little training of new employees in these areas. The event that all supply valves will be open in a fire was assigned a probability of $P(\text{vaa}) = 0.9$.

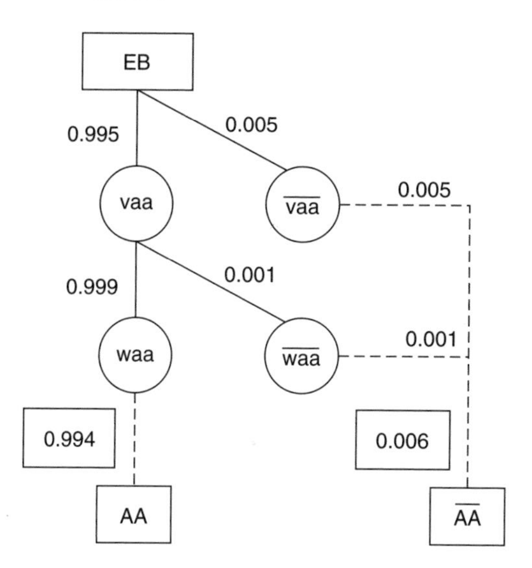

Figure 13.6

The accelerator for the dry-pipe system is a type that has a history of operational problems. Records indicate that the sprinkler system has not been flushed during the past 10 years. A probability that water will reach the sprinkler was assigned a value of $P(\text{waa}) = 0.9$.

Determine the probability of success for the agent application component of the A curve. Designate this Building B.

Solution

The values are shown in the SVN of Figure 13.7 and $P(\text{AA})$ is calculated as 0.81.

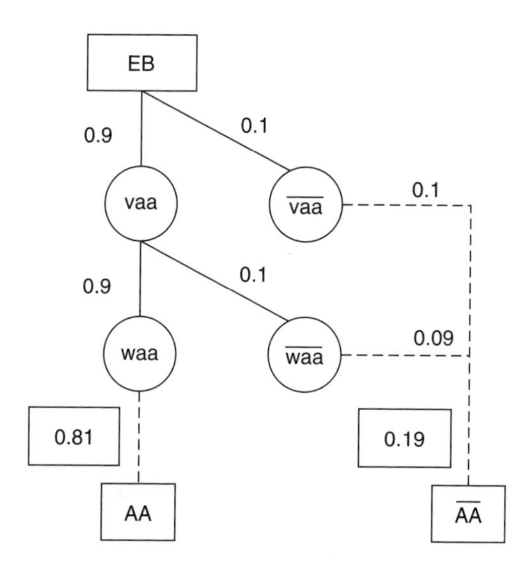

Figure 13.7

13.7 Operational effectiveness

Given that water reaches an open sprinkler, an A curve performance evaluation next examines the system's operational effectiveness. This involves the interaction of the fire and sprinkler water discharge. The design fire identifies the fire's rate of heat release, plume momentum, and speed of fire growth for a series of successively larger floor areas. These values would be given in the design fire of Section 1, Figure 5.1.

An SV analysis examines the success of sprinkler control (or extinguishment) for a specific fire size. The analysis concentrates on a single cell of line 4B of Figure 5.1 and involves examining the cells of rows 4B a, b, c, and d. These represent the following events:

- All sprinklers that protect that floor area will fuse (fac) before the fire can grow to a larger size.
- Water flow for the number of open sprinklers has sufficient quantity, pressure, and duration to control (or extinguish) the fire (cac).
- Water density will be sufficient to control the fire (dac).
- Obstructions will not shield the fire from water-spray discharge pattern enough to allow fire extension (wac).

For example, assume that we wish to evaluate the likelihood that four sprinklers will control a design fire. The first step is to identify the heat release, plume momentum, and speed of growth for that size of design fire (lines 1 a, b, c, d of Figure 5.1). The next step estimates the likelihood that the four sprinklers will fuse before the fire grows to a larger area (fac). Then, are the quantity, pressure, and duration of water delivered to the four open sprinklers sufficient (cac)? Is the actual delivered water density sufficient to control the fire (dac)? Finally, are there obstructions that may influence the water distribution pattern to the fire (wac)? Because this is an analysis rather than a design, the evaluations reflect a value judgment for each of these events.

Sprinkler systems often control a fire by preventing its growth or propagation beyond certain sizes, even though the sprinklers may not extinguish the fire completely. The inability to extinguish is often due to water-spray obstructions that prevent water from penetrating to the seat of the fire. Fires that are effectively controlled within an area can be extinguished relatively easily by occupant or fire department manual actions. When a sprinkler system controls a fire for subsequent manual extinguishment, the suppression is described as successful for analysis purposes. The symbol AC is used to describe this evaluation.

Sprinkler systems could be evaluated for their ability to extinguish rather than to control a fire. We use the symbol AE (rather than AC) to describe this analysis. Sprinkler system control (AC) has a higher probability of success than complete extinguishment (AE). Often the building's fire protection is relatively unaffected whether the sprinkler system controls or extinguishes the fire. If complete sprinkler extinguishment without auxiliary manual extinguishment is important for a particular application, the analysis examines that outcome. The evaluations are similar, with only the event likelihoods adjusted to address control or extinguishment.

13.8 Sprinkler fusing

A sprinkler performance analysis usually starts with a study of initial sprinkler actuation because fire sizes are easier to estimate for the first sprinkler. This analysis gives a sense of proportion for sprinkler sensitivity and the fire's behavior. If the first sprinkler does not control the fire, the cumulative suppression effectiveness improves with more sprinklers opening. However, the interaction of water, the fire, and additional sprinklers makes a technical analysis very difficult and judgment must augment fire and sprinkler information to estimate performance.

There are literally hundreds of different sprinklers available. All of them use essentially the same concepts. The differences are associated with the specialized features for the components. Some of the more significant features are described below.

The link-and-lever mechanism is the most common type of releasing device. The links are held in place by a solder that has a lower melting point than the parts that are joined. For sprinklers the most common temperatures at which the solder will fuse are 135–170°F (57.2–76.7°C). However, to accommodate different environmental conditions, sprinklers can have a range of temperature classifications up to 650°F. Temperature ratings are stamped on the sprinklers with associated color codes for visual recognition. Different temperature ratings for the solders reflect needs for the ambient temperature conditions in which the sprinklers will be placed. Most ordinary hazards, such as offices or residences, use 135°F or 165°F sprinklers. Industrial sprinklers may require higher temperatures because of operations in heated environments.

Frangible bulbs are another common heat actuating element. A liquid that does not completely fill a small glass bulb expands when heated. When the bubble disappears and the pressure rises, the glass bulb breaks, releasing the cap on the nozzle. When the cap is released, the sprinkler operates in the normal manner. The operating temperature is determined by the volume and type of liquid in the tube. Color codes indicate temperature classifications.

Although a temperature rating can give a sense of the fire size at fusing, the temperature rating alone can be misleading. Sprinkler sensitivity also varies with the mass, size, and shape of the solder; the temperature difference between the sprinkler and the surrounding environment; and the velocity of the fire gases that flow by the sprinkler. Frangible glass bulb sprinklers also vary in a similar manner. In other words, all 165°F sprinklers do not fuse at the same fire size, other conditions being equal. The response time index (RTI) gives a measure of the sprinkler link sensitivity to its actuation response. The RTI is a constant for a sprinkler. It is directly proportional to the mass and specific heat of the heat-actuating element and inversely proportional to the convective heat transfer coefficient and surface area of the element. The larger the mass of the solder, the less sensitive it is to fusing, regardless of the temperature rating. Fast-response sprinklers have a RTI of less than 50. The RTI for standard-response sprinklers may vary from 80 to 350, with 100 being a more common value.

The reason for an interest in sprinkler actuation technology is its relationship with fire development. Sprinkler systems can have a relatively broad range of performance characteristics. An A curve evaluation attempts to answer the question, What is the probability that the sprinkler system will control a fire of *xx* kW involving *yy* square feet of floor area? One of the parts of this question is, Will the links fuse before a fire grows beyond this size? The fire size and growth speed are compared with the response sensitivity of the sprinklers.

Sprinkler actuation is affected by factors associated with the fire size and room geometry. The hot layer takes longer to develop with high ceilings. This gives the fire more opportunity to grow before sprinklers actuate. High-RTI sprinklers will not fuse as rapidly as lower-RTI sprinklers. Sprinklers with inappropriate link temperatures will not fuse fast enough to apply agent before the fire grows to a larger size. Sprinklers that are corroded, painted, taped, or bagged have longer actuation times. Sprinklers that are protected from the heat because of location or obstructions that prevent the fire gases from reaching the heat-sensitive element have a delayed actuation. Figure 13.8 identifies the major factors that are associated with this assessment. Fortunately, most of these can be recognized in existing systems by observation. If a new design is anticipated, one may include appropriate requirements in its specifications.

13.9 Water flow

A water flow analysis determines the quantity, pressure, and duration of water for the number of open sprinklers selected for evaluation. The component considers the source of the water

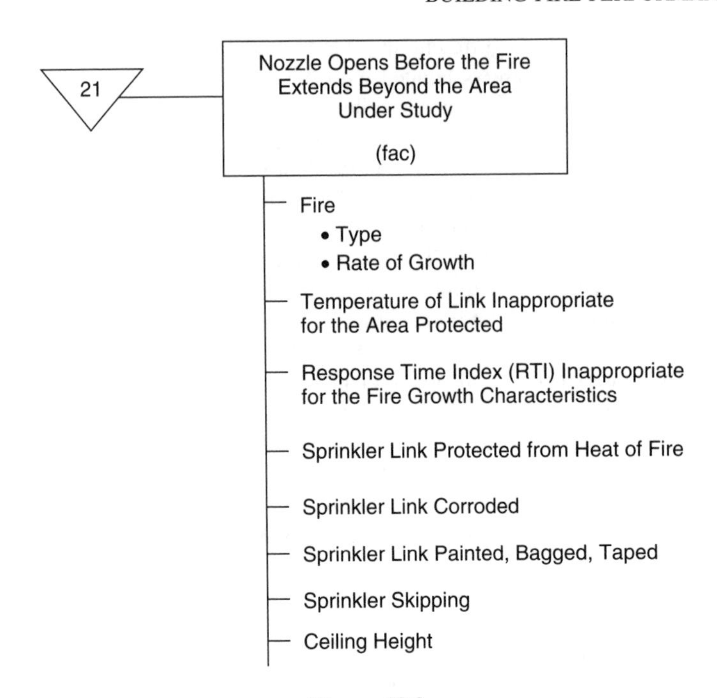

Figure 13.8

and the quantity available at the site. In the schematic of Figure 13.2, the water quantity and pressure available at the first water control valve is the base from which to start. Then the pressure changes due to control valves, backflow preventers and check valves, pumps, manual fire streams, fire department connections, risers, feed mains, cross mains, and branches can be incorporated to determine the pressure and quantity of water delivered to the number of sprinklers being evaluated. The collection of this information gives an insight into the water supply and the sprinkler system for the building being evaluated.

WATER SUPPLY

The water supply and external distribution pipes to a building influence the number of sprinklers that can be supplied and other demands for manual extinguishment water supply and the duration of water flow. The water supply may be broadly classified as public or private. A public water supply uses a system of underground pipes to transport the water from the source, such as a reservoir or well to the point of use, which is defined here as the boundary to the site. A private water supply processes and stores water for normal or emergency use on the site and moves it from the site boundary to the first building water control valve. In some cases an independent public water supply may be supplemented by private water supplies.

Public water sources and their underground piping determine the quantity and pressure that can be delivered at the time of use. Community water quantity and pressure will vary with the quality, size, and layout of the pumping and underground distribution system, pipe elevations, and other local demands at the time of use.

Private water sources provide quantity and pressure for normal and emergency use at the property. There are several types of water system, all of which have provisions for producing, treating, and storing a sufficient quantity of water for normal and emergency use. The water source may be a well, pond, lake, or stream. Elevated tanks are frequently used for the dual purpose of storage and maintaining adequate pressures.

The piping distribution on the site is a part of the complete sprinkler system. This piping system transports the water from the public or private supply to the building. The piping system can have a single feed to the building or a looped feed that provides multiple entry points and supplies. This can influence the dependability of the flow continuity (cac). For now, we merely recognize that the site's piping system is part of an evaluation.

CONTROL EQUIPMENT

Water control valves, backflow preventers, quick-opening devices, and pumps were discussed in Section 13.5. The focus at that time was their reliability in delivering water to an open sprinkler. Now the focus changes to determine the quantity and pressure of water for the number of open sprinklers that are being examined.

Many devices produce friction losses that reduce system pressure. Pumps do the opposite by increasing pressure and volume into the piping system. The capacity of the pumps, their controllers, the zones for which they pump, and the fire's duration are parts of a water flow evaluation.

When the water volume and pressure for sprinklers and fire department hose streams are inadequate, fire department connections are provided. Fire department connections (FDC) enable the fire department to pump water directly into a sprinkler system. This water can augment low-pressure conditions and override closed valves or failed quick-opening devices. The FDC availability, locations, and fire department water needs are part of water flow evaluations (cac).

AUXILIARY EQUIPMENT AND OTHER CONDITIONS

Although this chapter discusses automatic sprinkler suppression, one should have an awareness of other equipment and considerations that relate to a more complete understanding of a sprinkler system. One of the more valuable auxiliary devices is a water flow alarm with connections to a supervising service or to the local fire department. This device serves as an automatic fire alarm as well as a way to reduce potential water damage by enabling the fire department to shut down the sprinkler system after extinguishment. This is particularly useful in installations that are unoccupied for periods of time.

Test connections to perform maintenance should be located to reflect the system's performance demands. Valves and equipment necessary for operation must be protected against freezing, explosions, flooding, earthquakes, and windstorm. Also, when water is applied to a fire by sprinklers, by the fire department, or by both, it is a good idea to design a drainage system to allow the excess or unwanted water to be removed with a minimum of ancillary damage.

A continuous flow of water is necessary until the fire is extinguished. For those conditions where the fire is extinguished quickly by the sprinkler system, water supply needs are modest. When a large fire is controlled but continues to burn, the water supply must be maintained. Figure 13.9 shows factors that influence the continuous flow of water (cac).

13.10 Water discharge

When a sprinkler fuses, water is discharged through the nozzle and impacts on the deflector. The water pressure and the deflector design determine the size of the water droplets and the distribution pattern of the water spray. Factors that influence the operational effectiveness of the water discharge component include the sprinkler type, water droplet sizes, the water density, fire characteristics, and the room size.

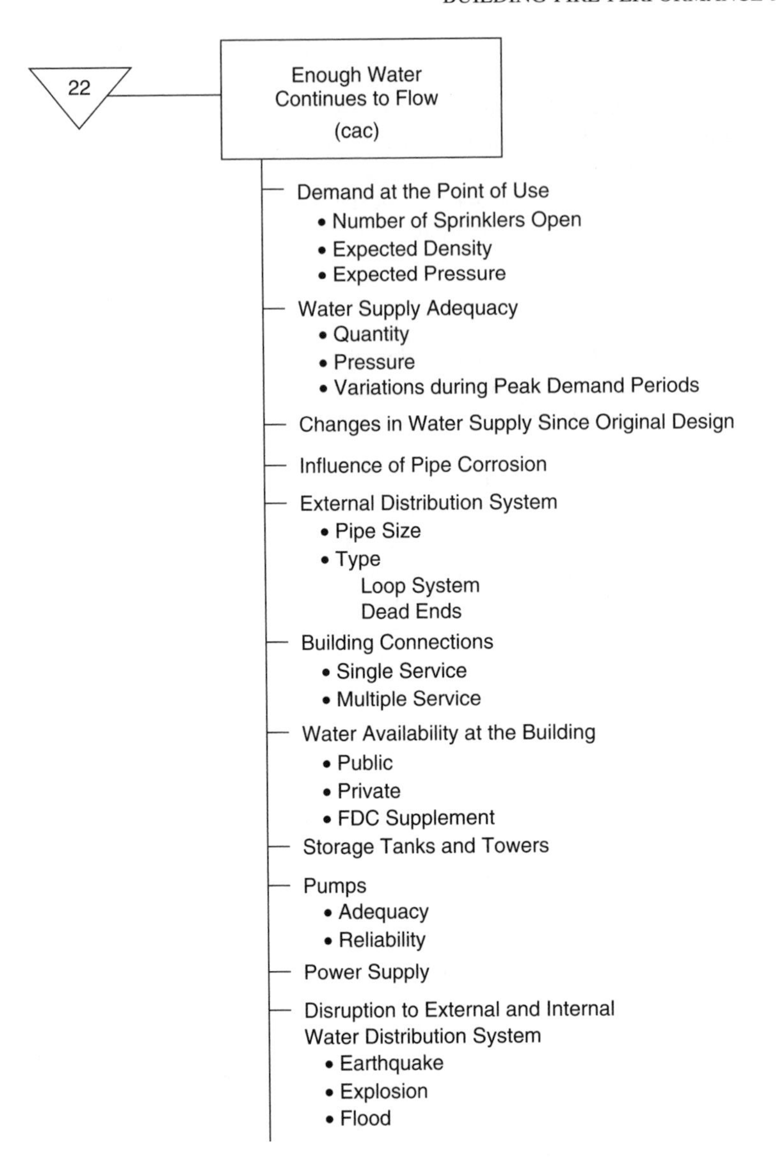

Figure 13.9

Sprinklers may be upright, pendant, or sidewall. Figure 13.10 illustrates each type. When upright sprinklers are used, the piping system is also visible. Water is directed upward to the deflector, which redirects a percentage of this water downward in an umbrella spray pattern to the fire. A sprinkler is often used in the pendant position. This is sometimes to hide the piping system in the ceiling and sometimes because of design needs. In the pendant position, water is directed downward to the deflector, and a percentage is redirected upward to strike the ceiling and again fall down to the fire in a spray pattern. Upright and pendant sprinklers should not be interchanged, because their designs are distinct and the spray patterns will differ.

Sprinklers extinguish a fire by several actions. One is the absorption of heat from the fire. Fire is an exothermic chemical reaction. When the heat generated by the fire is less than the heat lost

Figure 13.10

due to heat transfer and water evaporation, the fire will recede. The opposite occurs when the heat generation is greater than the losses. Heat absorption by the phase transformation of converting water to steam is a major heat loss from the fire and a significant contributor to fire control. In small, confined rooms, large amounts of steam contribute the displacement of air (oxygen) from the fire. Small water droplets are more efficient for steam generation than large water droplets because the water has more exposed surface area per volume.

Water from a sprinkler can also prewet unburned fuel adjacent to the burning materials. Before a fire can pyrolyze that fuel, the water must be removed by evaporation. The time needed for the fire to dry the fuel contributes to the delay in fire spread. In addition, the water spray cools the burning fuel to cause slower fire growth.

Any sprinkler water application to a fire will produce each of these mechanisms to some extent. It is possible to design the sprinkler to produce water droplets appropriate for the expected fire conditions. For example, a water mist may be appropriate for small rooms in which the escape of gases is limited. The small droplets evaporate more easily, creating steam and choking the fire. On the other hand, a water mist would be ineffective in a large room with fuels that produce a rapid heat release, because the fire pressures would drive the steam away before oxygen depletion could occur. In this type of situation it is more effective to use a fast-response sprinkler and discharge of large droplets that can penetrate the fire plume to the seat of the fire. For most situations the appropriate choice is a standard sprinkler that bridges between these extremes.

This discussion introduces the concept that all sprinkler systems do not have the same quality and performance effectiveness. The system design should be appropriate to the hazard. Room size, the rate of heat release, and the fire plume momentum can be important in selecting the appropriate sprinkler response for the hazard. The water discharge event (dac) includes this type of assessment.

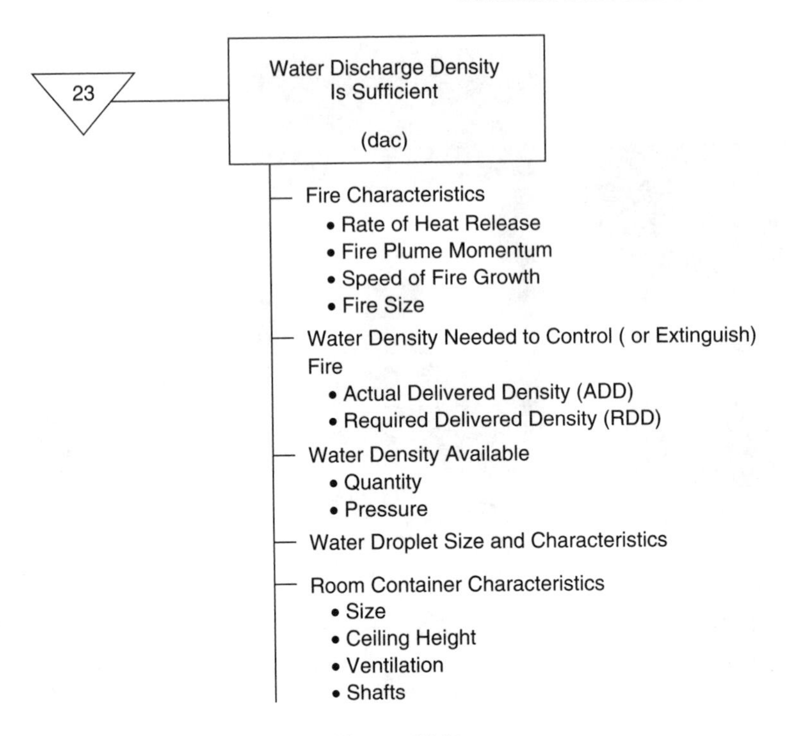

Figure 13.11

The NFPA sprinkler standard [1] provides guidance for automatic sprinkler design. Briefly, in the usual design process one identifies the type of hazard in the space. Then a preliminary piping layout for the building is selected and the hydraulically most demanding area is identified. Each sprinkler in this design area must have a discharge flow rate at least equal to a selected, defined water rate application. This rate is defined as the water application density and is expressed in gpm/ft^2 (lpm/m^2). The water supply must be sufficient to provide the required density and duration of flow. Water distribution is a principal component of the sprinkler standard.

The water density component (dac) estimates the likelihood that the water quantity and distribution will control the design fire. Figure 13.11 identifies the factors that affect this event.

13.11 Obstructions

The final part of operational effectiveness looks at the disruption of sprinkler discharge patterns by obstructions. The influence of obstructions can vary greatly. At one extreme, obstructions may provide a relatively minor interference to water distribution. At the other extreme, obstructions can have a major effect on the sprinkler system performance, including contributing to a sprinkler system failure. This event addresses the significance of obstructions to controlling a fire.

Perhaps a sprinkler system may not have had any major obstructions when it was installed. However, as buildings go through periodic, routine renovations, features such as new walls or air-conditioning ducts may be installed. Lack of care in construction and inspection can influence sprinkler spray patterns. In an existing building, conditions that shield discharge are relatively easy to recognize. In proposed new installations, the NFPA sprinkler standard incorporates requirements relating to installation practices when obstructions occur. Nevertheless, diligence is needed to ensure that obstructions are not introduced during the construction or renovation process.

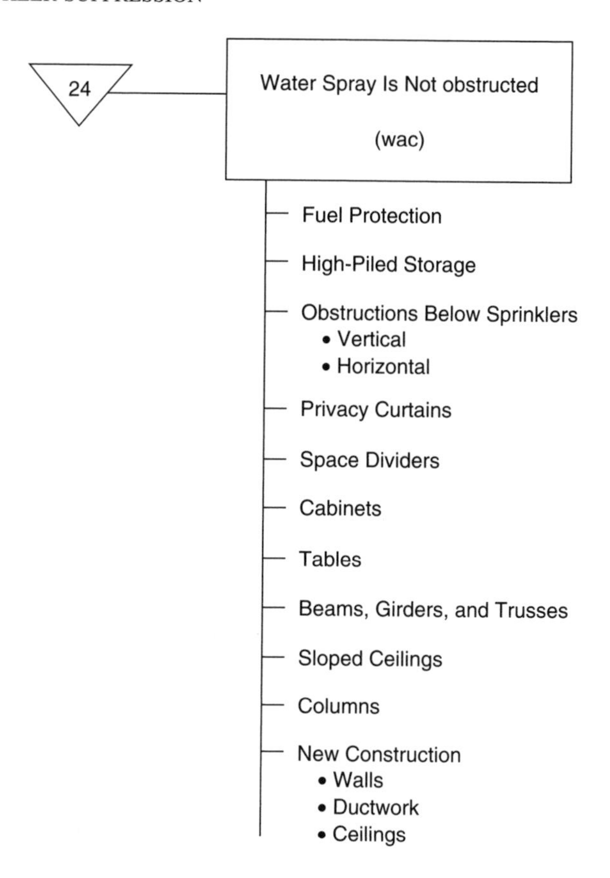

Figure 13.12

In addition to the more obvious obstructions of new partitions and ducts, other construction features, such as lighting fixtures, cable trays, beamed ceiling construction, columns, and soffits, can shield a potential fire from the sprinkler discharge pattern. Interior design involving items such as workspace dividers, privacy curtains, cabinets, and tables also affect performance by shielding a fire from the water spray. Shadow areas caused by high-piled storage introduced after the sprinkler system was installed also can shield a fire. Situations have occurred where a storage facility sprinkler system was originally designed for low-piled cartons in noncombustible containers. Newer technology using highly combustible container materials and different inventory practices introduced high-piled storage. These changes significantly affected the sprinkler system in two ways. First the design fire characteristics were changed dramatically. Secondly, by shielding a fire from the sprinkler water spray, patterns were different from those initially anticipated. These differences in condition affect the A curve.

The obstruction influence event (wac) provides an opportunity to assess the effect of shielding on sprinkler suppression. Figure 13.12 identifies common shielding conditions.

13.12 Framework for analysis: operational effectiveness

The four major components of a sprinkler system's operational effectiveness are shown in Figure 13.13. These events are conditional and their order should be sequential. The following examples illustrate the process for calculating design effectiveness.

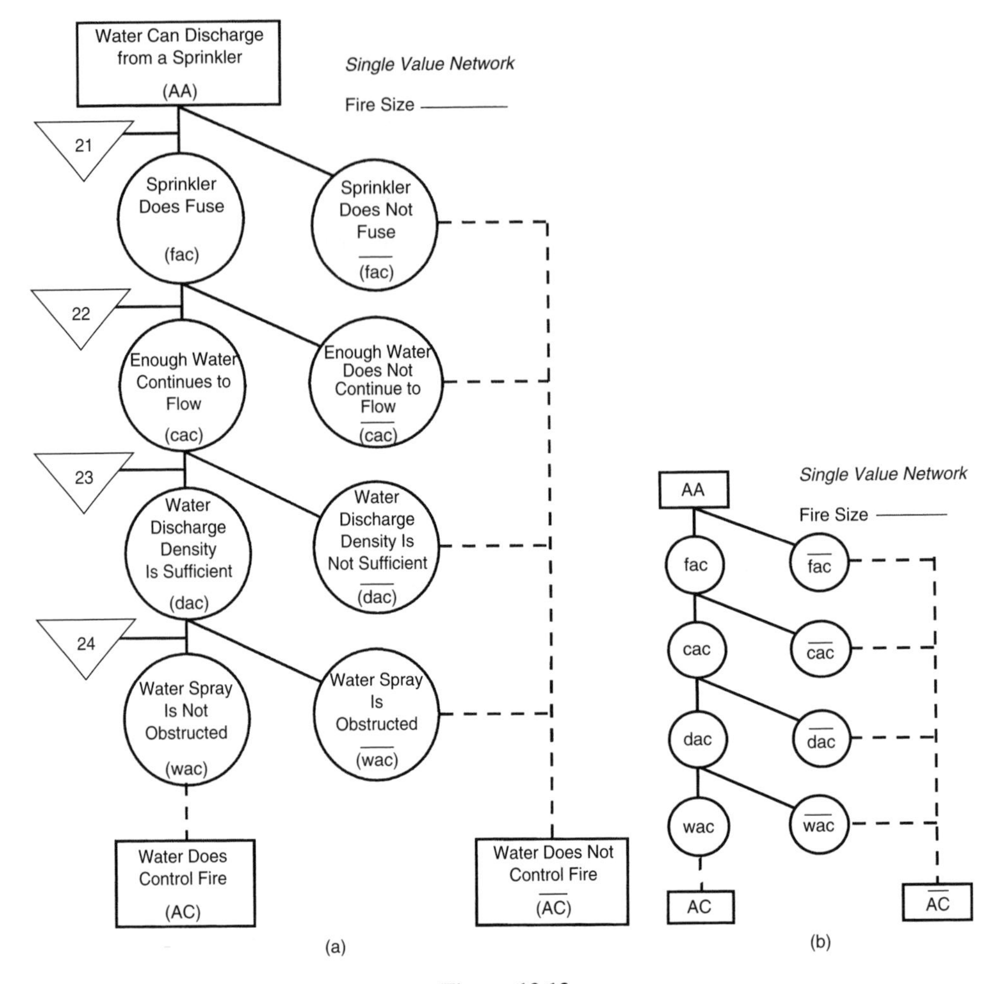

Figure 13.13

Example 13.3

A sprinkler system in a large room is evaluated for its design effectiveness. Initially it was decided to assess the likelihood that the first sprinkler will control the fire to get a sense of the expectations. Conditions for this assessment are as follows:

- A representative location and a design fire scenario was selected.
- The system is reliable and water can flow from the adjacent sprinklers.
- The RTI of the sprinkler was estimated. The fire plume characteristics, ceiling height, and sprinkler location guide the probability estimate for the first sprinkler to actuate. A value of $P(\text{fac}) = 0.7$ is selected. The minimum fire size for actuation of the first sprinkler is 800 kW.
- Given that the first sprinkler actuates, the water quantity, pressure, and continuity are certainly sufficient to extinguish a fire of approximately 800 kW. A value $P(\text{cac}) = 1.0$ is selected.
- Given that the first sprinkler actuates and the water flow is sufficient, the design density and distribution for this one sprinkler to extinguish this fire before it extends is a certainty. Therefore $P(\text{dac}) = 1.0$.

- Given that the first sprinkler actuates, enough water continuity is available, and the water density is sufficient, obstructions to the spray pattern are examined. Some minor obstructions were present that can prevent the water from being applied effectively to this small fire. These obstructions introduce some uncertainty in the sprinkler effectiveness. Considering the observable conditions, a judgment was made that there is only a moderate likelihood that the sprinkler will control the fire at 800 kW before it grows to a larger size. This estimate is expressed numerically as $P(\text{wac}) = 0.8$.

Solution

The probabilities are incorporated in the SVN of Figure 13.14, and $P(\text{AC}) = 0.56$ for this scenario.

Example 13.4

Evaluate the room of Example 13.3 for a fire area that would involve approximately six sprinklers. The design fire for this condition was estimated to be 8 MW. The design effectiveness will be reevaluated for this larger fire. The evaluation description is as follows:

- Considering the sprinkler RTI and the expected fire development, it is estimated that the probability of all six sprinklers actuating to control a fire of the expected size before it spreads to a larger area is $P(\text{fac}) = 0.98$.
- Given that the sprinklers fuse, it is estimated that the water quantity, pressure, and duration for this installation are marginally adequate. The value selected is $P(\text{cac}) = 0.95$.
- Given that the water pressure and quantity are adequate, the suppression density is judged to be very high, although not a certainty. A value of $P(\text{dac}) = 0.99$ is selected to reflect this belief.
- Obstructions to the discharge spray patterns were then examined. Obstructions still exist, although they are not as significant for this larger fire. There is only a small likelihood that the

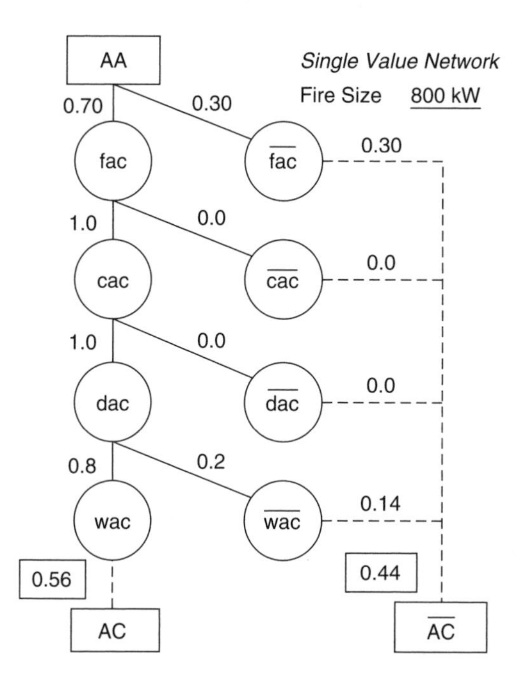

Figure 13.14

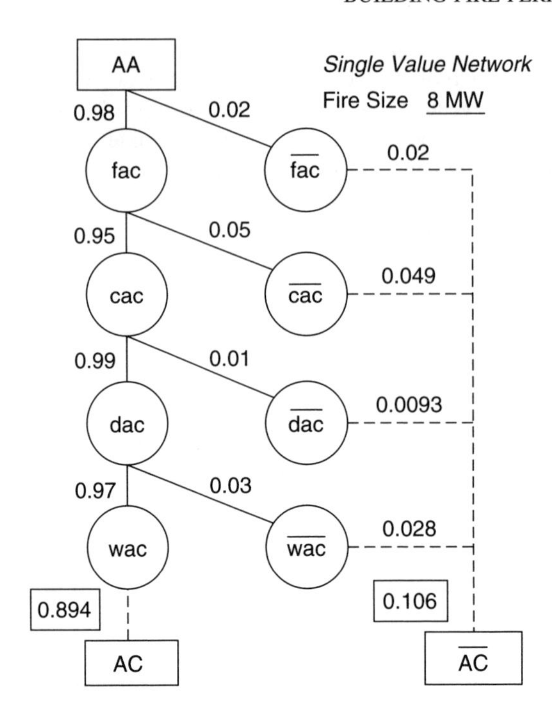

Figure 13.15

obstructions will have an adverse influence on the fire suppression before it grows to a larger size. This estimate is expressed numerically as $P(\text{wac}) = 0.97$.

Solution

These probabilities are shown in Figure 13.15. The design effectiveness is calculated as $P(\text{AC}) = 0.894$.

13.13 Calculating A curve coordinates

The combination of sprinkler reliability (AA) and operational effectiveness (AC) provides the likelihood of success for the sprinkler system at the fire size selected. Figure 13.16 shows the SVN for this relationship.

Figure 13.17 shows the segment of Figure 5.1 that relates to automatic sprinkler performance. Heat release rates for the design fire are shown in line 1a. Results from Examples 13.3 and 13.4 are included with Example 13.5 to calculate coordinates for fire sizes of 800 kW and 8 MW. Sprinkler reliability is a constant, as shown in line 4A. Values for fac, cac, dac, and wac are shown for fire sizes of 800 kW and 8 MW. Figures 13.14 and 13.15 show the calculations for row 4B. Values of row 4 are determined with the network of Figure 13.16 and illustrated by the calculations of Figure 13.18.

Example 13.5

Assume that the building of Examples 13.3 and 13.4 had the system reliability described in Example 13.1. Determine the probability that the sprinkler system will control a fire of these sizes.

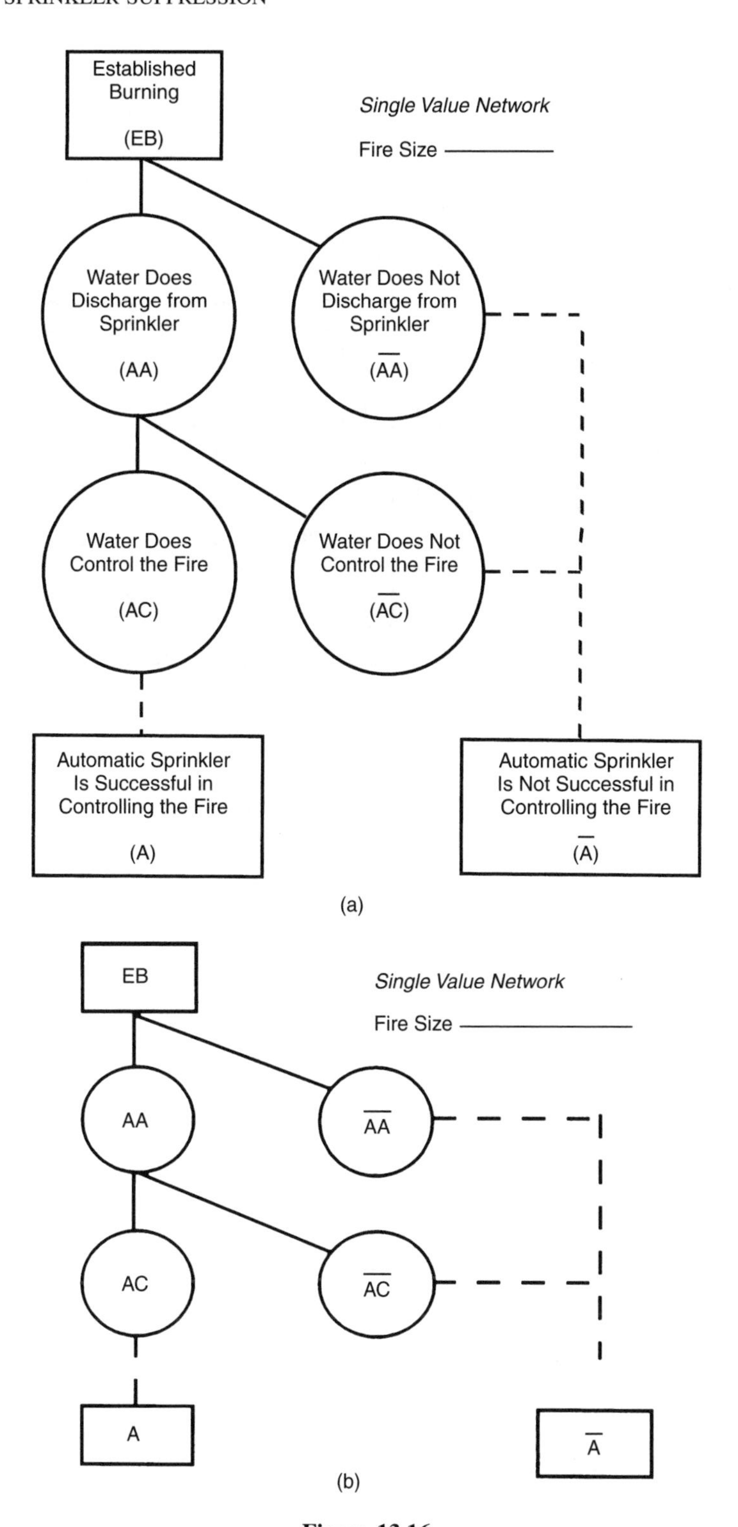
Established
Burning
(EB)
Single Value Network
Fire Size
Water Does
Discharge from
Sprinkler
(AA)
Water Does Not
Discharge from
Sprinkler
$(\overline{AA})$
Water Does
Control the Fire
(AC)
Water Does Not
Control the Fire
$(\overline{AC})$
Automatic Sprinkler
Is Successful in
Controlling the Fire
(A)
Automatic Sprinkler
Is Not Successful in
Controlling the Fire
$(\overline{A})$
(a)
EB
Single Value Network
Fire Size
AA
$\overline{AA}$
AC
$\overline{AC}$
A
$\overline{A}$
(b)

Figure 13.16

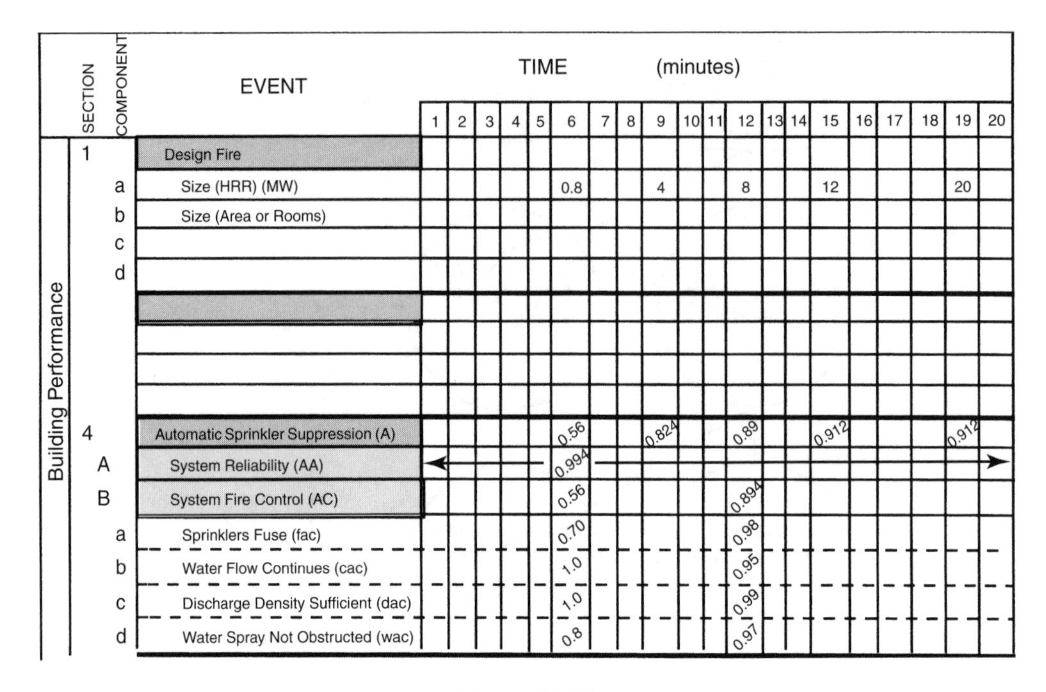

Figure 13.17

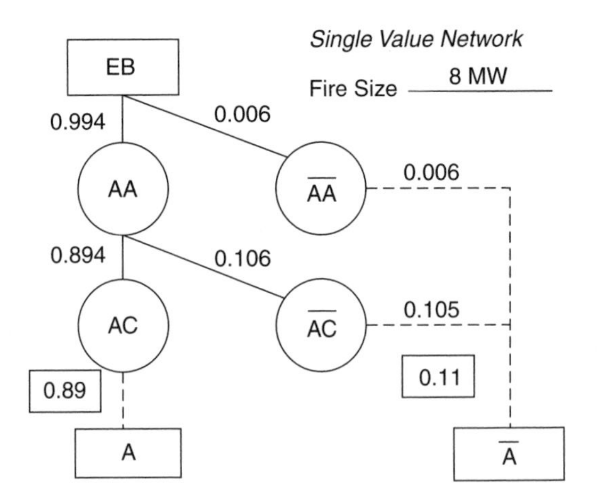

Figure 13.18

Solution

Because the reliability of the system needs to be evaluated only once for each zone, $P(\text{AA}) = 0.994$. The operational effectiveness for an 800 kW fire is $P(\text{AC}) = 0.56$ and for an 8 MW fire, $P(\text{AC}) = 0.894$. Incorporating the values of $P(\text{AA})$ and $P(\text{AC})$ for the 8 MW fire into Figure 13.18 gives a value of $P(\text{A}) = 0.89$ for this scenario. The probability of success in suppressing an 800 kW fire is $P(\text{A}) = 0.56$.

13.14 The A curve

All of the ingredients for level 2 SV sprinkler performance evaluations have been presented. A few additional observations about the A curve will be made in the context of an example and its discussion.

Example 13.6

Assume the building of Example 13.5 had SV assessments of $P(\mathrm{A})$ for a few additional fire sizes. The results were as follows. Plot these values on evaluator B and construct the A curve.

Fire size (MW)	P(A)
0.8	0.56 (Example 13.5)
4	0.824
8	0.89 (Example 13.5)
12	0.912
20	0.912

Solution

Figure 13.19 shows the coordinates. Connecting them produces the A curve.

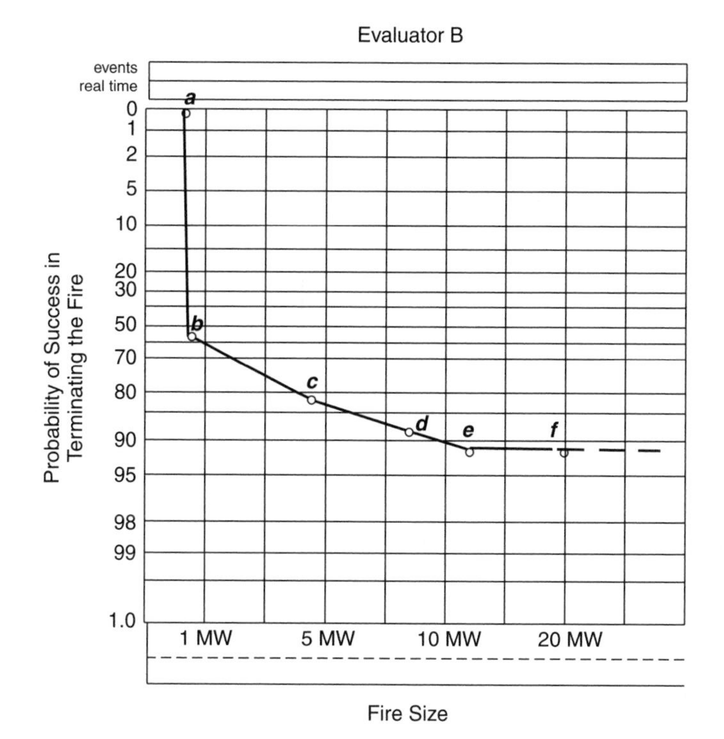

Figure 13.19

Discussion

The single values in the horizontal cells of Figure 13.17, Row 4 define an A curve. If one or a few coordinates are selected and evaluated with some care, the remaining coordinates can be extrapolated with confidence. A few carefully selected evaluations provide enough knowledge about the sprinkler system to predict its performance.

Figure 13.20 shows the first few events of a CVN that describes the thought process for an A curve. This CVN uses conditional events. Each sequential event is based on the condition that the fire has not been suppressed by a previous sprinkler size, $\overline{A}_{n-1}$. Thus, the link between $\overline{A}_{4\,MW}$ and $A_{8\,MW}$ is read as the probability that the sprinkler system will extinguish a fire of 8 MW given that it did not extinguish it at 4 MW. In symbols this is $P(A_{8\,MW}|\overline{A}_{4\,MW})$.

Figure 13.21 shows a CVN with the values of this example. The numbers in the column headed Cumulative Probability of Success show the single values for $P(A)$ that were described in the problem statement. The values in the body of the network are obtained from back calculation using the two rules of network calculations (Section 6.4):

- Multiply the probabilities along a continuous path to determine the probability of that path.
- Add like outcomes to obtain a result.

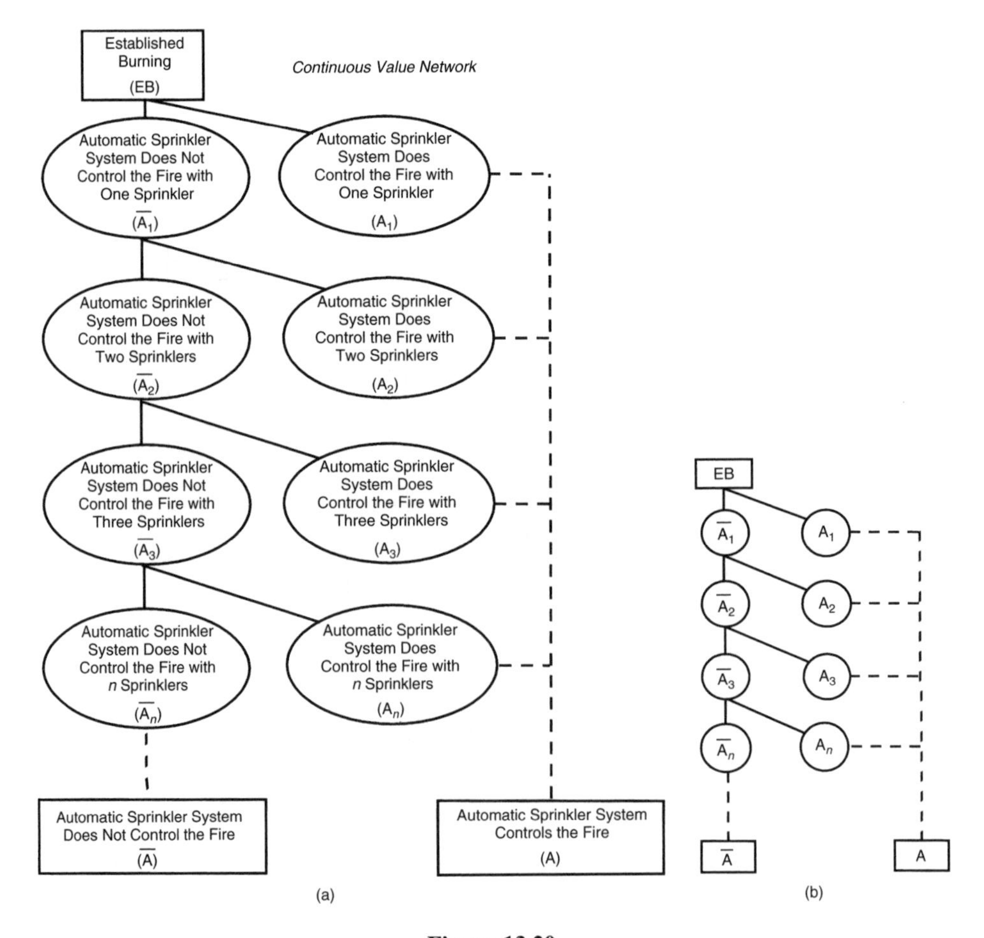

Figure 13.20

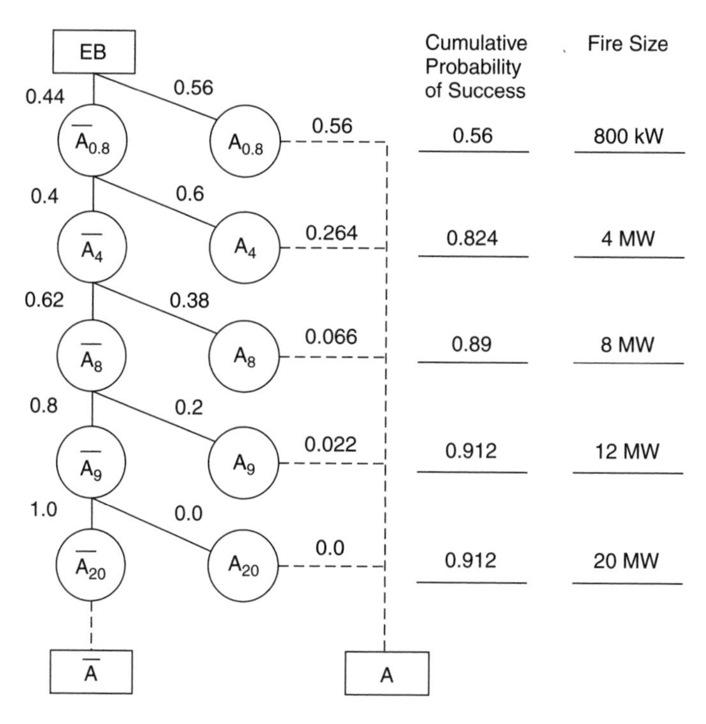

Figure 13.21

Because the network shows the universe of outcomes after each event, the sum of these outcomes must be 1.0. Therefore the initial value for 800 kW of 0.56 is made up of the single link $P(\mathrm{A}_{800\,\mathrm{kW}} = 0.56)$. Its complement $P(\overline{\mathrm{A}}_{800\,\mathrm{kW}}) = 0.44$ because the sum of all outcomes for 900 kW must be 1.0.

The difference between the probability of success at 800 kW and at 4 MW is 0.264. Therefore the continuous path from the start is $P(\overline{\mathrm{A}}_{800\,\mathrm{kW}})P(\mathrm{A}_{4\,\mathrm{MW}}) = 0.264$. Knowing $P(\overline{\mathrm{A}}_{800\,\mathrm{kW}}) = 0.44$, the conditional probability $P(\mathrm{A}_{4\,\mathrm{MW}}|\overline{\mathrm{A}}_{800\,\mathrm{kW}}) = 0.6$. The other conditional probability values can be back-calculated in this manner. Figure 13.22 shows these relationships.

The determination of conditional probabilities is unnecessary in practice. This exercise illustrates the relationships between a CV curve and its associated SV assessments.

Conditional probability is a useful concept for envisioning the sequential opening of sprinklers and the successively improving A curve. Figure 13.20 shows the thought process CVN. Unfortunately, we can't establish a technical rationale for evaluating conditional probabilities, such as $P(\mathrm{A}_{10\,\mathrm{MW}}|\overline{\mathrm{A}}_{4\,\mathrm{MW}})$. It currently seems beyond practical achievement. Fortunately, these conditional probability values give no advantage, because the technical basis for SV analyses satisfies the needs of performance evaluations.

All sprinkler systems have capacity limits. When too many sprinklers fuse, the water quantity and pressure get lower, density and coverage are reduced, and effectiveness decreases. Eventually the sprinkler system reaches its capacity, and if the fire has not been extinguished, the probability of success decreases, as shown by the dashed line in Figure 13.23. Up to this size, the cumulative curve (solid line) and the actual curve coincide. Because an A curve is cumulative, it remains horizontal after the system's capacity has been reached.

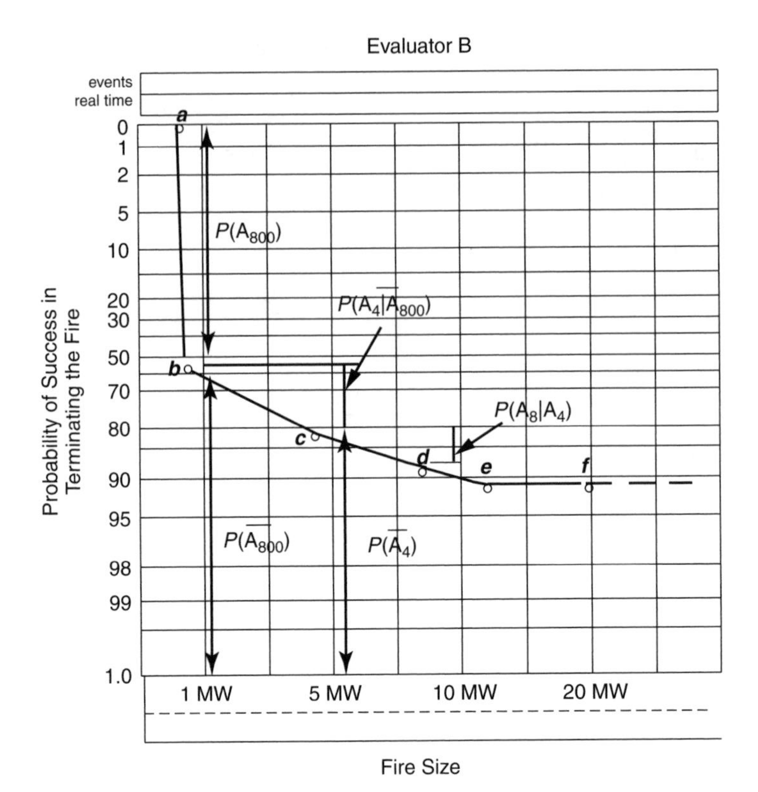

Figure 13.22

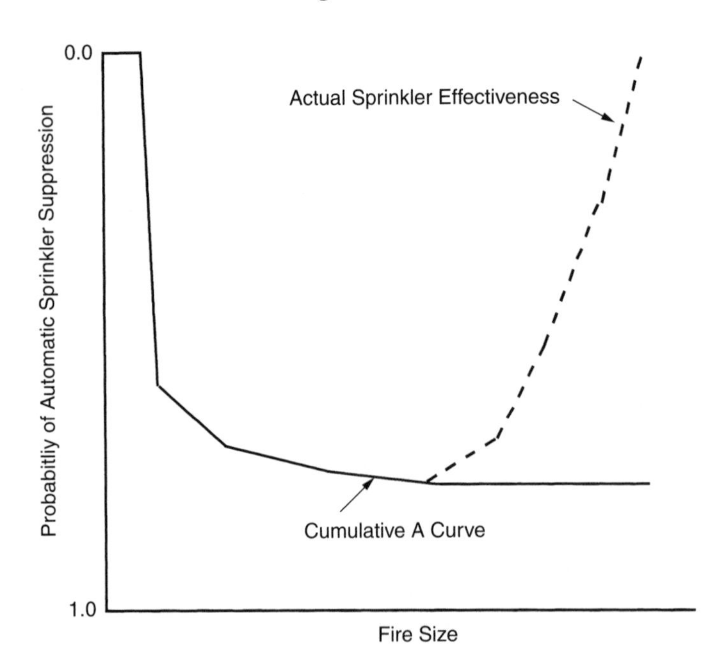

Figure 13.23

13.15 Quality comparisons

The A curve enables one to visualize sprinkler performance. An advantage of performance evaluations is the ability to compare different systems or alternatives. Example 13.7 illustrates how one may show performance comparisons.

Example 13.7

The operational effectiveness described in Example 13.6 is independent of the reliability of the sprinkler system. Compare differences in sprinkler performance of Example 13.6 for the reliability of Buildings A and B in Examples 13.1 and 13.2.

Solution

The A curve for Building A of Example 13.1 was constructed in Example 13.6. The effect of reduced reliability for Building B of Example 13.2 may be calculated by recognizing $P(\mathrm{A}) = P(\mathrm{AA})P(\mathrm{AC})$. For example, at 8 MW, $P(\mathrm{A}) = (0.81)(0.894) = 0.72$. Figure 13.24 shows the A curves for these two buildings.

13.16 Level 3 analysis

A level 3 analysis enables one to understand the envelope of expected performance for factors and variability in a building. For example, it could describe differences for factors such as changing the sprinkler RTI from 50 to 200; the effect of corrosion that reduces water main diameter by

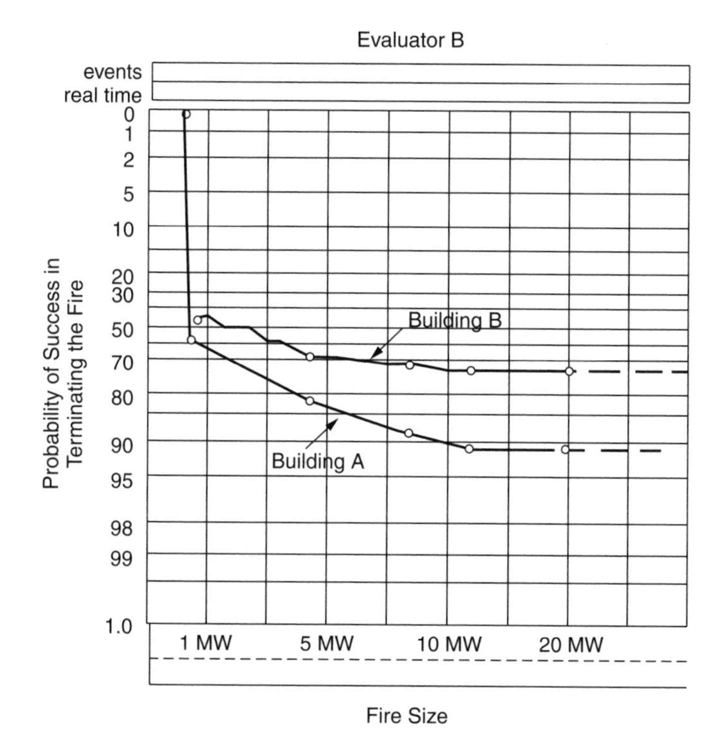

Figure 13.24

1 in; or reducing pump capacities. Differences can be described by variations in the A curve that reflect conditions of interest.

13.17 Level 1 evaluations

A level 1 evaluation is based on information that may be deduced from observations, simple calculations, and any documentation that may exist. Often performance inadequacies can be observed in existing buildings or they may be anticipated from the plans of proposed buildings. For example, commercial buildings undergo major renovations on average about every seven years. Fire protection components can be overlooked during reconstruction when new partitions or other construction features obstruct sprinkler spray patterns or contents change. Level 1 evaluations for a proposed building explores conditions in which potential weaknesses can develop. Evaluations tell the story about the expected sprinkler system performance for the site-specific building being studied.

Figure 13.25 shows the substance of an SV A curve coordinate analysis. One evaluates the reliability (AA) only once for the system or zone being studied, because it does not change with sprinkler operations. Figure 13.26 notes topics that influence this evaluation. The operational effectiveness evaluation involves the matching of design fire characteristics and operational effectiveness for each fire size selected for analysis. The design fire characteristics envision the speed of fire growth (αt^2 fire), area of coverage, cumulative heat release, and fire plume momentum. Sprinkler performance estimates the likelihood that the sprinklers will fuse and the sprinkler system will control the fire. Figure 13.27 identifies factors that influence this performance.

Descriptive terms enable one to describe quality from factors that are observed or anticipated from the written specifications of proposed new systems. Because sprinkler systems are typically very effective in controlling a fire, an irregular scale is more useful to communicate effectiveness (Figure 13.28). If desired, these verbal descriptions can be augmented with numerical values to enable system components to be compared more easily. As with all judgmental selections, one can adjust the position of a curve to communicate better.

Necessary information is not always available for level 1 evaluations. Nevertheless, much information can be acquired by observing conditions and envisioning operations. For example, design

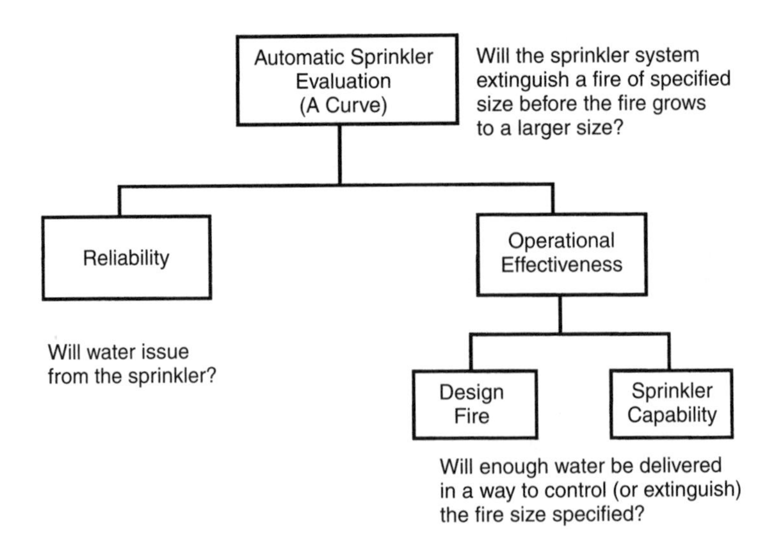

Figure 13.25

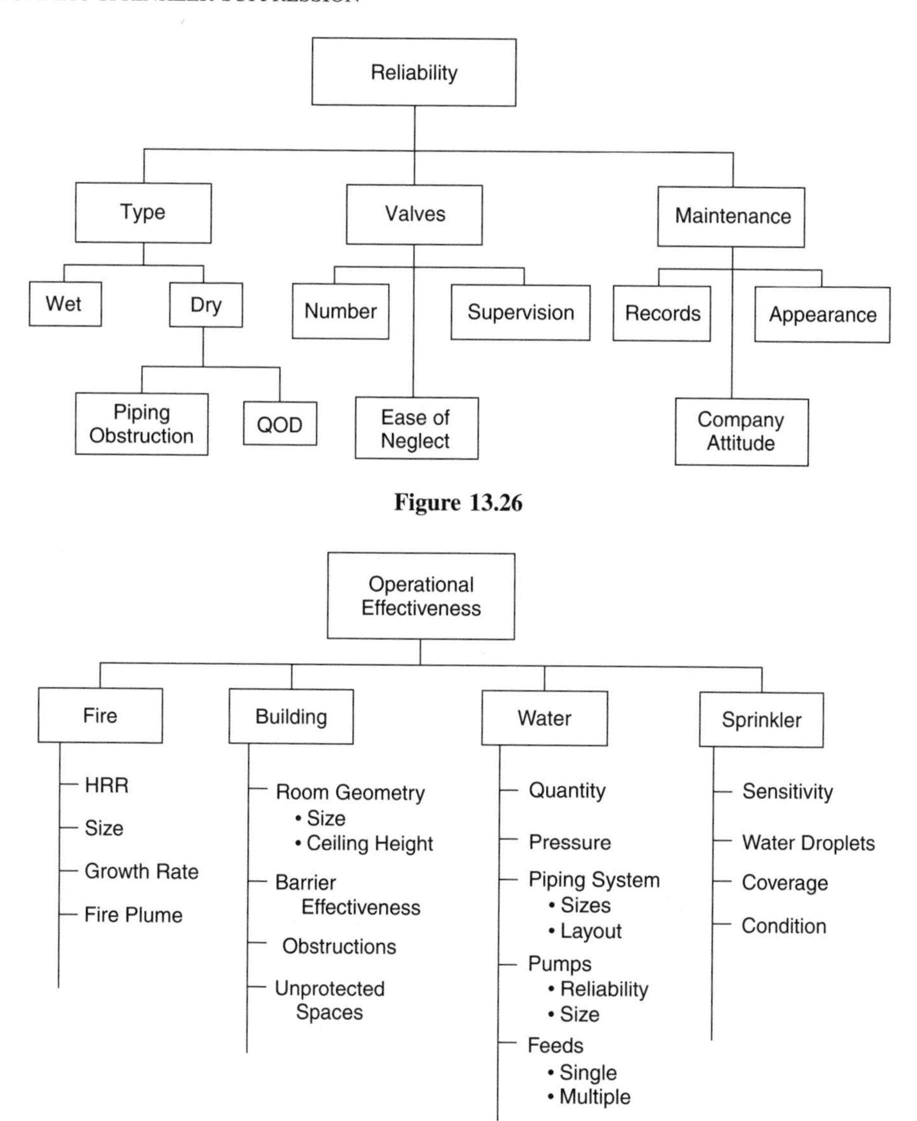

Figure 13.26

Figure 13.27

fire characteristics, fire plume momentum, and the effect of obstructions on distribution patterns can be envisioned. Pipe sizes, sprinkler characteristics, ceiling heights, distribution patterns, pump sizes, and barrier contributions can be recognized. Records may be available to document water pressures and capacities. Observations can give a sense of the system's operational effectiveness. The important concept is that an evaluation is based on what is known and observed, and the assessment reflects conditions that the individual understands.

Tables 13.1 and 13.2 illustrate factors that can guide reliability (AA), quality, and operational effectiveness (AC). Not all of these factors can be assessed, and some may not apply to the building being evaluated. However, they give a sense of considerations that affect performance and might be determined if enough time or information were available. As with all guidelines, one must select a different quality when reason indicates that a different rating is more appropriate.

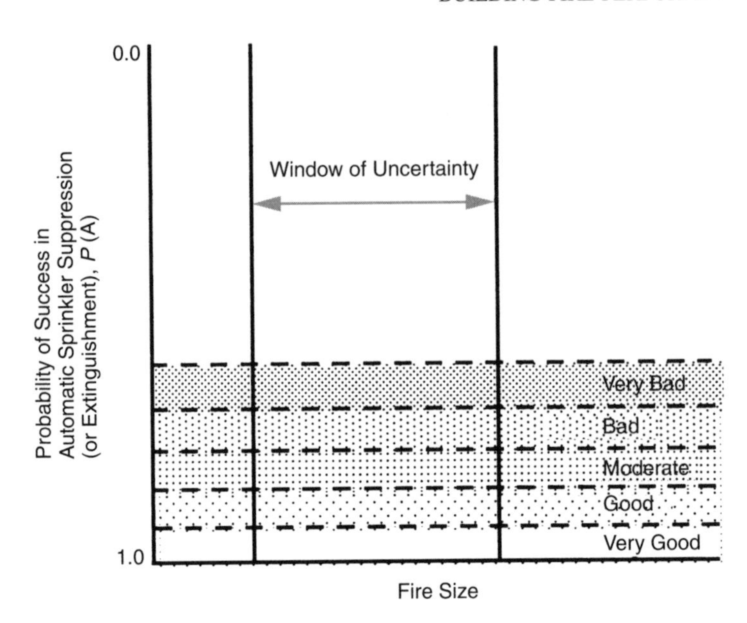

Figure 13.28

Table 13.1

Reliability rating	Major factors
Very good	Wet-pipe system; central station supervision of all control valves; regular maintenance and testing of valves, pumps, and piping system, and tanks (time duration?); automatic backup electrical power; sprinkler system was tested after initial installation; designed by a certified sprinkler designer
Good	Wet-pipe system; control valves chained and padlocked or electrically supervised; regular maintenance and testing of valves, pumps, and piping system, and tanks (time duration?); automatic backup electric power; designed by a certified sprinkler designer
Moderate	Wet- or dry-pipe systems; sporadic maintenance and testing (time duration); no corrosion protection
Bad	Wet- or dry-pipe systems
Very bad	Dry-pipe systems; no recorded testing and maintenance; no controlled valve supervision

13.18 Partially sprinklered buildings

Many buildings are constructed with sprinkler systems in some spaces rather than complete coverage. For example, corridor-only and exit-only sprinkler systems are often found in buildings. Sometimes much of a building may be sprinklered except for small areas that have contents deemed sensitive to water damage. There may be reasons for not installing a sprinkler system in a part of the building. Whether the logic and arguments are good or bad is immaterial, because a building evaluation assesses what is, not what should be.

The goal of evaluating a partially sprinklered building is the same as that of any building, i.e., to understand its fire performance and tell the story of what will happen for design fires located in different rooms. A partially sprinklered building requires at least two and sometimes a few more

Table 13.2

Operational function	Factors that alter effectiveness
Nozzle opens before the fire extends beyond the area under study	Fire (type, rate of growth); temperature of the link inappropriate for the design fire; response time index (RTI) inappropriate for the room size, ceiling height, or design fire; sprinkler link protected from the heat of the fire; sprinkler link corroded, painted, bagged, or taped; sprinkler skipping
Enough water continues to flow to control the fire	Demand at point of use (number of sprinklers open, expected density, expected pressure); water supply adequacy (quantity, pressure, peak demand period variations); changes in water supply since original design; influence of pipe corrosion; external distribution system (pipe size, type (e.g., loop or dead ends); building connections (single service or multiple service); water availability at the building; storage tanks and towers; pumps (adequacy, reliability); power supply; disruption to external or internal water distribution system (earthquake, explosion, flood)
Water discharge density is sufficient	Design fire characteristics (HRR, fire plume momentum, speed of growth, fire size); water density needed to control or extinguish the fire (actual delivered density is ADD, required delivered density is (RDD); water density available (quantity, pressure); water droplet size and characteristics; room size and container characteristics (size, ceiling height, ventilation, shafts)
Water spray is not obstructed	Fuel protection; high-piled storage; obstructions below sprinklers (vertical, horizontal); privacy curtains; space dividers; cabinets; tables; beams, girders, trusses; sloped ceilings; columns; new construction (walls, ductwork, ceilings)

rooms of origin for analysis. If a fire starts in a nonsprinklered area adjacent to a sprinklered area, the analysis looks at the barrier effectiveness and capacity of the sprinkler system for a fire that would move (often massively) into the sprinklered space. The sprinkler system must contend with a very different fire than would occur if the fire originated in the sprinklered room. Its suppression effectiveness may be significantly eroded. The questions in the evaluation process remain the same, but the answers and expected performance may be substantially different.

13.19 Fire department mutual aid

The fire department can be a valuable asset for sprinklered buildings. In sprinkler system fire control, the fire department can complete the extinguishment process and turn off the water after extinguishment to prevent excess water damage. A water flow alarm connection to a supervisory service or the fire department is an important part of this activity. In addition, the fire department can assess the situation and provide assistance to occupants or to other needs of the building.

In those cases where the sprinkler system may not control the fire, the fire department assumes the function of manual fire fighting. Major fire-fighting efforts may be required if the sprinkler system did not function or if it was substantially inadequate. Fire department connections to the sprinkler system can augment sprinkler water pressure and quantity, and the number and locations of fire department connections become part of a building evaluation.

13.20 Automatic suppression

The term "automatic suppression" is often used to mean different types of automatic systems. They may be described as follows.

GENERAL AREA EXTINGUISHING SYSTEMS

These systems protect a large area, usually an entire building. The automatic sprinkler system is the most common type of general area fire protection. Total flooding foam systems may also be considered a general area extinguishing system.

SPECIAL HAZARD EXTINGUISHING SYSTEMS

These systems are sometimes called spot protection. Their function is to protect identified special hazard areas within a building. For example, areas that involve

- flammable liquids in dip tanks, oil quenching tanks, or metal cleaning equipment;
- spray booths using flammable paints and finishing materials;
- test facilities where flammable liquids or gases are used or stored;
- commercial cooking involving deep fat fryers or grease accumulation in ducts and equipment.

BARRIER DEFENSE EQUIPMENT

Automatic equipment, usually involving sprinkler hardware, may be used to prevent fire extension through a barrier or opening. Illustrations of this type of equipment include

- water curtains;
- exposure cooling systems.

Almost all automatic suppression systems can operate without the need for human intervention. However, there can be situations where it may be desirable to have fixed manual extinguishing systems activated. This is relatively rare in buildings, but more commonly in ships. A fixed extinguishing system that requires human activation may be described as an automated extinguishing system, and its analysis combines automatic and manual extinguishing components.

The term "automatic suppression" in this book is used to mean general area automatic sprinkler system fire protection. If the sprinkler hardware covers the entire space (e.g., room or zone), the suppression is evaluated by an A curve analysis, as described in this chapter. The operation and framework for special hazard protection is described in Chapter 20.

Reference

1. NFPA 13, Standard for the Installation of Sprinkler Systems.

14 PUTTING IT TOGETHER: THE L CURVE

14.1 Introduction

The limit of flame movement describes the extent and severity of fire damage in a building. Its graph, called the L curve, identifies the probability of terminating a fire of any given size. One can understand and describe multiple-path fire spread by adapting the procedures of Chapter 9 to include manual and automatic suppression.

Calculating coordinates and constructing an L curve for a room of origin and a single propagation path is simple and direct. Calculating the limit for a barrier/space module is straightforward. However, one needs the bookkeeping capability of a computer program to manage the minute-by-minute local and global time integration for multiple-room fire propagation. The US Coast Guard R&D Center successfully demonstrated such a program in evaluating ships to provide an insight into understanding local and global fire performance [1]. The program was also used for several buildings to evaluate its strengths and weaknesses for practical applications. Unfortunately, the programs were not maintained and are no longer available.

This chapter describes L curve calculations for a room of origin. It also describes calculations to link multiple-room propagation involving barriers and spaces along a single path. Fortunately, one can satisfy most practical needs with the analysis of one or a few carefully selected barrier/space modules that use hand calculations or spreadsheet analysis.

14.2 Fire in a space

In any space, a fire can go out in only three ways: (1) by itself through self-termination, (2) by an automatic sprinkler system, or (3) by the local fire department.

THE FIRE

A performance evaluation starts with the fire. Every room can have a large number of potential fires depending on contents, arrangement, and ignition location. A selection of design fire characteristics becomes an early need for all evaluations.

Design fire characteristics are the loads that the fire defenses must resist. Philosophical and practical questions arise when selecting design fire characteristics. For example, should one be

Building Fire Performance Analysis R. W. Fitzgerald
 ISBN: 0-470-86326-9 (HB)

conservative and use the worst credible fire conditions? What should they be? On what are they based? How well does this design fire represent future conditions when an occupant changes materials and arrangements? How will major building renovations affect assumptions of the current design fire? These questions are major professional judgmental decisions.

This book describes fire performance analysis, not design. Therefore philosophical justification need not be as rigorous, and practical selections are easier. We must give some attention to variability in uses and materials and provide documentation of the rationale for selections. Describing the design fire carefully places performance expectations into context.

The definition of the model room is the most significant decision in a performance evaluation. A model room describes the conditions on which the design fire is based. The room interior design includes geometry and dimensions, contents, their materials and arrangement, and the interior finish. Although ventilation can be important to fire behavior, we assume that sufficient openings exist for efficient burning. This simplification allows one to incorporate ventilation conditions later, if necessary, although changes are rarely needed for practical applications. Level 2 and level 3 evaluations would document the model room with sketches and descriptive specifications. For most buildings, one or a few carefully selected model rooms are sufficient for most performance needs.

THE I CURVE

One may use the model room to analyze specific fire scenarios or to describe a general hazard classification with associated fire characteristics. Scenario analysis is useful for fire investigations or to get a sense of proportion for the way fires will behave in the model room. A self-termination (ST) curve (Chapter 8) describes a scenario outcome with a continuous value network (CVN) based on fire growth realms.

Fire growth hazard potential classifications are more useful for performance evaluations. They represent a relative hazard classification for the room's interior design. The classification may be expressed in descriptive terms (e.g., bad, moderate, good) or by a number. An I curve is a numerical representation of the room's fire growth potential. An I curve depicts a blended hazard classification rather than a specific fire scenario It provides a way to order the general hazard of a room interior design.

Any room has many potential ST curves, but only one I curve to represent the hazard classification. Fire characteristics that relate to model room conditions can be associated with I curve classification. We test fire defense performance against these design fire characteristics.

THE A CURVE

An A curve represents the quality and effectiveness of the automatic sprinkler system in suppressing a fire. The objective is to understand the sprinkler well enough to express its strengths and shortcomings clearly. Chapter 13 describes procedures to examine details that influence sprinkler reliability (AA) and operational effectiveness (AC). Performance knowledge is based on a scenario analysis that poses a significant threat to the sprinkler system. Although an A curve describes the sprinkler performance for a specific fire growth threat, the knowledge acquired in an evaluation, enables one to discriminate the performance for other threats.

THE M CURVE

An M curve (Chapter 12) represents the building's ability to work with the local fire department to manually suppress a fire. One acquires performance knowledge by a scenario analysis that compares an expected fire propagation with local fire department suppression activities. While the focus may be on fire propagation and manual suppression activities, the objective is to understand the building and site features that help or hinder fire ground operations.

14.3 Performance descriptors

A level 1 analysis describes a building's behavior with descriptive terms, performance graphs, and associated documentation. Observations, mental estimates, and judgment blend with information and knowledge to estimate I, A, and M curves that discriminate and describe quality and performance. A level 2 evaluation examines a component more carefully to acquire a deeper understanding of its behavior. Level 3 evaluations develop a performance sensitivity for important parameter or building variations.

A level 1 evaluation of component performance develops a sense of proportion for quality and component behavior. A few "standardized" classification terms and visual descriptors often help to discriminate and express performance categories more easily.

Figures 14.1 to 14.3 show representative shapes, standardized curves, and useful groupings for I, A, and M performance in a space. Classifications using the categories shown in Figures 14.1 to 14.3 provide an adequate sensitivity to discriminate among conditions. One can change the curve numbers to reflect personal value judgments of existing conditions.

14.4 The L curve for the room of origin

Figure 14.4 shows the network that provides the logical combination of I, A, and M events within a space. This network relates to the fault and success trees of risk analysis discussed in Appendix A.

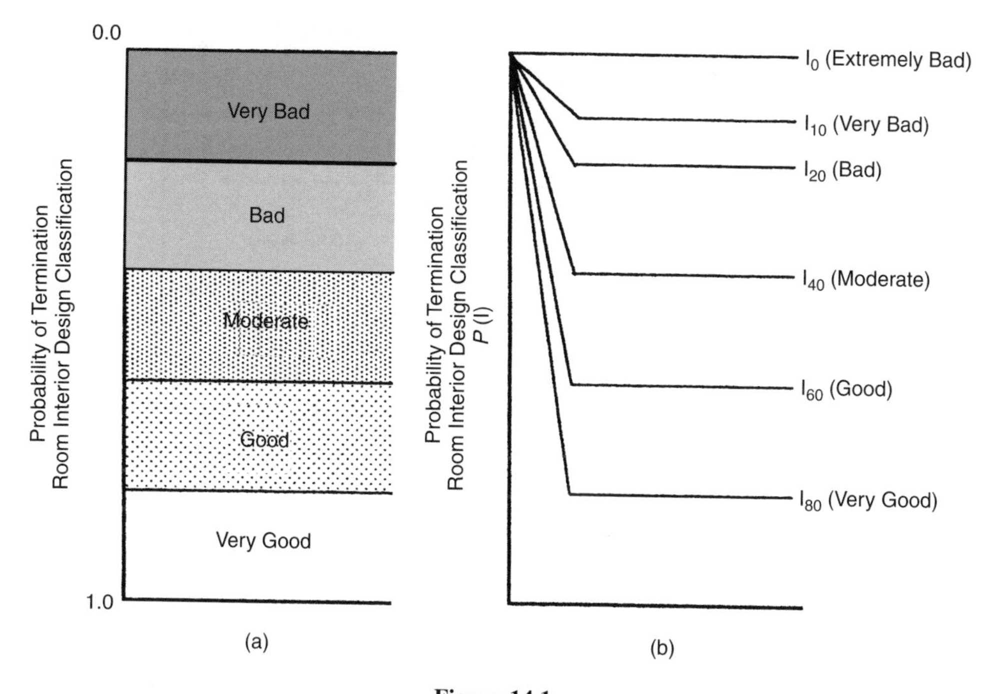

Figure 14.1

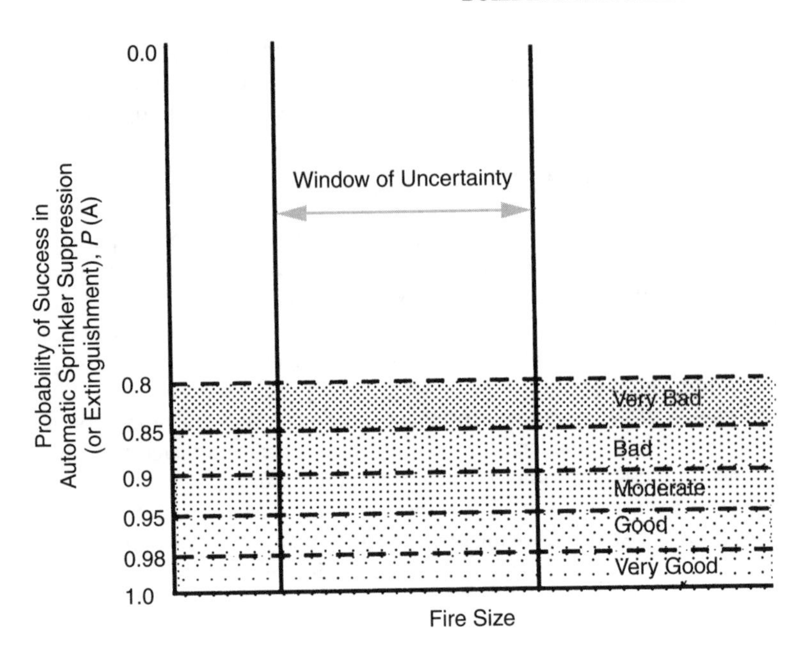

Figure 14.2

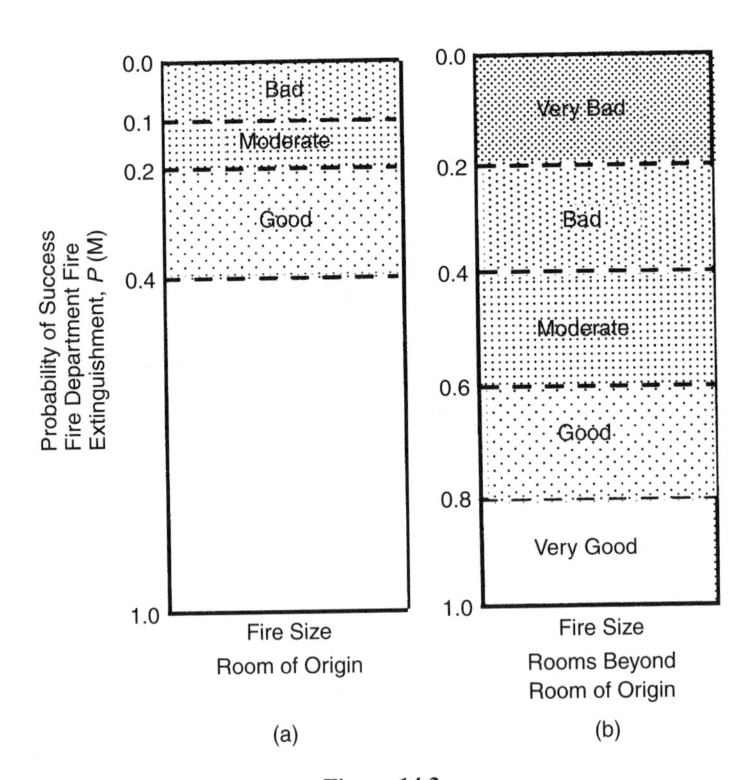

Figure 14.3

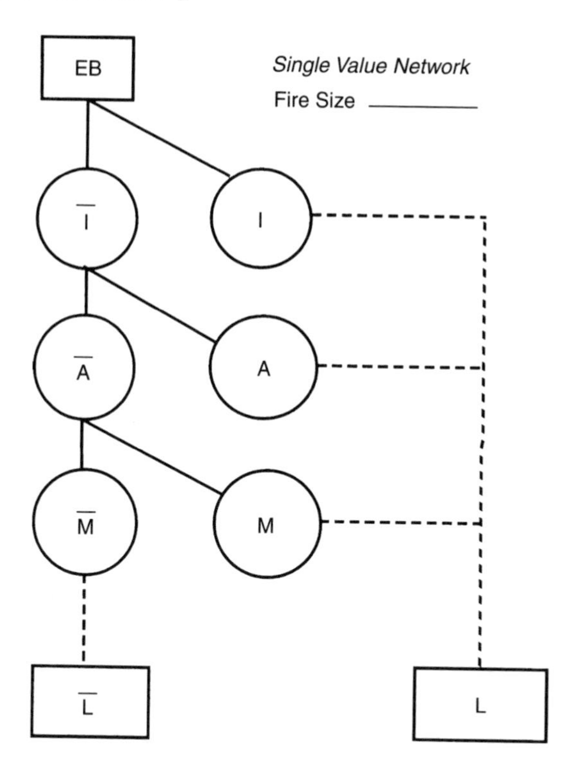

Figure 14.4

After one selects I, A, and M curves for the room of origin, they may be combined to produce an L curve. Figure 14.5 shows I, A, and M curves for a room of origin. One may calculate an L curve coordinate for any fire size with the SVN of Figure 14.4. Example 14.1 illustrates the process.

Example 14.1

The I, A, and M curves of Figure 14.5 have been selected for a room of origin. Calculate L curve coordinates at 1.5 MW and at FRI. Construct the L curve.

Solution

We note the I, A, and M coordinates at the two fire sizes from the curves as follows. Incorporating these values into the SVNs of Figure 14.6(a) and (b), the L values are calculated as 0.965 and 0.9762, respectively.

	1.5 MW	**Full room**
I value	0.3	0.3
A value	0.95	0.96
M value	0	0.15

The L curve is merely the continuous function defined by the calculated coordinates. One could calculate enough coordinates to develop the continuous curve. However, one can observe from

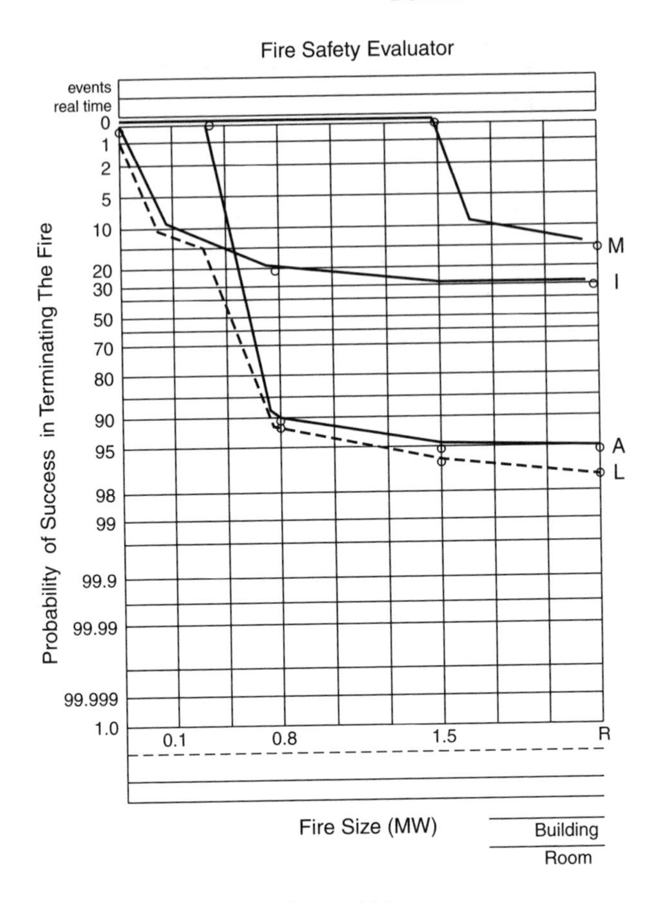

Figure 14.5

the network mathematics demonstrate that any L coordinate will always be equal to or better than the best of the three component curves. Consequently, one can calculate a single point and sketch the rest of the curve. This technique saves time and provides both the visual and numerical perspective of fire performance. The dashed line of Figure 14.5 shows the results.

Evaluator 3 is used when combining different components because the added scale allows extremes to be visually discriminated more easily. Evaluator 2 is truncated from evaluator 3 to avoid visual distortion in individual component behavior and maintain a sense of a proportion for envisioning performance.

14.5 Building L curve networks

The L curve for the room of origin is easy to construct after one evaluates the I, A, and M curves. The major issue in evaluating I, A, and M curves for spaces beyond the room of origin involves modes of barrier failure and time relationships for the fire defenses. Fortunately, one may extrapolate I, A, and M curves from room of origin information, and field performance $\overline{T}$, $\overline{D}$, and B barrier curves can be developed. The task is to put the pieces together in a logical and practical way.

Figure 14.7 shows a three-room building (or a three-room path of a larger building). Figure 14.8 describes a space/barrier network in which L_n represents the limit for a space, and barrier failure

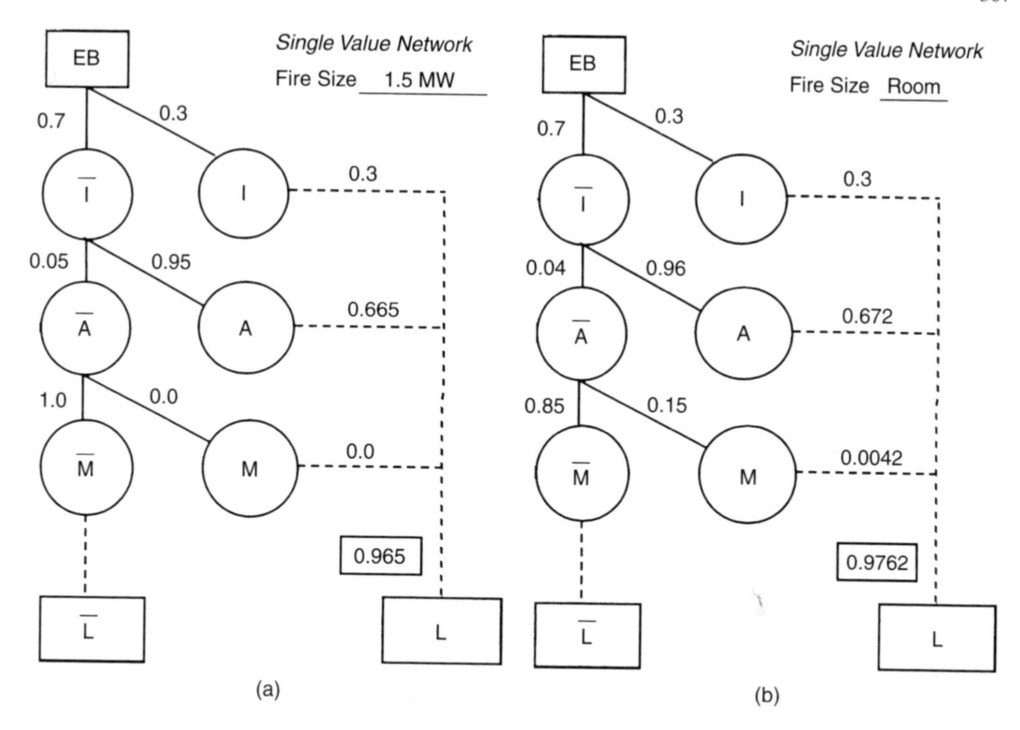

Figure 14.6

is described only for success B or failure $\overline{B}$. Section 6.10 introduced this network to describe propagation concepts and paths on which to focus attention. However, a more complete picture of what can happen during fire propagation requires some augmentation.

The two "building blocks" for a more in-depth analysis involve the space network of Figure 14.9 and the barrier network of Figure 14.10. Beyond the room of origin, the behavior in a space depends on the barrier failure mode. The fire defense performance is very different for a $\overline{D}$ failure than for a $\overline{T}$ failure. Figure 14.11 shows all possible outcomes for this three-room path.

It may appear that a complete analysis beyond the room of origin is a monumental task. Fortunately, we can simplify the apparent complexity. Consider first what happens when a $\overline{D}$ barrier failure occurs. A massive influx of fire gases enters the space very rapidly. The room reaches FRI quickly, making $P(\mathrm{I}) = 0.0$, no matter what it may have been before the fire. If a sprinkler system is present, the speed of fire gas movement is likely to overwhelm and move past

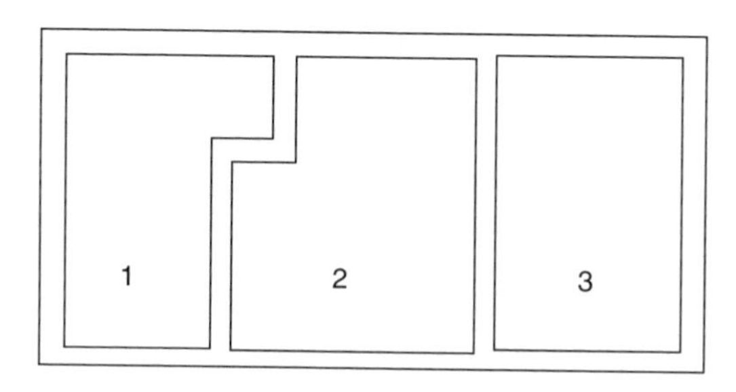

Figure 14.7

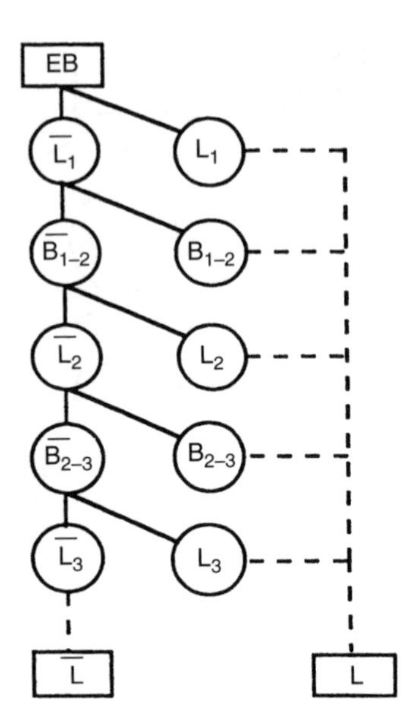

Figure 14.8

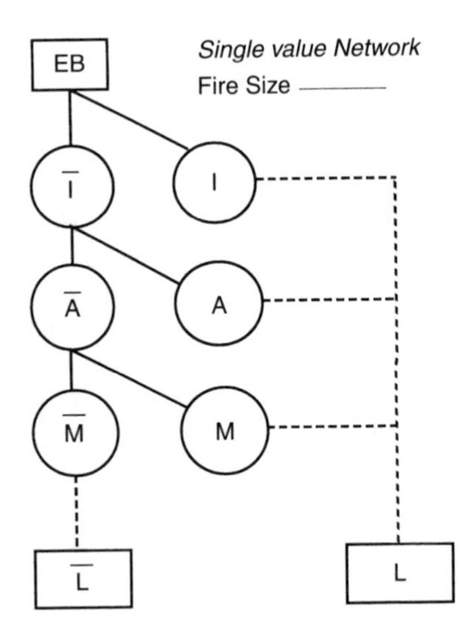

Figure 14.9

the sprinklers so they will not fuse early enough. Similarly, if the fire department does not have charged hose lines in place when the $\overline{D}$ failure occurs, it will have almost no chance to stop the fire before FRI. We can assign $P(\mathrm{A})$ and $P(\mathrm{M})$ success values of zero for these conditions.

Consider next what happens when a $\overline{T}$ failure occurs. A hot-spot ignition would cause an almost normal fire development in the room. The environment would be smokier and hotter, so fires grow more easily than if it were a room of origin. An I curve for the room as a sequential room with a $\overline{T}$ barrier failure mode would be slightly reduced. Thus, one can estimate $P(\mathrm{I})$ for the room and environmental conditions as the fire progresses.

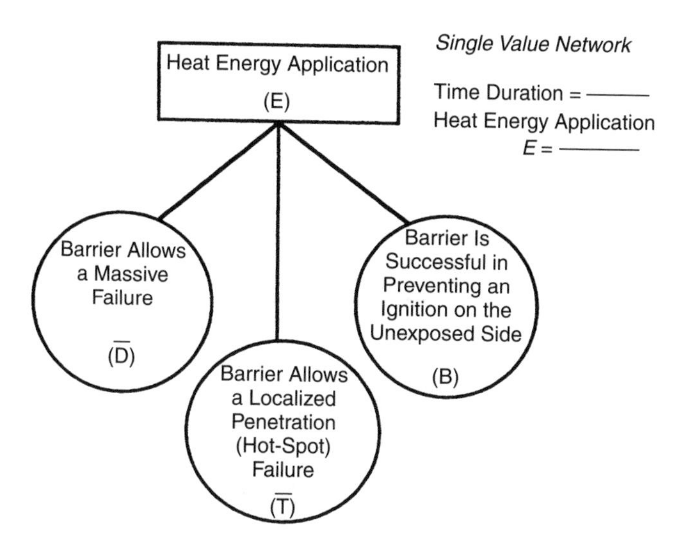

Figure 14.10

The status of the A curve can fall into three situations. If no sprinkler system is in the building, $P(\mathrm{A}) = 0.0$ for the room of origin and all other rooms. However, if the building is fully sprinklered and the fire was not put out in the room of origin, something is significantly wrong. We assign $P(\mathrm{A}) = 0.0$ to the sequential room because the system must be impaired. A third case arises when the building is partially sprinklered and the fire is moving from an unsprinklered room to a sprinklered area. Here $P(\mathrm{A})$ could have some value. Its assessment would be affected by a variety of factors such as sprinkler siting relative to the barrier failure location, fire growth speed, and system characteristics. Although some value for $P(\mathrm{A})$ may be appropriate, it would probably be derated substantially. Any value for $P(\mathrm{A}) > 0$ should be carefully scrutinized.

The final component for a $\overline{\mathrm{T}}$ failure is the M curve. Here fire department suppression can be very effective when barriers delay fire propagation to allow engine companies to set up and fight the fire. Depending on barrier effectiveness and time durations, values for $P(\mathrm{M})$ can vary from 0.0 to 1.0.

When these values are incorporated into a detailed analysis, the network of Figure 14.12 emerges. Here one can recognize that the sprinkler system is very effective in the room of origin and ineffective beyond the room of origin. Time delays make M curve values in the room of origin very small or nonexistent. However, the local fire department can be very effective beyond the room of origin when barriers provide enough time to mount a fire attack. The building performance of nonsprinklered buildings is dominated by the barrier effectiveness and local fire department resources.

14.6 L curve calculations

The simple space/barrier description of Figure 14.8 is related to Figure 14.11 which shows all possible outcomes. Depending on needs for technical analysis or simplified communication of results, one may move between them. Example 14.2 illustrates the relationships.

Example 14.2

Example 6.6 described calculations and graphing for a three-room building using the simplified space/barrier description of Figure 14.8. Figure 14.13 reproduces the network of Example 6.6.

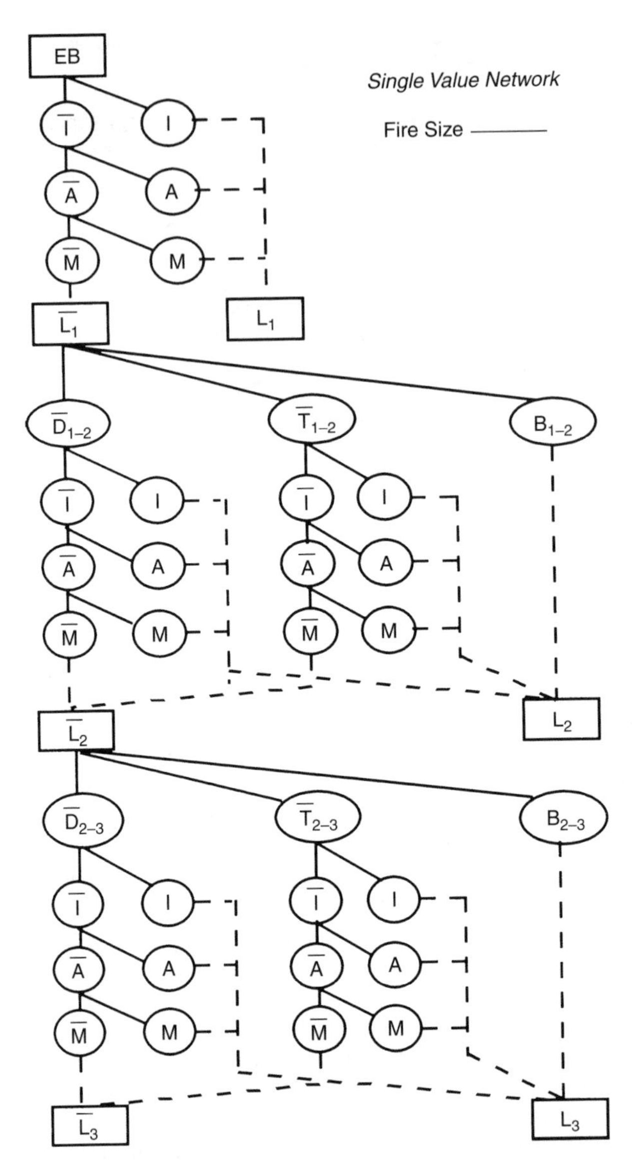

Figure 14.11

Figure 14.14 shows the complete network with hypothetical values assigned for the events of rooms 1 and 2. Calculate L curve values using Figure 14.14 for the second room and plot the results on evaluator 3.

Solution

Figure 14.15 shows calculations and some explanation for the network. The rules of network calculation are (1) multiply probabilities along a path (or partial path) to calculate an outcome; and (b) add like outcomes. Calculations for the room of origin in Figure 14.15 show $L_1 = 0.3$ and $\overline{L}_1 = 0.7$. These values correspond to point f in Figures 14.13 and 14.16.

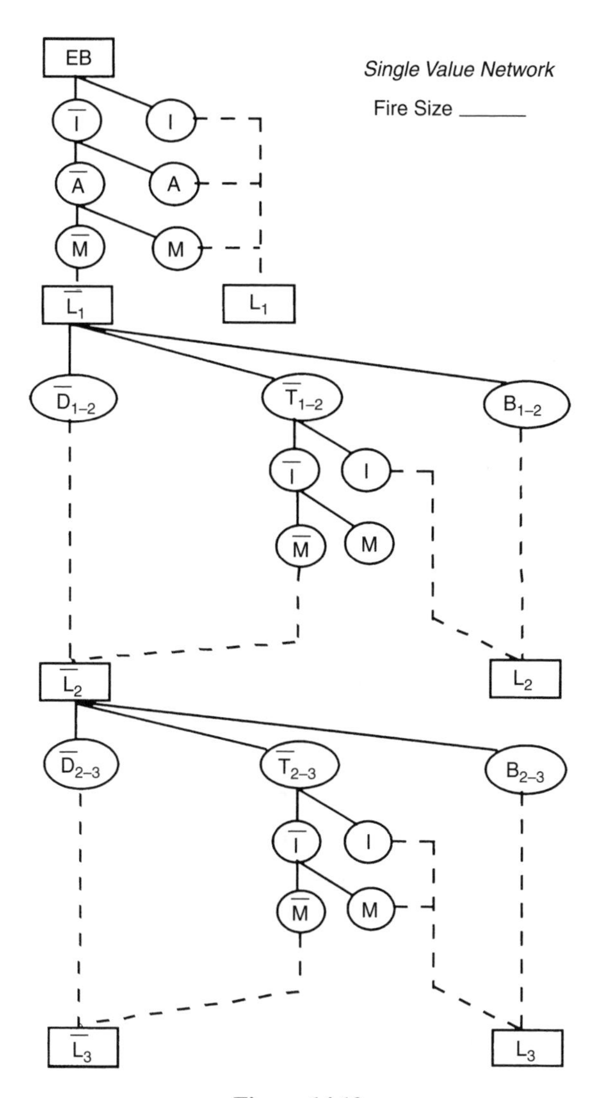

Figure 14.12

The barrier success of $P(\mathrm{B}) = 0.2$ is multiplied by $P(\overline{\mathrm{L}}_1) = 0.7$ to give the value $P(\mathrm{B}|\overline{\mathrm{L}}_1) = 0.14$. This is one part of the barrier/space module and is shown as value b in Figure 14.13. We can treat I, A, and M events as a package. These become subgroups for $\overline{\mathrm{T}}$ and $\overline{\mathrm{D}}$ failures in a room. To illustrate the procedure, we temporarily assign $P(\overline{\mathrm{T}}_{1-2}) = 1.0$ and call the room 2 success subgroup $\mathrm{L}_{2\mathrm{T}}$ and the failure group $\overline{\mathrm{L}}_{2\mathrm{T}}$, as shown in Figure 14.15. Their values are

$$\begin{aligned}
\mathrm{L}_{2\mathrm{T}} &= P(\mathrm{I}) + P(\mathrm{A}|\bar{\mathrm{I}}) + P(\mathrm{M}|\bar{\mathrm{I}}, \overline{\mathrm{A}}) \\
&= 0.2 + 0.0 + 0.44 = 0.64, \\
\overline{\mathrm{L}}_{2\mathrm{T}} &= P(\bar{\mathrm{I}})P(\overline{\mathrm{A}})P(\overline{\mathrm{M}}) \\
&= (0.8)(1.0)(0.45) = 0.36.
\end{aligned}$$

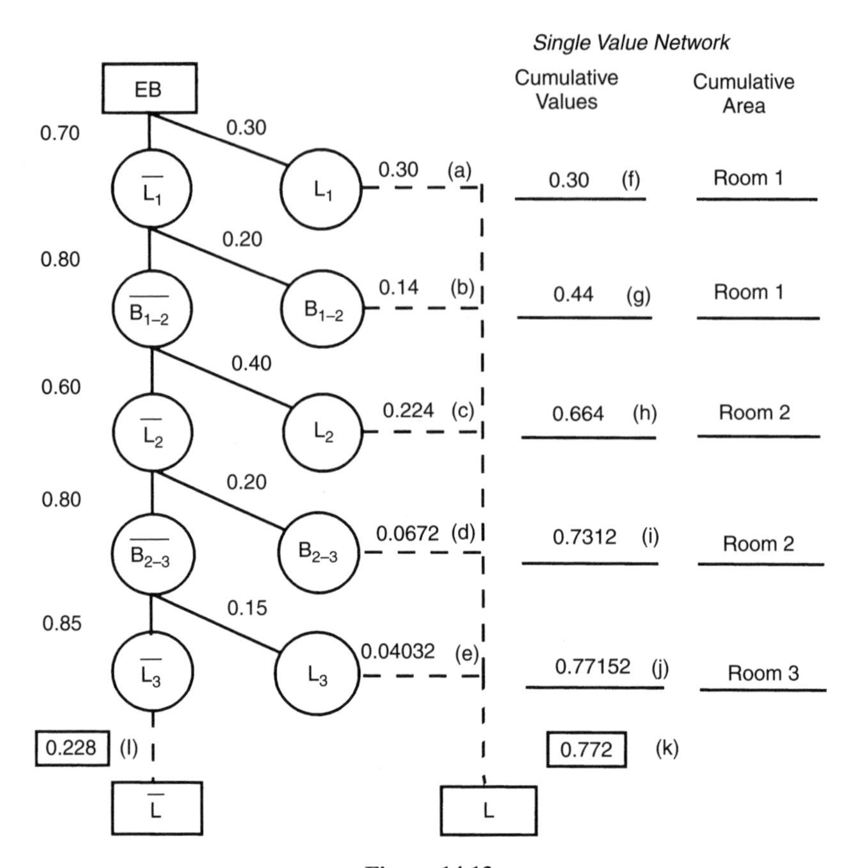

Figure 14.13

Similarly, the subgroups for a $\overline{\text{D}}$ failure would produce $\text{L}_{2\text{D}} = 0.0+0.0+0.0 = 0.0$ and $\overline{\text{L}}_{2\text{D}} = (1.0)(1.0)(1.0) = 1.0$.

The limit for Room 2 is obtained from Figure 14.15 as

$$\begin{aligned} P(\text{L}_2) &= P(\text{L}_1)+P(\text{B}_{1-2}|\overline{\text{L}}_1)+P(\overline{\text{L}}_1)P(\overline{\text{T}}_{1-2})P(\text{L}_{2\text{T}})+P(\overline{\text{L}}_1)P(\overline{\text{D}}_{1-2})P(\text{L}_{2\text{D}}) \\ &= 0.3+0.14+0.224+0.0 = 0.664. \end{aligned}$$

This value corresponds to point h in Figures 14.13 and 14.16.

The value for $P(\overline{\text{L}}_2)$ is

$$\begin{aligned} P(\overline{\text{L}}_2) &= P(\overline{\text{L}}_1)P(\overline{\text{D}}_{1-2})P(\overline{\text{L}}_{2\text{D}})+P(\overline{\text{L}}_1)P(\overline{\text{T}}_{1-2})P(\overline{\text{L}}_{2\text{T}}) \\ &= (0.7)(0.3)(1.0)+(0.7)(0.5)(0.36) \\ &= 0.336. \end{aligned}$$

Figure 14.15 shows all the events and complete calculations to illustrate the process. L curve calculations for rooms beyond the room of origin are systematic when one packages the limit for a space (Figure 14.4) for $\overline{\text{T}}$ and $\overline{\text{D}}$ failure conditions.

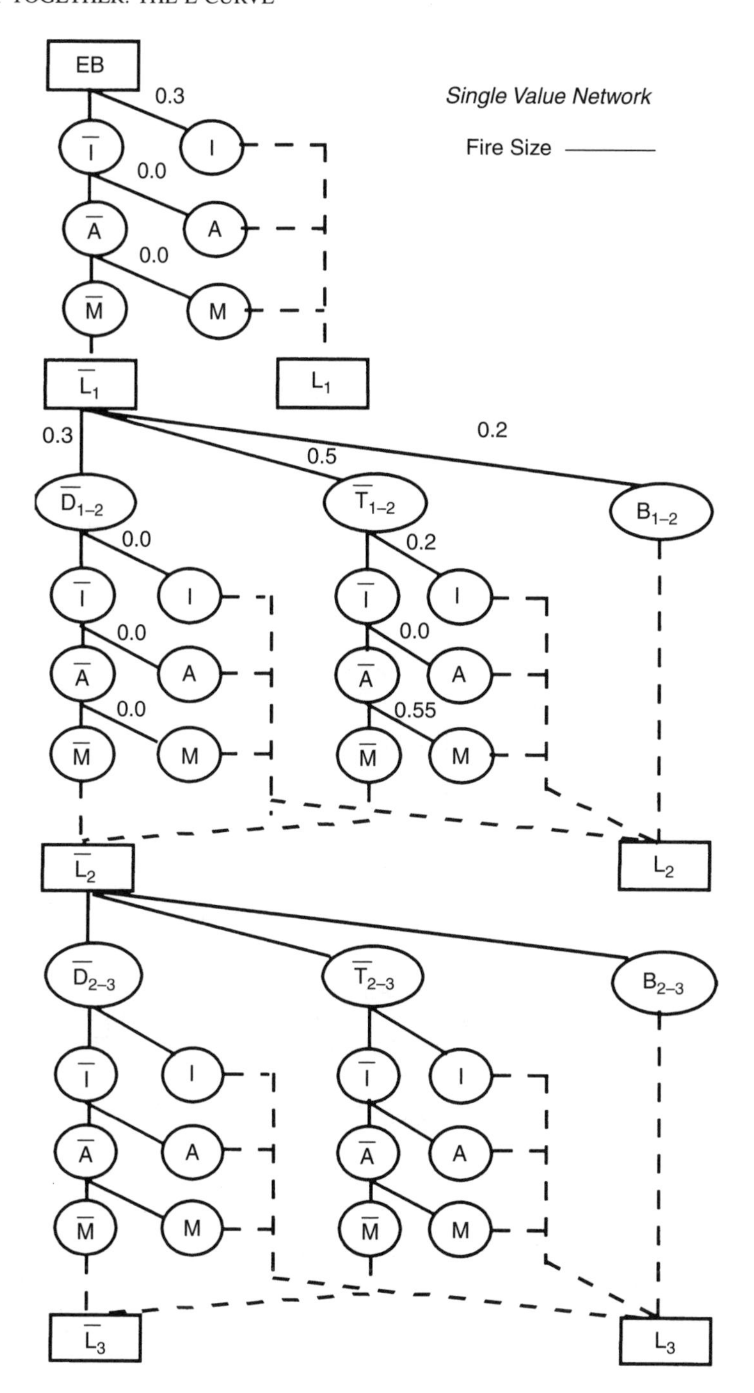
Single Value Network
Fire Size
EB
0.3
I
0.0
A
0.0
M
L1
0.3
0.5
0.2
D1–2
T1–2
B1–2
0.0
0.2
0.0
0.0
0.0
0.55
L2
D2–3
T2–3
B2–3
L3

Figure 14.14

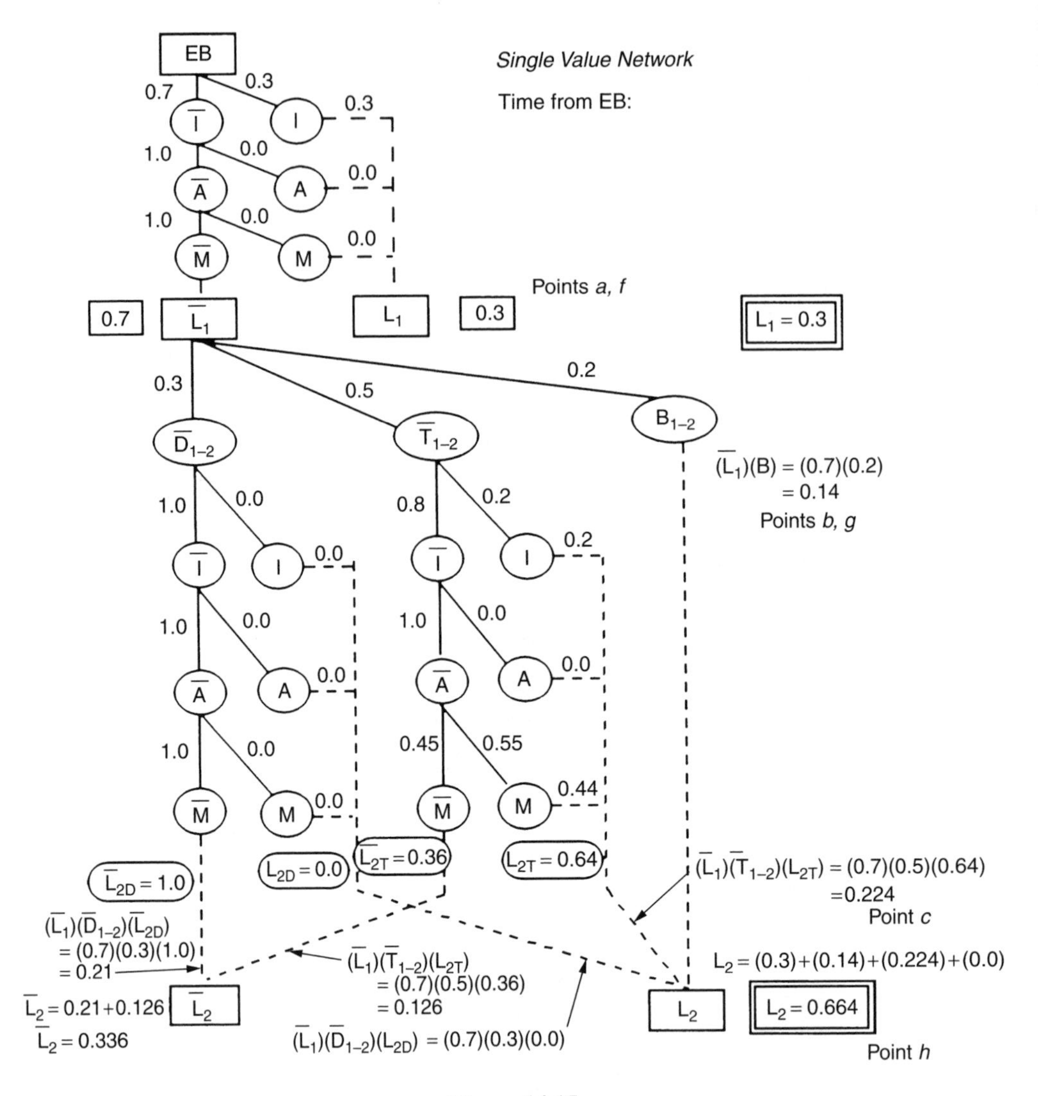

Figure 14.15

14.7 L curve discussion

One may move from one space/barrier network to another. Figure 14.11 provides the analytical logic for the interrelationships and Figure 14.8 simplifies communication. Spreadsheets simplify calculations and plotting tasks. However, the responsibility for input data, value judgments, and decision making remains with the user.

Note that calculations such as in Example 14.2 provide only a single value "photograph" associated with an instant of time. The L curve gradually shifts upward as time elapses until room burnouts occur. At room burnout, the L curve value reaches its most pessimistic level and does not deteriorate any further. This value becomes a worst bound for the flame/heat analysis.

Professional practice focuses on thinking, understanding, and deciding rather than calculations that can be provided by a computer program. Here the program serves as an accountant that

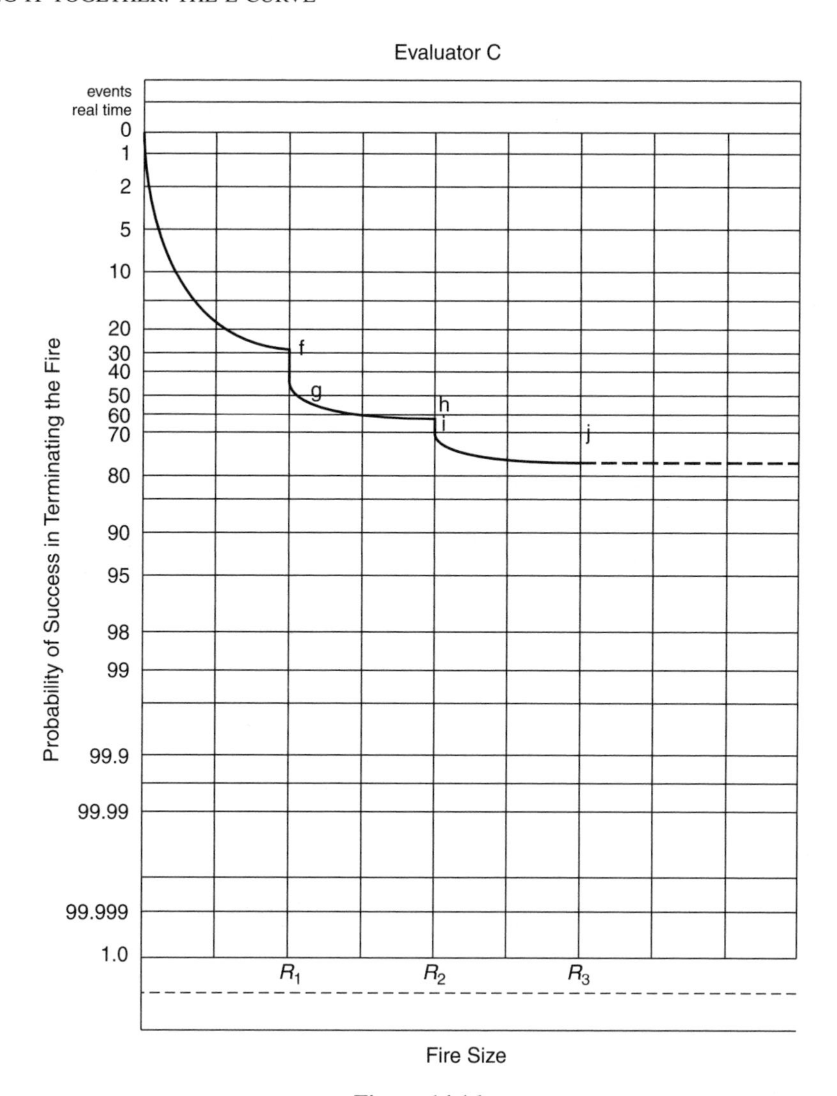

Figure 14.16

calculates values and integrates global time accumulations with local time outcomes for subcomponents. It can produce selected outcomes on a minute-by-minute basis to identify times or conditions when important events occur.

Understanding time relationships for critical events enables an individual to quickly develop a clear picture of performance and to examine a variety of "what if" situations easily. Computer-generated information for risk management applications can identify graphically on floor plans the impact of different alternatives on the building performance. Vulnerable locations for life safety, property protection, and operational continuity may be easily recognized by management. The impact of changes and costs may be incorporated into plans for visual recognition by management decision makers.

Rather than to note what "can be" in computational sophistication, we wish to point out that one may handle important assessments without the use of a computer. When level 1 sketches provide an overview, a few carefully selected evaluations will provide sufficient understanding to describe with some confidence a building's performance. The analytical structure is modular and integrated to enable one to move back and forth between a macro (level 1) view and micro (level 2) studies of performance.

Barrier effectiveness and the barrier/space module are the fundamental performance units. They enable an individual to focus on a single subcomponent at a time. Then one can put components together in a way that is technically defensible and provides a transparent understanding of performance. These units are structured analogous to free body diagrams that are the fundamental building blocks of analytical mechanics.

14.8 Summary

This chapter describes calculations to integrate the components of fire suppression. The evaluation of any individual fire propagation path is neither difficult nor time-consuming. Complications can arise because fire propagation is in three dimensions and is time dependent. Also, there will be six or more barriers that surround each space. Each of these barriers can fail at different times because their performance effectiveness will be different. This creates complexity in ordering the relationships between the different potential fire paths. Nevertheless, one can conduct a valuable and reasonable building evaluation by judiciously selecting the room of origin and the barriers, and extrapolating other rooms using reasonable technical judgment.

One may get an impression that the evaluation of rooms beyond the room of origin and the calculation of an L curve becomes a lengthy and complicated process. That is not the reality. Many techniques can streamline the process so evaluations can be completed in a reasonable amount of time. The illustrative calculations in these examples describe a process and provide an insight into the relationship between physical performance and mathematical representations.

Reference

1. Fitzgerald, R., Richards, R., and Beyler, C. Firesafety analysis of the polar icebreaker design. *Journal of Fire Protection Engineering*, Vol. **3**, No. 4, October 1991.

15 CONCEPTS IN STRUCTURAL FRAME ANALYSIS

15.1 Introduction

Tools to analyze structural behavior in fires are better defined and understood than any other component in the fire safety system. Yet structural analysis and design principles for fire conditions are rarely incorporated into building analysis and design. Instead, fire endurance test ratings dominate structural fire protection. While fire endurance times are valuable for comparisons, they do not describe performance adequately. The advantages and limitations of the standard fire endurance test for barriers discussed in Chapter 10 are generally appropriate for structural frames.

A performance evaluation attempts to understand and describe the way in which the specific structural system will behave during a building fire. This requires a basic knowledge of structural analysis and design procedures, building framing, and an awareness of the conditions that can cause failure. The type, installation, and reliability of structural protection is a major part of the evaluation.

This chapter describes structural framing systems and conditions that can lead to collapse or excessive structural deformation. It discusses concepts of structural behavior in fires (Fr curve) and identifies potential sources of weakness. The focus is on level 1 awareness and a way to describe the dynamic changes in condition as the fire continues to burn. The technical analysis for level 2 and level 3 evaluations forms the background to the way of thinking, but is not presented here.

15.2 Structural design

Routine structural design starts with schematic building plans. Potential vertical support locations are identified and a series of alternate designs are considered. The alternatives are evaluated for functionality, economy, constructibility, compatibility, aesthetics, and safety. The structural system is selected to provide the best solution to the spectrum of needs while maintaining structural safety and serviceability requirements.

Structural design arranges and sizes the framing system to support gravity, wind, and earthquake loads. The weight of the structural frame and the fixed building elements comprise the dead loads. Live loads include gravity loads associated with the use, occupancy, and location, as well as high winds, earthquake, and any unusual conditions that may reasonably be expected during

Building Fire Performance Analysis R. W. Fitzgerald
 ISBN: 0-470-86326-9 (HB)

the building's use and life. Structural drawings and specifications provide information for cost estimates and construction.

Minimum requirements for live loads, wind loads, and earthquake forces are specified in the building code. Professional standards identify allowable control values and serviceability requirements. The professional registration stamp on the drawings indicates the engineer of record, who is responsible for using state-of-the-art technology. The framing is designed to provide safe, functional performance.

The role of the code and structural engineering is different for fire. Although fire endurance requirements may have been acknowledged in the early stages, they are not a conscious part of structural design. Efforts are directed toward ensuring structural safety during routine operation and certain abnormal conditions. After the structural design has been completed, standard fire test information is incorporated into the specifications. For fire, the responsibility for safety shifts from the engineer to the building code and its enforcement. The code prescribes fire test endurance requirements. Structural analysis for fire is not a part of the usual design process nor is it within the scope present of expectations.

15.3 What are we looking for?

The prescriptive building code process that relates fire endurance requirements with occupancy, height, and area has been a part of codes for about 75 years. This process works well to provide a framework for a variety of prescriptive requirements and alternatives associated with fire. There is no question that the good practices incorporated into modern prescriptive codes have enormously improved the level of fire safety in buildings. Unfortunately, the level of risk associated with code compliance remains unknown.

A performance analysis estimates the likelihood of structural collapse during a fire. It gives a sense of proportion for the building's risk levels. One of the functions is to recognize situations in which collapse is likely to occur. Sometimes observations may suggest a potential failure. Other situations may require an in-depth analysis to describe expected behavior. Level 1, 2, and 3 evaluations enable one to tailor the value of information to its cost.

15.4 Structural frame and barriers

The function of a structural frame is to support itself and the applied loads without collapse or excessive deformation. A sudden loss of stability in part or all of the building results in collapse. Collapse can also occur when initially gradual deformations begin to accelerate and the affected section of the building falls down. Excessive deformation without collapse can cause barriers to distort and crack, creating $\overline{T}$ failures that allow ignitions in an adjacent space.

Chapter 10 notes that sometimes the structural frame is an integral part of a barrier and other times it provides support. When a structural frame is a part of the barrier, the performance becomes a part of the $\overline{T}$ and $\overline{D}$ analysis. When the structural frame supports a barrier, the frame behavior becomes a limiting value that affects the $\overline{T}$ and $\overline{D}$ performance. In either case the collapse or excessive deflection potential is important information for fire fighting and risk characterizations. An individual must clearly understand the function and construction details to describe performance.

15.5 The structural frame

A structural frame is an assembly of framing elements that provide the skeletal support for the building. The elements of a structural frame may be organized in the following manner:

1. Horizontal elements
   a. Slabs or deck
   b. Joists or purlins
   c. Beams
   d. Girders
2. Vertical elements
   a. Columns
   b. Bearing walls

These terms are routine in the building industry, and their use in describing the structural system can convey information about its potential behavior in a fire. Figure 15.1(a) shows a floor system with a concrete slab, beams, girders, and columns. Figure 15.1(b) illustrates a corrugated steel deck with concrete fill, open-web bar joists, and a bearing wall. The slab or deck is a horizontal wearing surface that carries its own weight and the contributing weight of the building (dead load) as well as the live loads of contents and people. It is structurally designed and its thickness is related to the spacing of the supporting beams.

When the slab or deck supports are relatively widely separated (e.g., approximately 4 ft or more), they are called beams. When the spacing is closer and evenly spaced, they are called joists. Roof support members are usually called purlins or rafters. Beam, joist, and purlin are common terms used for structural steel, reinforced concrete, prestressed concrete, and wood construction.

Girders support beams or joists. Columns support girders. In some low-rise buildings, bearing walls support the beams or joists and girders are omitted. In the hierarchy of structural framing, the slab or deck is supported by beams or joists; the beams or joists are supported by girders; girders are supported by columns; when girders are omitted, the members are supported by a bearing wall.

The structural materials are most commonly wood, structural steel, reinforced concrete, prestressed concrete, and masonry. Materials and framing systems can be mixed and matched in a variety of ways. The selection of materials and the manner in which they are put together has a major influence on the structural frame performance. Although there are other structural systems available, the vast majority of buildings use these elements.

Another important structural element that can replace a beam or girder is the truss. A truss is an articulated structure composed of slender, lightweight members connected in the form of triangles. The members act either in tension or compression and are positioned to carry loads very efficiently. Truss members are most commonly structural steel or wood. When steel trusses are very light and evenly spaced, as in Figure 15.1(a), they are called open-web joists or bar joists.

Figure 15.2(a) and (b) show two examples of truss arrangements. There are a wide variety of ways to arrange members, but their names and forms are unimportant for fire safety. However, it is important to be able to recognize a truss because its stability during a fire can be very different from that of more substantial floor or roof construction systems. Truss characteristics are long spans with lightweight, slender members arranged in a series of triangles.

15.6 Performance concepts

Section 10.14 describes material behavior at elevated temperatures. Some of the factors that affect structural behavior in a fire are as follows:

- reduction of load carrying capacity due to structural member elevated temperatures;
- insulation protection that delays heat penetration into structural elements;

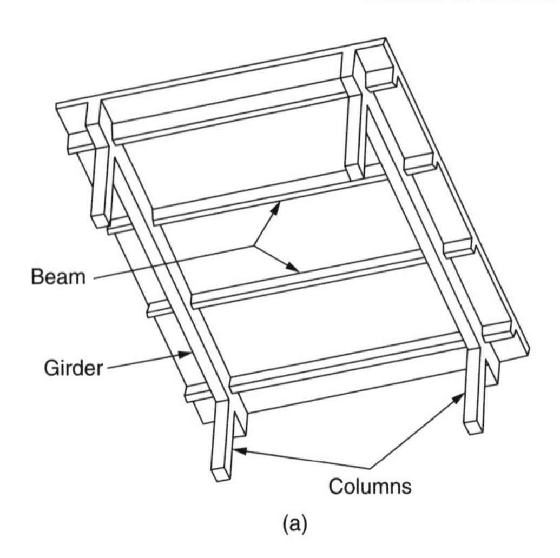

(a)

- Floor deck spans joist spaces.
- Floor deck may consist of:
  - Metal decking w/ concrete fill
  - Precast concrete planks
  - Plywood panels or wood planking, requiring a nailable top chord or nailer bolted to top chord

Bearing wall

Girder

Column

(b)

Figure 15.1 (Building Construction Illustrated 3e; Ching & Adams; © 2001 John Wiley & Sons, Inc. This material is used by permission of John Wiley & Sons, Inc.)

- magnitude of applied loads;
- effect of thermal expansion;
- influence of connections and frame construction;
- potential for progressive collapse.

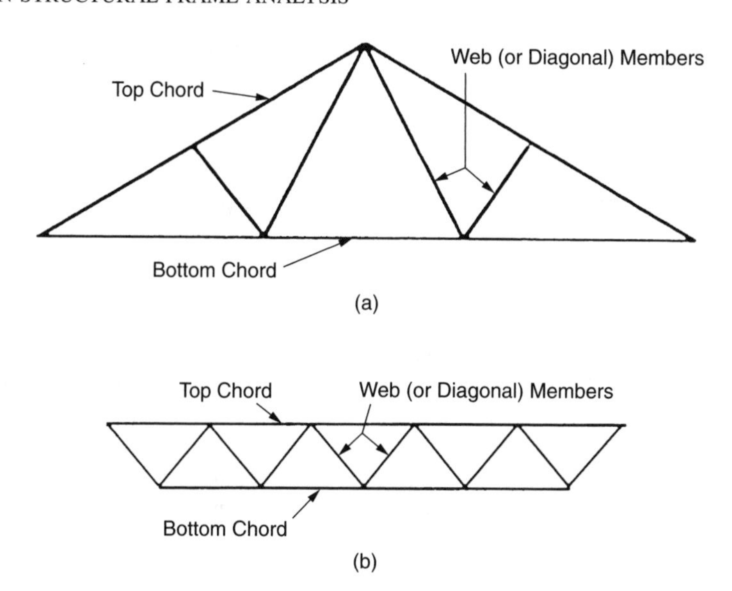

Figure 15.2

All structural materials deteriorate and lose strength when subjected to the heat of a fire. The way in which the strength is reduced depends on materials, structural member temperatures, and how the frame is constructed. The general structural behavior is illustrated by Figure 15.3, where increases in heat produce reductions in load-carrying capacity.

The type and amount of insulation protection are important to predicting the temperature increase in structural members. Observing the completeness insulating materials is also important. Missing sections in spray-on coatings or weaknesses in concrete or gypsum coverings enable heat energy to compromise structural strength.

Suspended ceilings, often called membrane ceilings, provide a barrier to the passage of heat energy into the plenum space containing structural framework above. When the suspended ceiling remains infact, heat energy movement from the fire into the void space is delayed. The protection

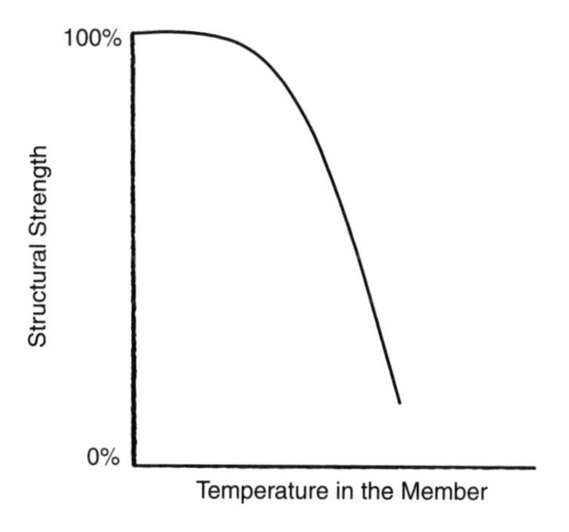

Figure 15.3

is compromised if the suspension system (as a separate structural system) fails or if tiles are missing or fall out. Ceiling suspension systems should have expansion joints to allow the suspension supports to expand without excessive deflection.

Concrete cover in reinforced concrete and prestressed concrete structural members acts as an insulation to protect the steel reinforcing bars. Concrete spalling exposes reinforcing bars to temperature increases. This may or may not be significant, depending on their location and other construction details.

Wood members burn in a fire. Charring of the exposed surfaces reduces their size and load-carrying capacity. The speed of char layer development and the resulting load-carrying reduction depends on the wood species and the number of surfaces exposed to the fire.

The magnitude and type of applied loads influence performance. The structural member must support the dead loads all the time. Live loads from stationary contents such as furniture, machinery, and storage materials will not move during the fire. On the other hand, live loads caused by people would be expected to move away from a fire on the floor below their location. The magnitude of live loads can vary greatly, and this can affect the structural frame performance.

Heat causes structural materials to expand. This action can have different effects on structural behavior. When a horizontal member expands and movement is not constrained, only the strength reduction due to the temperature increase will occur. However, when elongation is prevented, additional axial stresses are induced. Initially this axial thermal force increases compressive stresses, decreases tensile stresses, and increases deflections. In trusses the net effect will reduce load-carrying capacity. On the other hand, the load-carrying capacity in steel and concrete flexural members will probably increase due to the prestressing action. When the stiffness of the member is substantially reduced and the deflected shape approaches a catenary, the axial forces change character and become tensile.

Connections are important to composite structural behavior. Connections must support the combined effects of gravity and thermal loads. Not only can the combined stresses increase over the levels expected during normal loading conditions, but connections must handle both tensile and compressive loads at different times during the fire. Continuous construction and moment connections influence the collapse potential for thermally induced conditions.

When bearing walls support horizontal framing systems, the lateral expansion will tend to push the wall out of line. This thermally induced force can cause premature collapse in masonry walls by inducing flexural stresses and eccentricities.

Thermal stresses can affect vertical columns in several ways. When the expansion is unrestricted, no additional axial loads are induced. However, when expansion is constrained by other parts of the building or by the framing system, additional axial loads are induced, as shown in Figure 15.4. These thermally induced axial loads, combined with reduction in the modulus of elasticity, reduce column capacity.

A second influence on column strength is caused by thermal movement that pushes columns out of plumb. Eccentric loading induces a moment Pe, as illustrated in Figure 15.5. The stress magnification may be calculated from

$$\sigma = P/A \pm Pec/I, \tag{15.1}$$

where

σ = stress,
P = applied axial load,
A = cross-sectional area,
e = load eccentricity,
I/c = section modulus.

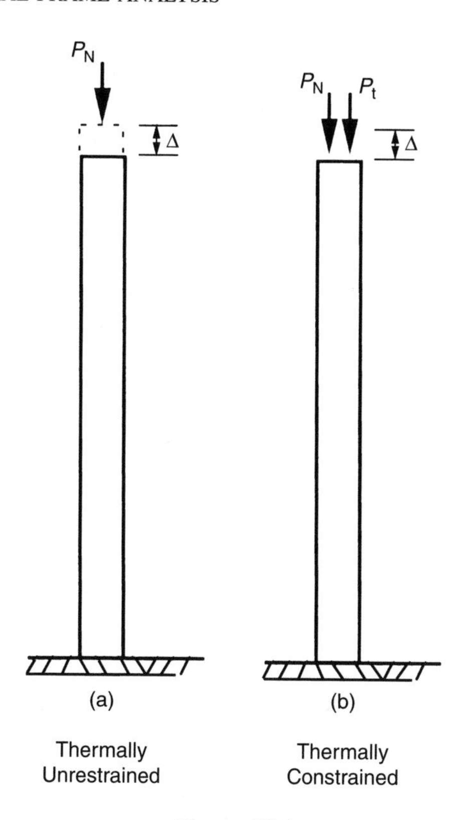

Figure 15.4

Figure 15.5

The details of construction influence structural strength during a fire. Continuous construction is inherently stronger than simple (statically determinate) construction. Structural theory enables one to recognize the necessary conditions and calculate the increase in strength for loading and support conditions. The collapse of upper floors onto lower floors can trigger progressive structural collapse. The structural framing influences the likelihood of local floor failure and the dynamic effect of more extensive failure.

15.7 Structural performance description

Structural theory enables one to calculate deflections and collapse mechanisms at normal temperatures. Even with this well-defined theory, dimensional variability, lack of precision for physical property values, and inaccuracy in theoretical equations provide some uncertainty as the loads approach the limit state. One can describe the structural performance at normal temperatures by a characteristic curve that includes three regions:

(a) the load at which failure will certainly not occur;
(b) the load at which failure certainly will occur;
(c) a window of uncertainty for a likelihood of failure.

The structural frame (Fr) curve in Figure 15.6 labeled "normal temperatures" would describe this performance up to the limit state.

Figure 15.3 shows that structural strength decreases as the temperature increases. Figure 15.6 illustrates the effect of different temperatures on the Fr curve. The abscissa of Figure 15.6 describes applied load-carrying capacity. When the dead load is known, one can estimate the live load that will cause failure. Fr curve construction provides an understanding of how the structural frame will behave in a fire. Example 15.1 illustrates Fr curve construction. This system was used to illustrate the role of structural capacity on $\overline{\text{D}}$ performance in Example 10.7.

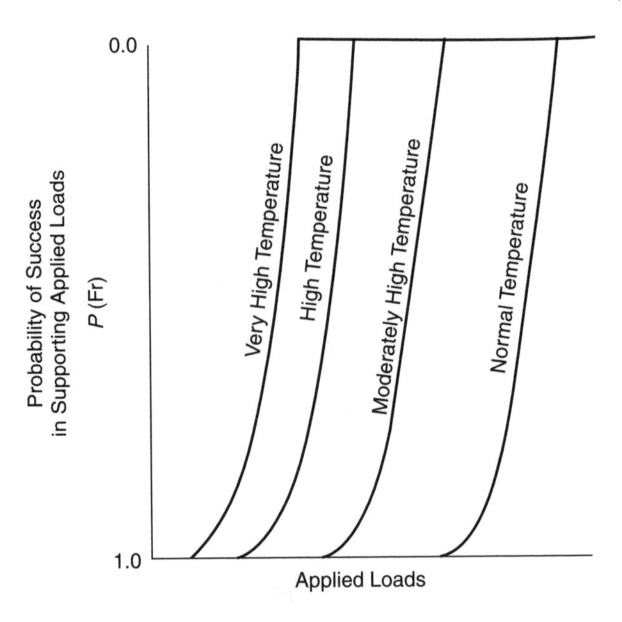

Figure 15.6

Example 15.1

A structural floor system is composed of W 12 × 14 beams spaced 4 ft 0 in c/c supporting a 3 1/2 in concrete slab on cold-formed steel deck. The beams span 17 ft 0 in and the steel's yield strength is $F_y = 36$ ksi. Construct Fr curves for this beam for the following conditions:

- unprotected steel beam;
- steel beam protected by 1/2 in of spray-on insulation;
- steel beam protected by 1 in of spray-on insulation.

Solution

A W 12 × 14 provides a plastic section modulus of $Z_x = 17.4\,\text{in}^3$. At normal temperatures the ultimate moment is $M_u = F_y Z_x = (36)(17.4)/12$, so $M_u = 52.2$ ft-k. The resistance factor $\phi = 0.9$ indicates that variations in dimensions, properties, and professional practice (i.e., equations) could reduce this ultimate moment to $M_u(\text{min}) = (0.9)(52.2) = 47.0$ ft-k. It seems reasonable to estimate that M_u would certainly not be less than 47.0 ft-k. Factors such as strain hardening and other conservative factors in selection of ϕ indicate that the resisting moment would certainly not be more than about 56 ft-k.

These calculations and reasoning are the basis for the Fr curve at normal temperatures shown in Figure 15.7. Increased temperatures reduce the ultimate moment, as indicated in Figures 15.3 and 10.18. This influence of ultimate moment reduction shifts the Fr curve to the left, as illustrated in Figure 15.7.

One can associate the shift in the Fr curve with the time when a critical moment is reached. However, for a specific structural size, insulation and load, it is more useful to identify a time duration before collapse. The ultimate moment at elevated temperatures can be calculated with a spreadsheet analysis using normal structural relationships and heat transfer analysis from the room time–temperature relationship. Figure 15.8 shows the results for three protection conditions (see Figure 10.18) for room conditions that produce a standard time–temperature relationship.

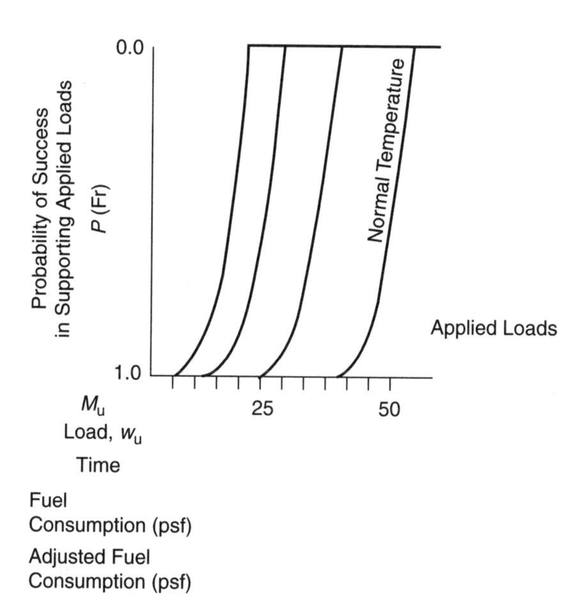

Figure 15.7

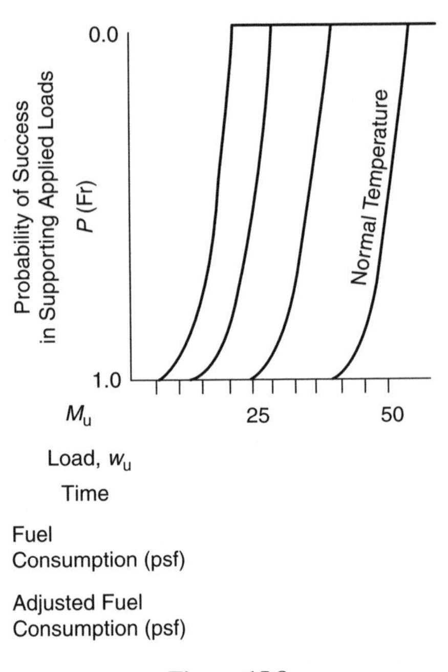

Figure 15.8

One can show the results in terms of ultimate moment, M_u, or as a ratio of limit load to dead load, w_u/w_{DL}. Figure 15.8 uses the latter format to show the results of the three insulation conditions for the standard time–temperature relationship. The dead load is as follows:

$$\begin{aligned} &\text{4 in concrete slab, 50lb/ft}^2 \times \text{4ft} &= \text{200lb/ft} \\ &\text{Steel beam weight} &= \text{14lb/ft} \\ &w_{DL} &= \text{214lb/ft} \end{aligned}$$

These values would indicate that collapse under dead load alone would be about 8 min for the unprotected beam, but 97 min for a beam protected with 1 in of spray-on insulation.

If the live load were at the full design value (50 psf), the value for $w_{LL} = 50\,\text{lb/ft}^2 \times 4\,\text{ft} = 200\,\text{lb/ft}$. The critical load is $w_{TL} = w_{LL} + w_{DL} = 200 + 214 = 414\,\text{lb/ft}$. If this were to cause collapse, the ratio of $w_u/w_{DL} = 414/214 = 1.9$, and failure could be expected in about 48 min. If the actual loading were 20 psf, $w_u/w_{DL} = 1.4$ and collapse could be expected in about 52 min. If no live loads were on the floor, $w_u/w_{DL} = 1.0$ and collapse could be expected in about 57 min.

If the rate of mass loss were related to the time–temperature curve, the fuel load consumption could be estimated. For a live load of 50 psf, the Ingberg conversion would indicate a fuel consumption of about 8.7 psf.

15.8 Discussion

The construction of the Fr curves of Section 15.7 illustrates how one might develop a level 2 performance estimate. Unprotected steel construction using smaller-sized members does not give much time for frame stability. Even accounting for the time for initial fire growth, collapse of this small unprotected steel beam could be expected in about 10–12 min after EB. Chapter 10 describes ways of constructing catalog curves and estimating real time, fuel consumption, and the percentage of design fire heat applied to the barrier. These procedures may also be used with structural frame analysis.

The time to failure for the beam of Example 15.1 can be increased by insulating the steel. The common insulation methods are spray-on mineral fiber, encasing the beam in gypsum board, or installing an approved membrane ceiling. For example, if the beam is protected by mineral fiber insulation of 1/2 in or 1 in, the time to failure can be extended, as seen in Figure 15.8.

Estimating structural frame performance uses the state of knowledge available. For structural framing, one can calculate the limit state load at normal temperatures to establish a bound and sense of proportion for performance. Then, for selected time intervals after EB, the following questions are evaluated:

- Will enough heat be generated and reach the surface of the structural member to reduce its strength to the point of failure?
- Will enough heat penetrate into the structural member to weaken it to the point of failure?

Failure may be defined as collapse or excessive deformation.

A level 1 estimate uses observations and mental estimates to recognize potential weaknesses and get a sense of proportion for performance. A level 2 structural performance analysis starts with a design fire that identifies a time–temperature relationship and calculates structural behavior. Level 3 analyses investigate the sensitivity of important parameters to get a better in-depth understanding of performance.

15.9 Level 1 evaluations

A level 1 evaluation attempts to identify the structural framing system and deficiencies of construction or structural protection that could lead to collapse in a fire. Level 1 evaluations use observational recognition of important factors that can easily be overlooked in design and construction yet become very significant in the performance during a fire. The objective of a level 1 analysis is to recognize potential structural weaknesses. This results in better decisions regarding occupant and fire-fighter safety or identifies a need for additional examination.

Important information may be inaccessible in an existing building. Therefore evaluations are based on information and associated deductions that can be gathered within the available time. The types of information that provide a basis for observational evaluations are shown below.

STRUCTURAL FRAME

Trusses of any material; exposed steel or wood trusses, beams, and girders; exposed steel or wood columns; prestressed concrete construction.

INSULATION PROTECTION OF STRUCTURAL MEMBERS

Missing sections of spray-on or boxed-in insulation to expose structural members; concrete spalling; membrane ceilings (e.g., holes, questionable suspension strength, unusual lighting fixtures or diffusers); holes, cracks, or questionable weaknesses in coverings.

APPLIED LOADS

Large dead load presence; large stationary live loads (e.g., machinery, piled storage, book stacks).

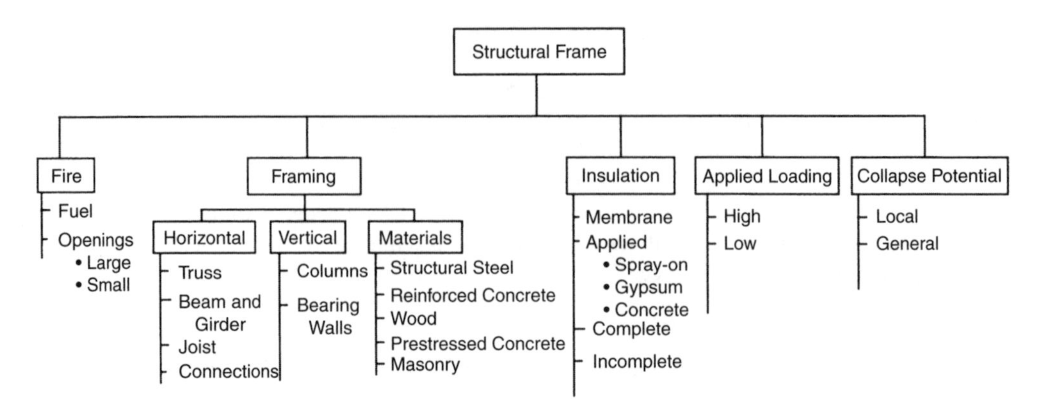

Figure 15.9

CONSTRUCTION FEATURES

Simple and continuous construction locations; thermal expansion effect on support and wall movement; rigidity of movement restraint. Figure 15.9 shows the major factors of interest. The major observable factors that pose questions about adequate structural performance include

- trusses, regardless of the materials of construction; exposed trusses are a particular concern;
- exposed, unprotected wood or steel joists or studs;
- exposed structural steel or wood beams and girders;
- exposed wood or structural steel columns;
- prestressed concrete beams;
- connections that are exposed or seem susceptible to changes in carrying capacity due to heat absorption;
- holes in membrane ceilings that will permit heat to move easily into the space between the ceiling and the structural members;
- sections of fire protection insulating materials that have fallen off structural members;
- potential for thermal expansion to push walls out of plumb;
- large live loads on some or all of the frame.

15.10 Summary

The function of a structural frame is to support the applied dead and live loads for the duration of a fire. Collapse presents a danger to occupants and fire fighters as well as to property and the operational continuity of the building. The structural frame is evaluated as a separate component, even though it may also be a part of a barrier. Structural floor systems illustrate a structural frame that is simultaneously part of a barrier. Certainly structural collapse will be recognized as a barrier failure. However, excessive deflection without collapse can cause cracks that will allow flame movement to penetrate into the space above.

As with all performance evaluations, Fr curves reflect your understanding of how the structural frame will behave during a fire. A level 1 Fr curve can recognize the likelihood of collapse. However, this is complex because of interactions between loads, heat energy application, framing materials, and arrangements. These interactions can be estimated relatively accurately by level 2 calculation methods and shown on an Fr curve if the needs of the application warrant this added study.

16 SMOKE ANALYSIS

16.1 Introduction

The flame/heat component dominates a building's performance because it defines the fire behavior and gives a clear picture of strengths and weaknesses for most of the fire safety system. A structural frame analysis amplifies this understanding to include a perception of the building's stability during the fire. A smoke/gas evaluation expands knowledge even further and is the final performance assessment needed to develop risk characterizations.

The goal of a smoke evaluation (Sm curve) is to understand the building's behavior regarding smoke generation and movement. The analysis provides a sense of proportion for the smoke movement in important building spaces. Conceptually, one can envision smoke generation and its movement into target spaces. The analysis blends theory and technology with judgment to make performance estimates.

One must make several interrelated decisions to construct a Sm curve. The first describes the smoke generation rate and cumulative quantity for a sufficient duration to understand the time that critical building spaces will remain tenable. The second defines tenability in ways to numerically estimate performance and in a manner that can be comprehended by the layman. A third selects a strategy to estimate the time that selected target spaces will remain tenable.

This chapter provides a way of thinking about smoke tenability and building analysis. The goal is to develop an awareness for building factors that affect the time that critical target spaces will remain tenable. The way of thinking will focus on level 1 evaluations derived from level 2 analyses. One may expand the understanding with 3 evaluations.

16.2 Overview of the process

Smoke and toxic gases adversely affect humans and also some equipment, property, and information storage. A building performance expressed by a smoke analysis (Sm curve) estimates the tenability time for critical target spaces. Figure 16.1 illustrates a representative Sm curve that describes a target space's transition from tenable to untenable conditions.

A smoke analysis starts with a description of the design fire smoke generation. Some of this smoke will be pushed outside, and the remainder will move through the building. The internal paths through which the smoke will travel are identified, and one distributes the fraction of

Building Fire Performance Analysis R. W. Fitzgerald

ISBN: 0-470-86326-9 (HB)

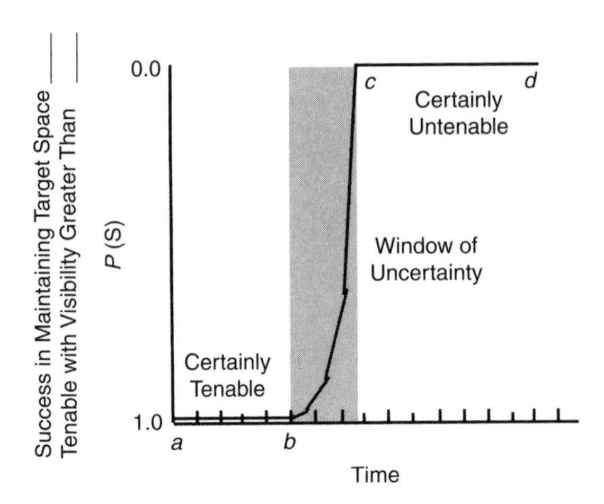

Figure 16.1

generated smoke that remains in the building through this space/barrier network. One estimates the accumulation of smoke and the likelihood that the target spaces will remain tenable at each time step.

Smoke movement is relatively easy to conceptualize, but difficult to analyze. A building fire is a complicated happening with many forces that influence smoke movement. The fire generates smoke and acts as a pump with driving forces of buoyancy and fire pressures of the heated air. Natural and man-made pressures influence smoke movement within the building. Temperature differentials and wind cause the natural pressures. Openings caused by the fire, fire ground ventilation activities, leakage, and opening and closing of doors may be augmented with mechanical air pressures. These actions occur during the fire, making the actual smoke movement a dynamic process. Nevertheless, a systematic, methodical evaluation can give a good sense of the building's smoke behavior.

A level 2 analysis uses a computer program to simulate smoke movement from the room of origin through a group of barrier/space paths. One uses the output as an initial base to estimate the tenability in target spaces. Sm curves translate the deterministic output to express a degree of belief that target spaces will remain tenable at each time increment. The probability estimate enables an individual to associate computer assumptions, limitations, and simplifications to relate the outputs to the specific building conditions. It provides an added tool to give computer applications more credibility. A level 3 analysis examines the influences of variability on computer results to establish an envelope for performance sensitivity. The representative shape of the performance curve in Figure 16.1 remains the same. Confidence in results relates to the quantity and quality of information used for degree-of-belief estimates.

A level 1 estimate envisions smoke generation within the room of origin and movement outside room. One may envision movement through interior openings and along the major paths that transport the smoke. The estimate may include the influence of natural and man-made pressures. Observations and mental estimates can give a sense of proportion for the time during which a target space will remain tenable.

16.3 Smoke

We have all experienced friendly or hostile fires and can envision a rising smoke cloud. Smoke quantity, quality, characteristics, and behavior can vary enormously among fires. Although both smoldering and flaming fires generate smoke, performance evaluations use flaming fires. An analysis considers the quantity of smoke generated, its characteristics of temperature, visibility, and toxicity, and the forces that move it in a building.

Smoke obscuration is due to soot, which is an unburned product of combustion. Some materials produce more soot than others, and inefficient combustion (e.g., poorly ventilated fires) produce more soot than efficient combustion. Smoke may be nonirritant where visibility reduction is due to light attenuation by soot particles. Smoke may also be irritant when the eyes sting and tear. One may have a long ocular visibility in irritant smoke, but may be unable to keep eyes open because of the pain and discomfort. This reduces movement speed and causes anxiety. Sometimes a fire may produce irritant smoke and low visibility, although the two characteristics do not necessarily accompany one another. Although irritant smoke is a common problem, we do not know how to recognize or measure its conditions in an evaluation. Therefore tenability criteria will be based on visibility reduction due to smoke obscuration.

Entrained air in the fire plume becomes the medium in which smoke is transported, and smoke movement is essentially the transport of the smoke/air mixture. The rate of smoke production is nearly the same as the rate of air entrainment in the combustion process. Fuels, fire size, and room geometry are the most important influences on air entrainment. The geometry includes room volume, ceiling height, and the size and location of the ventilation openings. The generation rate produces a volume of smoke that has characteristics of temperature, visibility, toxicity, and corrosion. Although one can roughly estimate the volume of smoke and its characteristics, these attributes can vary greatly in building fires.

A description of the time-related volume of smoke production is needed for building evaluations. The rate of air entrainment before full room involvement (FRI) may be calculated for growing compartment fires as

$$M = 0.188 P_\mathrm{f} y^{3/2}, \tag{16.1}$$

where

M = rate of air entrainment, i.e., smoke production (kg/s),
P_f = fire perimeter (m),
y = distance between the floor and the underside of the smoke layer below the ceiling (m).

The mass rate of smoke production can be converted into a volume rate by calculations by using the ideal gas law for, at different temperatures. However, after FRI, smoke production increases greatly above that predicted by equation (16.1). Computer programs can provide a more accurate estimate of smoke production. User supplied computer inputs include:

- heat release rates;
- heat of combustion;
- fire position (center of room, near walls, corners);
- limiting oxygen index;
- ratios for estimating species concentrations (H/C, C/CO_2, CO/CO_2);
- type of fire (unconstrained means burning takes place inside the plume; constrained means burning takes place where there is a sufficient supply of oxygen).

Computer programs are discussed briefly in Section 16.15. Their use requires considerable time and interpretation of results, which is not available for a level 1 evaluation. Default smoke production curves provide a base for examining features that affect building performance beyond the room of origin. One could associate default curves with the room classifications described in

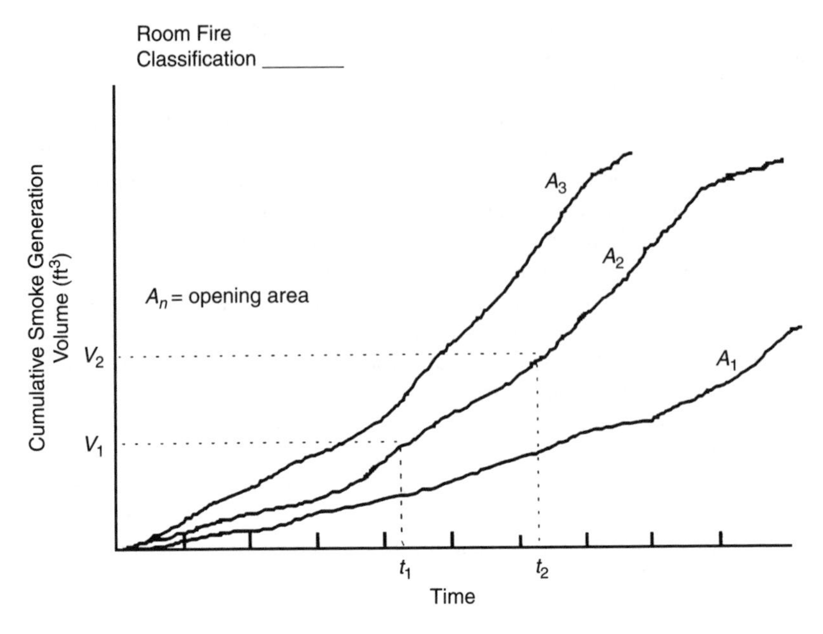

Figure 16.2

Chapter 8 to provide a rapid level 1 estimating basis to start the building analysis. Although this has not been done here, Figure 16.2 shows hypothetical curves for a model room.

16.4 Smoke movement

A growing fire must have ventilation for a sufficient air supply to reach FRI. Therefore at least one opening will exist in the room of origin to exchange the smoky effluent with new air. If the openings are in exterior walls and no interior openings exist, most of the smoke will be released to the outside; the building interior, although not the room of origin, will be relatively smoke free. On the other hand, if openings are present to the interior of the building, smoke can migrate rapidly through the available paths.

A Sm curve describes the probability that a target space will remain tenable for time durations after EB. Figure 16.1 illustrates a representative Sm curve. Each time duration is evaluated by answering the following two sequential questions:

- Will enough smoke reach the boundaries of the target room to make it untenable?
- Will enough smoke enter and accumulate in the target space to make it untenable?

FIRST QUESTION

The first question addresses the following topics.

Time and design fire smoke generation

For the time duration, has enough smoke been generated to make the target space untenable?

Smoke transport

For the time duration, will the smoke reach the target room? This involves factors such as

- smoke characteristics (e.g., temperature, pressure, smokiness, air entrainment, deposition);
- path routes from the room of origin to the target space;
- path length;
- number of branches that allow smoke to separate and distribute to other paths (including the outside);
- number and influence of barriers to movement along the path;
- existence and effectiveness of a smokeproof tower;
- barrier opening sizes;
- natural air pressures to influence smoke movement (e.g., stack effect);
- influence of wind pressures;
- mechanical air pressures to influence smoke movement.

SECOND QUESTION

The second question examines the building features that affect smoke entering, leaving, and accumulating in the target space. The following features influence this event.

Barrier effectiveness

This involves observations such as

- existence of a physical barrier;
- doors and other openings and their status (open, closed, time durations);
- leakage.

Space protection

This investigates the effectiveness of maintaining a tenable atmosphere with features such as

- space size and shape;
- barriers to smoke movement;
- venting capabilities and effectiveness;
- space pressurization and its effectiveness.

This two-phase analysis allows one to separate the smoke movement analysis from its accumulation in the target space. This enables one to systematically recognize and organize features that affect building performance. Observations, recognition of physical influences to smoke movement, and use of a mental model to trace movement along the path to the target room become the basis for the estimates. Often one may expedite the process by using the barrier/space network to estimate the fraction of smoke reaching each target space. Estimates for each new time increment may adjust this value.

Many forces influence smoke movement from the room of origin to a target room. One involves buoyancy characteristics changes due to the cooling of hot smoke. Also, natural air movement and the air currents that are in buildings can be important, as can mechanical air movement. The opening and closing of doors, fire department ventilation practices, wind direction and speed, and automatic venting also influence smoke migration during a fire.

16.5 Buoyancy forces

Smoky air that flows out of the room of origin is hot, and its temperature is one of the design fire characteristics. Hot air is a fluid that behaves in accordance with Boyle's law, $PV = RT$. In this expression the pressure, P, does not change much because of the room sizes and the openings that are present in all buildings. The constant, R, is consistent for any specific gas. Therefore a given air mass will change volume in a building fire in proportion to its absolute temperature, T. The absolute temperature of fire gases in the room of origin is about three times the normal temperature. Therefore, immediately upon flowing out of a room, the hot smoke can have as much as three times the volume of the same mass of cold smoky air.

It is convenient to envision smoky air as "magic" balloons that have unusual characteristics. Their original composition can change temperature, mix with entrained air, and deposit soot as it moves away from the fire. Consequently, one may track the properties of a smoke unit as it moves through a building. For example, a 1 ft^3 (0.0283 m^3) volume at room temperature will expand to about 3 ft^3 (0.085 m^3) in a fire, and also incorporate soot and the other constituents of smoke. This decreased density (buoyancy) allows colder units to push the hot balloons upward until they reach the ceiling and then out through the room openings.

Several things happen to the smoke units as they leave the room of origin and move through the building. They will lose some heat to the building surfaces, and additional cool air is entrained (substituted) into the units. These actions cause the smoke to cool, reducing the total volume and diluting particulates and gases. Also, some soot will deposit on the surfaces along which the smoke flows. At a greater distance from the fire, the smoke may approach the temperature of ambient air. This cold smoke contains particulates that reduce visibility along with toxic and corrosive fire gases, while having the flow characteristics of the ambient air.

The transition from hot smoke to cold smoke depends on the cooling effect of air entrainment and heat losses to boundary surfaces. The characteristics of hot smoke and cold smoke are very different. Hot smoke is very buoyant, causing an upward movement unless constrained by building surfaces. It is often possible to recognize a stratified, clear plane of demarcation between the hot smoke layer and the cold layer in a building fire. The buoyant pressures caused by heat cause substantial flow of smoke. As air moves away from a fire, it cools relatively quickly, causing a loss in buoyancy and greater diffusion. The condition of a uniform smoke mixture from floor to ceiling is described as diffusion. Diffuse, cold smoke moves like the air in which it is mixed.

16.6 Natural air movement

After identifying potential paths from the room of origin to the target rooms, one assesses the smoke movement through those paths. Besides the buoyancy forces, natural air currents influence the movement. These natural air forces involve the stack effect and wind.

Air moves through a building by fluid forces caused by differences in pressures. The slight increase in pressure caused by higher temperatures is enough to cause smoke to migrate from a space of higher temperature to one of lower temperature. Boyle's law for two different spaces may be expressed as

$$R = \frac{P_1 V_1}{T_1} = \frac{P_2 V_2}{T_2}, \tag{16.2}$$

where

P = fluid pressure,
V = volume,
T = temperature.

Pressure differences across barriers may be calculated by knowing the temperature and volume between two rooms. This pressure differential creates the forces that cause air to migrate into an adjacent space.

The stack effect, sometimes called the chimney effect because it relates to vertical air movement, also causes natural air movement in buildings. The stack effect is caused by differences in temperature between the inside and the outside of a building. Assume initially that the air outside a building is cold relative to the inside. The cold air is more dense and its outside pressure is greater than that inside the building. A building normally has many openings (leakage) throughout its height. Also, doors are often open on the lower floors. Cold air moves into the building through these low openings, causing the warmer inside air to be forced upward.

The inside air moves upward through elevator shafts, stairwells, other vertical shafts, and by floor to floor leakage. The inside air is forced outside toward the top of the building through leakage and other openings that may be present. All buildings have leakage around openings and penetrations. The stack effect causes air to circulate upward in this situation. Somewhere near the middle of the building, depending on the opening areas throughout the building, is a neutral pressure plane. At this level, the pressure to move from the outside into the building is equal to the pressure to move from the inside to the outside. Figure 16.3 shows the general pressures and air movement caused by the stack effect. When outside air temperatures are higher than the inside air, the movement is inverted and a reverse stack effect is created.

The location of the fire with respect to the neutral pressure plane will influence the smoke movement. For example, assume a fire occurs on a lower floor of a tall building. Buoyancy caused by the fire gas temperatures will cause upward pressures. As the hot gases cool due to air entrainment and heat loss to the boundaries, the buoyancy reduces until it reaches ambient temperature. However, the stack effect also helps to move smoke to the upper floors. The largest pressures occur near the top of the building and those locations may be more threatened than floors

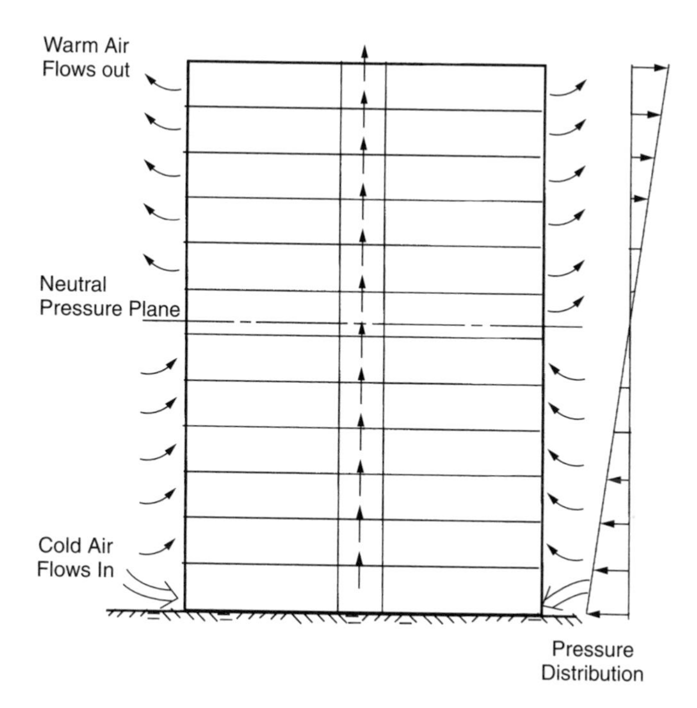

Figure 16.3

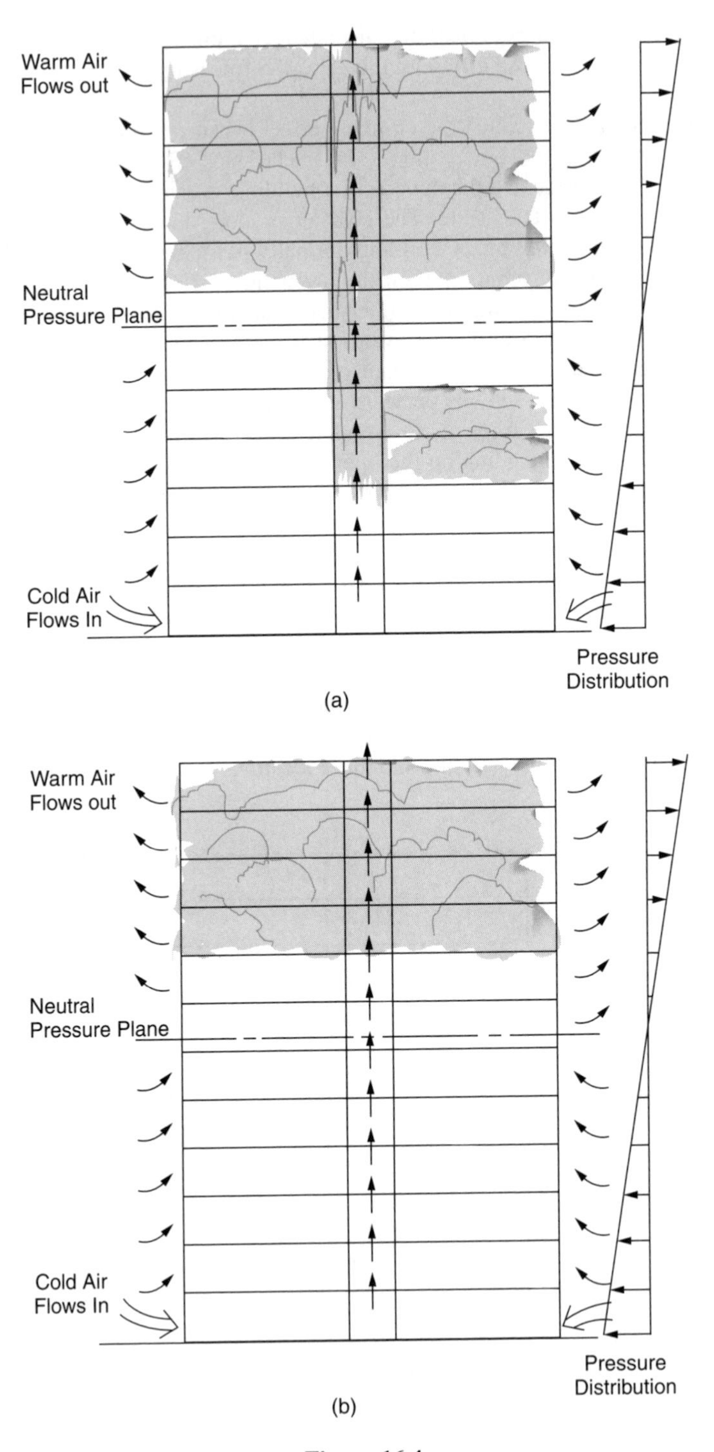
Warm Air
Flows out
Neutral
Pressure Plane
Cold Air
Flows In
Pressure
Distribution
(a)
Warm Air
Flows out
Neutral
Pressure Plane
Cold Air
Flows In
Pressure
Distribution
(b)

Figure 16.4

closer to the fire. The floors near the neutral pressure plane will have some smoke migration, but the concentration will not be as great as at the top floors. A fire that occurs above the neutral pressure plane will also have some smoke migration above the fire due to leakage and some greater smoke at the top floors. Smoke will also move outside a building due to leakage when the inside pressures are greater than outside. Figure 16.4(a) and (b) illustrate these conditions.

Wind is another force that causes air movement in buildings. When wind blows on a building, an array of positive and negative (suction) pressures are established. Positive pressures develop on the windward side. Negative pressures will occur on the adjacent and leeward sides. Figure 16.5(a) shows these pressures. Either a pressure or a suction will develop on the roof, depending on the slope. Flat roofs develop a suction caused by the winds, as shown in Figure 16.5(b). Sloping roofs develop a positive pressure on the windward side and a negative pressure on the leeward side, as shown in Figure 16.5(c).

Wind pressures can complicate a smoke analysis. This can be important when either automatic or fire department ventilation occurs. For example, in a flat-roofed building, as in Figure 16.5(b), an open vent with a wind will create larger negative forces at the roof and help remove air from inside the building. On the other hand, if the roof is sloped, as in Figure 16.5(c), the side on

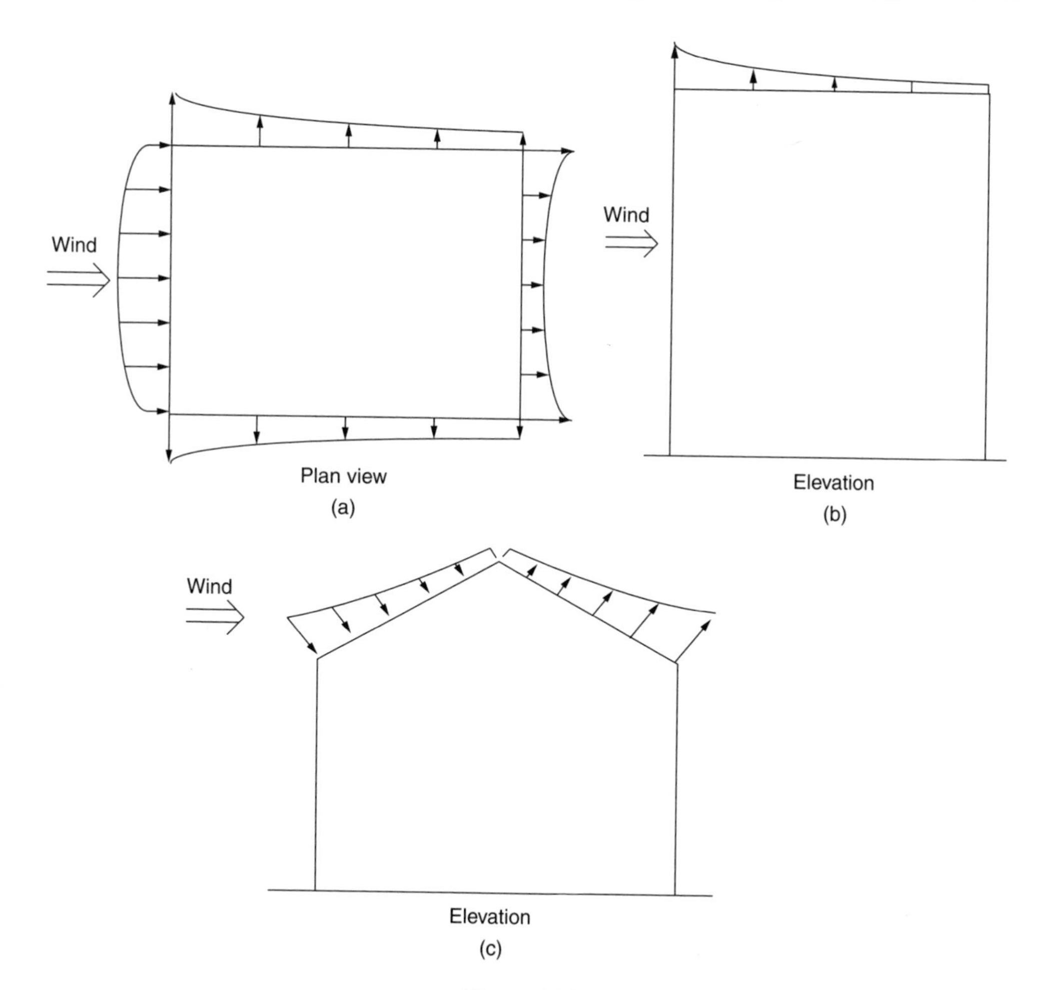

Figure 16.5

which venting occurs can affect internal smoke movement. If the windward slope is vented, wind will force escaping smoke back into the building. If the leeward slope is vented, the negative pressure helps remove the inside air faster.

A fire may vent itself by breaking a window. Smoke from a fire on the windward side will be driven toward the building interior. A fire on the leeward side lessens interior smoke conditions. We do not know the direction of the wind until the time of the fire. Therefore different scenarios become useful to develop an understanding of conditions that can affect smoke movement.

16.7 Influences of mechanical air pressures

When a building has mechanical air-handling equipment, its operation to pressurize or exhaust can influence potential smoke movement. During a fire, an air-handling system may

- continue normal operation;
- shut down;
- change to a defined emergency mode of operation.

Whatever the operational mode of the mechanical air-handling equipment, its influence and reliability are part of smoke tenability evaluation.

If the air-handling controls of an HVAC system continue in normal operation after a fire starts in a space, the smoke will spread throughout the zone. The rapid smoke spread may cause anxiety among occupants and reduce visibility in rooms far removed from the fire. This smoke may also cause toxic reactions and increase the likelihood that the rooms will become untenable more quickly. A role for air duct smoke detectors is to close dampers so that smoky air cannot be forced into rooms. Air systems can also feed fresh air to the fire, causing it to burn faster with greater intensity.

This forced air movement is prevented when HVAC systems are shut down during a fire. However, shutting down a system does not prevent ducts from transporting smoke to these rooms if dampers are not present or inoperative. Smoke may be forced through the supply and return air ducts or plenum spaces by fire gas buoyancy, the stack effect, and wind pressures.

The third condition for HVAC system is that of emergency operation. It is possible to install fans to influence airflow. These fans can be programmed to place some rooms in an exhaust mode and others in a pressurization mode. Operations of this type are described as smoke management systems.

HVAC emergency operations can affect target space tenability. Pressure differentials created by mechanical fans can prevent smoke from entering into a space. When a pressure differential is combined with a physical barrier, a target space may remain tenable for a long time. The time during which doors may be open influences target space tenability.

16.8 Fire department ventilation

Fire department ventilation operations can affect the smoke tenability in certain spaces. The usual function of ventilation is to raise the smoke level in fire areas so that suppression forces can find the fire and extinguish it more easily. Ventilation also helps in extinguishment for some situations and in smoke removal.

Intentional fire department actions or unintentional fire venting can cause vent openings. Heat and smoke release, and pressure differential changes within the building can affect smoke and heat conditions in building spaces. While these influences are rarely incorporated into a smoke tenability evaluation, a series of performance curves could describe "what if" scenarios.

16.9 Tenability

A fire produces heat, soot particles of different sizes, water, and a variety of gases, none of which are good for human health. In addition, these products can damage the operation of some equipment and data storage. Factors that influence tenability criteria for people or property are discussed below.

HUMAN TENABILITY

A wide variety of toxic gases accompany the combustion process and are transported by the same air that carries soot particles which reduce visibility in smoke. Carbon monoxide and carbon dioxide are in all fires. Depending on the fuel, many other gases that damage health are present in fires.

Fire tests can identify toxic gases from building and content materials. Toxicity measures and performance criteria have not been identified. While lethal concentrations for some gases can be established for laboratory animals, levels of exposure doses (including the cocktail effect of gas mixture synergism) that produce human impairment are not known. Although one can recognize that the toxic gases are bad for humans, we currently have no measure to quantify building performance for that toxicity.

Occupants and construction industry professionals can understand visibility reduction. One can envision and roughly approximate visibility in building fires. However, smoke visibility reduction usually precedes the effects of toxic gases on humans, often by an ample time. Consequently, if we base tenability criteria on visibility distance in smoke, the results will be conservative and one can evaluate building performance.

Visual impairment can be affected by the density of soot particles that constitute smoke or by irritants that sting and cause the eyes to tear. It is possible to have low visibility due to dense smoke and not have eye irritation. Conversely, it is possible to have light smoke with long ocular visibility and be unable to see because the eyes close due to the stinging irritants. Ocular visibility is currently a more practical measure because we do not know how to estimate eye irritants.

Normal ambient light affects visibility and so does background lighting. When the lights are out and the building is dark, one cannot see into the distance whether or not smoke is present. We base performance visibility distance through smoke on conditions of normal interior lighting. This criterion does not identify whether an individual will be hospitalized or killed by the products of combustion. The criteria identifies only a smokiness level that can quantify performance. The life safety tenability criteria may be stated as the ability of an individual to see x feet (meters) in a building fire.

Humans can tolerate a mild increase in temperature and some smokiness and toxicity for a short time during building egress or while remaining in a room. Tenability selection will reflect differing life safety needs and vary with factors that are unique to the building. Values for an occupant of a defend-in-place room will differ from values for an individual in transit during a building evacuation. A building with uncomplicated circulation patterns will differ from one with a complicated architectural plan where direction finding indicators are more difficult to recognize. The tenability criteria may change for occupants who are unfamiliar with the building or who may have physical or psychological impairments.

A tenability selection will consider factors such as

- building height, floor size, and architectural layout;
- number of occupants and initial location of the representative occupant;
- potential egress routes and their number and complexity;

- familiarity of occupants with the building architectural layout;
- occupant age, mobility, physical and mental condition, and expected activity for the fire scenario;
- time of day and day of week;
- prefire instructions, education, and training.

Some smoke is common and expected in building fires. Light smoke conditions can generally be handled. However, too much smoke in the target spaces is unacceptable. Smoke tenability defines the question of how much is too much. The individual responsible for the evaluation must select appropriate values for tenability. Some codes have identified visibility distances ranging from 2 to 10 m (6 ft to 33 ft) and some researchers recommend values between 3 and 25 m (10 ft to 80 ft) for various conditions.

EQUIPMENT AND DATA STORAGE

Although life safety is normally a primary focus of attention, fire gases can affect equipment, instrument functionality, and data storage. It is difficult to identify the particular corrosive products and exposures that affect performance. An additional problem is measuring the corrosive products and estimating their residual amount in rooms distant from the room of origin.

Equipment manufacturers are the best source to identify the effect of corrosive gases on their products. Fire tests can indicate if these gases will be released during combustion. If corrosive gases are released, they will be fellow travelers with smoke. If one models the corrosive gases of interest as a mixture with smoke, the onset of a specified visibility level may be used to define equipment tenability. Although correlations may be weak, they do provide a way to describe tenability for equipment or data storage.

16.10 Visibility in smoke

The performance measure for smoke tenability is the onset of a defined visibility level. Visibility is related to soot yield and optical density, which can be predicted by a computer model. The relationships are identified briefly in this section.

The extinction coefficient, which describes the smoke's ability to absorb or scatter light, depends on the optical properties and concentration of soot. Beer's Law predicts the attenuation of light by an absorbing or scattering media. This is expressed as,

$$I_L = I_O e^{-kL}, \tag{16.3}$$

where

I_L = intensity of a light beam after traveling a path length L,
I_O = original intensity of a light beam,
k = spectral extinction coefficient,
L = path length.

Equation (16.3) can be converted from base e to base 10 using

$$I_L/I_O = 10^{-DL}, \tag{16.4}$$

where D is the optical density. Optical density (OD) can be related to visibility by using the extinction coefficient

$$k = D/2.3. \tag{16.5}$$

Visibility through smoke is the maximum distance an individual can see an object. The visibility distance times the extinction coefficient is a constant:

$$Lk = C \qquad (k < 0.25) \tag{16.6}$$

or

$$Lk = C(0.133 - 1.47 \ln k) \qquad (0.25 < k < 1.1). \tag{16.7}$$

The constant C varies depending on the brightness of the object being viewed. Here are values of C for three types of object:

- Illuminated sign: C varies from 5 to 10.
- Reflective sign: C varies from 2 to 4.
- Walls, floors, doors: $C \approx 2$.

Figure 16.6 shows the relationship of visibility distances and optical density using these equations. If one obtains an optical density from a computer program, that value can be converted into an extinction coefficient and then into visibility distances.

16.11 Building geometry

The sensitivity of an architectural design to fire locations and target spaces is a critical part of several performance analyses. Flame spread, smoke spread, occupant egress, and fire attack routes all involve a space/barrier path connectivity. In uncomplicated buildings, appropriate locations for the room of origin and target spaces are relatively straight forward. However, when architectural plans become complicated, performance analyses involving movement paths for fire, smoke, people, and fire fighting become more difficult. One needs a tool to help organize space/barrier paths needed for analysis.

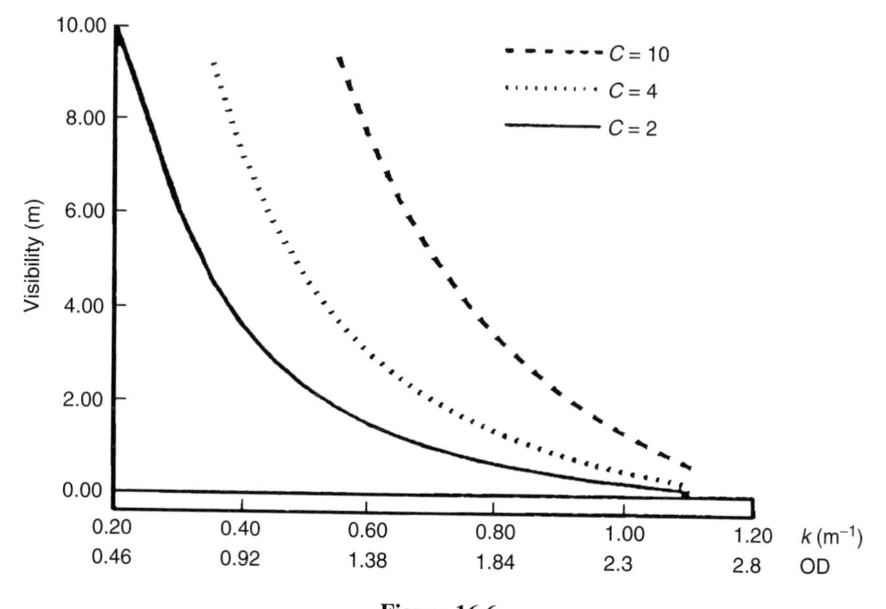

Figure 16.6

A barrier/space connectivity network provides a way to recognize architectural plan sensitivity. Figure 9.5 illustrates a network constructed by visual recognition that could be used for level 1 evaluations. However, when the architectural plans are complicated or when level 2 or level 3 evaluations examine sensitivity to different locations, adjacency matrix construction becomes a useful tool. Data-structuring programs [1] can help to construct interconnectivity networks for complicated architectural layouts.

To illustrate the process, Figure 16.7(a) and (b) show a simple floor plan and a space/barrier connectivity network for occupant movement. Table 16.1 shows an adjacency matrix that identifies room connectivity. Here a 1 identifies rooms through which people or smoke can easily move and a 0 identifies when movement cannot occur. The network of Figure 16.7(b) is constructed

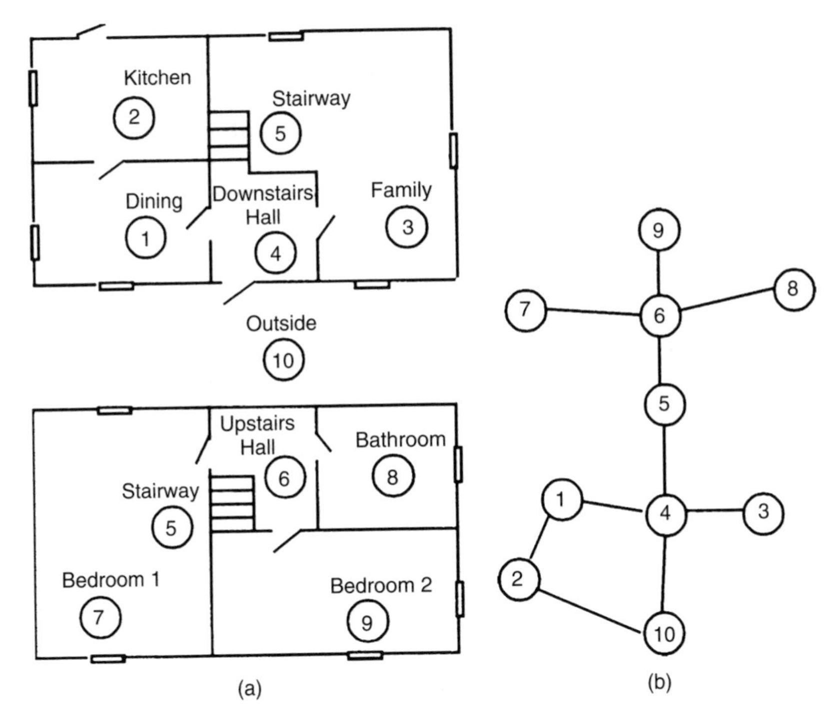

Figure 16.7

Table 16.1

Room	Module	1	2	3	4	5	6	7	8	9	10
DR	1	1	1	0	1	0	0	0	0	0	0
Kitchen	2	1	1	0	0	0	0	0	0	0	1
Family	3	0	0	1	1	0	0	0	0	0	0
L Hall	4	1	0	1	1	1	0	0	0	0	1
Stairs	5	0	0	0	1	1	1	0	0	0	0
U Hall	6	0	0	0	0	1	1	1	1	1	0
BR 1	7	0	0	0	0	0	1	1	0	0	0
Bath	8	0	0	0	0	0	1	0	1	0	0
BR 2	9	0	0	0	0	0	1	0	0	1	0
Outside	10	0	1	0	1	0	0	0	0	0	1

from this matrix. Complicated plans can use connectivity networks to systematically examine a variety of performance analyses.

The adjacency matrix of Table 16.1 shows building paths for smoke or egress. A similar matrix can identify potential fire propagation paths when appropriate zeros are changed to ones. This produces networks similar to Figure 9.5 (and more complete). The networks may be simplified to study propagation paths of interest.

16.12 The Sm curve

The Sm curve describes the building's performance in maintaining tenability in the target space. Figure 16.1 shows a representative smoke curve. Although we advocate tenability as a visibility distance, one may use any measures that can be quantified to a performance event.

We can envision the construction of an Sm curve as a CVN. However, because we cannot determine conditional probabilities directly, we use SV estimates for selected time durations to provide the coordinates for an Sm curve. Figure 16.8 shows the SVN to calculate an Sm curve coordinate. The factors that influence the events are shown in Figures 16.9 and 16.10.

At each time interval, the first event estimates whether enough smoke can reach the boundaries of the target space, B_S. The second event estimates whether enough smoke can enter the target space and accumulate to make the space untenable, E_S. Because this network is single value, the process is analogous to taking a photograph inside the target space at each time interval.

One can use Figure 16.8 to calculate Sm curve coordinates whether the deterministic analysis is a level 1 observational estimate or a level 2 computer analysis.

16.13 Sm curve construction

Smoke generation is a part of design fire characteristics. Smoke movement evaluations are initially based on a fire that continues to burn. After one understands this performance, the effect of fire termination may be incorporated to provide a comparison. The evaluation involves the following steps:

1. Identify the room of fire origin.
2. Identify the target rooms (specific building spaces) for which the Sm curve will be constructed.
3. Specify the acceptable level of tenability. This may be expressed as X_1 feet (meters) of visibility Y feet (meters) above the floor for hot smoke; or X_2 feet (meters) of visibility in diffuse smoke.
4. Select the design fire for the room of origin. This fire describes the "load" or smoke generation characteristics to analyze the building. The design fire characteristics include the rate of smoke generation, the temperature of the effluent, and the soot concentrations that influence visibility. If toxic or corrosive gases are a tenability consideration, their nature and concentration also become part of the design fire characteristics.
5. Identify the possible smoke movement paths from the room of origin to a target room. The technique described in Section 16.11 is useful to sort out the rooms and scenario conditions.
6. Identify possible paths of smoke movement through duct systems to the target room. One can adapt the adjacency matrix technique for these additional paths.
7. For each time selected, determine the smoke visibility in the target space. If a computer program is used, this relationship may be plotted directly. If level 1 estimates are used, the process assesses the following for each time step:
   - What was the total volume of smoke generated?
   - What fraction of the total generated smoke will reach the target room boundary?

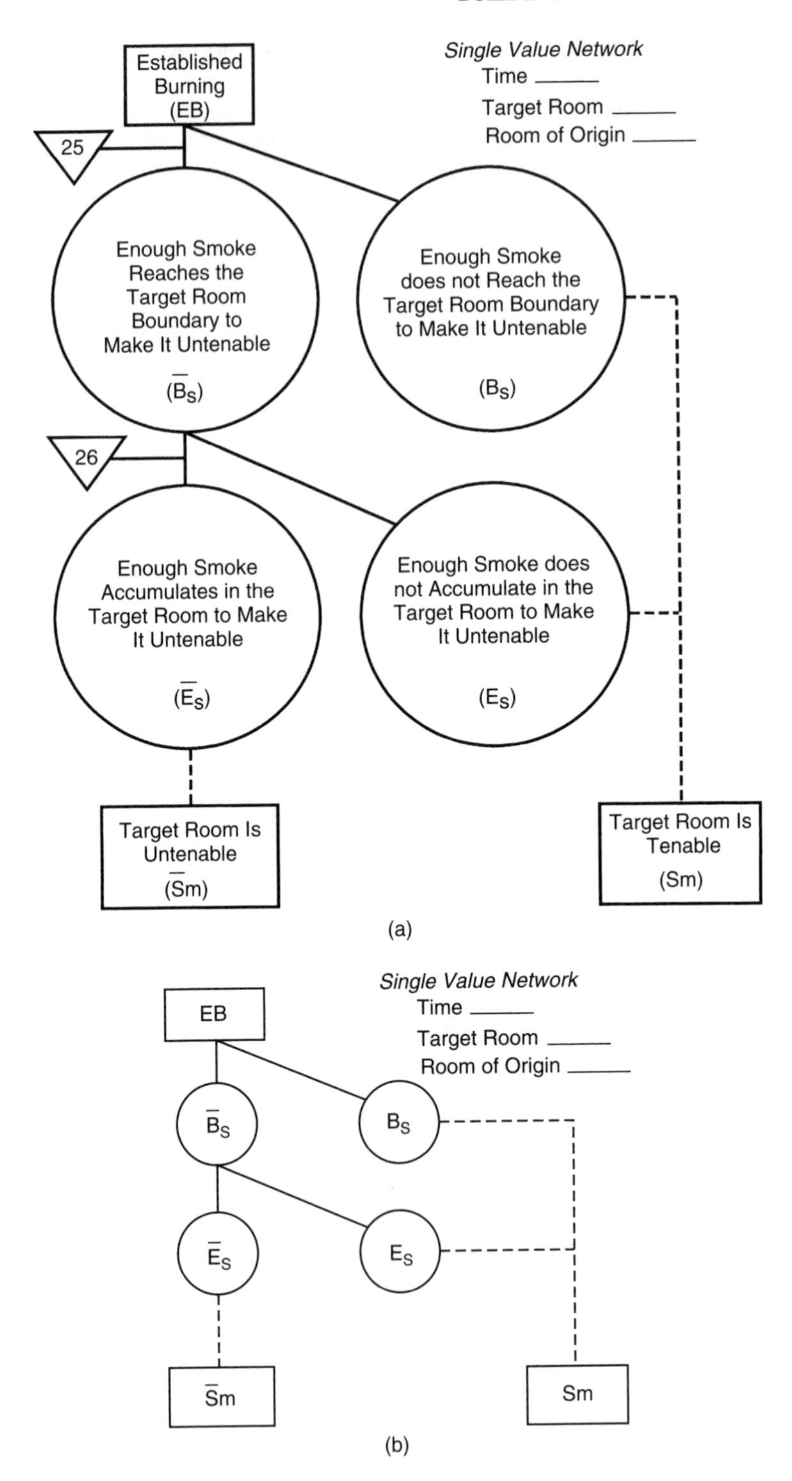

Single Value Network
Time ______
Target Room ______
Room of Origin ______
Established Burning (EB)
25
Enough Smoke Reaches the Target Room Boundary to Make It Untenable
($\bar{B}_S$)
Enough Smoke does not Reach the Target Room Boundary to Make It Untenable
(B_S)
26
Enough Smoke Accumulates in the Target Room to Make It Untenable
($\bar{E}_S$)
Enough Smoke does not Accumulate in the Target Room to Make It Untenable
(E_S)
Target Room Is Untenable
($\bar{Sm}$)
Target Room Is Tenable
(Sm)
(a)
Single Value Network
Time ______
Target Room ______
Room of Origin ______
EB
$\bar{B}_S$
B_S
$\bar{E}_S$
E_S
$\bar{Sm}$
Sm
(b)

Figure 16.8

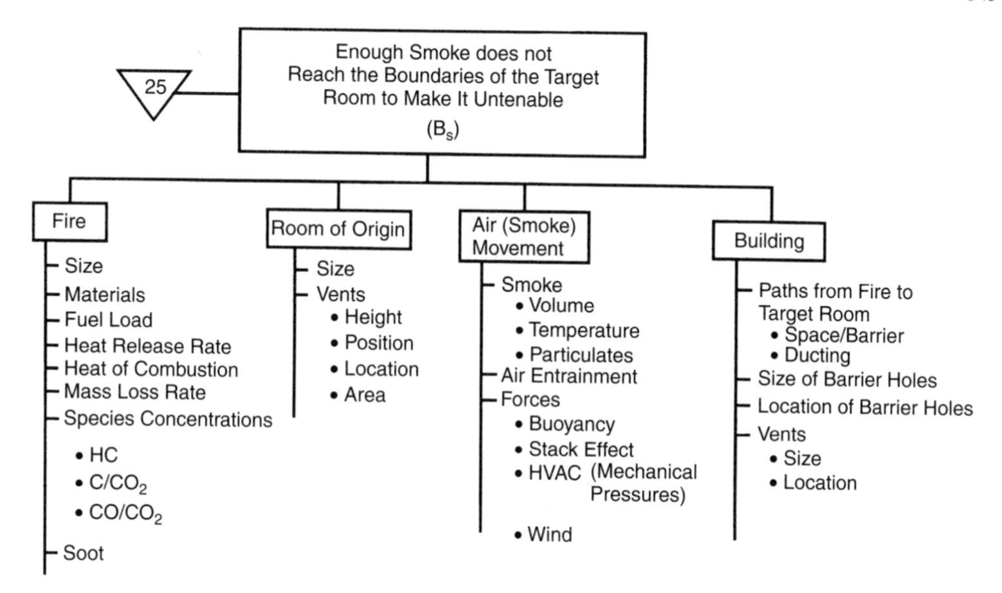

Figure 16.9

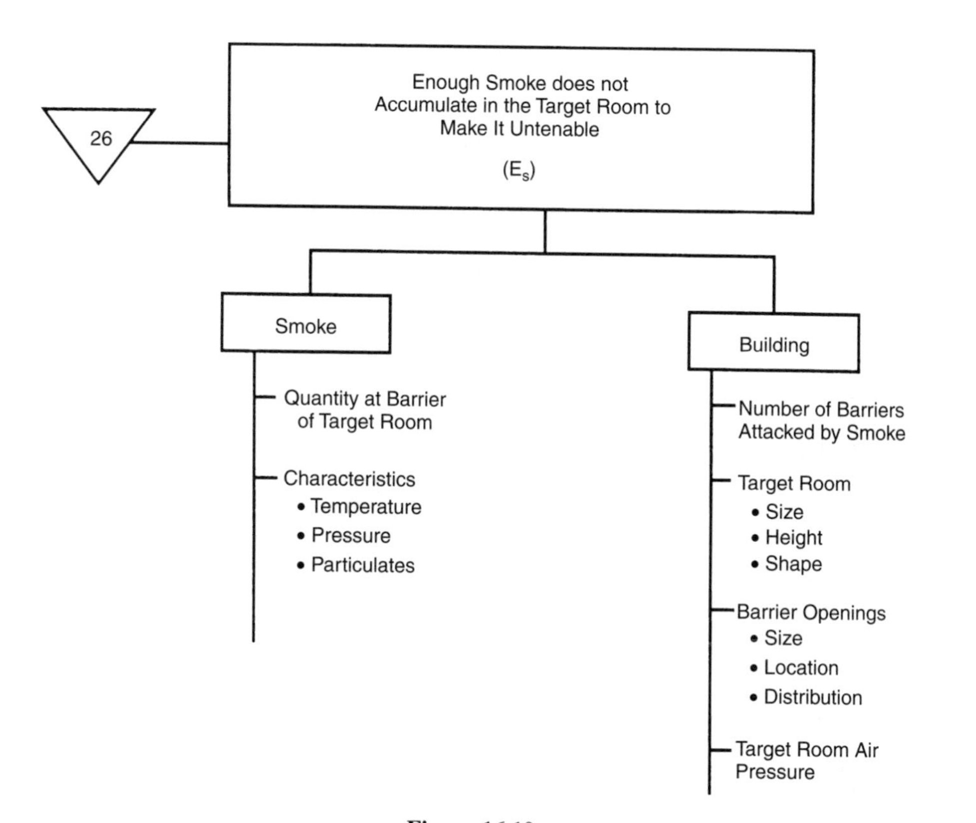

Figure 16.10

- What fraction of smoke that reaches the target room boundary will enter and accumulate in the target room?
- What is the visibility level in the target room?

8. Using the information from step 7, establish value judgments for the events of Figure 16.8 to identify a degree-of-belief probability that the target spaces will remain tenable for that time step. These values combine the basic smoke analysis of step 7 with knowledge of natural and man-made forces that affect smoke movement.
9. Repeat steps 7 and 8 for as many other time intervals as necessary to understand the building's performance for smoke movement, and construct an Sm curve for the target room.

16.14 Design fire smoke characteristics

The room of origin acts as a smoke generator and a pump to push the smoke to other parts of the building. If the fire propagates rapidly to several rooms, the smoke generator enlarges. Estimates that use only the room of origin and a space/barrier distribution network are usually sufficient to understand the building's response to smoke.

The design fire characteristics for a smoke movement analysis include

- a relationship between time and the volume of smoke generation;
- the smoke temperature;
- an obscuration measure to predict visibility;
- if needed, corrosive or toxic gases to define tenability for a specific object.

The design fire for smoke is more difficult than other performance evaluations. Ventilation is important to smoke volume, and the status of doors and windows can change. Fuels and materials can change. More important, the available time for decision making precludes an in-depth computer analysis unless the information value is greater than its cost.

A building's smoke performance is dominated by features that influence smoke behavior outside the room of origin. Building fires can produce enormous amounts of smoke very quickly. Smoke management features become a primary focus of attention. In many ways, the smoke performance is really an examination of how long a smoke management system will provide protection for occupants before the onset of untenable conditions. An accurate prediction of smoke production is not unimportant, but it is secondary to the physical building smoke defenses. A ball-park estimate for the room of origin smoke production and characteristics is usually sufficient to examine the rest of the building.

Room interior design classifications described in Chapter 8 can be used to identify default values for level 1 smoke generation. Defined model room conditions can be a basis for a small group of time–smoke generation curves. Computer analysis and thoughtful judgment can be integrated to document the standardized smoke characteristics. Selection of a default value enables one to make initial level 1 Sm curve approximations. Computer scenarios can be used with level 2 estimates to describe the design fire smoke characteristics.

16.15 Computer smoke modeling

Although one may envision smoke movement easily, the sense of proportion for time durations to untenable conditions is somewhat more difficult to develop. Increased experience and the use of smoke movement computer programs provide greater confidence in estimating building performance. The cost for this greater confidence is the time to apply the programs to a building and to translate the outputs into probabilistic Sm curves.

A series of masters theses were produced at Worcester Polytechnic Institute that studied smoke theory and applications to building analysis. Davoodi [2] examined calculation of extinction coefficients for compartment fires and Chutoransky [3] developed an insight into estimating smoke movement phenomena with a concept for generating and tracking smoke. Ghosh [4] and Nadeau and Wojcik [5] demonstrated the suitability of these techniques by using computer programs for smoke movement and egress to analyze building performances. Dolph [6] examined uncertainty and application variability for computer-based Sm curves in greater detail. Collectively these theses demonstrate the feasibility of translating deterministic computer outputs into probabilistic performance curves.

The major advantage of using computer analyses for smoke movement is having the ability to calculate and graph the visibility of target spaces over time. Varying the scenario by opening doors or windows or by changing other variables provides a sense of proportion for smoke behavior. There are two major disadvantages with computer analyses. The first is the time needed to analyze scenarios. Several working days may be needed to do basic analyses for enough variations to understand the building's smoke performance. The second disadvantage may also be viewed as a distinct advantage and strength. To translate deterministic results into probabilistic degree-of-belief calibrations, the user must understand the computer program and application well enough to take the uncertainties and limitations into account. Experience in making calibration estimates increases professional competence.

Having completed the deterministic analysis, a graph such as in Figure 16.1 can be drawn to represent the time during which the target space will be tenable. There is still a window of uncertainty when using computer programs, although one may have a greater confidence in judgments. Whatever the method of analysis, the graph represents an individual's judgment, understanding, and degree of belief in the behavior.

16.16 Building analysis

The goal of a building analysis is to understand building behavior and tell a story about what may be expected during a building fire. Chapter 4 describes a four-story apartment building. Although one may recognize the inherent problems associated with the building, the fire scenario will be continued to illustrate a level 1 Sm curve construction.

Example 16.1

Construct level 1 Sm curves for the fire scenario of Chapter 4.

Solution

The following information is obtained using Sections 16.12 and 16.13. Figure 16.11 shows the floor plan with apartment 310 as the room of fire origin. Table 16.2 identifies the target spaces.

A visibility of 15 ft through diffuse smoke was selected as the criterion for describing tenability of the space. This short distance is based on occupants being very familiar with the building. On the other hand, this value may be unreasonably low because the occupants are elderly, move slowly, and are likely to be more frightened of heavy smoke when they cannot see to an exit.

Figure 16.12 shows the design fire for smoke volume. The bedroom window will break a few minutes into the fire. One could change the smoke curve to incorporate that action. However, the group of smoke curves in Figure 16.12 are based on three different opening sizes. Rather than specifying a single fire, we will interpolate and select what appear to be the most appropriate values for different vent openings.

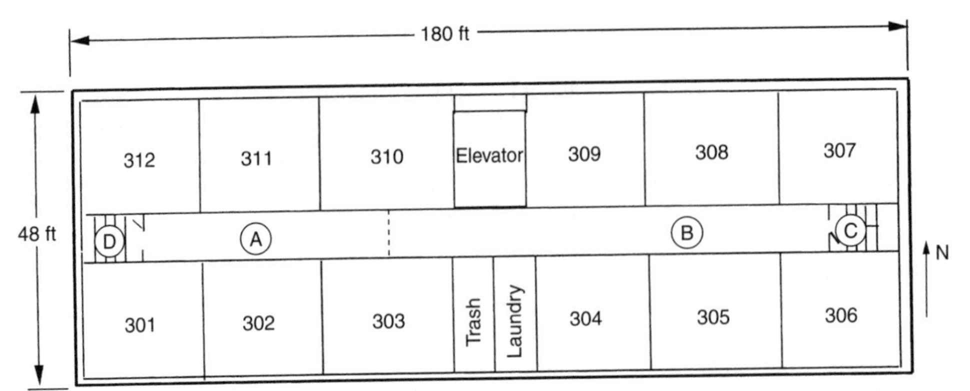

Figure 16.11

Table 16.2

Target space	Description
A	Third-floor corridor from room 310 to the west stairwell
B	Third-floor corridor from room 310 to the east stairwell
C	East stairwell from third to fourth floor
D	West stairwell from third to fourth floor
E	Fourth-floor corridor
F	East stairwell from third to second floor
G	West stairwell from third to second floor
H	Second-floor corridor

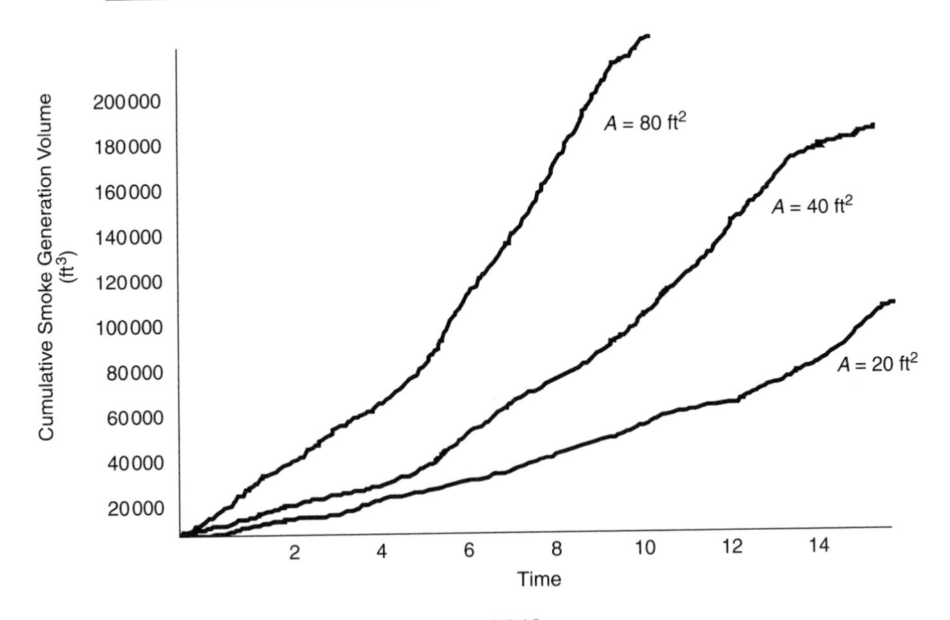

Figure 16.12

A computer analysis for this building would produce scenario information such as time durations, smoke layer temperatures and dimensions, and optical densities. A level 2 analysis would utilize the computer output to establish a spread sheet of information from which to establish subjective probability estimates and construct the Sm curve. Although confidence in these performance curves may be greater than for level 1 curves, the cost in professional time and expenses may be greater than its value. A level 1 analysis may be sufficient to "tell the story" and give a sense of proportion for building expectations involving different states of door and window openings.

A level 1 analysis focuses on the building. The primary attention is directed to potential smoke movement paths and the status of doors, windows, and large ducts. One examines the potential

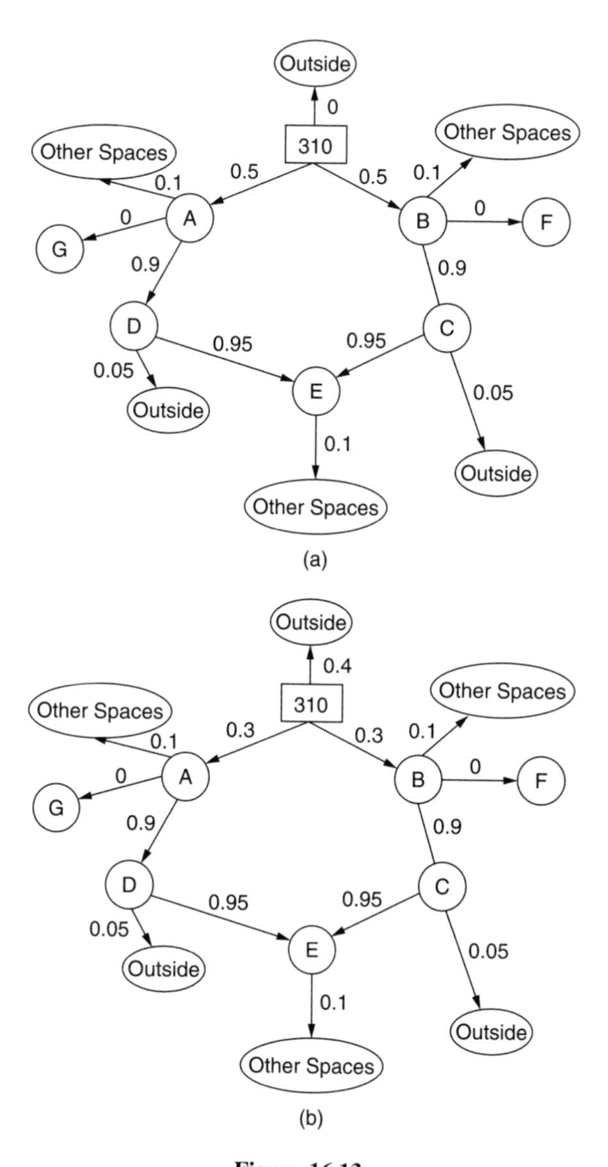

Figure 16.13

distribution of smoke and the forces that affect movement along the major paths. This fundamental information may be augmented by adjustments for room conditions, air pocket reservoirs that collect smoke, and effects such as air entrainment, heat losses, dilution, and deposition. The estimate of the time a target space will remain tenable uses all relevant information within the time frame available for study. In this example, no smoke management measures have been used, so the Sm curve will largely reflect the status of doors and windows with the design fire.

Figure 16.13(a) shows the network of interconnected spaces that distribute the smoke for an open corridor door to room 310 with no outside openings. Figure 16.13(b) shows the same network with a bedroom window broken.

The time steps for smoke volume distributions and visibility in target spaces may be shown on a spreadsheet. As an alternative, we show the results in Figure 16.14. This uses only the design fire and smoke tenability segments of Figure 4.7. The smoke tenability cells of Figure 16.15 would probably use descriptive terms (e.g., high, moderate, low) for a level 1 estimate and probability numbers for a level 2 estimate.

Figure 16.16 shows level 1 Sm curves for several target rooms. Here are the major factors that influenced the selection of calibration descriptors for each time step:

- The door to room 310 remains open.
- Links to the fire doors at each stairwell are unlikely to fuse. These doors are likely to remain open for a time. There are no barriers to smoke propagation along the paths.
- The wall segment from the door head to the ceiling is 1 ft deep. This will delay smoke movement briefly at the third- and fourth-floor barriers between the corridors and the exits.
- Distances for smoke travel through corridors and exits are short.
- No active smoke management systems have been installed.
- Negligible wind conditions exist.

Example 16.2

Construct level 1 Sm curves for Example 16.1 if the third-floor fire door to target space C is closed.

		EVENT	TIME (minutes)															
				2		4		6		8		10		12		14		
Building Performance	1	Design Fire																
	a	Size (HRR)																
	b	Size (Area or Rooms)																
	c	Cumulative Smoke Generation (100 ft³)		10		20		60		90		140						
	7	Smoke Tenability (Sm)																
	a	Room 310			Bad													
	b	3rd Floor Corridor A						OK	M	Bad								
	c	3rd Floor Corridor B						OK	M	Bad								
	d	East Stairwell C							OK		Bad							
	e	West Stairwell C							OK		Bad							
	f	4th Floor Corridor E								OK	M	M	Bad					

Figure 16.14

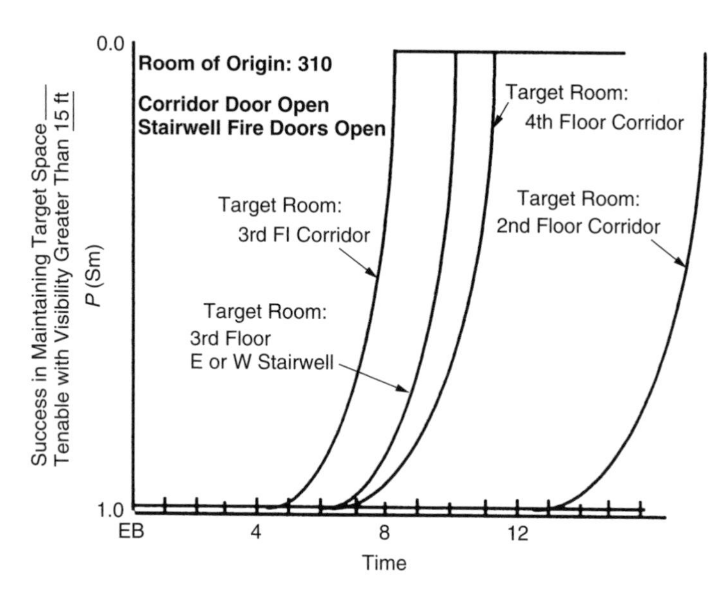

Figure 16.15

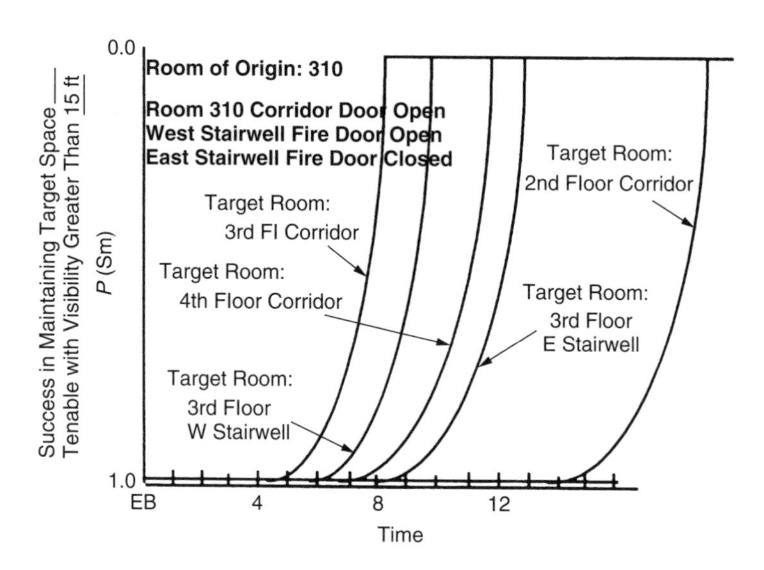

Figure 16.16

Solution

Smoke leakage will occur into target space C, but not enough to make it untenable. However, the smoke that enters target space D will move to the fourth floor and down the corridor. The smoke will have lost much of its buoyancy when it reaches stairwell C. As smoke continues to enter the stairwell and does not have a way to vent, it will move down to contaminate the east stairwell (target space C). Figure 16.16 shows level 1 Sm curves for this scenario.

16.17 Closure

Smoke tenability must be identified in a way that provides a recognizable measure that allows one to evaluate the building performance. Currently, we define tenability for life safety as a visibility level in a target space. However, the process is structured to be independent of the measurement system. One can use any defined measure of smoke, toxicity, or heat.

Life safety depends on more than levels of visibility, even though smoky conditions do contribute to anxiety and the danger of disorientation. State-of-the-art fire science cannot yet identify lethal or hazardous levels of toxicity for airborne products of combustion. Because loss of visibility due to smoke normally precedes physical impairment from toxic gases, an analysis that uses visibility as the tenability criterion is conservative.

These simple examples illustrate a process for evaluating smoke tenability for target rooms. Large buildings with complex air path movements are more difficult to analyze than small buildings. However, the process enables one to evaluate smoke tenability for building spaces and describe the results to others.

A technique of construction project management uses most pessimistic, most optimistic, and most likely estimates for time durations. This technique can be adapted to express a range of time for which the target room may be tenable. It provides a better understanding of a building's smoke performance.

References

1. Skiena, S. S. *The Algorithm Design Manual.* Springer-Verlag, 1998.
2. Davoodi, H. Time dependent calculation of visibility in compartment fires. MSc thesis in fire protection engineering, WPI, 1987.
3. Chutoransky, R. An engineering method for predicting smoke movement in buildings. MSc thesis in fire protection engineering, WPI, 1988.
4. Ghosh, G. Lifesafety analysis in the building firesafety engineering method. MSc thesis in fire protection engineering, WPI, 1994
5. Nadeau, D. R. and Wojcik, M. J. The Influence of building codes on fire performance. MSc thesis in fire protection engineering, WPI, 1998.
6. Dolph, B. J. Quantitative target space smoke curve analysis and procedures. MSc thesis in fire protection engineering, WPI, 1995.

17 BUILDING PERFORMANCE

17.1 Introduction

Buildings are constructed to perform a function, and each building is unique. Each building has associated needs, uses, resources, and constraints. The ability to use existing or new buildings or to move to other buildings is central to a free market. Decisions are judgmental involving costs, functionality, safety, convenience, and personal preferences. Major determinants in the decision involve functional and economic comparisons for day-to-day operations.

The primary attention in building design and use focuses on routine activities and operations. Safety and functional conditions for structural, heating and cooling, light, and power are integral to a design. Abnormal conditions that rarely occur include earthquakes, high winds, and fire. Codes, laws, insurance, and professional engineering address natural disasters such earthquakes or hurricanes. Fire is a ubiquitous but rare event for a specific building. Laws, codes, and insurance provide guidance to make a building safer from fire.

Most people do not think about fire design in buildings unless they have experienced one. People are happy to insist on safety as long as others pay for it, so fire safety becomes a legislated solution rather than a designed solution. The primary concern is to meet the code well enough to get necessary building and occupancy permits. Fire performance is not a routine design issue, because it is assumed that the code treats this adequately. It has rarely been considered by owners that fire should be a routine design consideration, because fire is an abnormal load that no one plans to have in their building. It is often perceived that unnecessary code requirements increase costs, and those resources can be better spent on improving other quality features. Why are they necessary if the building has never had a fire?

An awareness of fire as a design function rather than a code issue is slowly attracting attention. Economics is the primary incentive for this transition. The question of need is being considered differently by enterprises whose survival depends on avoiding catastrophic losses. Awareness of data and equipment that is vital to operational survival is gaining greater notice. Formerly unforeseen accidental fires, civil disobedience, terrorism, and disgruntled employees are raising different questions about safety and security. Risks, costs, and performance are closely intertwined.

Thinking about fire in terms of performance rather than as a code issue gives a different perspective on building functionality. Before addressing the issues of risk, cost, and decision making, we shall pause briefly to review some concepts about performance analysis.

Building Fire Performance Analysis R. W. Fitzgerald
 ISBN: 0-470-86326-9 (HB)

17.2 Scenario analysis

Performance evaluations develop an understanding of the way a building will work in a fire. Although risk is often a focus of attention, it is the building's fire behavior that establishes the understanding needed for making good decisions. Fire growth and propagation, as defined by the L curve, dominate performance.

We acquire knowledge of a building's behavior from scenario analyses. The selection of a design fire and its location becomes the most important professional decision in a building evaluation. The building layout, its contents and operations, and the passive fire defenses all have a role in defining design fire characteristics.

There are only two types of building: sprinklered and nonsprinklered. These conditions identify the active fire intervention methods. The effectiveness of the dominant fire intervention method provides a major part of the knowledge about the building.

The role of the design fire is to test the effectiveness of active fire intervention defenses in limiting the fire size and damage. Design fire location is as important in a scenario analysis as the other fire characteristics.

17.3 Building performance summary

The goal of a building evaluation is to understand the building well enough to describe clearly its performance and the associated risk characterizations. Communication is the ability to tell a story about the building and what is likely to occur with different fire conditions or scenarios. The story must be credible and relate to expectations for the specific building of interest. Evaluations are never averaged or blended. However an envelope of behavior that bounds the most likely expectations may help to tell the story. Performance descriptions are based on specific design fire conditions and details of design and installation.

Quality is discriminated by comparing performance differences for details of design and installation. Relative performance relates to site-specific features of the application. The ability to select appropriate details and conditions is a skill. A complete building performance analysis may be summarized as follows.

FIRE DEFENSE INTERVENTIONS

Buildings are either sprinklered or nonsprinklered. Partially sprinklered building evaluations combine performance expectations with different scenarios.

DESIGN FIRE

The design fire is the most important part of a performance analysis. The design fire becomes the load in a load and resistance analysis that is the basis of technical evaluations. Fire defense functions establish the combustion product selection.

BARRIER EFFECTIVENESS AND FIRE DEPARTMENT SUPPRESSION

Flame movement analysis for a nonsprinklered building is dominated by the effectiveness of barriers to contain the fire. Barrier fire endurance (fire rating) is not the significant measure. Rather, it is the condition of the doors, windows, and other openings that affect the ease of fire propagation into adjacent rooms. The design fire for fire department suppression analysis usually includes a room of origin that is sensitive to multiple-room involvement.

Estimating the time to first water application involves building design features and community fire resources. Detection and fire department notification are exclusively the building's responsibility. Response relates to fire station location and fire department procedures. Suppression resources of staffing, equipment, and water supply are predominantly community resource functions. The M curve analysis shows how the site and architectural features help or hinder the local community resources.

AUTOMATIC SPRINKLER SUPPRESSION

Flame movement analysis for a sprinklered building examines the quality and reliability of the sprinkler system design and installation. The design fire is an important part of every performance evaluation. The system reliability provides the initial water flow. Its design effectiveness provides enough water, discharges it in an appropriate form, and distributes it over the area.

BUILDING BEHAVIOR

The flame movement analysis provides a way to understand the expected building behavior and extent of flame/heat damage. The building fire behavior is different from the design fire. A design fire defines the dynamic products of combustion that test the building's fire defenses. The building fire behavior combines the design fire characteristics with the fire defenses to describe fire performance.

STRUCTURAL FAILURE

Design fire characteristics define the environment that affects the structural system. The structural frame analysis develops an understanding of deflection or collapse expectations. Results from a structural frame analysis may be used as a stand-alone description or they may provide feedback for the flame movement analysis. Structural movements can change barrier behavior, and the differences may be important to a flame movement analysis. When the structural analysis shows deformations that affect barrier effectiveness, a reexamination of the flame movement analysis may be appropriate.

SMOKE TENABILITY

After the flame movement and structural frame analyses have been completed, the smoke tenability is evaluated. The design fire defines the smoke load for evaluating the building's ability to maintain target space tenability. The time that target spaces remain tenable provides the building knowledge.

SYSTEMS BEHAVIOR

Although the normal goal is to give a systems understanding of the complete building performance, individual components may be examined as a part of a more narrow study.

ONE SIZE DOES NOT FIT ALL

The type of analysis is tailored to the needs of the problem. Three different levels of evaluation for building evaluations all use the same framework and thought process. The differences relate to time and information levels. Level 1 provides a rapid macrounderstanding of performance. It is a rapid analysis based on knowledge, observations, and estimates. Level 2 analyses produce a detailed understanding of building performance. These analyses follow a level 1 evaluation and use network diagrams to direct the process. One uses a level 3 evaluation to understand the building's sensitivity to uncertainty and variability. These sensitivity analyses enable if-then questions to be addressed.

Risk characterizations developed after performance evaluations provide an understanding of building fire behavior. These risk characterizations provide information for a variety of applications. Among these, the development of comprehensive risk management programs is the most versatile.

17.4 Information organization

The building originally described in Chapter 4 has been used throughout this book to illustrate performance evaluations. Time-based integration of component performances enables one to understand the building's behavior. This provides an informational base to develop risk characterizations for the building. Risk characterizations for an existing building may be presented to an owner or occupant in a brief stand-alone report. However, more often risk characterizations become a starting point to develop risk management programs or identify design alternatives to change the risk.

Figures 17.1(a), (b), and (c) show the building plans and elevation. Additional information is available in Chapter 4. Although the building is small and does not offer a broad scope of challenges, it does provide a setting to organize and portray performance information. While the building's fire weaknesses may be obvious at this point, these features often exist unrecognized in normal use. This example is a composite of features from case studies and actual buildings.

Example 17.1

Describe the performance expectations of the building in Figure 17.1. Organize the discussion around the analytical framework.

Solution

The building features and their effect on performance are outlined below.

Building fire performance

I Flame/heat movement (L curve)

A Fire growth potential of the room interior designs (I curve)

1. Apartments and rooms are relatively small.
2. Ventilation is sufficient for fire growth to FRI. After FRI, windows in bedrooms and at the balcony will break. The absence of corridor door closers will provide an additional opening if occupants do not close corridor doors when leaving the apartment.
3. There is little or no control over contents of the occupants; furniture is likely to be constructed of high-HRR materials that will cause FRI easily. The fire growth hazard potential classification for the room interior design will be in the range *very high to high* for all apartments.
4. Combustible exterior cladding can propagate flames over the outside building surface.
5. Exposed rafters and wood attic framing will allow fire to grow in that space. An unknown (although unlikely) amount of storage is in the attic. Nevertheless, the exposed wood surfaces suggest that a high fire growth potential is appropriate for the space.

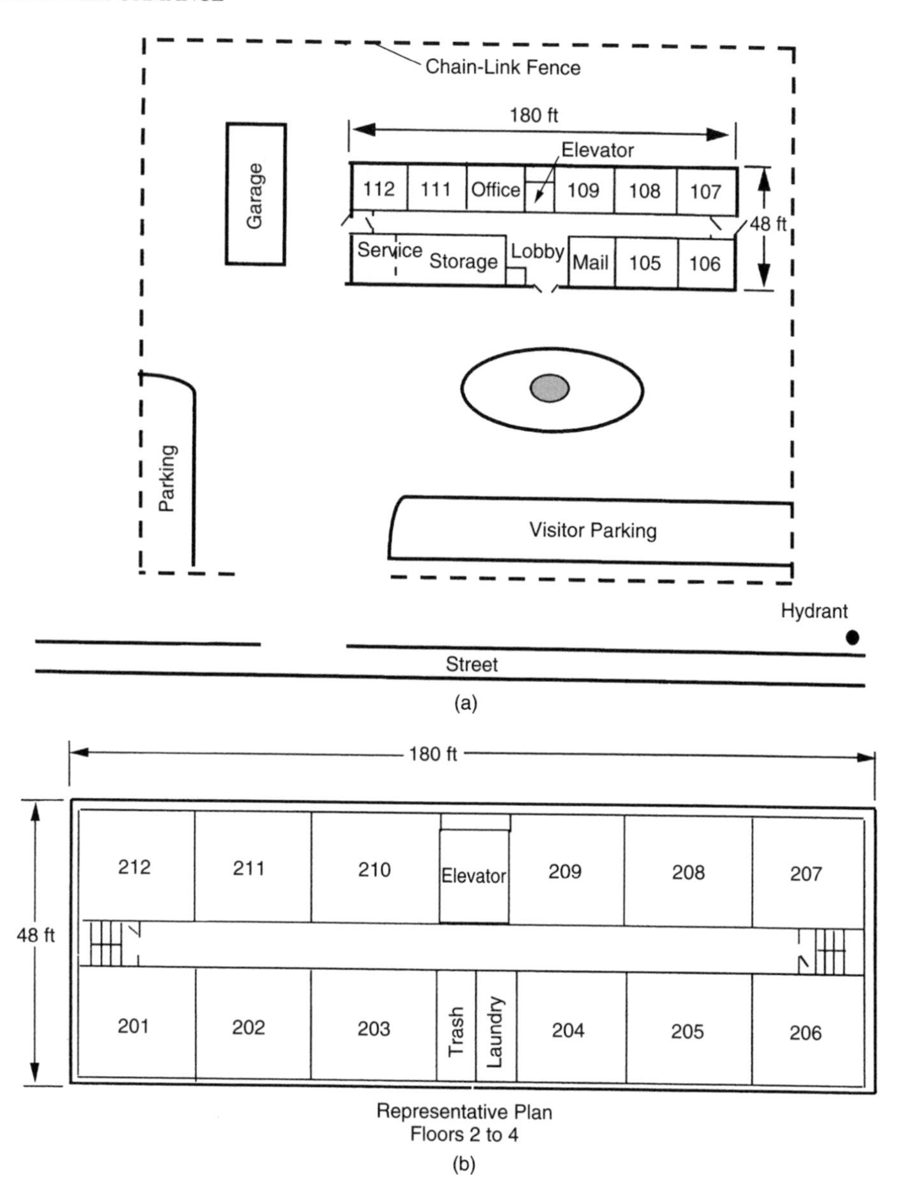

Figure 17.1

B Automatic suppression (A curve)

1. No sprinkler system is provided.

C Fire department extinguishment (M curve)

1. Detection.
   a. Single-station smoke detectors are in each apartment.
   b. Human detection in an apartment will occur when the occupant is awake, alert, able, and present to recognize a fire. The fire size at detection and the detection reliability vary with occupant conditions.

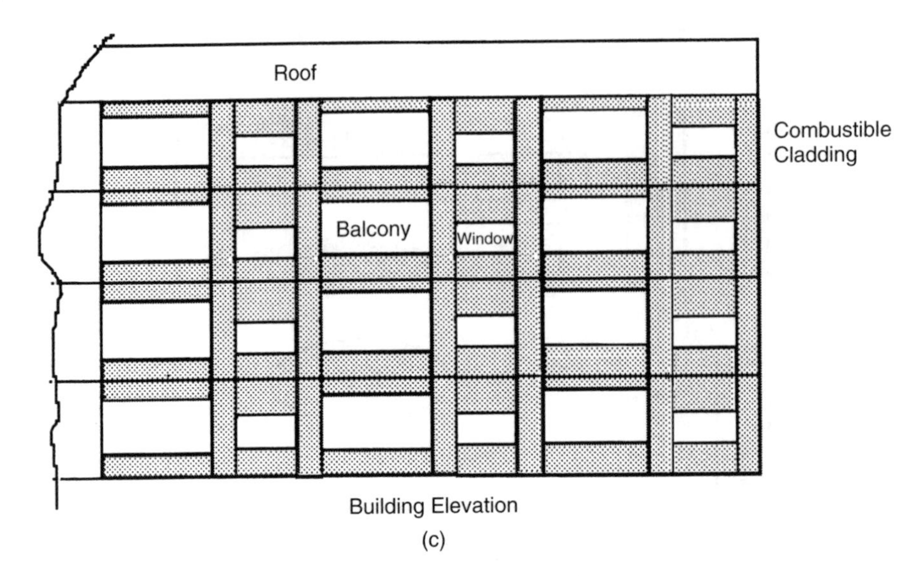

Figure 17.1 *(continued)*

c. Detection by an individual outside the apartment of origin will not occur until the fire has grown substantially.

2. Fire department notification.
   a. A human provides the link to notify the fire department.
   b. The individual must locate a telephone, call the correct number, and deliver the message correctly. Many distractions (e.g., attempt to extinguish, alert others, flee) will delay the event of notification (MN curve) or make it more unreliable. Early fire department notification is unlikely and unreliable.
3. Design fires.
   a. *Apartments*: Two types of fire can occur in an apartment:
      - Small, slow burning in materials separated from others that allow growth.
      - Established burning in furniture that has a high HRR.

      This condition is used for the design fire for this building evaluation.
   b. *Service Rooms*: A fire can occur in mechanical, trash, and laundry rooms:
      - Rooms are relatively small.
      - Ventilation will be adequate for early fire growth, but perhaps restricted for larger fires.
      - Combustible contents will vary. Select the very high fire growth classification for the trash and laundry rooms and the high classification for the mechanical room.
4. Barrier effectiveness.
   a. Fire resistance of interior walls and floor ceiling barriers should be very adequate. No concerns about fire endurance ratings.
   b. The absence of door closers require that an occupant must consciously and physically close the door on leaving. This is very uncertain for normal usage and in an emergency situation.
   c. Fire doors at the ends of the corridors are held open by fusible links. Expected delay in operation will allow heat and smoke to move into the exits. If wedges maintain an open door, the resistance is lost.
   d. Windows to the exterior walls are weak. FRI would probably break windows to the bedroom and balcony.

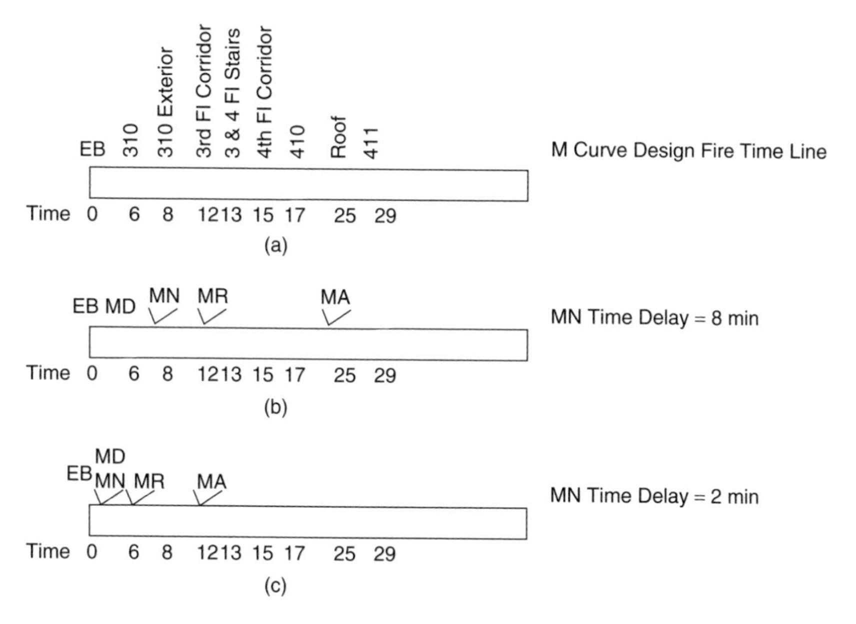

Figure 17.2

e. No spandrel fire break exists at the bedroom windows. The balcony does provide a separation to impede external fire spread at those locations.

f. The barrier at the soffit of the roof overhang has poor resistance. At locations of roof vents, the barrier will be breached very quickly. In sections where the plywood is attacked, expect about 20 min before a massive failure.

5. Design fire for fire department extinguishment.
   a. The design fire time line in Figure 17.2(a) seems reasonable for fires that start in apartments. Exterior fire propagation is very likely. The attic is likely to become involved in most apartment fires.
   b. Fires in trash and laundry rooms will reach FRI quickly, but will not cause a large fire area. Lack of exterior windows will avoid exterior flame propagation.
   c. Fires in the mechanical and storage rooms can be fast and large. They will cause some difficulty in extinguishment. However, they will not be as demanding on the fire ground commander and the resources as an apartment design fire with exterior flame spread and attic involvement.
6. Fire department response.
   a. After notification, responding fire forces should be in attendance within approximately 4 min. Response is two engines, one ladder truck, and a battalion chief with a total staffing of 12 people.
   b. Chain-link fence and landscaping will delay placing vehicles at the back of the building. Exterior access to the back of the building is difficult.
7. Initial water application.
   a. If the time for detection and fire department notification involves a delay (from EB) of 8 min, when the first-in company arrives the fire will involve apartment 310, the third-floor corridor, and the exterior cladding in the back of the building. Within the next six minutes, apartment 410 will become involved, major parts of the back exterior wall in flames, and fire would be close to or already in the attic. The stairs and fourth-floor corridor would be involved. The time line of Figure 17.2(b) illustrates this condition.

b. If the time for detection and notification involves a delay of only 2 min from EB, only apartment 310 will be involved. Within the next six minutes, the exterior cladding in the back will begin to be involved and the third-floor corridor will be involved. It is a very different situation for the small fire-fighting force. The time line of Figure 17.2(c) shows this condition.
c. The incident commander will note that the fire is large at arrival and there is a major threat for occupant life safety. The officer may or may not know about the exterior flame spread at the back until more investigation. A rapid decision must be made on allocating the responding forces between fire suppression and search and rescue. Often approximately one-third of the first alarm is allocated to fire suppression, although this can vary substantially. Multiple alarms will be sounded immediately. The evaluation examines who will respond and how long it will take.
d. The single hydrant requires a long supply lay around the chain-link fence into the parking area in front of the building. This will involve a delay in supplying enough water for this fire. In winter, with cold or snowy weather, digging out the hydrant or hydrant freezing will require a search for alternate sources of water. Lack of hydrants close to the building could cause major difficulties.
e. The fire is on the third floor. No standpipes are available. Fire suppression forces must stretch a line from the pumper to the position of fire attack. This requires about 300 ft per line plus extra for slack. Three building turns and a long charged line make movement very staffing intensive. External attack from the rear is difficult because of access difficulties. Internal attack is difficult because of corridor fire involvement and visibility difficulties in finding the fire. At the time of initial water application, apartment 410 will be involved, and the attic is likely to be involved. The fire doors at the exits may have some influence on fire extension, although their effectiveness for this scenario is uncertain.
f. A design fire on the first or second floors would cause greater fire damage and spread. Also, it would smoke-log more corridors, making fire suppression movement difficult.
g. In this scenario for apartment 310, a time duration of 9 min is estimated to apply first water to an internal attack for the conditions of an 8 min delay in notification from EB. A time duration of 6 min is estimated if the delay in notification is reduced to 2 min from EB. Figure 17.2(c) shows the time lines for these conditions.

8. Fire department extinguishment.
   a. When the conditions at the time of first water application are those of Figure 17.2(b), there is little chance that the fire can be extinguished before losing much of the top two floors. Roof collapse is almost certain. Progressive floor collapse can occur. There is a small likelihood of losing the entire building.
   b. When the conditions at the time of first water application are based on the earlier fire department notification of Figure 17.2(c), the likelihood of extinguishment is greatly improved. An internal attack can push the fire out of apartment 310. An external hose line can extinguish external spread. Eventually the roof can be opened and fires in the attic space can be extinguished. Fire damage can be expected to involve all of apartment 310. Some or all of apartment 410, exterior siding, and some of the attic space.
9. The M curve.
   a. Figure 17.3 shows M curve for several fire scenarios.

II Structural frame behavior (Fr curve)

1. Design fire.
   a. Interior apartment and other rooms have a normal fire development. Ventilation will be adequate. The time for a fully developed fire will be about 8–12 min.

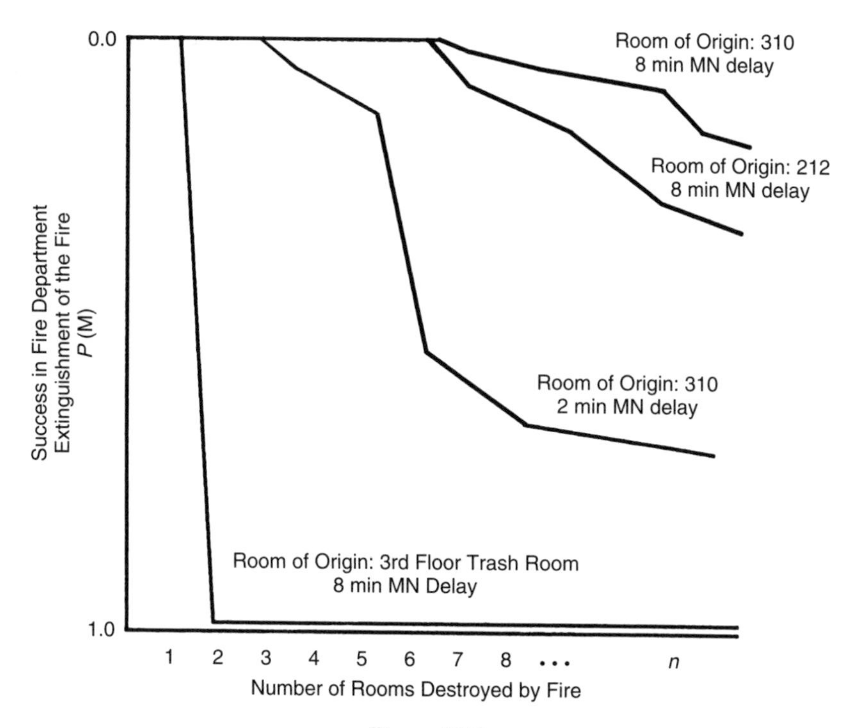

Figure 17.3

b. Exposed wood surfaces in the attic augment the little other fuel that is likely to be in this space. However, the M curve design fire shows that an EB and fire development can easily occur under conditions of an apartment fire and delayed fire department notification.

2. Structural performance.
   a. The structural frame for the interior rooms is protected by gypsum board. Few, if any holes and penetrations are expected. The gypsum board fire protection should be adequate to prevent ignition of the structural frame.
   b. The vulnerability of roof construction is not from barrier deterioration in the apartments below, but from the exterior flame spread that breaches the overhang soffit. The roof system is truss construction, and after an established burning in the attic space, collapse may be expected to occur within about 20 min.
   c. Attic collapse will probably cause failure of the fourth-floor ceiling joists, and this can trigger a progressive collapse if other floors are weakened by the fire. Otherwise, we may expect only a collapse of the roof to the fourth-floor supports.
   d. Figure 17.4 shows an Fr curve for the walls, floors, and roof trusses of the attic. A dashed line is used to give a global time Fr curve performance that is delayed until EB occurs in the space.

III Building performance for smoke tenability

1. Conditions.
   a. The status of the apartment corridor doors and the fire doors at the stairwells will have an effect on the time the corridor and stairwells remain tenable.
   b. After FRI a large volume of dense smoke is available to fill the corridors and stairs. Even fire doors have cracks and provide only a short delay in smoke filling.

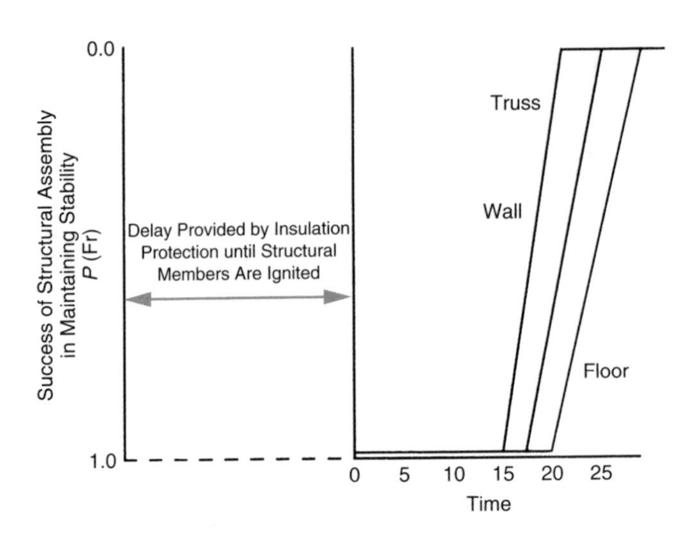

Figure 17.4

c. Occupants who may try to evacuate the building will open doors. Fire fighters may keep stairwell doors open to pass hoses through. Doors can be expected to be open during all or part of the time.

2. Figure 17.5(a) and (b) show smoke tenability curves for different scenarios.

This level 1 analysis shows expected performance for the system components. The principal value is in the understanding that comes from the evaluation because the process enables one to recognize the important factors. Rarely does a single element dominate performance. Rather, it is the synergistic influence of a combination of events that often leads to the system outcome. The understanding comes from scenario evaluations, and this knowledge can be portrayed in a way that enables the effectiveness of potential changes to be compared.

One may observe that a disaster will not occur when a small fire is put out by an occupant. Nor will it occur when the fuel packages are separated and the fire department is notified in a timely manner. However, these scenarios would not expose the weaknesses of this building.

17.5 Alternative comparisons

Section 17.4 illustrates a general building performance description. One would expect this story to enable most individuals to understand the weaknesses of this building and what may be expected. However, some applications may involve decisions for component parts. Example 17.2 illustrates a way to structure a more localized decision.

Example 17.2

During an M curve analysis for a proposed new building, detector and notification alternatives were discussed. A question was raised concerning the cost and quality of the detection and notification system and its value to building protection. The following preliminary analysis is a response to this query:

1. A scenario was developed to identify a room of origin and reasonable expectations of potential fire propagation. A design fire on which to base an M curve analysis was identified and described in Figure 17.6(a).

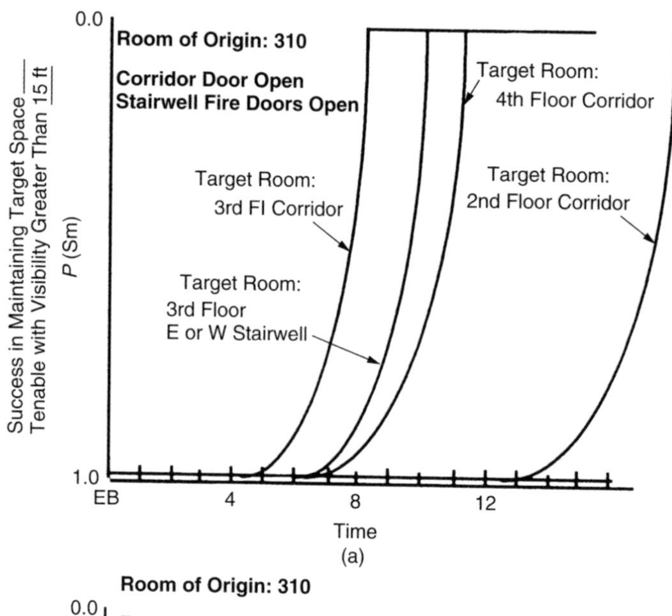

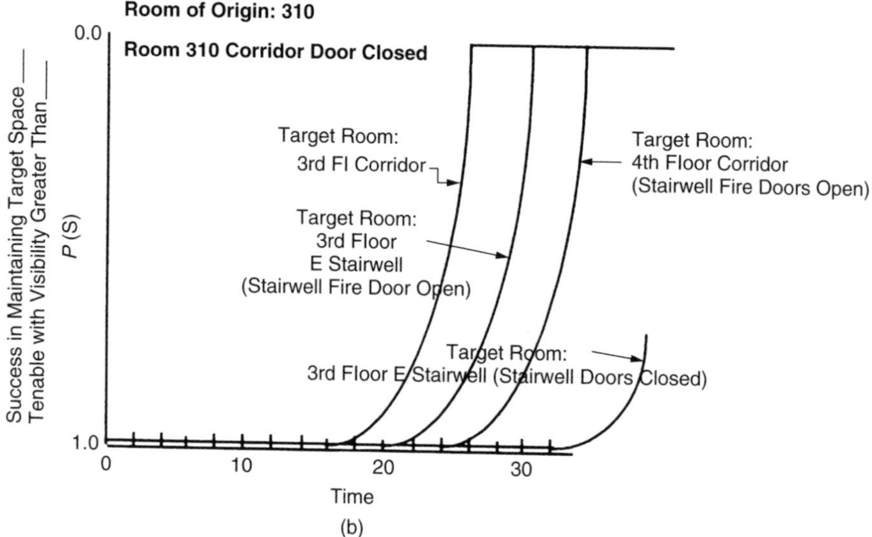

Figure 17.5

2. Two possible detection and notification designs are considered. Both comply with all applicable codes and standards.
   - System A is expected to detect the fire size in Figure 17.6(b) with a reliability of 96% and to notify the fire department at the fire size shown with a reliability of 98%. The annunciator will give clear directions for the fire location.
   - System B is expected to detect the fire size in Figure 17.6(c) with a reliability of 96% and to notify the fire department at the fire size shown with a reliability of 92%. The annunciator will give general directions for the fire location.
3. Fire department response and the time to establish supply water will be the same for each scenario. The time to find the fire and the environmental conditions of visibility and heat will

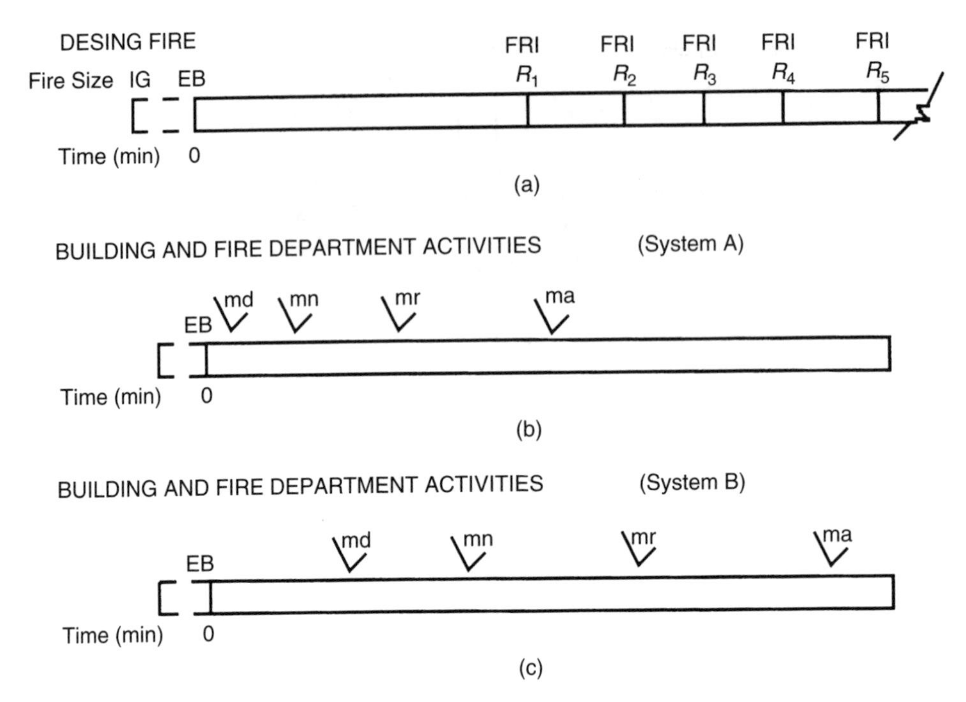

Figure 17.6

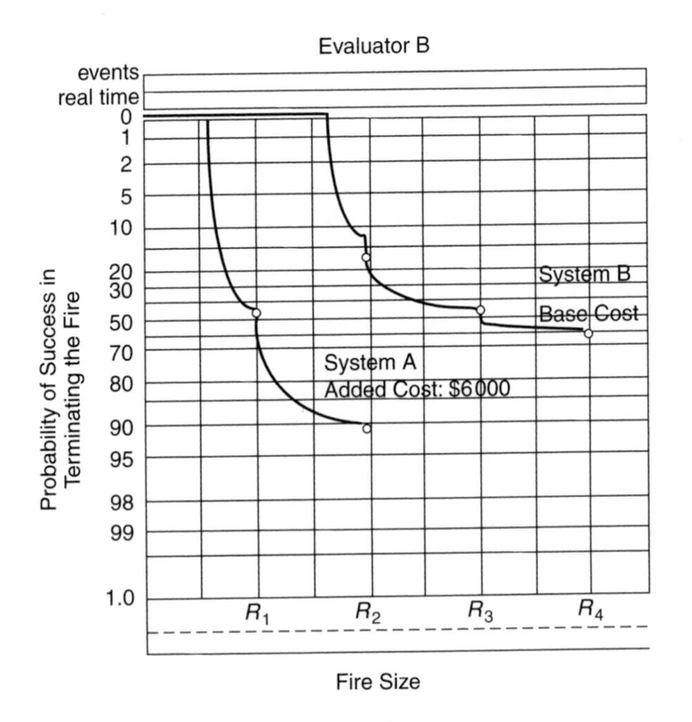

Figure 17.7

be greater with system B. This will result in a somewhat greater time duration to apply first water to the fire. The fire sizes at first water application are shown in Figure 17.6.

4. The following estimates were made with respect to fire extinguishment:
   - System A: the first alarm response should be adequate to control and extinguish a fire of two rooms.
   - System B: the first alarm response will have difficulty with a fire of four rooms. A second alarm response is expected, and possibly even a third alarm will be needed to control and extinguish the fire.
5. A building analysis examined M curve events for each scenario. The results are shown by the M curves in Figure 17.7.

System B is selected as a basis for comparison. The initial capital cost to install the higher-quality system A is \$6000 and the annual operating maintenance cost is \$200. The costs and effectiveness of the two systems are shown in Figure 17.7. The owner can make a decision within the context of the building, its operations, and the management conditions.

18 RISK CHARACTERIZATIONS

18.1 Introduction

The building evaluation process provides an understanding of what to expect regarding fire damage, smoke tenability, and structural behavior. Although an analysis may examine only specific scenarios, the acquired knowledge enables one to understand a building's performance regardless of where a fire may start. Often pockets of comparatively moderate fire conditions may pose a greater threat to certain contents, operations, or people than more severe fire conditions at other locations.

Risk characterizations identify what can be expected for the building's occupants, contents, and operations from a fire in any location. The merging of knowledge of the building's fire behavior with its functional operations provides an insight into understanding the way the building will work during a fire. Risk characterizations tell the story in a way that users and decision makers can understand and take appropriate actions.

This chapter describes risks and ways to represent them for a building. The base of knowledge is a functional analysis, and one describes risks in a way that the layman can understand and use to make better decisions. As with performance evaluations, three levels of risk characterization can tailor available time to the needs of the problem. Level 2 or level 3 characterizations augment the level 1 knowledge and understanding.

18.2 Risk characterizations

Buildings are built to perform a function. That function may be to house its occupants, conduct office management operations, sell merchandise, educate students, store materials, feed and entertain people, or provide space and facilities for a variety of other activities. Normally a combination of activities occur in the same building.

Some level of risk is present in all activities related to and independent of hostile fire conditions. There can be a risk of too much exposure to the sun or too little; eating too much food or too little; driving too fast or too slow. One can extend this to recognize that the risk of too many safety requirements may make a project uneconomical to construct or operate, whereas too few (or the wrong type) of safety measures provide unsafe conditions. A window of tolerable conditions exists with almost all endeavors.

Building Fire Performance Analysis R. W. Fitzgerald
 ISBN: 0-470-86326-9 (HB)

A risk characterization identifies what is in danger and gives a description of the expected threat. As with a performance analysis, communication with others not in the fire business involves telling a story about what can happen and why. The goal is to provide a clear understanding of the type and nature of threat to the occupants and operations.

A risk characterization does not identify corrective actions to change the threat, nor does it identify appropriate or acceptable levels of risk. Those are functions of risk management. However, a clear understanding of risks associated with the building's functional operations and fire performance is central to developing an effective risk management program.

The basis for a risk characterization is the building's fire performance. Often one becomes aware of risks during a performance evaluation. Risk characterizations may be presented in a brief stand-alone report. This information is a starting point for risk management programs. Risk characterizations address the following:

1. Human safety
   a. Occupants who are expected to leave the building in a fire emergency
   b. Occupants who are restrained, incapacitated, or unable to leave the building
   c. Occupants who may prefer to remain in the building
   d. Fire fighters conducting emergency operations
2. Property protection
   a. Ordinary contents
   b. Valuable contents
   c. Heritage
3. Operational continuity
   a. Equipment
   b. Information and data storage
   c. Material supplies
   d. Functional spaces
4. Neighboring property
   a. Exposure losses to neighboring structures
   b. Exposure losses to other enterprises within the building
   c. Community losses (e.g., jobs, taxes)
5. Environment
   a. Groundwater contamination
   b. Surface runoff contamination
   c. Air pollution

PART A: HUMAN SAFETY

18.3 Human safety

Life safety is a major concern in building behavior. People may be threatened and react to building fires in a variety of ways. One building may provide superior safety for occupants and fire fighters and another may exhibit very poor qualities. A risk characterization identifies the nature, relative significance, and consequences of the building's performance to fire.

People–building interactions can assume three different roles. One is conventional occupant egress in leaving the building in a fire emergency. A second involves a defend-in-place strategy. This is appropriate for individuals in hospitals, nursing homes, prisons, or other situations where movement is inappropriate or difficult. A third strategy involves moving to a designed area of refuge. Rescue or assistance by emergency personnel can be a part of each option.

The basic principle of a life safety risk assessment is that people and too many harmful combustion products should not occupy the same space at the same time. Target spaces are rooms that may contain individuals on a temporary or semipermanent basis. For example, corridor segments, stairwells, hospital rooms, and prison cells are potential target spaces for human risk characterizations. Tenability is the threshold for "too much" combustion product. One describes tenability functionally in a way that building performance can be estimated and individuals can understand its meaning.

The most useful and common definition of tenability specifies a maximum distance through which an occupant can see in smoke conditions during normal building lighting conditions. These define visibility conditions in which most occupants will stop moving along a travel path and attempt to turn back. Much anxiety may be present, and if one remained in those smoke conditions long enough, the effect of toxic gases would cause serious harm. One cannot estimate the probability of death with this definition, although one would note conditions where the likelihood of survival is poor. The understanding of risk for humans is gained by evaluating the available time an individual may have before experiencing untenable conditions.

Buildings do not provide unlimited time for egress, and prescriptive codes do not identify any available times. Codes specify good practices to help individuals leave a building safely in a fire emergency. The level of risk is indeterminate, safety is not assured, and alternatives cannot be compared.

18.4 Overview for life safety analysis

Occupants have several alternatives during a fire emergency. Figure 18.1 describes actions that may or may not lead to avoidance of death and injury (life safety). Events 0, 1, 3, 17 identify the traditional (successful) evacuation from a building. This may involve either a code-complying means of egress or simply a way out by normal circulation routes.

Events 0, 2, 5 or 6 identify a defend (remain) in place situation. Here the occupant remains in the room and is either safe from the combustion products (event 5) or is not safe (event 6). If the occupant is not safe, rescue may or may not take place, as noted by events 7 and 8. Here "rescue" means by the fire department or by trained nursing or security staff. Rescue by another occupant or passerby is not considered a part of life safety evaluations.

Although designed and effective areas of refuge are rare, in theory they may be appropriate for protecting people in a fire. It is possible in building fires to encounter an accidental area of refuge. That is, an occupant accidentally may reach an area that provides enough protection to enable the individual to survive the fire's duration. The accidental area of refuge that may provide coincidental survival conditions is not considered in a building evaluation. This process does not recognize an area of refuge unless it is a part of a conscious design and becomes the focus of a tenability evaluation.

Paths 0–1–4–9 or 0–1–4–10–11 or 0–1–4–10–12 identify the area-of-refuge situation. If the occupant is not safe, rescue may or may not take place, as noted by events 13, 14 and 15, 16. If rescue is a part of a conscious building performance design, only the fire department or trained building staff are considered. Other occupants may provide rescue aid, but they are not included in an analysis. If a family member is expected to rescue very young, elderly, or disabled individuals, this situation must be identified and those actions become an explicit part of the evaluation. For example, if building occupants are expected to provide rescue aid, and this is ineffective, one should not fault the building performance. However, one may question the criterion that occupants must provide rescue aid.

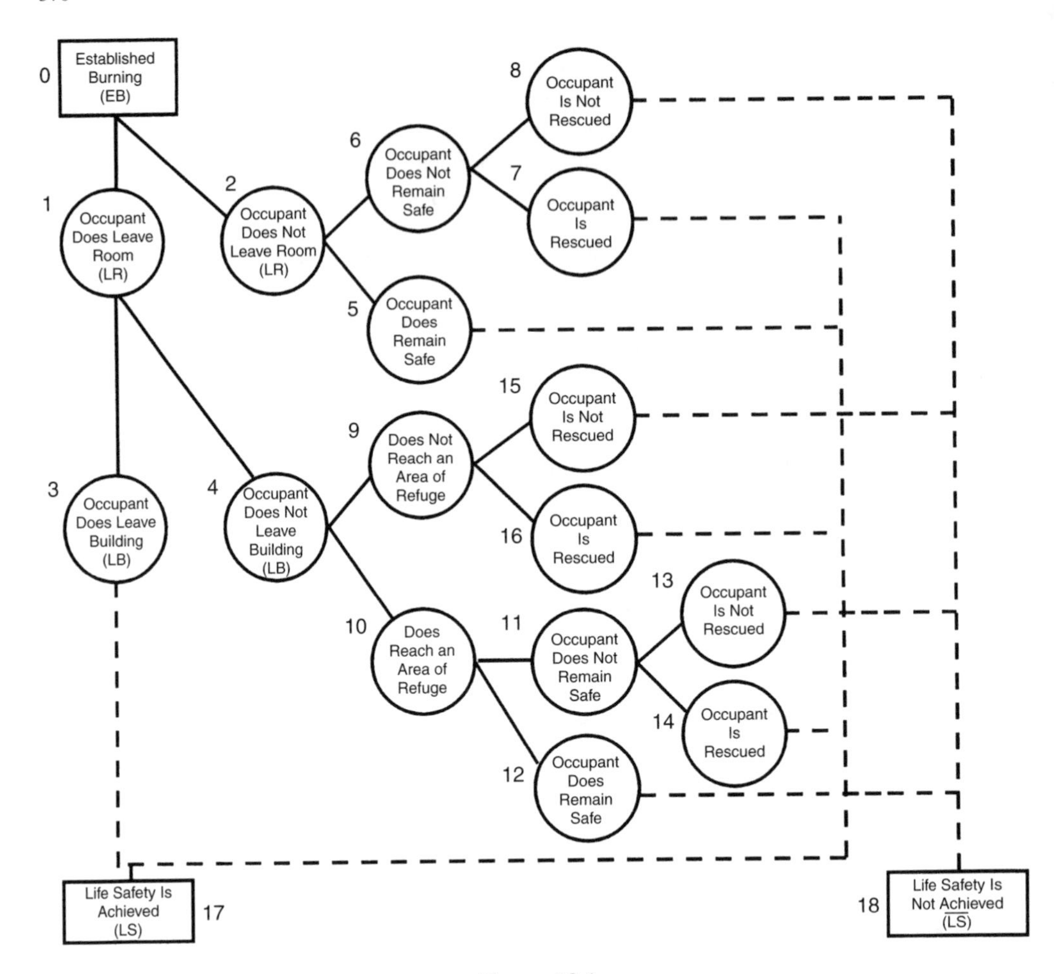

Figure 18.1

18.5 Human reaction to products of combustion

The physiological and psychological effects of building fires on humans can be dreadful. Phillips [1–6] has written on these effects, and this section is adapted from her writings and with her help.

Humans are fragile beings when subjected to products of combustion in a confined space. The major hazards for an occupant of a burning building include the effects of a diminished supply of oxygen, inhalation of smoke and toxic gases, the physical effects of heat and flame, and dangers relating to structural collapse. Although panic is rare if individuals believe a way out exists, anxiety and emotional shock cause added problems.

A continuous, ample supply of oxygen is necessary for human survival. Air contains 21% oxygen. When the oxygen is reduced to about 16%, clear thinking becomes difficult and muscular coordination for skilled movements is lost. At levels less than 14%, judgment becomes faulty and behavior is irrational. At levels less than 8–10%, collapse occurs and death may result within 10–15 min unless reoxygenation is rapid.

Oxygen deprivation is a common hazard in fire. The fire itself competes for the available oxygen, although this is not a significant cause of oxygen deprivation. Instead it is the effect of the products of combustion on the human respiratory system. One of the chief culprits is carbon monoxide. Carbon monoxide is the result of incomplete combustion and is present in all building fires. It is a major cause of fire deaths because the hemoglobin in the red blood cells has a much greater affinity for carbon monoxide than for oxygen. Consequently, carbon monoxide replaces oxygen that the red blood cells normally carry from the lungs to the other parts of the body. The resulting hypoxia (low blood oxygen) can cause death unless prompt treatment is available.

Toxic gases that permeate the building can affect occupants a long distance from the fire. These gases can travel through ventilation systems and other openings and channels that exist in all building construction. Doors, stairwells, pipe chases, elevators, and service ducts are but a few of the avenues through which smoke and fire gases can travel. An exposure to concentrations of carbon monoxide alone at 0.4% will be fatal in less than one hour. However, much smaller amounts may be deadly when combined with other toxic products. This cocktail effect of gas mixtures is clearly evident, but exposure time and species have not been established.

In addition to the theft of oxygen by carbon monoxide, oxygen can be blocked from the body tissues in other ways. The inhalation of irritant gases in the smoke can cause an outpouring of fluid into the air passages. This causes swelling of their walls, restricting or obstructing the inflow of air to the cells from which it normally passes into the bloodstream. Victims may suffer a wide range of symptoms such as headaches, nausea, fatigue, and difficult breathing. In addition, individuals may experience confusion, impaired mental functioning, and diminished coordination. This can cause what may be perceived as irrational behavior, such as clawing at a door rather than turning the knob.

Heat in a burning building can cause serious physiological problems. Humans can tolerate temperatures of 150–200°F (65–90°C) for only short periods of time. The tolerance time drops rapidly above 200°F (90°C), and in temperatures above 300°F (150°C), survival is but a few minutes. These temperatures are easily attained in building fires where temperatures within 10 ft of the flames can exceed 300°F (150°C).

The secondary effects of temperatures well below 200°F (90°C) cause other physiological problems. People cannot work in an atmosphere of 120°F (50°C) without protective clothing. Even then, the heart rate increases, causing increased stress. Fatigue, dehydration, heat exhaustion, and heat shock commonly accompany these temperatures. Both climatic conditions and a fire environment can cause temperatures of this level. These conditions contribute to fire-fighter fatigue.

Burns due to heat and flames are often incorrectly assumed to be the primary cause of death and injury. Smoke and toxic gases remain the major threat. Besides smoke casualties on the fire scene, nearly half of those who reach the hospital die of respiratory tract damage due to irritant gases. Although fewer fatalities result from burns, the pain and disfigurement cause serious long-term complications. Bacteria are *everywhere*, and bacterial infections from burn wounds and damaged lungs are major killers. Shock results from fluid losses out of the burned tissues. This formerly major killer has declined with vigorous intravenous replacement of plasma and other fluids.

Structural collapse of building elements can cause physical injury and restrict movement through the building. Collapse is a concern for building occupants and a particular danger to fire fighters. Deaths and serious injuries occur each year because of unanticipated structural failure. Some of these failures result from inherent building weaknesses, but many are the result of renovations to existing buildings that materially, though not obviously, affect the structural integrity of the support elements. A building should not contain surprises of this type.

Besides being accompanied by toxic and irritant gases, smoke contributes indirectly to death. Dense smoke obscures visibility and irritates the eyes. Consequently, an occupant may become disoriented and be unable to identify escape routes and use them. Visibility reductions cause occupants to turn back from an intended escape route. At a visibility of about 10 ft (3 m), 90% of occupants will have turned back in a building with which they are very familiar.

Knowledge of fire in a building causes anxiety among its occupants. The combination of emotions, conditioned responses, training, rational action, and physiological and mental impairment due to products of combustion all influence human behavior. To these factors should be added the size, shape, and function of the building, the physical and mental capacity of its occupants, and interior design features that influence decision making.

Behavior patterns of individuals subjected to sudden, unexplained danger cover a wide range of action. Some people will correctly assess the danger and take prompt action whereas others may withdraw from the situation. Some may deny the situation altogether and take no action or totally unrelated actions. Still others will take no initiative, but will follow a leader and obey orders if and when they are given.

It is not possible to predict actions of an individual, but group behavior patterns can be statistically quantified. However, this assumes that the occupants are capable of reacting physically and mentally. Hospitals, nursing homes, prisons, and mental hospitals all house occupants who cannot react with freedom of movement. Part of the risk characterization is an analysis of combustion products that can enter target spaces occupied by these individuals.

Special building fire situations often arise. In high-rise buildings, for example, occupants cannot evacuate within a reasonable time. Unless effective areas of refuge can be provided or the fire is extinguished quickly, occupants may be exposed to the toxic combustion products for significant portions of the fire duration. A symptom of oxygen deprivation is that a normally rational individual behaves irrationally. Low levels of carbon monoxide, for example, reduce levels of oxygen available to the brain and other tissues, increasing the potential for irrational behavior.

18.6 Prescriptive code egress

A major part of building codes is a collection of good practices relating to egress. A *means of egress* is defined as a continuous, unobstructed path of travel through which an individual can move to a public way. A public way is a street, alley, or other land outside the building that leads to a street.

A means of egress has three major components: the *exit access*, the *exit*, and the *exit discharge*. An exit access is the part of a means of egress that enables an individual to move from a location within the building to the exit. In most buildings the exit access is a corridor leading to the exit stairwell. However, in buildings such as assembly rooms, stores, and galleries, the exit access may lead through aisles or other paths. The exit access must provide free and unobstructed travel. Therefore intermediate rooms that can be locked are not a part of the exit access.

Depending on the occupancy type and certain building features, such as the presence of sprinklers, the codes prescribe maximum travel distances. Codes prescribe a minimum width to accommodate the number of persons and the character of the activity. The building code allows a dead-end distance for most occupancies. This dead end is restricted in length because two avenues of possible escape are not available. If the dead end were excessively long, an individual could be trapped and prevented from returning to find another route to an exit.

The *exit* is the second component of a means of egress. In the perception of the layman, an exit is synonymous with the way out. However, from a code perspective, the exit is a special part of the building that connects the exit access with the exit discharge. Although an individual may not have protection from combustion products when traversing the exit access, the exit is viewed as a temporary refuge area that protects an individual for the time needed to move through it. As a temporary area of refuge, the exit has special requirements to maintain its integrity and tenability. This includes fire-resistant rated enclosures, opening protectives, and, for some buildings, smokeproof enclosures. The code includes additional safety requirements, such as dimensional limits, guards and handrails, and walking surface specifications. Exit requirements are so distinctly defined that many "exits" to the layman should be considered as "ways out" under a code, because they do not meet prescribed conditions.

Codes permit a *horizontal exit* in many buildings. A horizontal exit permits passage from one part of a building to another part on the same level. There are many specialized requirements such as fire resistance, refuge area requirements, and egress requirements from the building section that serves as the refuge area. The reliability of the door closer that separates the fire area from the refuge area is an important requirement.

The *exit discharge* is the third component of an egress system. An exit discharge enables individuals to reach a position of safety outside the building when leaving the exit. Some buildings discharge into a lobby or other building area. When this occurs, additional requirements, including automatic sprinkler protection, are often required for those spaces.

A concern with respect to an exit discharge is the assurance that an individual will not inadvertently travel past a ground floor when a building has floors below ground level. To avoid this, codes normally require stairwells to be discontinuous at the ground floor to force individuals to leave one part of the exit before entering the other part. In addition, stairs are required to distinctly identify the ground-floor discharge.

In addition to dimensional and other special requirements for a means of egress, codes identify the *capacity* of each egress component. The capacity of a component is determined by dividing its width by a factor. For example, the factor for a level component such as a door, corridor, or horizontal exit is usually 0.2, whereas the factor for stairs is usually 0.3. Therefore the capacity of a 34 in door is 34 in / 0.2 = 170 persons. The capacity of a 44 in stairway is 44 in / 0.3 = 146 persons.

The final code factor is the *occupant load*, which is the minimum population for which to design the means of egress. The minimum number of people is calculated by dividing the gross area (or sometimes the net area) by a factor based on the building's use and occupancy. For example, the floor area per occupant for a business area is 100 ft^2 gross. Therefore the minimum occupant load for a gross office floor area of 30 000 ft^2 is 30 000 / 100 = 300 persons. If the actual population is larger than this, the larger number is used. If the actual number is smaller, the minimum calculated number is used. The sum of each egress component must have a capacity large enough to accommodate the occupant load.

Sometimes one observes signs that state "The maximum number of occupants shall not exceed *xxx* people," or words to that effect. The capacity of the egress components determines this number. For example, if a room has a capacity of 300 people, but the exit doors have a capacity of 180 individuals and the stairways a capacity of 120 persons, the room capacity is only 120 individuals.

18.7 Plans approval for prescriptive code egress

Building egress is a major consideration that requires approval by the authority having jurisdiction (AHJ). The process normally involves the following steps. The building size and occupancy are established and the occupant load is determined for the different areas on each floor. Exit locations are noted, and the AHJ accepts their remoteness from each other. Exit access distances are checked regarding occupancy and other building features, such as sprinklers. The widths of doors, corridors, stairs, and other parts of the egress components are determined, and the occupant load must be smaller than any component limitation. Finally, other specialized requirements for dimensions, fire resistance, and special characteristics are checked. When all code requirements are met, the egress system is in compliance and legal.

The building egress system is part of a total architectural circulation plan. A variety of considerations besides fire requirements are a part of a circulation pattern. The earlier discussion referred to code compliance rather than safety. It is often assumed that a code-compliant design is a safe design. However, the quality of egress systems can vary, and a performance-based risk analysis requires sensitivity about the occupants of the building and what they will be doing at different times of the day.

18.8 Overview of egress performance evaluations

A prescriptive code egress system gives an inventory of good practices. However, the quality of egress systems can vary greatly, and life safety encompasses more than good practices and routes out of a building. We need to understand and characterize life safety risks in a way that helps occupants and others to make better decisions.

A performance evaluation for egress relates two separate components. The first describes the time from EB during which each target space along an egress path will remain tenable. The second identifies the travel time for an individual to move from an occupant room through the egress path. A missing link is the preevacuation time between EB and the instant that the occupant opens the door to start escape movement. The performance measure is the available preevacuation time delay that an individual can have before encountering untenable conditions along the way out.

Building features such as detection and occupant alerting are parts of this preevacuation delay. Other factors may contribute to occupant delay in deciding to open the door and start leaving the building. This performance analysis does not require one to predict actions or decisions that an individual might make when exposed to one or multiple fire cues. It merely identifies the time from EB to when movement must be started to avoid untenable conditions during egress.

18.9 Tenability

The definition of tenability is a central issue in a life safety analysis. The tenability criterion is a major decision in a performance analysis that requires a clear rationale tailored to the building use and architecture and occupant characteristics. Here we define tenability as a visibility through smoke under normal ambient lighting conditions. Individuals attempting to evacuate a building will reevaluate their intent and turn back to seek another way out when visibility falls below a certain level. Toxic gases are fellow travelers in the same air and are more hazardous to people than smoke. However, because obscured visibility normally occurs before the toxic gases affect humans, this selection is conservative. Perhaps more importantly, it is more readily perceived as an imminent danger by occupants.

A performance evaluation identifies the likelihood that people and untenable smoke or other fire conditions will occupy the same space at the same time. *Note that this analysis does not determine the likelihood that an individual will be injured or killed in a fire*. It merely identifies an expectation that people will be subjected to untenable smoke (or other fire) conditions as they move through the building. This enables one to gain a clear understanding of the building's effect on egress.

18.10 The analytical process

Identification of potential egress routes to the outside is a necessary but not sufficient part of evaluating a building's egress performance. Only the normal circulation patterns are considered for performance analyses. This includes the usual ways out of a building by ambulatory means

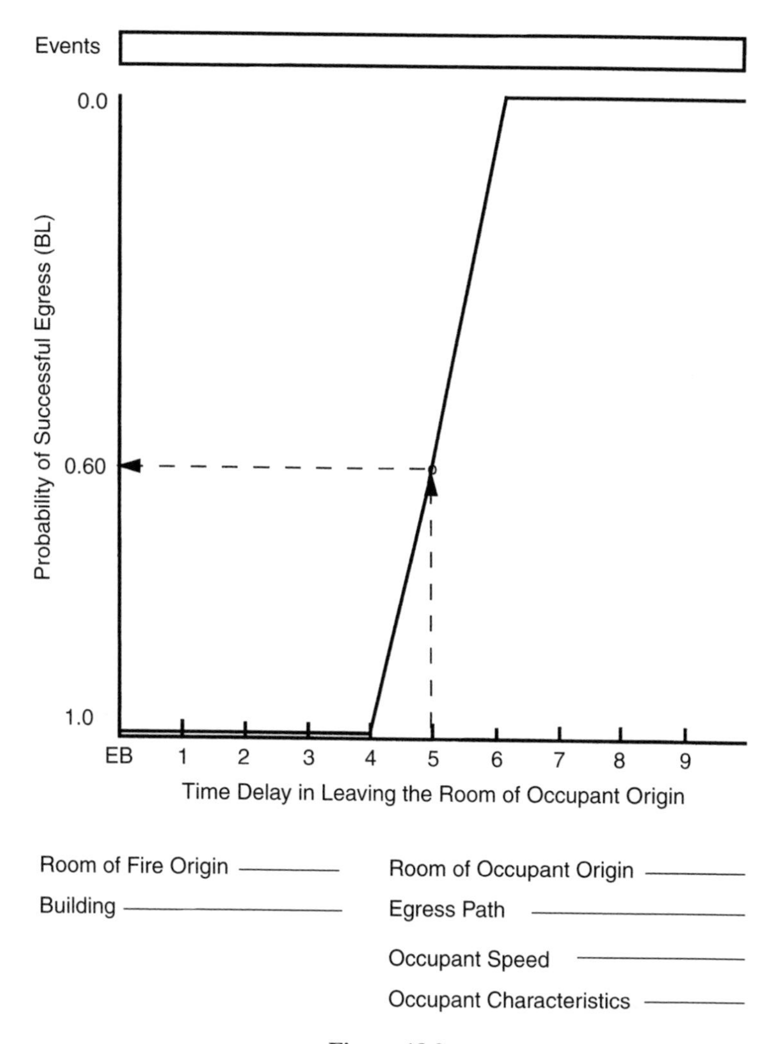

Figure 18.2

as well as the use of elevators and escalators. Evacuation through windows, roofs, and fire department assisted rescue is not included in the analysis.

Figure 18.2 illustrates a representative life safety (LS) by egress graph (LS curve) of building performance. The ordinate indicates the probability that an individual can travel the egress path without encountering untenable conditions in the spaces through which he or she must move. The abscissa shows the available preevacuation time delay. In Figure 18.2 if an occupant leaves the room within 4 min of EB, safe egress can be expected. If the delay is 5 min, the probability of avoiding untenable conditions along the egress path is reduced to 60%. If the evacuation start is delayed to 6 min from EB, the occupant will certainly encounter untenable conditions.

A procedure to estimate performance is described below:

1. Select the room of occupant origin. Although this procedure is appropriate for rooms with few occupants as well as for assembly areas with many occupants, initially we focus on a single representative occupant in a specific room.

2. Identify the representative individual's characteristics involving the speed of movement and agility to traverse building obstacles, such as corridors, stairs, and doors. These characteristics define occupant mobility and any impairments. For example, an athletic, unimpaired individual exhibits one type of movement. An impaired individual unable to move easily along corridors and stairs exhibits another. Blind, deaf, young, elderly, or frail individuals are other categories. Each characteristic group exhibits a different building performance.
3. Identify a specific egress route to study. An individual usually has choices of routes both initially and as the evacuation evolves. A performance curve is based on each specific egress route of interest.

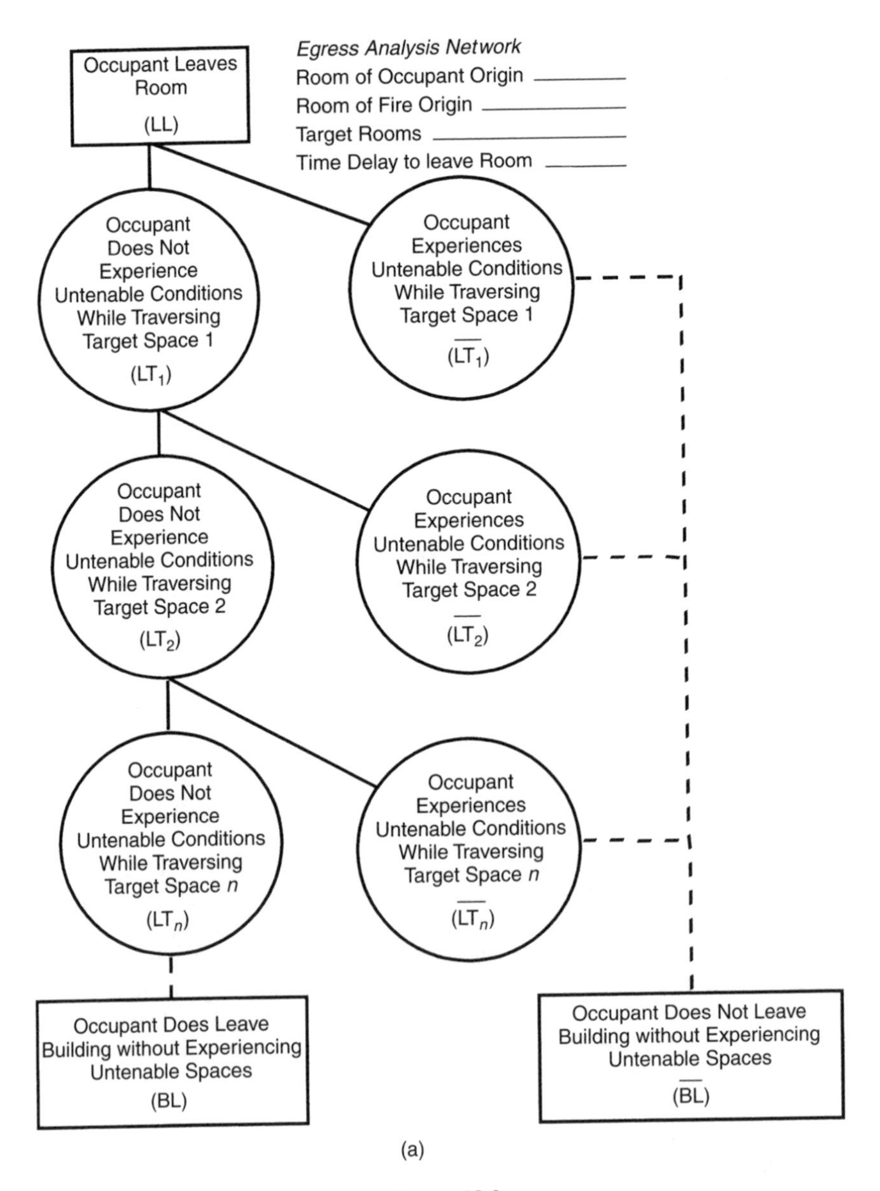

(a)

Figure 18.3

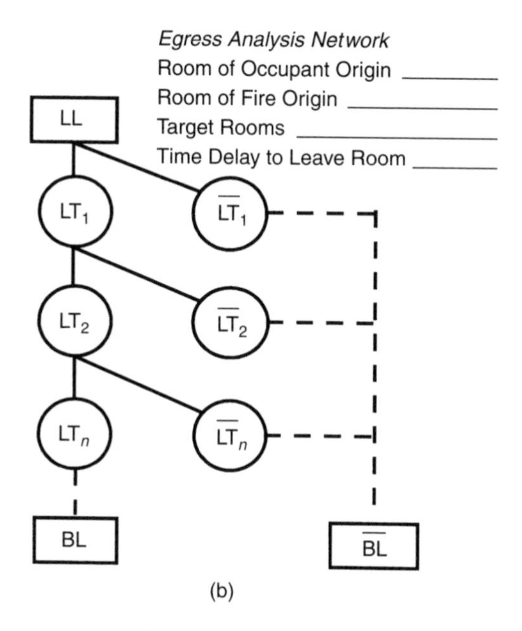

Figure 18.3 (*continued*)

4. Select target rooms along the egress route that are important to evacuation success.
5. Identify a room of fire origin that provides a representative condition or a specific threat for the target rooms. This room of fire origin is not necessarily the same as that for the limit (L curve) analysis.
6. Identify tenability criteria for the occupants. The criteria depend on the characteristics of the individuals and the building use. Construct Sm curves for the target spaces. One may wish to construct most optimistic and most pessimistic bounds to provide a better insight into the ranges of risk.
7. Calculate time durations for the individual to reach each of the successive target spaces along the route. This is a simple calculation of distance / speed = time elapsed. Because the route is segmented, one may use different speeds to reflect differing conditions along the path.
8. Add the time sequence of step 7 to selected occupant preevacuation time delays. Each preevacuation time delay provides an LS curve coordinate.
9. Compare the time lines of step 8 with the Sm curves of step 6. The probability for each life safety coordinate is the probability of Sm curve tenability for the occupant positions. Figure 18.3 provides a framework to calculate the likelihood that an occupant will traverse an egress path without encountering untenable conditions.
10. Examine enough time periods and other egress paths to get an understanding of building performance.
11. Describe the risk characterizations for individuals who may use building egress paths during a fire emergency and document the rationale.

While this procedure may appear to have many activities, steps 2 through 6 would have been completed as a part of the Sm curve performance analysis. Steps 1 and 7 through 11 become the basis for identifying relative risk characterizations for different egress paths, occupants, and and occupant locations. Although this analysis focuses on occupant safety, a sense of proportion for the outcomes is likely to have been recognized during the performance evaluations. Component interactions help to form the understanding that enables one to portray accurate and convincing risk characterizations.

Example 18.1

Construct an LS curve to describe a building's performance for the following scenario. A representative room was selected for an occupant's location. Three target rooms along an egress route were chosen for Sm curve analysis. A tenability criterion of 10 ft visibility through smoke was selected as the limit of acceptable performance. The Sm curves for these rooms are shown in Figure 18.4.

Building occupants have a wide range of ages and physical and mental abilities. This analysis will assume occupants are unimpaired by medication, alcohol, or other drugs. The population in the building is small, allowing unrestricted movement. The analysis studies two representative occupants. Person A is a young, unimpaired individual who moves easily. Person B is elderly with some impairment who walks with somewhat greater difficulty. All occupants are familiar with the building. The time that an individual will remain in each target space after leaving the room of occupant origin was calculated and is shown in Figure 18.5(a) and (b).

Solution

The abscissa of the life safety (LS) curve identifies the time delay in leaving the room of occupant origin. This case focuses only on travel through the egress path, given that the individual has

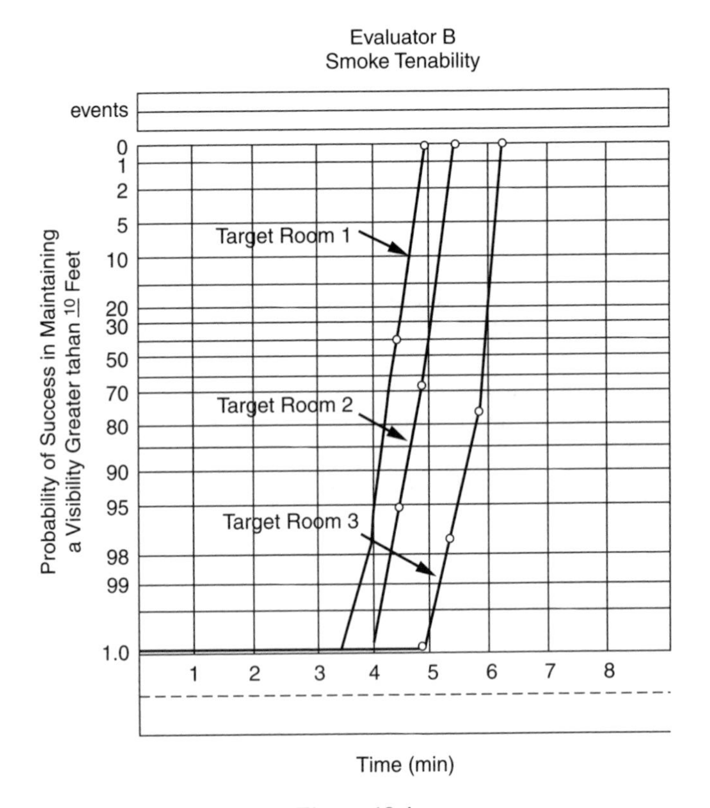

Figure 18.4

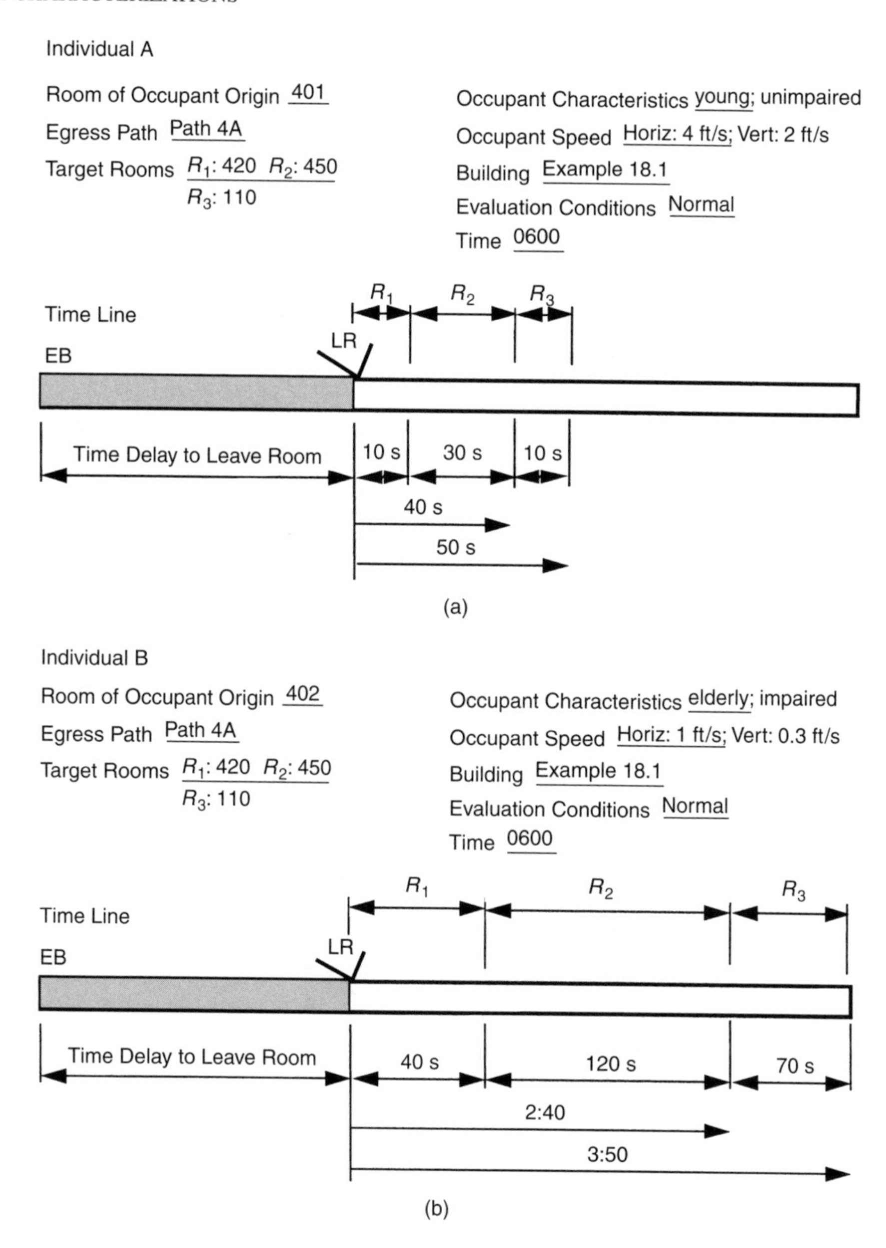

Figure 18.5

decided to leave the room of occupant origin. Here calculations for an individual successfully leaving the building (LB) will be the same as the life safety (LS) curve.

Table 18.1 shows values for the probability of moving through the target spaces without encountering untenable conditions. These values are obtained by matching the time that an individual will be in a target space with the smoke tenability (Sm) curves for the space (Figure 18.4). Figure 18.6 shows a SVN to calculate the likelihood that an individual will traverse egress path 4A without encountering untenable conditions along the path. This illustration uses values for

Table 18.1

Time delay	Time to enter R1	$P(LT_1)$	Time to enter R2	$P(LT_2)$	Time to enter R3	$P(LT_3)$	$P(BL)$
Individual A: Young, not impaired							
0:00	0:10	1.0	0:40	1.0	0:50	1.0	1.0
3:20	3:30	1.0	4:00	1.0	4:10	1.0	1.0
4:00	4:10	0.95	4:40	0.85	4:50	1.0	0.81
4:20	4:30	0.4	5:00	0.7	5:10	0.99	0.28
4:30	4:40	0.1	5:10	0.1	5:20	0.98	0.01
Individual B: Elderly, slightly impaired							
0:00	0:40	1.0	2:40	1.0	3:50	1.0	1.0
1:10	1:40	1.0	2:50	1.0	5:00	1.0	1.0
1:40	2:20	1.0	4:20	0.95	5:30	0.95	0.90
2:00	2:40	1.0	4:40	0.85	5:50	0.8	0.68
2:20	2:50	1.0	4:50	0.8	6:10	0.0	0.0

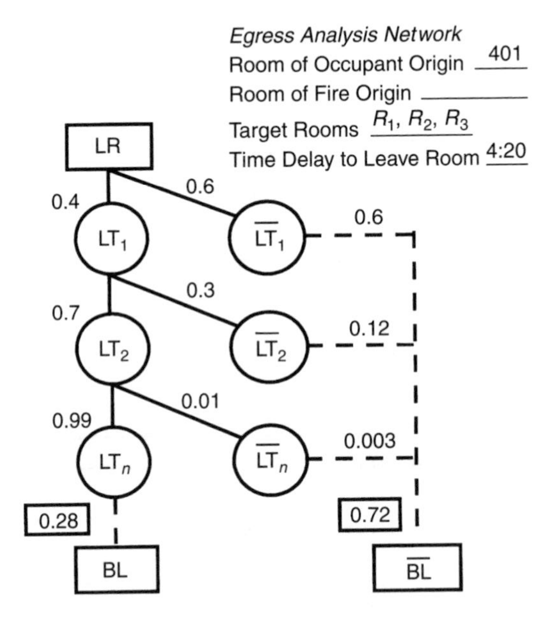

Figure 18.6

individual A with a preevacuation time delay of 4:20. The values for Table 18.1 provide the information for the life safety (LS) curves of Figure 18.7.

Discussion

Figure 18.7 depicts the likelihood that individuals A and B can traverse egress path 4A without encountering untenable conditions. Individual B must leave the room within about 1 min after established burning (EB) to leave the building without encountering untenable conditions. However, these untenable conditions were defined as a visibility of 10 ft. This is an extremely small distance for most individuals. Individual A can delay leaving the room for nearly 3 1/2 min after

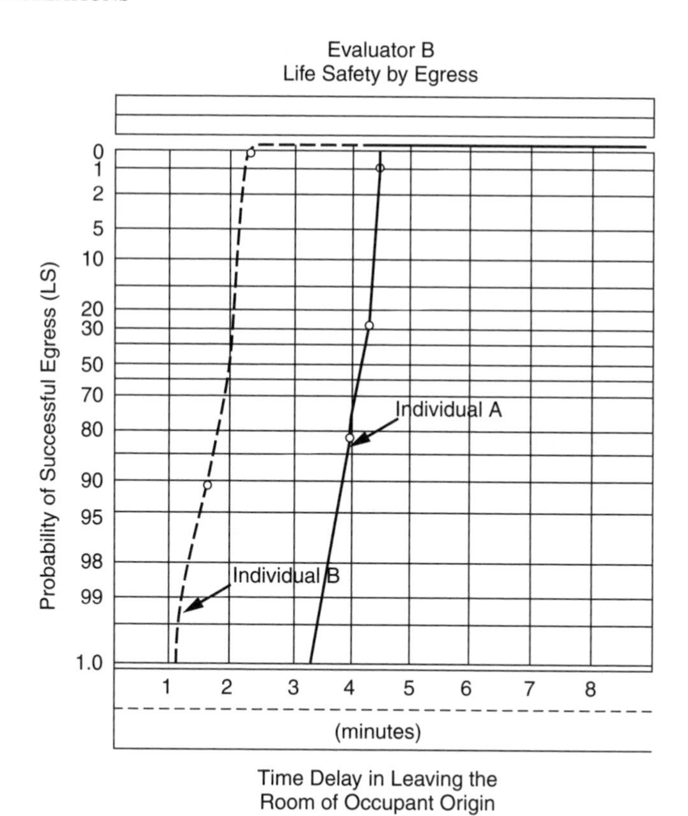

Figure 18.7

EB. Besides this luxury, the age and physical characteristics of individual A may give a greater incentive to move further through dense smoke.

Note that the building performance described by the (LS) curve does not depend on detection, alerting, and occupant decisions. These factors are important in understanding the complete life safety system. However, because the time delay does not use these factors, one can uncouple them from the fundamental building behavior. This makes it easier for all parties to recognize performance needs of detection and occupant alerting. A risk management program can inform occupants of potential problems and expected actions in prefire instructions.

18.11 Preevacuation activity analysis

A building evaluation for egress considers the building performance and the occupant preevacuation activities. Figure 18.8 shows an SVN for these components. The occupant traversing the egress path without encountering untenable conditions (leaves building, LB) was discussed above. We now look at the decision to leave the room (LR).

Figure 18.9 identifies critical preevacuation events of fire detection, occupant alert, and occupant decision. Detection would have been evaluated earlier as a part of the M curve. Figure 18.10 shows the factors that influence detection (MD). Occupant alert could be part of the M curve evaluation, but often occupants are not involved in the fire department notification process. Some buildings may have intentional or unintentional delays in alerting occupants. The occupant decision component is very uncertain and variable, although it is possible to combine "what if"

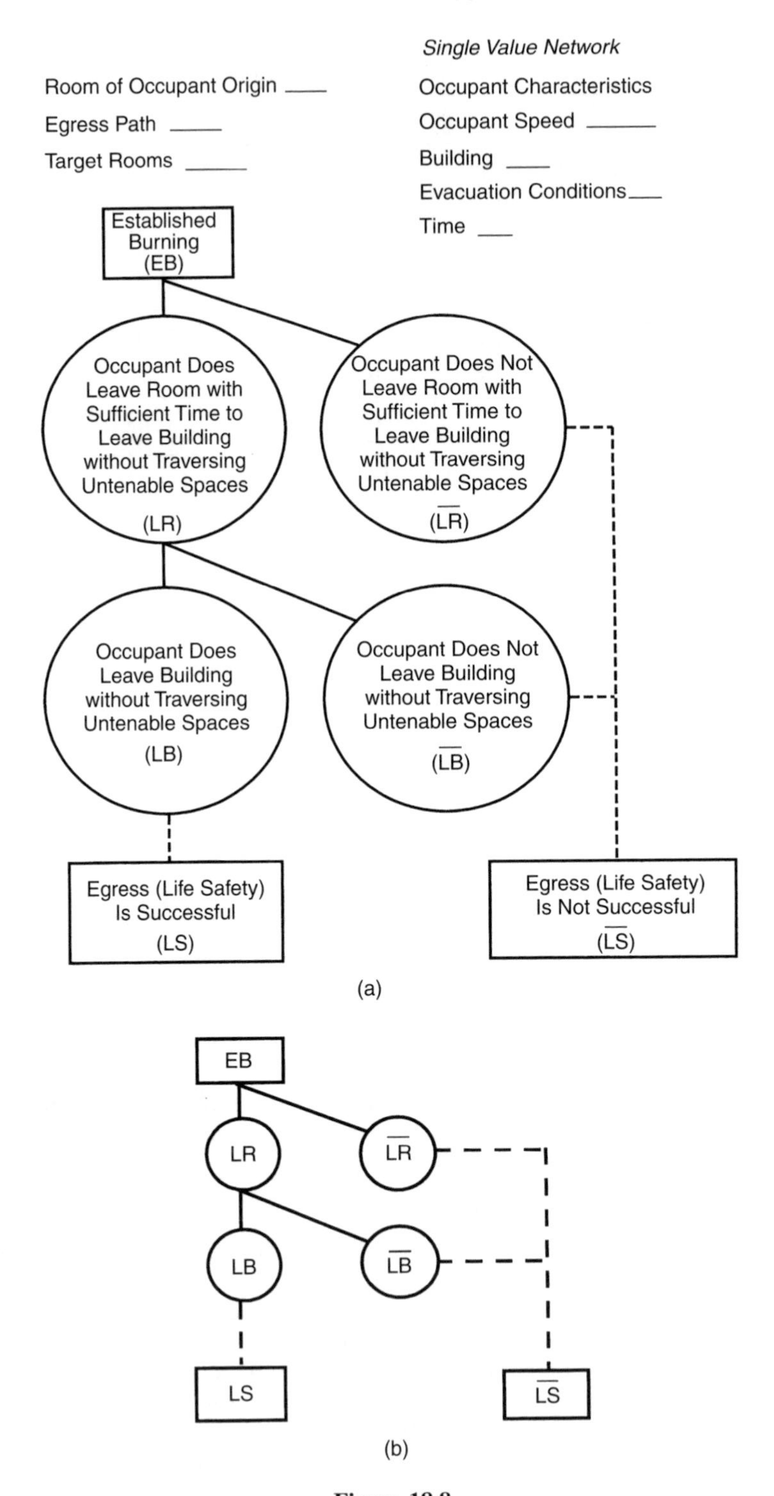
Single Value Network
Room of Occupant Origin ____
Egress Path ____
Target Rooms ____
Occupant Characteristics
Occupant Speed ____
Building ____
Evacuation Conditions____
Time ___
Established Burning (EB)
Occupant Does Leave Room with Sufficient Time to Leave Building without Traversing Untenable Spaces (LR)
Occupant Does Not Leave Room with Sufficient Time to Leave Building without Traversing Untenable Spaces ($\overline{LR}$)
Occupant Does Leave Building without Traversing Untenable Spaces (LB)
Occupant Does Not Leave Building without Traversing Untenable Spaces ($\overline{LB}$)
Egress (Life Safety) Is Successful (LS)
Egress (Life Safety) Is Not Successful ($\overline{LS}$)
(a)
EB
LR
$\overline{LR}$
LB
$\overline{LB}$
LS
$\overline{LS}$
(b)

Figure 18.8

scenarios with research data to describe possible outcomes. Constructing a time line for these events provides an insight into building and occupant conditions.

The occupant alert event (LA) requires attention to the characteristics, location, and activities of the representative individual on whom the assessment is focused. For example, if the alerting process involves sounding a noise, the actions of deaf occupants would be evaluated as a separate scenario from unimpaired hearing individuals. Also, the alarm decibel level is affected by the location of the sounding device and intervening architectural obstacles, such as closed doors. Note that the awareness of a signal or cue does not include the decisions that an individual may make after that event. This event considers only the means by which the building alerts an

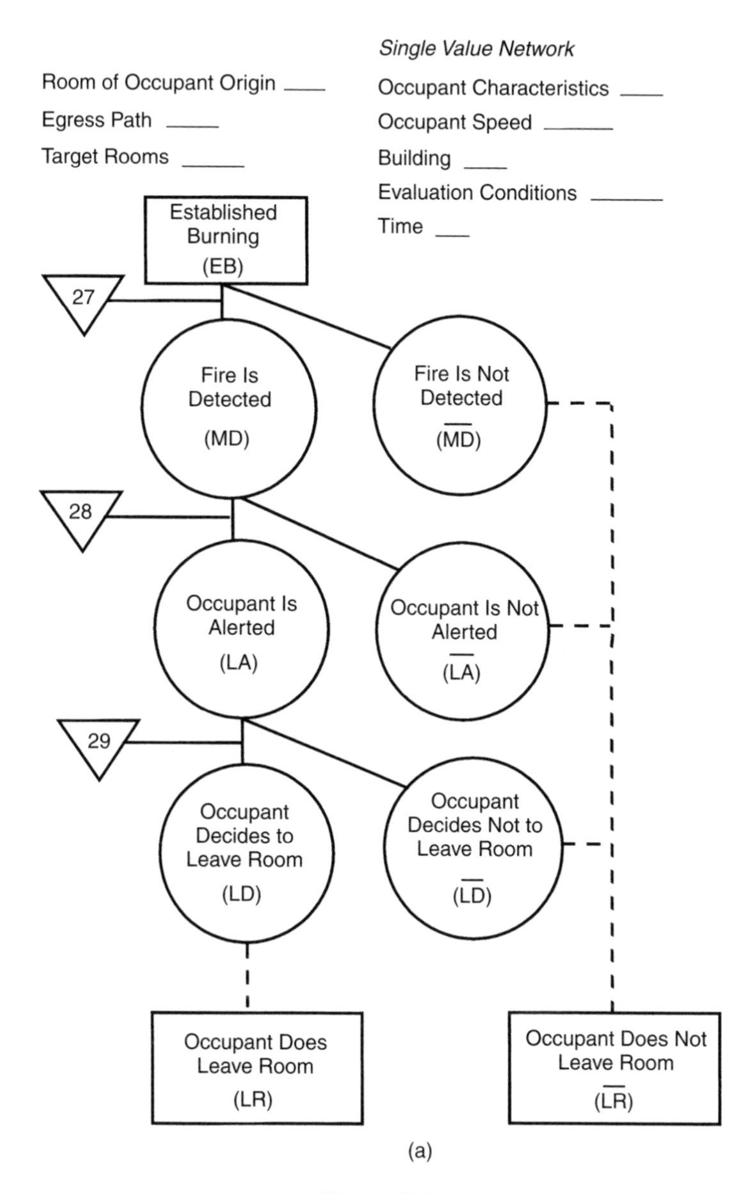

Figure 18.9

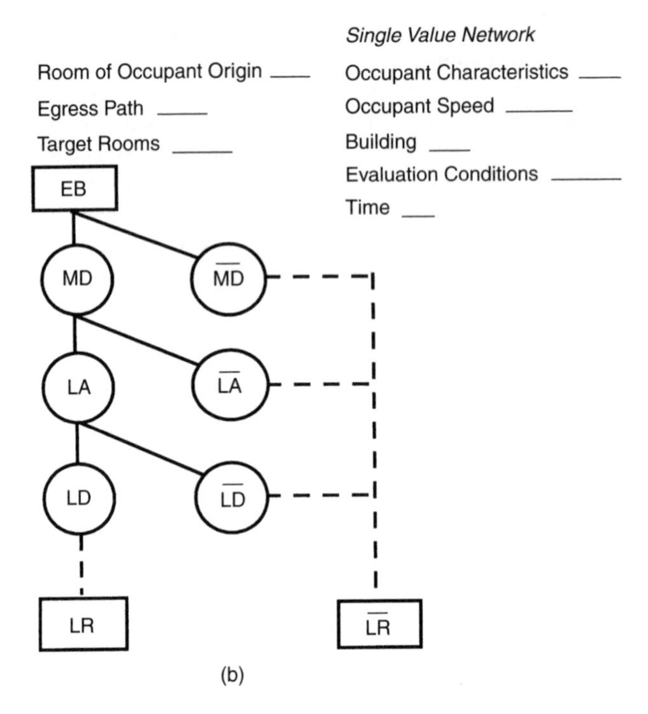

Figure 18.9 (*continued*)

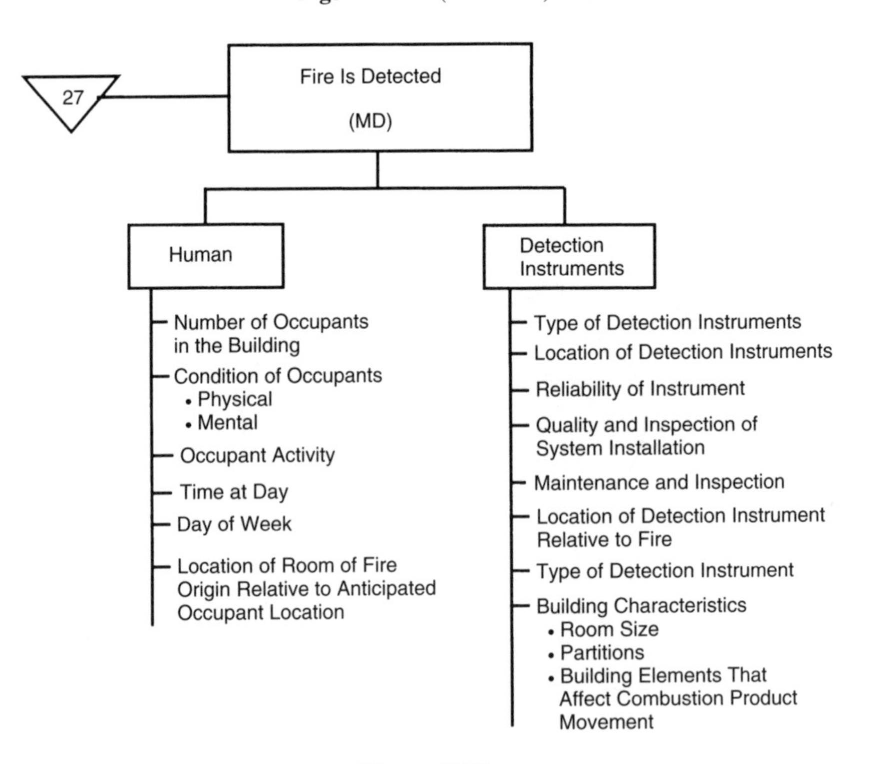

Figure 18.10

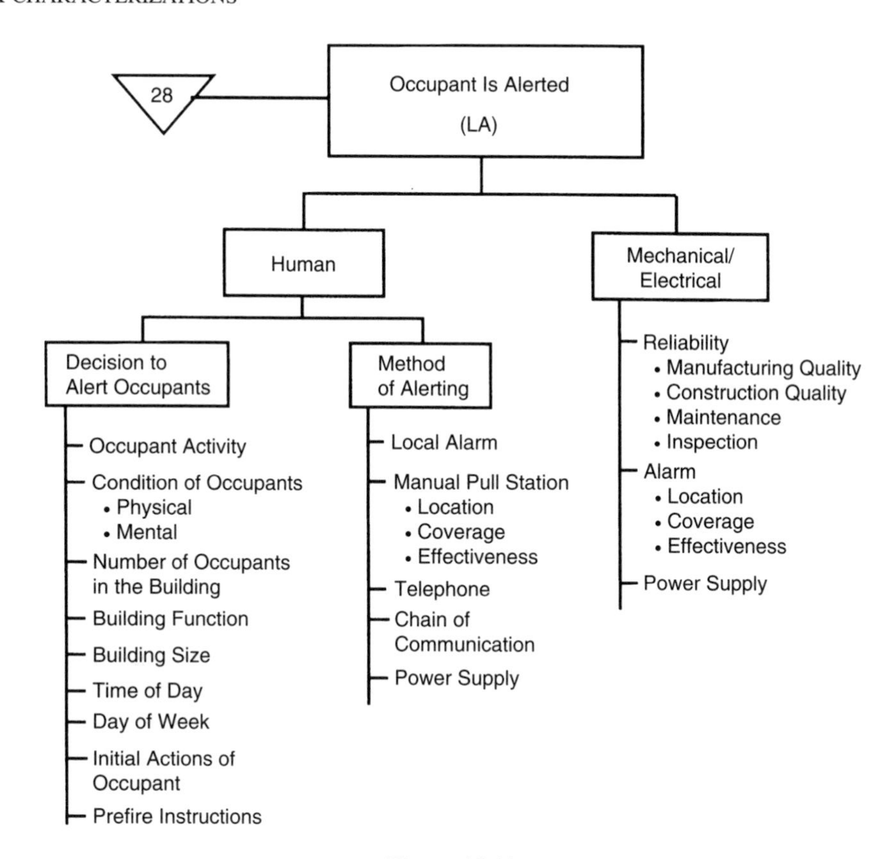

Figure 18.11

individual of specified characteristics of a fire emergency. Figure 18.11 describes the factors that affect this evaluation.

After an occupant has been alerted of a fire emergency, a variety of actions can occur. All involve time delays. Studies have shown this delay can be several minutes. The common assumption that building evacuation starts at the instant of detection and alarm is a myth. Time delays are common and often comparatively long.

Attitudes, experiences, cues, and prefire emergency training influence the decision to leave the room (LR). This decision is highly personal and is influenced by factors that involve interactions of people, the fire, and the building. Often the time duration is much longer than imagined by individuals who may give superficial assumptions about appropriate behavior. Examining this event for building occupants gives an insight to more effective risk management programs.

A variety of factors and conditions influence an individual's decisions after becoming aware of a fire emergency. For example, the manner in which an individual becomes aware of the fire emergency will influence his or her actions. The ringing of an audible alarm, particularly after a series of false alarms, is not as likely to produce the same response as seeing dark, acrid smoke or hearing an individual shouting, "Fire!" The intensity, volume, pitch, and earnestness conveyed by a voice alarm has a major influence in occupant actions.

The number and intensity of fire cues influences the perception of needs for action. A fire cue may be a level of seeing, hearing, smelling, or feeling the products of combustion, or it may involve the actions of others or of alarm systems. The perception of real danger in uncertain situations seems to be the most important factor. *Recognition* of a threat may be very ambiguous and unclear during early stages of a fire. As the volume and intensity of combustion products increases, the uncertainty declines. The time that remains to take effective action is also reduced. Experiences, a predisposed optimistic wish, concepts of personal abilities, and prefire training or instructions influence initial actions.

When an individual becomes aware of mild or ambiguous cues, most individuals attempt to obtain additional information. This process of *validation* may take the form of a physical investigation and surveillance, or it may involve questioning other individuals. Eventually the threat becomes defined in the mind of the individual. This *definition* stage involves the recognition of the meaning and structure of the threat, although not necessarily its seriousness.

After an individual defines the threat and knows the nature of the problem, the *evaluation* of personal actions begins. The size and perceived growth of the fire, the building size and function, availability of fire extinguishers, the activity and potential danger to others, and prefire instructions are among the factors that influence an evaluation of appropriate actions. The concern and actions of a group of people often influence decisions more than those of an individual in isolation. Possibly the evaluation will lead to a series of actions, usually involving variations of fight or flight. The initial decision of an individual to take specific action is completed with evaluation. Often the evaluation may include a series of if-then scenarios that are useful in establishing prefire instructions and fire program management plans.

After an individual decides an appropriate course of action, the *commitment* stage is entered. At commitment the individual carries out the decision made during evaluation. That action may be successful, such as leaving the building safely or fighting and extinguishing the fire. On the other hand, the initial commitment may be unsuccessful and the situation must be reassessed. This *reassessment* may result in continued efforts or in redirecting actions. For example, an individual initially may have attempted to fight the fire, but being unsuccessful, they attempt to leave the building.

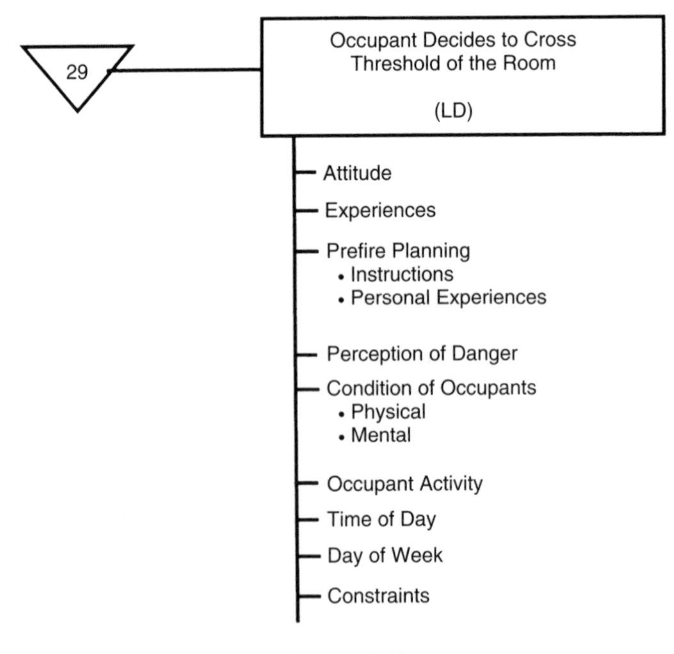

Figure 18.12

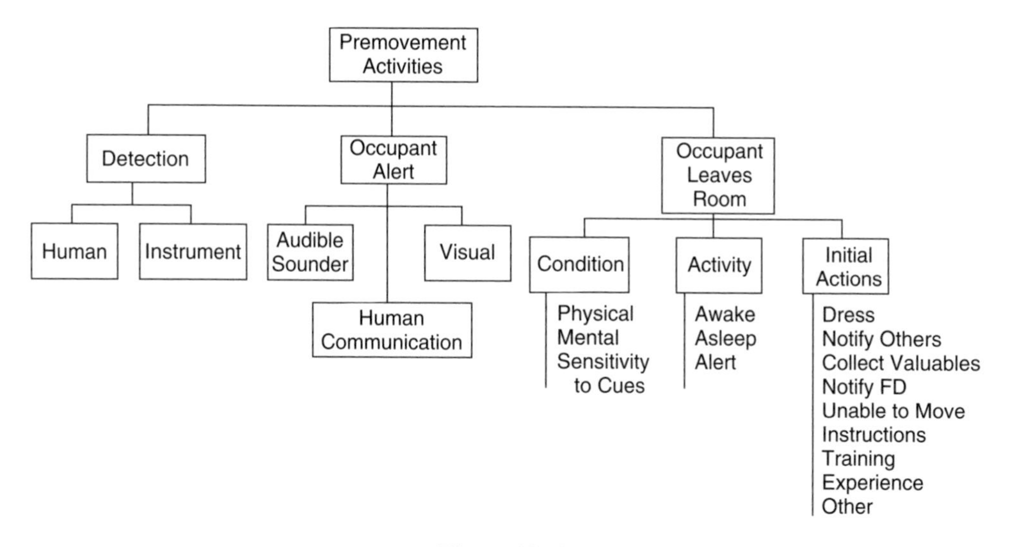

Figure 18.13

The time frame to complete the decision stages described above may be very brief or it may be longer. However, the process is dynamic. The fire grows at an exponential rate. Conditions change more rapidly than can be imagined by individuals who have never experienced a fire situation. Anxiety, stress, and levels of activity reflect the intensity and importance of the cues available. The time to define and evaluate decisions can be reduced to seconds when cues are unambiguous and intense. On the other hand, an individual may act in a relaxed manner until the movement of combustion products changes the environment so rapidly that plans must be reassessed and changed very quickly. Figure 18.12 shows factors that influence occupant decisions.

Premovement activities involve building performance and human decisions. Figure 18.13 describes the major elements of the process.

18.12 Movement times

After the occupant opens the door and starts to leave the room (LR), the local travel clock starts. We estimate travel time through escape routes and construct a time line. The time needed to traverse the egress route will depend on a variety of factors. Among these are people concentrations in the egress route, the distance, path recognition, amount of smoke, and the physical and attitudinal characteristics of the individual.

When the population density is low, individuals are essentially moving alone, and speeds relate to the physical capability of the individual. However, disorientation, anxiety, visibility, building familiarity, and recognition of an appropriate route can be major problems. The architectural layout for some buildings is so complex that it is possible to become disoriented under normal conditions. Under conditions of stress, reduced visibility, and unclear directional orientation, selecting an appropriate path may involve time and also wrong choices. Figure 18.14 describes the factors that influence the movement from a starting location to transit through the egress path.

When a building has a moderate concentration of people, leaders often provide instructions. Moderate crowds may reduce speeds somewhat, although not much. Major problems can arise from congestion at locations where different paths merge. In buildings having zones with large concentrations of people or with constraints to movement, such as narrow aisles and fixed seating, travel speed may be reduced drastically.

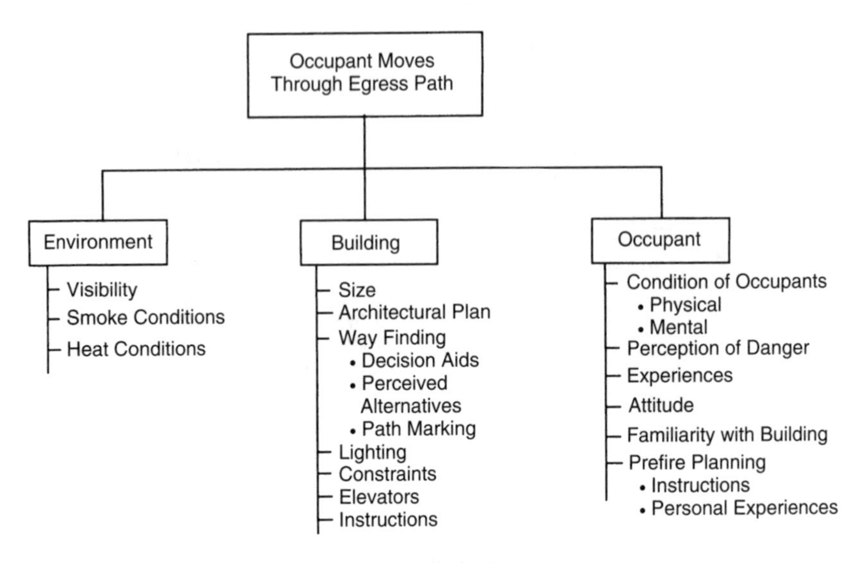

Figure 18.14

Table 18.2

Activity	Speed (ft/s)
Comparative movement	
100 yard dash in 10s	30
Mile run in 4 min	22
Mile walk in 20 min	4.4
150 feet in 90s	1.7
Normal Adult Walking speeds:	
Level passageway	4.3
Stairways (down)	2.5
Stairways (up)	1.9

Mobility characteristics must be defined. For example, is the occupant able to move 100 ft in 60, 90, or 300 s? Are there any constraints to movement exist, such as the need for wheelchairs or crutches? Are there any impediments to movement and decision making, such as blindness or deafness? Table 18.2 gives a sense of proportion on speed of ambulatory movement.

Life safety scenarios may use other characteristics. For example, risk characterizations are different for occupants who are asleep, intoxicated, or under the influence of drugs. They differ for occupants mentally incapable of making appropriate decisions in a fire emergency, as well as the very young or the infirm elderly. Should a building design be faulted when a parent leaves infants or small children alone for a short time while socializing elsewhere?

18.13 Defend in place

Event 2 in the organizational diagram of Figure 18.1 shows that the individual may not leave the room. This could be by intent because the individual may have decided that survival chances would be better by remaining in the room than by attempting to leave the building. Or perhaps the individual is asleep or under the influence of drugs or alcohol, hence unaware of the occupant alert. In other situations, occupants are expected to remain in their rooms for the duration of a fire. Nursing homes, prisons, and hospitals are examples of a defend-in-place strategy for the occupants.

A defend-in-place life safety analysis evaluates the likelihood that an individual will be protected from untenable conditions. The analysis assumes that the room under study is not a room of origin.

A defend-in-place evaluation has these steps:

1. Identify the room of fire origin.
2. Select the target space to be evaluated.
3. Identify the tenability measure for the occupants of the target space. The criteria will reflect the characteristics of the room occupants. The tolerable level of products of combustion for the room might be expressed as follows:

   X feet of visibility in smoke;
   Y_1 ppm of CO dose rate and a limiting acquired dose;
   Y_2 ppm of other specified toxic gases dose rate and a limiting acquired dose;
   Z degrees of temperature;
   Avoidance of structural collapse.

4. Construct target space tenability curves as described in Chapter 16. One may construct the most optimistic and most pessimistic bounds to identify segments of time for which the target space certainly will be tenable, may be tenable, or certainly will be untenable.
5. Document the basis for the conclusions.

Tenability curves describe the available time and relative danger for an occupant who remains in the room. The tenability curve for the target (defend-in-place) room becomes the building performance curve.

18.14 Areas of refuge

It may be impractical to provide for egress or defend-in-place for certain types of buildings. The concept of area of refuge is appealing for tall buildings, large buildings, hospitals, and ships.

An area of refuge is a designed and designated area in which occupants may find refuge for the duration of a fire. Few buildings have a designed area of refuge. However, if one were incorporated into the building design, its performance evaluation becomes merely a combination of egress analysis and defend-in-place analysis.

An individual must move from an occupied room to the designated area of refuge. This involves identifying paths to the area of refuge and constructing smoke tenability curves for the target rooms along the route. The analysis examines the available time for an individual to move into the area of refuge without passing through untenable target spaces.

The area of refuge is a defend-in-place room. One identifies tolerable levels for combustion products and constructs performance curves. One can compare tenable time in the area of refuge with the fire duration.

18.15 Fire department rescue

Saving lives is a primary responsibility of a local fire department. This activity in a fire emergency may involve several types of operation. In one case the fire department may assign fire fighters to guide or direct occupants to safety. If occupants are trapped, a search and rescue operation may take place where fire fighters methodically search the building to find trapped individuals.

If victims are discovered, the rescue may require substantial physical effort. A third way to save lives is to extinguish the fire quickly to stop the generation of combustion products.

Guiding occupants to safety is the least physically demanding and requires the smallest number of fire fighters. Sometimes fire fighters take charge of occupant movement and act as an authority to eliminate the need for occupants to evaluate cues. This operation can efficiently handle large numbers of people. Also included in this category is the traffic supervision and operation of elevators. A fire fighter can take control of an elevator and use it to manage the transportation of occupants in a building fire. Fire fighters may transport individuals from above a fire to a few floors below the fire, rather than to the ground floor, thus saving substantial time.

Search and rescue is a time-consuming, staffing-intensive activity. Two types of search operations are used, primary and secondary. The primary search is done on arrival, particularly if there is reason to believe that people may be in the building or if a bystander says that occupants remain in the building. A primary search is usually a very rapid exploration of rooms to discover if any victims are still in the building. On the fire floor, fire fighters usually move directly toward the fire then systematically search the rooms for victims. On floors above the fire floor, the search usually starts at the point of floor entrance and moves toward the fire.

The primary search often has conditions of poor to no visibility and may involve high heat conditions. Physical strength requirements are demanding because of the low visibility, heat, heavy protective equipment and self-contained breathing apparatus (SCBA), and the forcible entry tools needed to gain entry into locked or blocked areas. The mental stress demands are also significant. Room searches when there is no visibility are time-consuming and imperfect.

If a victim is discovered in a primary search, their removal can be difficult and time-consuming. The fire fighter must have the physical strength to carry the victim to safety. A general rule of thumb is that the maximum rescue capability of one fire fighter at any one fire emergency is two occupants. This staffing expenditure is for the entire incident. When search and rescue is an expectation of the fire department, enough personnel must be dispatched to the incident in sufficient time to conduct an effective search and to perform all the other activities that are demanded.

A secondary search is normally conducted after the fire is under control. A secondary search is more careful and is often combined with the overhaul operations. This detailed examination of the building for victims can take much time if smoke, clutter, and debris are present.

There is a demand for available staffing on the fire ground. The fire ground commander must decide on the allocation of fire fighters. Should the available staffing be allocated to search and rescue or to fire extinguishment? If an individual is lost because of a delay in search and rescue, criticism may be directed at the fire ground commander. On the other hand, if search and rescue efforts use up too much staffing and the fire attack and water application is delayed, the fire can propagate because of its exponential growth characteristics. This more extensive fire propagation could endanger even more individuals. When staffing is insufficient, decisions can be difficult.

18.16 Evaluating fire department rescue

Fire department rescue is identified by several events of Figure 18.1. Local fire departments regularly assist occupants to safety or rescue them. The presence of this organization in times of emergency is comforting to those who feel threatened. The value of a fire department in an emergency is clear. Risk characterizations associated with fire department rescue must identify potential problems before an emergency arises. Risk management identifies alternative measures to alleviate weaknesses.

Most fire ground commanders allocate the available (and frequently insufficient) resources to accomplish both tasks. However, fire ground operations are very labor-intensive. When insufficient staffing is available, the tasks must be accomplished by fewer individuals. This causes fatigue and delay. Fire grows exponentially and humans move linearly, so time is critical at a fire scene.

The analysis of fire department rescue involves two parts. The first identifies locations of occupants who may be involved in untenable conditions during the fire. The second examines emergency service response and estimates the number of fire fighters available for rescue.

Building performance and risk evaluations indicate problems that can be expected with the building. For example, life safety evaluations identify only the likelihood that people will encounter untenable conditions, rather than the likelihood of death or injury. Humans can certainly traverse some untenable conditions. But one can also recognize that people may be trapped or will turn back when confronted by untenable conditions. These individuals may require assistance in leaving the building or they may need to be rescued.

Estimating whether occupants may need assistance is a first step. Another involves the fire department response. The M curve analysis identifies the number and types of apparatus and the staffing. The major fire ground decision involves the allocation of staffing between fire suppression and search and rescue. The number of fire fighters responding at the scene becomes critical to the success in each part of the operation.

A 1993 survey concerning first-alarm staffing and allocation of resources received responses from 27 communities of widely varying populations [7]. The average number of fire fighters in a first response was 18–22 with the range varying from 8 to 37. Approximately 25% of responding fire fighters were normally used for primary rescue operations, although here too the range was quite large.

The goal of a building analysis is to understand what to expect before the fire. Building performance and life safety analyses, integrated with the local fire department resources and capabilities, give a good insight into potential problems in life safety and rescue. This enables one to discern the nature of life safety problems. Generally, buildings that have clear life safety problems also pose search and rescue difficulties.

18.17 Fire-fighter safety

Risk characterizations for fire fighters are often delegated to the fire service with little thought by others associated with the design, management, or operation of the building. However, the problem relates to community fire fighters and the building design.

Fire ground operations involve three types of hazard. One occurs during fire suppression operations. Another involves the search and rescue of occupants, and the third relates to overhaul operations. A knowledge of the building and its expected fire conditions is central to risk characterizations for fire fighters.

Risks associated with fire suppression operations can be recognized during an M curve analysis. The fire size and conditions that relate to finding the fire or to trapping fire fighters can be associated with the time line estimates. In addition, the potential for collapse can be assessed. Collapse may occur because thermal expansion can push frangible walls to the point of instability. Collapse can also occur when high-piled storage is supported by unprotected light steel frames. Heat can reduce strength and compression stability to the point of failure. Because these frames are not a part of the building structure, they may easily remain unrecognized.

A second part of fire suppression risk characterizations relates to hazards that are a part of routine building operations. For example, the presence of unprotected holes and shafts is particularly dangerous during operations where visibility can be almost nonexistent. Also, rooms associated with biological or chemical activities can have hazards that may create health problems in the near or long term after the fire is extinguished. A risk characterization for fire fighters includes these types of conditions.

The danger to fire fighters can be even greater during search and rescue operations. The culture of fire fighters has a strong inclination toward saving people in danger. They often take personal risks to rescue an individual. Results of a risk characterization for occupants should be shared with the local fire department to provide an awareness of potential rescue problems. When the building management does not have this type of risk characterization, the local fire department may wish to develop one of its own as a part of the prefire planning process.

Overhaul operations can provide unexpected hazards. Smoldering fires produce increased quantities of carbon monoxide and other toxic gases. Fire fighters who do not wear SCBA gear are in greater danger to the effects of carbon monoxide and the cocktail effect of other gases. The effects may be delayed until the fire scene is secured and the staff returns to the fire house.

PART B: PROPERTY SAFETY

18.18 Other risks

A building provides protection for property and operational functions besides life safety. Risk characterizations evaluate the likelihood that property or operational functions will survive a fire. The usual focus of attention is high-value property and operational continuity. Risk characterizations are specific to the building operations. In a technological society the storage, management, and transmission of information become an integral part of both areas. Other concerns of modern building fires involve threats to neighbors and the environment.

Potential risks may be addressed by risk management programs that tailor loss-reduction strategies to site-specific needs. A prerequisite is a clear understanding or the nature and substance of the risk. An analysis locates the property, mission, or functional features that are at risk and estimates the likelihood of loss associated with a building fire. The room of origin for these risks may be different from those for life safety. However, the knowledge gained from a performance analysis will allow extrapolations to characterize these risks.

18.19 Property protection analysis

Property, including the building structure and its contents, may be exposed to fire damage or destruction. Many buildings have individual rooms that house contents of high monetary value (e.g., information) or irreplaceable historical or sentimental value. The value of this property can be large enough to draw attention when describing a building's fire performance. Some owners do recognize a problem. Others are completely unaware of what may be at risk.

Electronic communications, computers, and controls plus their transmission equipment require a new awareness of property value. For example, fiber-optic interconnections between rooms can have substantial value. A minor incident may damage the cables, incurring a major repair expense. Equipment distant from a fire may be very sensitive to smoke or certain airborne products of combustion. Information and records are valuable commodities.

Protection of property from the flame/heat destruction, as identified by the L curve, is one part of an analysis. However, one may also express property sensitivity (tenability) in terms such as these:

X ppm of HCl (or other corrosive gas);
Y degrees of heat;
Z amount of smoke;
W amount of water application.

We evaluate high-value rooms in a similar way to a defend-in-place life safety analysis. That is, the sensitivity of the contents to the type and concentration of combustion products is determined. Then we construct a tenability curve for the space. This tenability curve, in conjunction with the L curve, gives a sense of proportion to the threat. Fire in another room may pose a major threat.

Risk characterizations include an assessment of the value of information, equipment, and other property in the building. Often only a few rooms of very high value may be important. A performance analysis gives an understanding of the risk to these pockets of high value.

18.20 Continuity of operations analysis

Disruption of operational continuity (business interruption) has been important historically and it is becoming even more important in this age of information technology and just in time operations. The building code does not address this issue. It is possible for a relatively small fire that does not impact life safety to greatly affect the well-being of a business enterprise. Functional downtime after a fire can have an important influence on economic health. Loss of market share through an inability to provide services or materials can be significant. A risk characterization includes an assessment of the survivability of operations to combustion products.

An operational continuity analysis is similar to that for property protection. That is, one identifies spaces that are mission-sensitive and determines the vulnerability of the equipment or contents to combustion products. The L curve and the relevant Sm curves provide an understanding of fire performance on business operations. Fires far removed from the room of interest can be a risk. For example, electrical wiring circuits often pass through intermediate rooms. Loss of a room distant from the mission-sensitive room may be insignificant in itself, but can interrupt electronic communication severely.

The influence of data storage is another concern for relatively small fires. Often owners and designers think of redundancy in terms of electronic redundancy. Location can affect equipment redundancy. For example, when backup copies are made, their storage location is important. The development of risk characterizations involves techniques for questioning owners and technical personnel to uncover the sensitivity of operations to products of combustion.

18.21 Threat to neighboring exposures

The threat of a building fire to neighboring structures is another risk characterization. The exposure threat from neighboring structures to the building being studied may also play a role in a risk characterization report. Neighbors can occupy adjacent buildings or they can be within the same building. Risk is not limited by corporate boundaries.

One may use an extension of the space/barrier concept (Section 8.10) to evaluate neighboring structures. Fire can propagate between two buildings by radiant heat energy transfer or flying brands. The intervening space between the buildings becomes analogous to fuel package separations in a large room where spreadover is the fire propagation mechanism. Probability values are calibration estimates for the physical conditions that are present. The process becomes one of answering the question, Given a fire of X_n rooms, what is the probability that the fire will ignite (1) the roof, (2) exterior cladding, or (3) interior contents of room *xxx* in the adjacent building? Besides the flame/heat propagation, smoke/gas or structural collapse can affect neighboring buildings.

Exposures may also be within a structure. For example, malls and offices house different business enterprises under the same roof. A fire in one firm can influence business operations in the same building or the same complex of buildings. Within a building with several tenants, the question becomes, Given a fire in tenant spaces X, what is the risk that can be expected in tenant spaces Y? Risk characterization considers appropriate scenarios between corporate boundaries.

18.22 Threat to the environment

Environmental concerns have gained attention in recent years because of a series of disastrous and expensive catastrophes relating to fire emergencies. Environmental consequence analyses can be included in a building evaluation, if needed. A risk characterization identifies the effect of airborne, runoff, or subsurface movement of toxic products produced by a large fire. The understanding is based on building performance evaluations. A risk characterization identifies the scope of the problem, whereas risk management identifies potential solutions and their effectiveness.

A building evaluation provides information about the likelihood of fire duration and damage. The L curve analysis, particularly knowledge obtained from an M curve, gives information about materials that can be involved in the fire. The M curve analysis provides a sense of the likely time duration and extent of the fire. Water or other agents used in the extinguishment process will be known and their impact may be examined.

The combination of extinguishing agent and building contents can identify potential environmentally toxic products. Some of the products will be airborne. Environmental sensitivity to airborne products of combustion can be evaluated by an environmental engineer with experience in air pollution. Chemical products created by combustion and extinguishing agent reactions can be evaluated. An environmental engineer with knowledge of toxic products, surface runoff and subsurface transport can evaluate potential environmental risks.

18.23 Closure

The key to understanding risk characterizations for exposed property is to know how the building works. Buildings are constructed to perform functions, and disruption by fire or other natural and man-made threats are not in routine operations. Less attention may be given to disruptive threats because of their infrequency and the immediacy of day-to-day crises. Consequently, the role of risk characterization is to communicate what is at risk and the potential severity. This aspect of a building fire analysis focuses on what can be lost, not what can be the loss.

Fire performance is integrated with the building's functional operations. One selects potential rooms of origin to relate fire performance to building operations. Level 1 evaluations are fast and informative. Observations, estimates, and judgment are integrated to give a picture of performance in different building locations. Combining normal building functions with level 1 evaluations gives a picture of performance and associated risks. That picture becomes the basis for selecting factors to examine more thoroughly. Information acquired through further examination can help to make better decisions.

References

1. Phillips, A. W. Are we playing with fire? Part I. *The International Fire Chief*, Vol. **38**, No. 9, p. 20, October/November 1972.
2. Phillips, A. W. Are we playing with fire? Part II. *The International Fire Chief*, Vol. **38**, No. 10, p. 1, December 1972.
3. Phillips, A. W. Are we playing with fire? Part III. *The International Fire Chief*, Vol. **39**, No. 1, p. 3, January 1973.
4. Phillips, A. W. and Cope, O. The revelation of respiratory tract damage as a principal killer of the burned patient. Burn therapy II. *Annals of Surgery*, Vol. **144**, p. 1, 1962.
5. Phillips, A. W. Beware the facial burn. Burn therapy III. *Annals of Surgery*, Vol. **156**, p. 59, 1962.
6. Phillips, A. W., Tanner, J, and Cope, O. Respiratory tract damage and the meaning of restlessness. Burn therapy IV. *Annals of Surgery*, Vol. **158**, p. 799, 1963.
7. Menke, W. K. Predicting effectiveness of manual suppression. MSc thesis in fire protection engineering, WPI, 1994.

19 INTRODUCTION TO RISK MANAGEMENT

19.1 Introduction

Risk management can mean many different things to different individuals. An insurance company, an investment broker, a fire officer, a building contractor, and an industrial company president all have different meanings, functions, and ways of dealing with the risk. Yet the thread of commonality that weaves through each type of application involves making decisions under uncertainty.

One part of risk management is to understand specifically what is at risk; another is to have a sense of its relative severity; a third is to make a decision about what to do about the risk. Sometimes these pieces are addressed with rigor, sometimes in a fuzzy, casual manner, sometimes in a mixed mode. Some view risk management exclusively as a decision of what type of insurance to purchase. Some risks are ignored (i.e., a passive decision to do nothing) in the hope that misfortune will not happen. If an accident were to occur, decisions of what to do would be handled at that time. Some risk management is formulated with a thoughtful, rational process that involves a partnership between technical experts in the exposure field and the team that must manage and operate the enterprise.

This introduction to risk management will suggest a way of thinking which structures a process that links technical competence in the fire field with management of the business enterprise. The goal is to make better decisions faster, easier, and more economically. Complex operations must deal with a variety of day-to-day decisions to keep the business going. Fire safety is normally low on the list of immediate needs. Yet there are times when fire safety moves up on the agenda to require decisions that can affect the physical and financial well-being of the enterprise.

19.2 Comparisons

The needs of the problem should determine the best way to compare alternatives. For example, decision-making needs by an insurance company on rate-setting policy are different from those of a manufacturing company deciding what type of insurance to purchase. The former deals with a large population of insured whereas the latter is involved with a unique, individual operation. Objective probability and statistical studies may be appropriate for an insurance company to compare alternatives. However, individual business decisions are more appropriately structured

Building Fire Performance Analysis R. W. Fitzgerald
 ISBN: 0-470-86326-9 (HB)

on understanding the unique operations that take place. Measurements and communication should be appropriate to the needs of the application.

Risk management focuses on the individual, site-specific building. It becomes a personal analysis for that building and its operation. Information has both cost and value. We operate on a principle that the value of information must be greater than its cost. Therefore an evaluation is tailored to the needs of the problem. When a level 1 basic evaluation is adequate to provide sufficient understanding of building performance and risk, no additional information is needed. When one needs a greater understanding of detailed features, we can use levels 2 or 3 to target specific information.

19.3 Process overview

Figure 19.1 shows a complete risk management process. There are many potential uses for this structure, and some applications will rearrange or even omit certain activities. The activities are distinct, but the information and knowledge have interactions. The intent is to provide an organized structure for the process.

The functions of each part of the process are summarized below. They are discussed in greater detail in this chapter.

UNDERSTAND THE PROBLEM

Understanding how the building and operations work during normal usage is the most important aspect of any risk management program. This has a minor influence on risk assessment, but a major impact on risk management. In gaining an understanding of how the building works, one identifies what is at risk and the sensitivity to combustion and suppression products.

IDENTIFY BUILDING FEATURES

This activity defines the building for the performance evaluation and its associated risk characterizations. Space utilization plans can note specific features important to an evaluation.

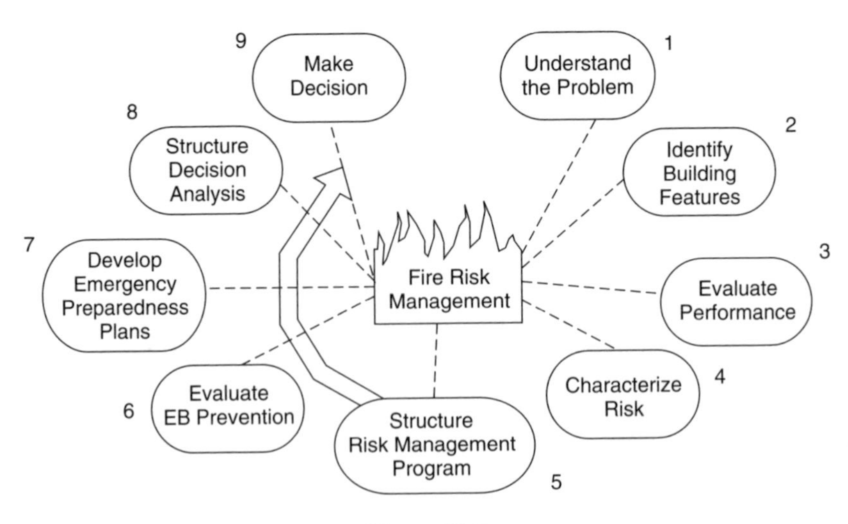

Figure 19.1

EVALUATE PERFORMANCE

One develops the understanding of technical performance by evaluating flame/heat damage, structural frame behavior, and smoke tenability.

CHARACTERIZE RISK

After one understands how the building will behave, threats to people, property, mission, neighbors, and the environment are analyzed to characterize the risks.

STRUCTURE A RISK MANAGEMENT PROGRAM

In theory this activity starts after characterizing the risks. However, much knowledge will have been gained during the evaluations and some ideas will have been formulated during that process. This activity devotes a conscious effort to a risk management program. The process may continue in parallel with other activities, because alternatives will depend on or be influenced by fire prevention, emergency planning, and decisions relating to actions. If a risk management program is not needed, the activity may be omitted.

EVALUATE EB PREVENTION

This first step of a traditional hazard analysis is intentionally delayed until after we understand the risk. It is more useful to clearly understand the consequences of an unwanted ignition in isolation. Then, if improving or maintaining an effective fire prevention program is an option, its importance and function can be recognized more clearly.

Preven established burning has two components. One is the traditional ignition prevention involving the separation of heat energy sources and available kindling fuels. The second follows an unwanted ignition by preventing flames from growing to established burning. This incorporates initial fire growth and fire suppression by building occupants.

DEVELOP EMERGENCY PREPAREDNESS PLANS

Emergency preparedness plans are based on understanding the building function and operations, the fire defense system, and the associated needs of the enterprise. One may organize plans into three sets of activities that may be instituted before, during, and after the incident.

STRUCTURE DECISION ANALYSIS

A decision analysis structures alternative courses of action in a way that enables a manager to compare the impact of decisions. One normally packages alternatives into a very few integrated plans of action. Cost and effectiveness comparisons are included with the alternative packages.

MAKE DECISION

The choice may be to do nothing, to select one of the alternatives, or to investigate other changes. Management decisions should be based on a clear understanding of the building's performance and associated risks. Costs and effectiveness of the alternatives are an integral part of a decision analysis structure.

19.4 Understand the problem

Four major areas of information assembly are involved in this activity. They include (1) to understand how the building functions and operates; (2) to identify what is at risk; (3) to learn what managers believe is important and are willing to accept regarding losses on a fire incident; and (4) to document the problem definition. Figure 19.2 shows a mind map of the major parts of the activity.

Understanding the functions, operations, and how the building works is a difficult yet important activity. This task is easier to do for existing buildings. Proposed new buildings may require more discussion with operations management and the architect. The information is essential to establish a workable, effective risk management program.

This activity includes identification of the sensitivity of people, contents, and operational continuity to the potential products of combustion and time frames for emergency actions. In cooperation with the building's operations management, tenable and tolerable limits of the products of combustion should be identified. Often management may not know suitable values and may ask what is appropriate. One may describe tenable limits for human protection with citations and a rationale for reasonable values. Tenability limits for information technology, data storage, and equipment can be more difficult to ascertain, and those limits may be delayed until one gains a better understanding of the building's performance. Sensitivity of neighboring properties to

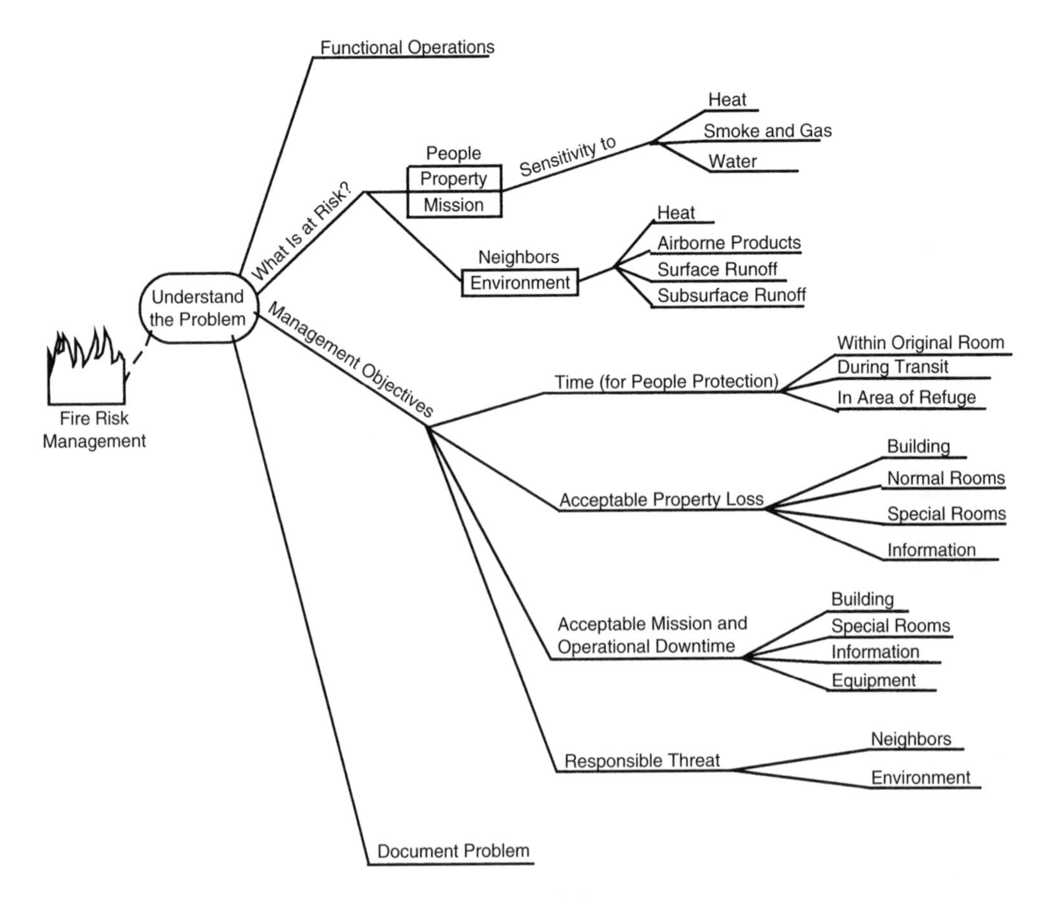

Figure 19.2

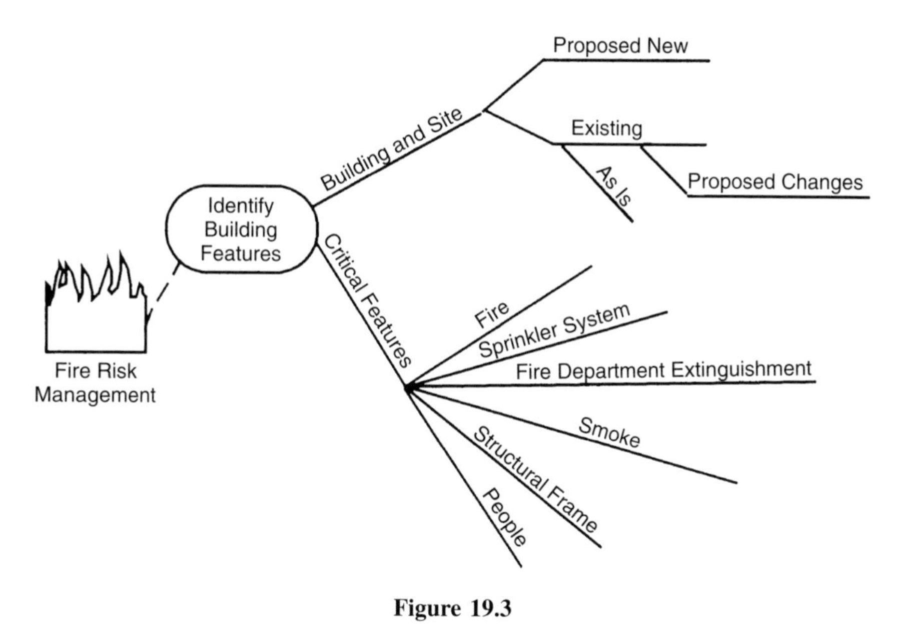

Figure 19.3

exposure fires and the susceptibility of the environment to air and water runoff may be noted. A better time to incorporate this aspect of risk management may be after the performance evaluation is completed. At that time, one has a better sense of the problems which actually exist.

Documentation of the information, results, and problems to be addressed is valuable in this activity. Space utilization plans become useful ways of documenting and retrieving the information. Broad observations may often be sufficient for the documentation. Data collection forms, such as those shown in Appendix D, illustrate a rapid way to collect and document information for this activity.

19.5 Describe the building

This activity identifies building features that are important to performance evaluations and risk characterizations. It becomes the visible object that links functions, risks, and performance. Its outcome defines the building's fire defenses and features that influence their performance and the associated risk characterizations. Figure 19.3 shows a mind map of factors for consideration.

19.6 Evaluate building performance

The flame/heat performance evaluations provide an understanding of the fire size, conditions, and damage that can be expected. Although a detailed analysis may focus on a single or few rooms of origin, extrapolation of that knowledge to envision fires in other locations enables one to describe differences in performance. Some fire locations will clearly cause less damage. Other locations may identify conditions where the fire could be worse even if it is in a small but critical pocket of operations. This may or may not indicate whether one should give more attention to this location. The single (or few) detailed analyses plus their extrapolations give an understanding of the performance variations that can occur in the building.

Knowledge of the building's response to fires becomes a basis for understanding how the building's structural frame and smoke tenability of target spaces are likely to perform. Again, one

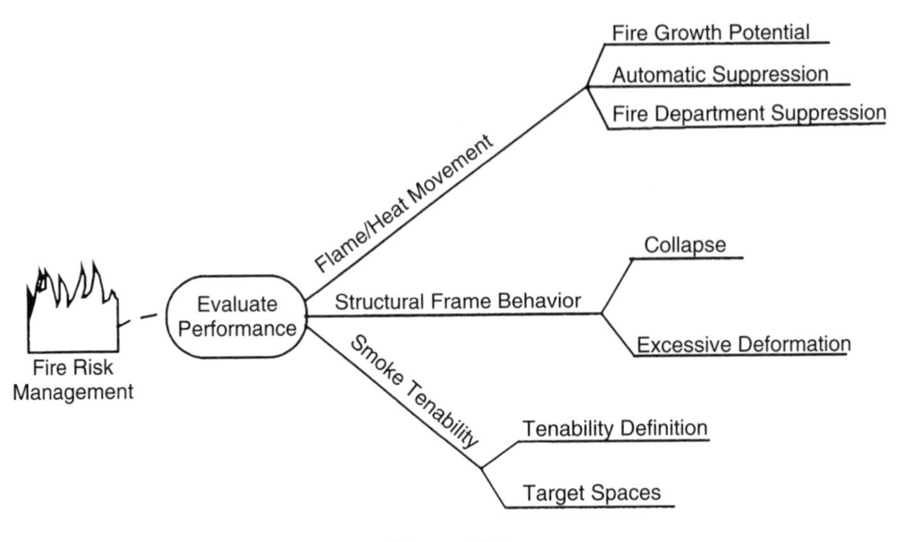

Figure 19.4

or a few detailed analyses can be extrapolated by envisioning fires in other locations and estimating building behavior. Logical extrapolation will provide an understanding of fire performance.

Figure 19.4 shows a mind map of the evaluations associated with this activity.

19.7 Characterize risk

Performance evaluations provide knowledge of the fire, the associated structural stability, and smoke tenability. The goal is to understand the range of performance to tell a credible story about the building's fire behavior. This understanding, combined with the information gathered in Activity 1, Understanding the Problem, provides a basis for characterizing the risks.

The risk for people can be expressed in terms of the time the building will provide to individuals before they encounter untenable conditions. Expected fire damage may be adapted to include the risk for important property contents or information and functions essential to the facility's operational continuity or mission. The impact on neighbors and the environment may also be described, if necessary. Regardless of how one measures risk, it must be described in terms that a business manager can understand quickly and accurately.

Figure 19.5 shows a mind map of the ingredients of a risk characterization for a building.

19.8 Structure a risk management program

The knowledge and understanding that emerges from completing the first four parts of the risk management process of Figure 19.1 is a central element of almost all applications. If the purpose is to describe building performance from the perspective of code plans approval, the process would be completed because enough information is available for a regulatory decision involving acceptable performance. However, the acquired knowledge can serve a variety of other applications such as corporate planning or fire ground operations. A risk management program looks at the question, How can I do better with my available resources?

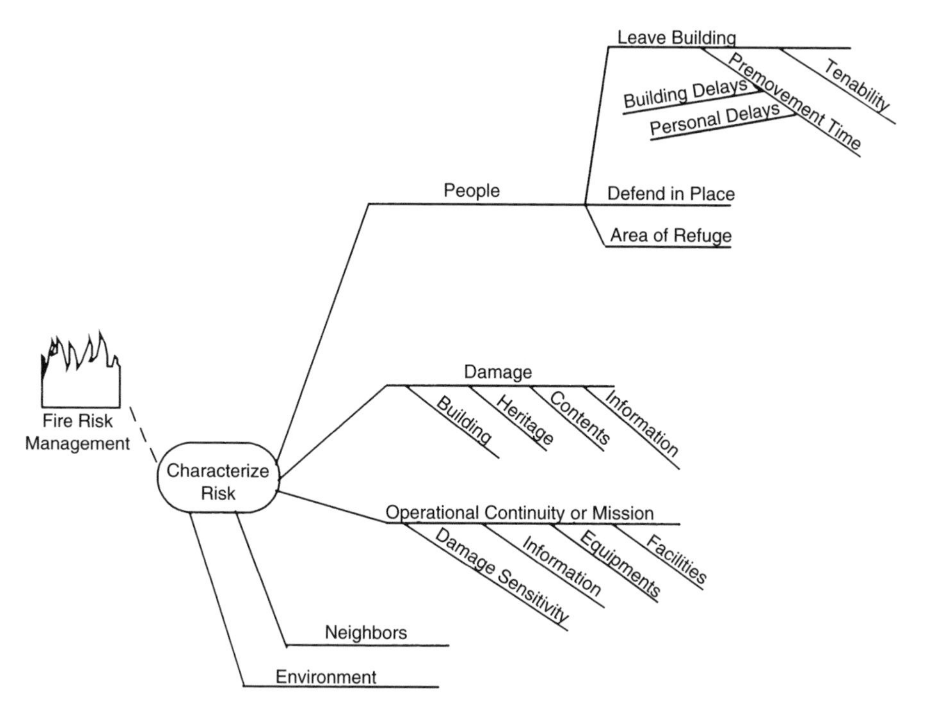

Figure 19.5

Parts 1 to 4 provide a foundation of knowledge from which to organize comprehensive services for a building or enterprise. For example, with this knowledge one may plan the general structure of a risk management program. A risk management program is an integrated organizational strategy that establishes practices and policy with which to guide the fire risk management for the building or the enterprise. The risk management program can be developed in parallel with information from a "prevent EB" analysis and the development of emergency plans. The entire process can be incorporated within a decision analysis that packages alternative courses of action and their costs and effectiveness.

The basis for a risk management program is the building fire performance. This analysis can give a good perspective of the consequences of an unwanted fire anywhere in the building. Sometimes a building's survival may rest only on the effectiveness of the "prevent EB" program. If this is the case, building management must be aware of the situation and give adequate attention to fire prevention, improve the building's fire defenses, or purchase appropriate insurance. Regardless of the performance of the building after EB, a knowledge of the quality of "prevent EB" activities is an important part of a comprehensive fire risk management program.

The development of emergency preparedness plans is another part of a fire risk management program. These plans consider actions for various fire conditions before, during, and after ignition and EB occur. Disaster plans may incorporate hazards beyond fire, such as hurricanes, earthquakes, floods, terrorism, and civil disobedience.

Figure 19.6 shows the factors of a comprehensive risk management program. The program is tailored to a specific enterprise, rather than forcing a fit into a standard format. The knowledge that is gained provides a basis for the decision analysis described in Section 19.11.

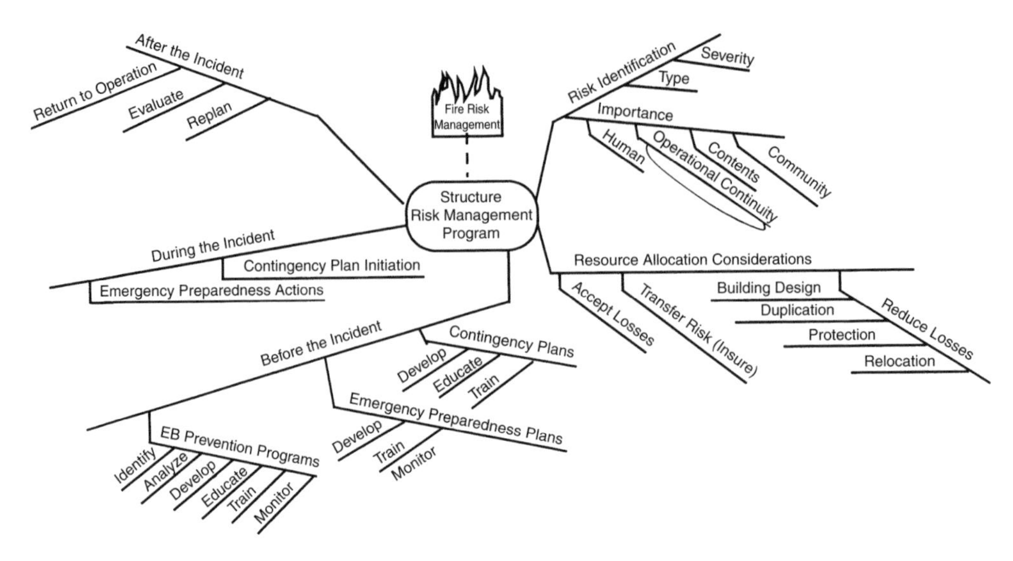

Figure 19.6

19.9 Evaluate EB prevention

The evaluation of fire prevention involves two events. One is the traditional "prevent ignition" analysis. Here the goal is to avoid ignition by evaluating the ability of building operations to separate heat energy sources from readily ignitable fuels. Heat fluxes, piloted flames, and their time durations are a part of this assessment. Level 1 evaluations would use traditional hazard assessment procedures to get a sense of trouble locations for unwanted ignitions.

The second part of a "prevent EB" evaluation analyzes the fire from that first fragile flame to established burning. This process considers the fire and early fire extinguishment. The ease and speed with which the fire grows during this period depend on the fuels and heat fluxes. Early fire extinguishment considers only occupants. Although occupant suppression is possible for fire sizes greater than EB, from a building evaluation viewpoint, occupant extinguishment is limited to small fires. If special training or conditions warrant other definitions of occupant suppression capabilities, they can be incorporated.

Figure 19.7 shows a mind map of the elements in a "prevent EB" evaluation.

19.10 Emergency preparedness

The Latin phrase *praemontis praemuntis* 'forewarned is forearmed' is appropriate for emergency planning. Although neither a specific date nor the exact scenario for a disaster is known, contingency plans can be developed before an anticipated event while enough time is available to provide a thoughtful response strategy. This book focuses on fire emergencies, but many contingency strategies are also appropriate for earthquakes, hurricanes, floods, terrorism, or civil disobedience.

An emergency plan is based on elements in Figure 19.2, Understand the Problem, and Figure 19.5, Characterize the Risk. The emergency plan provides a strategy to preserve the people, property, and operational continuity or mission of the building. The knowledge gained from the performance analyses provides a base to identify physical actions that can augment the building's capabilities.

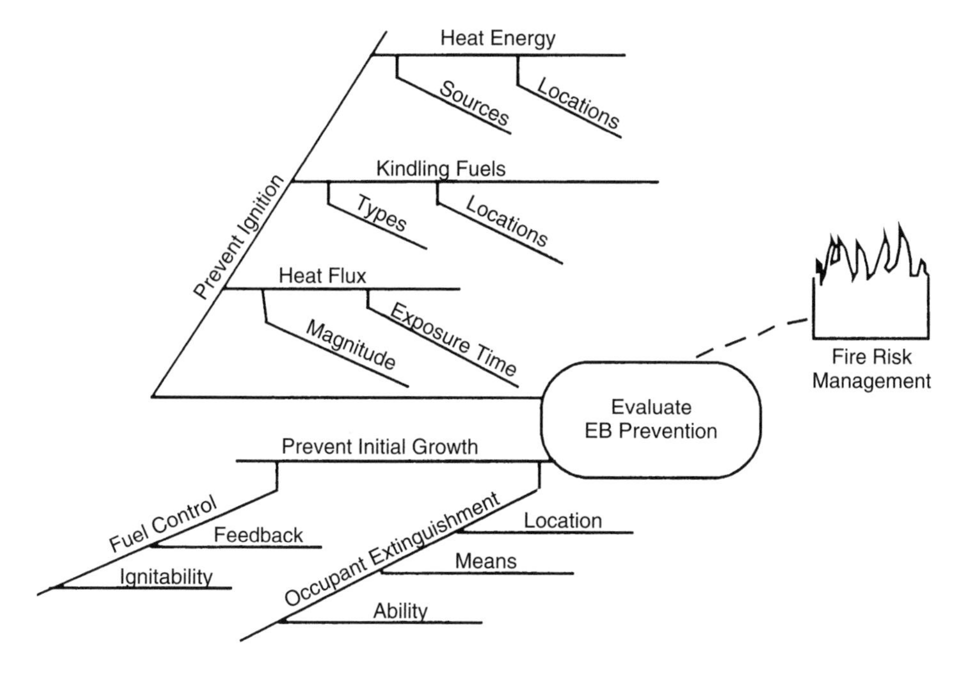

Figure 19.7

The plan may be organized into actions that take place before, during, and after an event. Each action is tailored to the site-specific needs of the building. One may wish to ensure that physical and other (e.g., electronic) redundancies are separated adequately. For example, important electrical equipment often has redundancy installations. When electronically redundant systems such as those of a telephone center are separated by a few inches of air space, a fire in one will probably cause fire in the adjacent gear. Fire is often not a consideration by others in designing functional protection. Individuals sometimes tend to overlook events that are beyond the scope of their expertise. Therefore a single risk manager must be responsible for coordinating the complete system into an integrated plan.

Figure 19.8 shows a mind map of the elements of an emergency preparedness plan.

19.11 Decision analysis

The primary goal of a decision analysis is to organize feasible alternatives to concisely inform a decision maker about courses of action. Advantages, disadvantages, costs, effectiveness, and potential consequences of each alternative are linked to the alternatives. The decision maker is rarely the individual who has done the analyses or understands details of the complete picture. One may organize information to compare alternatives. The presentation must be clear enough for a decision maker to recognize meaningful differences and implications of the alternatives. The time available to grasp the essence of the story is brief, and it must be told succinctly and accurately.

The development of feasible alternatives requires effort. It is a synergistic process that integrates performance and associated risk expectations, costs, and economic forecasting with a good understanding of the building operations and functions. The development of the risk management program becomes a basis for structuring a decision analysis. The alternatives considered in

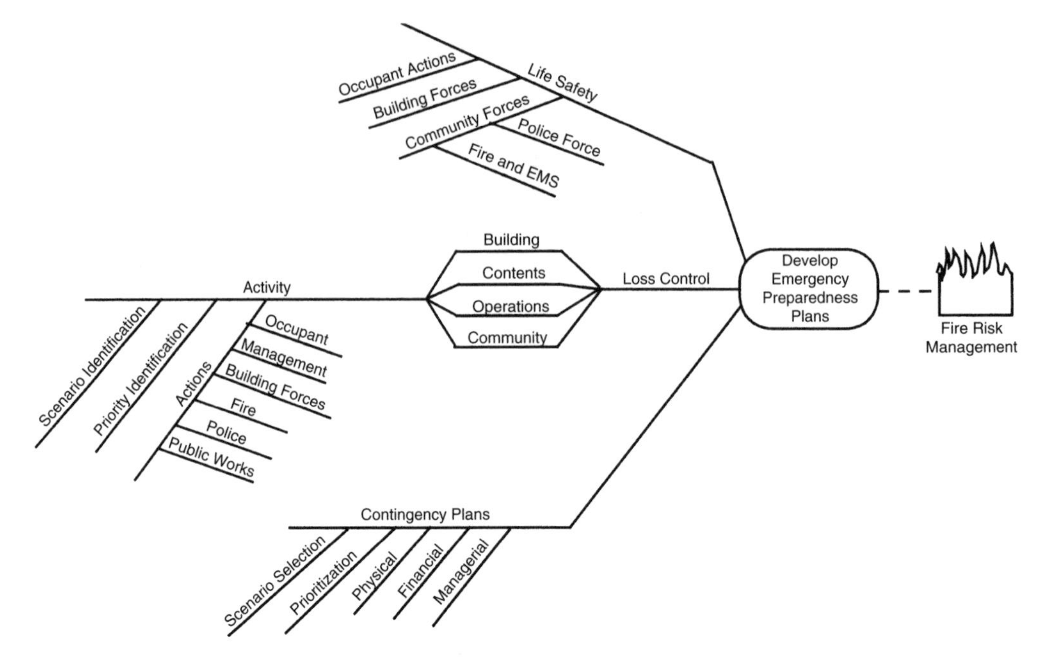

Figure 19.8

risk management programs become an integral part for the decision analysis structure. The mix among accepting potential losses, transferring risk by buying insurance, and changing the risk by installing fire defenses can be augmented with better "prevent EB" procedures and effective emergency preparedness plans.

One possible alternative is always the status quo. This involves no additional expenditures and the risk characterization identifies potential losses that can occur. It is often desirable to compare alternatives with this performance. For example, the effectiveness of fire prevention procedures and emergency preparedness plans may be superimposed to provide a picture of the overall risk as well to distinguish the role of each in the outcome. The cost and effectiveness of changes in alternative plans can be compared more easily to this base.

When one bases risk outcomes on the understanding that evolves from performance estimates, the effect of proposed changes can be described relatively quickly and easily. Therefore knowledge from a performance analysis enables one to recognize logical modification in the building and their effect on performance. These potential changes may be incorporated into candidate risk management programs to enable decisions to concentrate on the big picture and the associated costs.

Clear communication of outcomes and costs is fundamental to a decision analysis structure. Implications of risk, loss aversion, cost, and affect on the operations and stability of the enterprise are a part of the process. Although engineers often think in terms of success and avoidance of failure, risk managers are more often influenced by loss considerations. The analytical framework of this method focuses on success. In risk management descriptions business decisions it is usually better to express outcomes in terms of loss potential rather than success. The conversion is trivial, but the understanding may be greatly improved.

Figure 19.9 shows a mind map of the elements of a decision analysis.

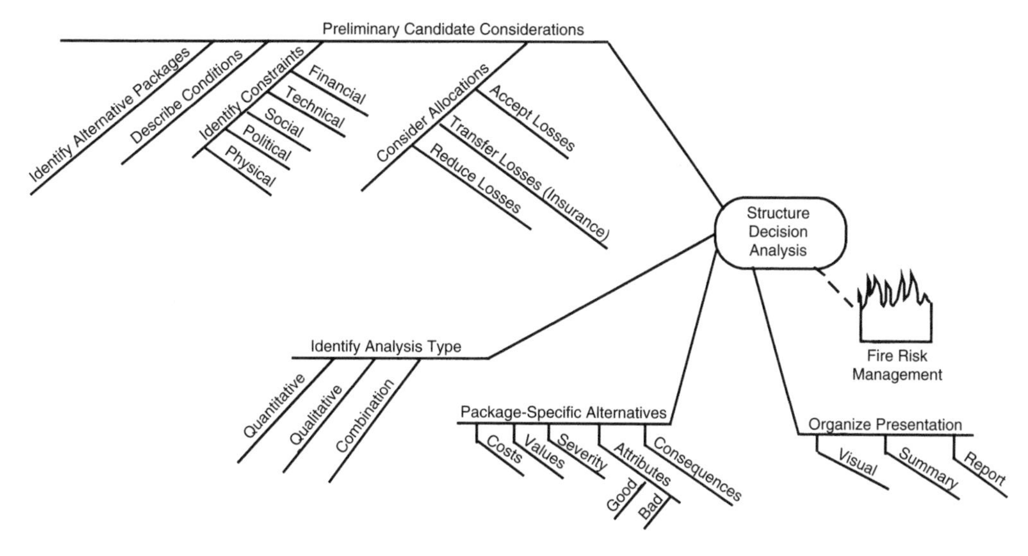

Figure 19.9

19.12 Decision making

Decision makers are rarely the individuals who do the performance analysis or develop the risk management program. Documents that form the basis of a decision must portray the existing situation and alternatives clearly and succinctly. Management rarely has time to read reports and investigate technical reasoning for recommendations. Consequently, the report must frame the alternatives in a way that management can comprehend the implications clearly. A decision may be selection of a prepared alternative or it may be a request to explore another solution not originally presented in the decision structure. Because performance is based on understanding from technical estimates, changes can often be evaluated relatively quickly. The final part of Figure 19.1 is making a decision.

20 PREVENT ESTABLISHED BURNING

20.1 Introduction

Identification of conditions that contribute to ignition and initial fire growth is an important part of traditional hazard analysis. The results are often blended with a study of fire propagation and fire defense operations to produce a comprehensive description of the peril that a building faces. One might expect fire prevention to be introduced early in a performance analysis because its failure is the triggering event that sets the entire process into motion. However, the delay in discussing this topic is intentional because fire prevention is so important to the risk management of business enterprises.

The primary focus of a performance evaluation is to understand how a building will behave after a fire starts. The frequency or ease of first ignition and the ability to reach established burning is irrelevant to the performance of the building after EB has occurred. Often it is difficult to grasp the true role of fire prevention when the two parts are commingled. Many buildings survive only on the success of fire prevention. It is important to recognize when this condition exists.

A risk management program can be developed more easily when one understands the building's performance after EB separately. The system modularity allows one to decouple any one element and examine it in relative isolation. In this way, the costs and effectiveness for any proposed change can be distinguished.

Fire prevention is often viewed by the public and by some fire professionals as only "prevent ignition." We define fire prevention to mean *prevent established burning* and separate it into two parts. The first part is the traditional ignition prevention. This evaluates the ease of producing the first fragile flame that initiates a fire. The second part is prevention of the first flame from growing to the size defined as established burning.

A fire prevention evaluation can start from a fire-free status or it can start from a defined ignition in any designated location. The separate but sequential study of these two parts provides a sense of relative threat. It also gives a measure of the effectiveness of occupant fire suppression. This chapter describes how to use either part as a starting point. The objective is to provide a

Building Fire Performance Analysis R. W. Fitzgerald
 ISBN: 0-470-86326-9 (HB)

logical analysis that uses available information to evaluate a building's fire prevention program in a way that is useful for decision making.

20.2 Prevent EB

The two distinct components of fire prevention are shown in Figure 20.1. Ignition (IG) is defined as the appearance of the first small flame. Before this event, the heat energy applied over a duration of time may discolor the fuel or it may cause smoldering. Smoldering can appear briefly and then a flame appears; or smoldering can last for a long period of time with smoke being generated in the process. During the early stages of a fire, flames can appear and then give way to smoldering; and then flames can reappear. The fire can go out by itself at any time. For our purposes, ignition is defined as the first appearance of any flame. The changes of state after that time, including the fire going out before reaching EB, are merely stages in the evolution of the fire.

During this often unstable initial period of fire growth, either the fire continues to develop and grow, or it goes out. When the fire goes out without any external intervention, that event is defined as self-termination. If the fire grows to the critical size that defines established burning, the fire will have moved to the next realm of development. The fire may also be extinguished by an occupant within this realm between IG and EB. Figure 20.2 shows the SVN that describes these events. Special situations discussed later in this chapter show how automatic suppression of special hazard equipment may be incorporated into an evaluation.

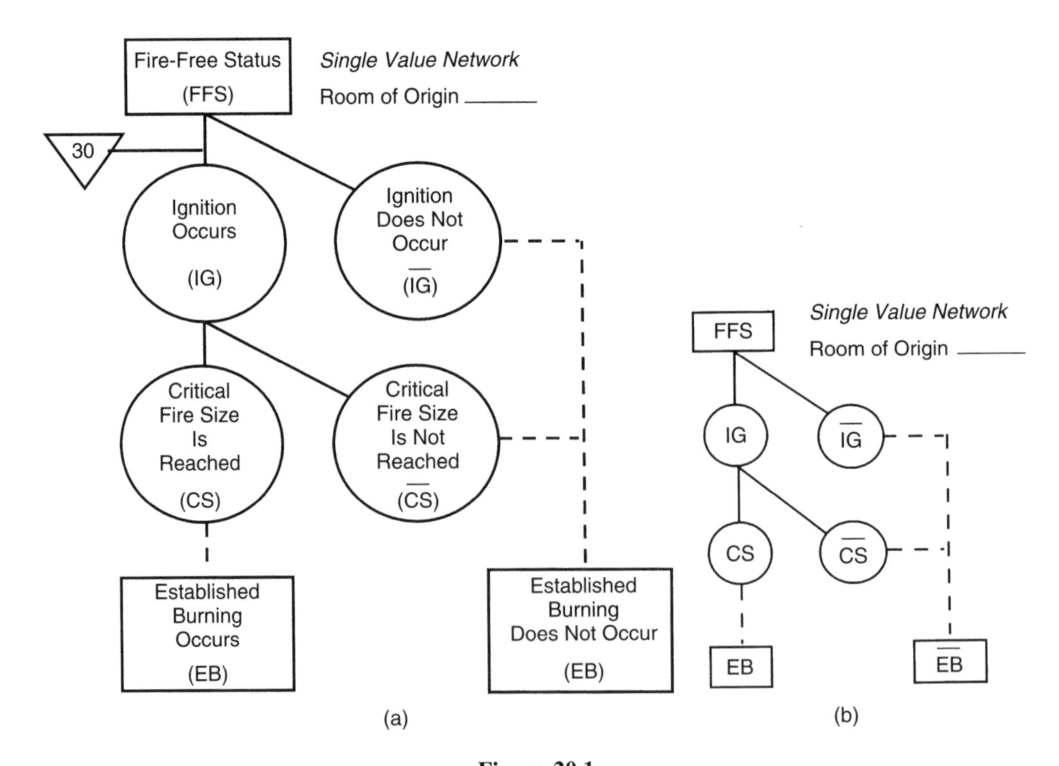

Figure 20.1

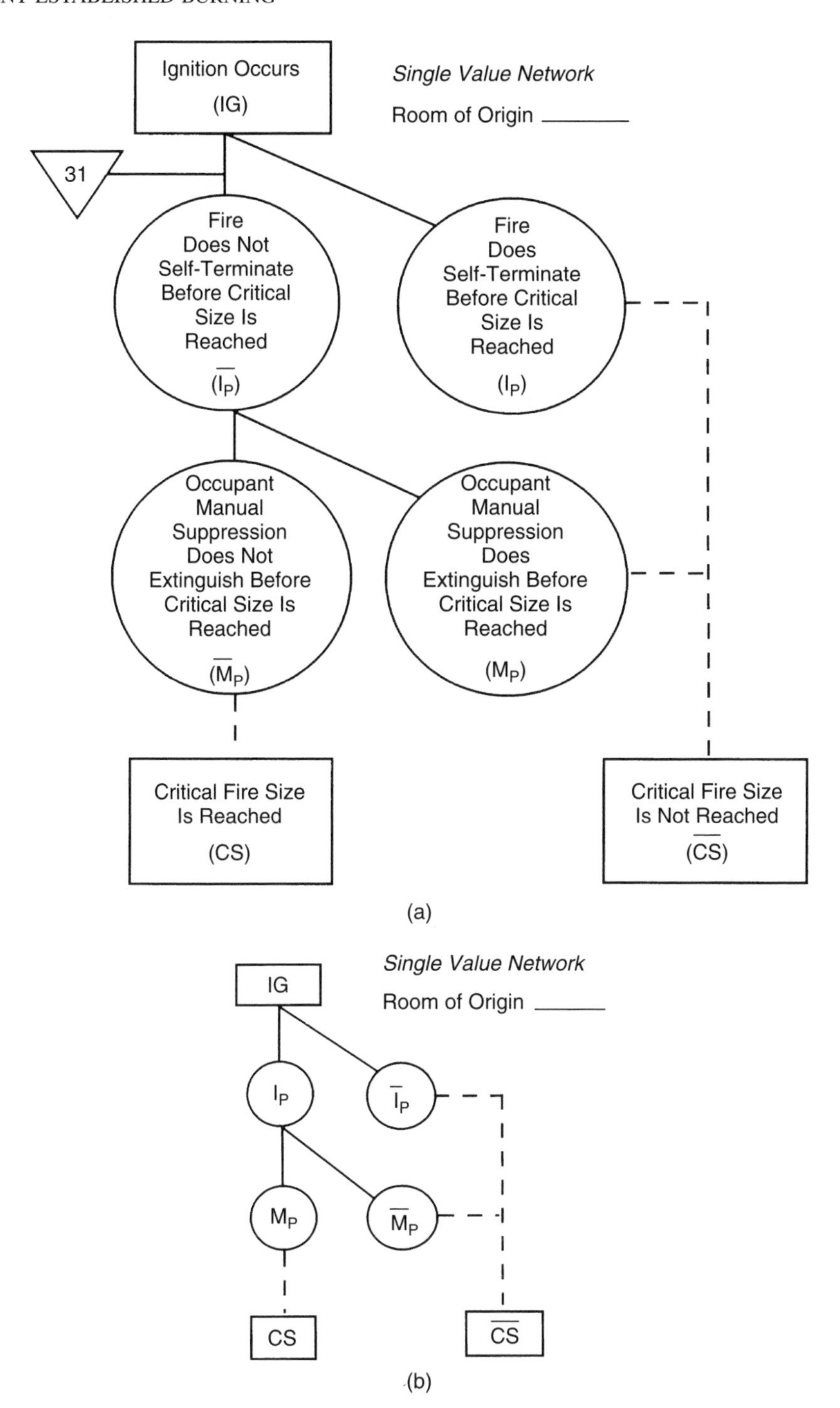
Ignition Occurs
(IG)
Single Value Network
Room of Origin
31
Fire
Does Not
Self-Terminate
Before Critical
Size Is
Reached
$(\overline{I_P})$
Fire
Does
Self-Terminate
Before Critical
Size Is
Reached
(I_P)
Occupant
Manual
Suppression
Does Not
Extinguish Before
Critical Size Is
Reached
$(\overline{M}_P)$
Occupant
Manual
Suppression
Does
Extinguish Before
Critical Size Is
Reached
(M_P)
Critical Fire Size
Is Reached
(CS)
Critical Fire Size
Is Not Reached
$(\overline{CS})$
(a)
Single Value Network
IG
Room of Origin
I_P
$\overline{I}_P$
M_P
$\overline{M}_P$
CS
$\overline{CS}$
(b)

Figure 20.2

20.3 Ignition potential

Ignition (IG) is commonly described in statistical measures. A statistical frequency may be appropriate for some specialized purposes, but it is not useful when the objective is to provide a clear performance understanding of a specific building. In this case an organized analytical process is needed to recognize problems and correct them, if necessary.

In assessing the likelihood of an ignition potential, the building is studied for the location, type, and intensity of

- potential sources of heat energy;
- kindling fuels.

Then one estimates the likelihood that heat sources and kindling fuels will come together closely enough for a long enough time to cause an ignition. This examination provides a good insight into operations and potential trouble spots that can cause ignition. Often information can be recorded on space utilization plans for documentation or communication.

Ignition potential is viewed here in terms of a process. The probability estimate is based on observation and judgment. In this way, if a risk management alternative involves ignition prevention, modifications will relate to the specific building operations. Factors that influence ignition are shown in Figure 20.3.

Level 1 evaluations are organized around the factors of Figure 20.3. The assessment is based on observing building operations to identify the ease and likelihood of combining enough of the factors to cause an ignition. Calibrations may use verbal descriptors, such as red, yellow, green, or a numerical scale. Conditions of certainly no and certainly yes are also valid. Figure 20.4 shows scales.

A range of additional sophistication is available to assess the likelihood of ignition for level 2 and level 3 evaluations. Calculations involving distance, heat flux, or other factors may be used. Analytical hierarchy procedures [1] can describe relative ignition potentials. Statistical data may be useful for some applications. It is important to tailor the analytical process to the needs of the application.

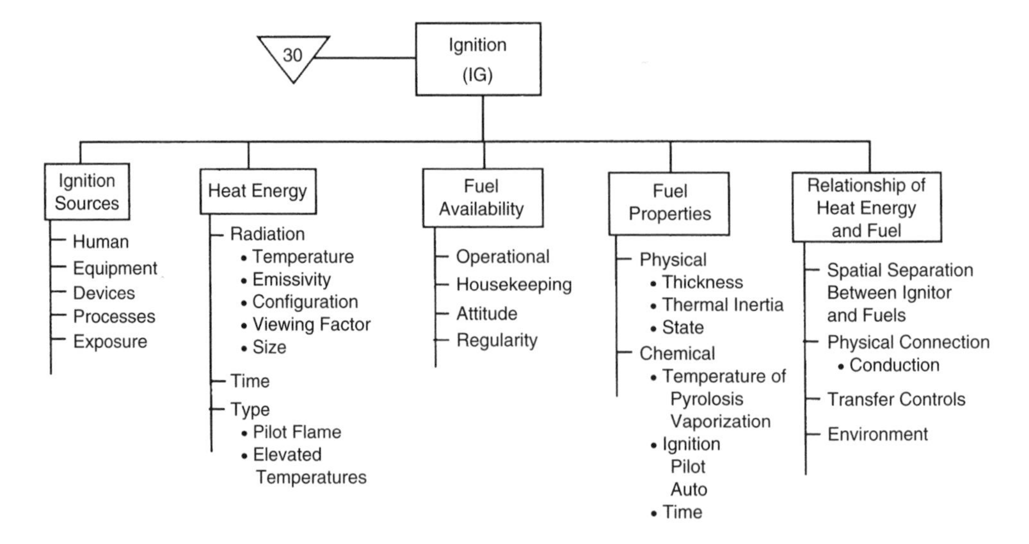

Figure 20.3

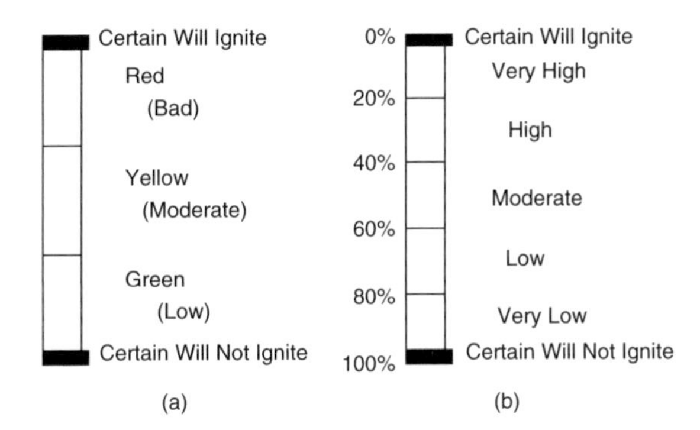

Figure 20.4

20.4 Initial fire growth potential

Given ignition, the analysis examines the initial fire growth potential (I_P) which estimates the likelihood that the fragile flames can grow to established burning. The critical flame size that defines EB is either

- a flame of about 20 kW, or 10 in (25 cm), or knee high, or a wastebasket size; or
- a size more appropriate to the building operation.

A knee high size (bucket of fire) is appropriate for EB in most compartmented buildings. This fire size becomes established in the sense that fire growth predictions can be based on physical principles.

Before reaching the critical size, the fire will either

- go out itself (self-terminate), or
- become stronger and grow in size and power.

This stage of fire growth is described in Section 8.5 as realm 2, initial burning between the ignition point and the radiation point. The assessment of the likelihood of a fire growing to EB is guided by the factors in Figure 20.5.

If the heat flux continues to impact on the target fuel after ignition, it will contribute to the fire growth. Fire retardant treatments that reduce combustibility can be particularly effective when the fire is small, such as in realm 2. However, fire retardant impregnations, coatings, and paints are not effective forever. Many have a limited life or durability and must be maintained regularly. Nevertheless, they can substantially reduce the likelihood of initial fire growth.

Continuity means that the fuel is located at the flame height or above it. For example, an ignition on a horizontal surface, such as a carpet, usually self-terminates because there is no fuel continuity above the flame height. However, if that same carpet is placed on a vertical wall, the flames can preheat the unburned fuel and the fire can grow more easily. Fuel positioning and orientation affect the ease of initial fire growth.

Fuel thickness and surface roughness influence small fires. Thin materials, such as loose paper and wood shavings, behave very differently than those same materials in the form of bundled paper, books, or solid wood blocks. Combustible fuels that are poor insulators (good conductors) allow the heat to penetrate through the depth of the material more easily than good insulators (poor conductors). Good insulators keep the heat near the surface and the material gasifies much more

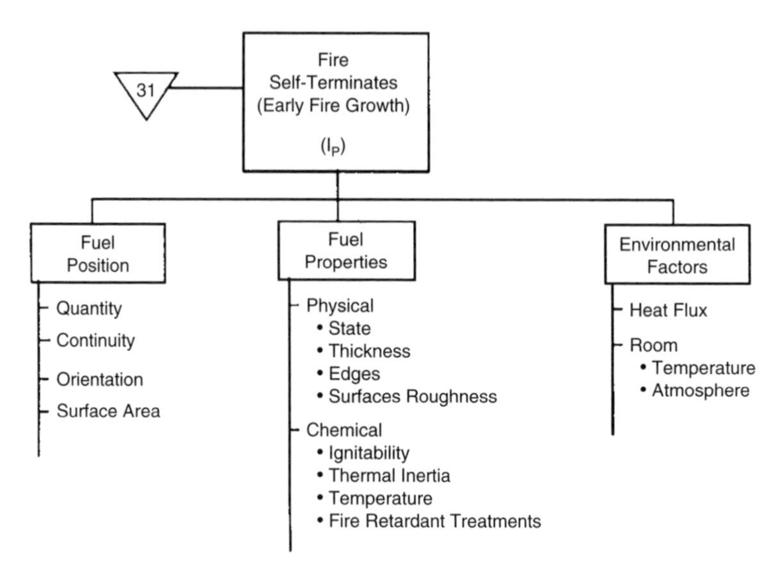

Figure 20.5

rapidly and readily. This allows the flames to grow and develop much more easily. Often materials like foam plastics exhibit rapid fire growth primarily because they are such good insulators.

Many ignitions go out shortly after starting. Others continue to grow easily. The EB estimate is based on judgment that enough factors will work together to allow continued fire growth.

20.5 Occupant extinguishment

Occupant extinguishment (M_P) is very dependent on human decisions. These decisions are strongly influenced by the presence or absence of experiences, training, and instructions. Although one may have skepticism about an untrained occupant's suppression skills, when individuals are in close proximity to the fire when it starts, they may extinguish it quickly. Often these ignitions are never reported, so statistical evidence is difficult to obtain.

Evaluating occupant extinguishment is analogous to fire department suppression. It compares the time for fire growth with the time to get agent on the fire and extinguish it. Figure 20.6 shows the events that comprise occupant extinguishment. They are (1) occupant recognizes the ignition, (2) occupant decides to attempt to extinguish the fire, (3) agent is applied, and (4) fire is extinguished.

Occupant extinguishment normally occurs within such a short time frame that an individual who is not in close proximity to the fire has less chance of extinguishing it. Although a glass of water or a shod foot may be used, the analysis focuses on an individual using a portable fire extinguisher. Figures 20.7 to 20.10 show the factors that influence these events.

The fire size that an occupant can extinguish is defined for analytical purposes to be established burning. Some skilled occupants may be able to extinguish fires slightly larger than the waste basket-sized fire. However, an untrained occupant using a fire extinguisher would normally have difficulty extinguishing a fire the size of a lounge chair.

This fire size for occupant extinguishment is set at EB for two reasons. The first reason is that if the occupant is near the ignition, he or she can take rapid, effective extinguishment action. This is important to a fire prevention program. The second reason enables the analysis to avoid commingling manual suppression actions of occupants with those of fire fighters. Established

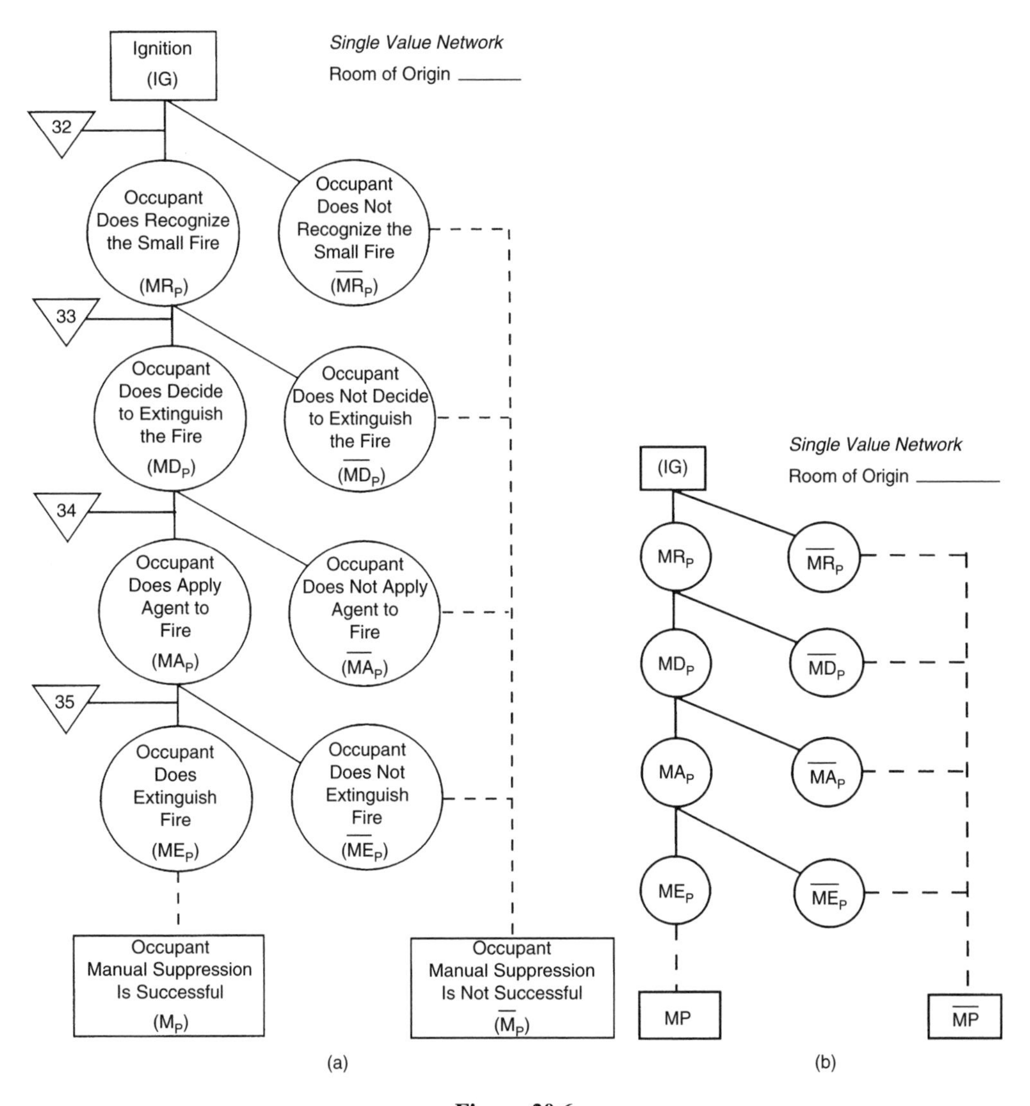

Figure 20.6

burning is a convenient, conservative demarcation that enables an evaluation to incorporate both forms of manual suppression in a realistic, yet disciplined manner. Analyzing fire prevention separately from the active and passive fire defenses provides a flexibility that does not compromise either segment.

20.6 Portable fire extinguishers

Portable fire extinguishers have a role in building fire safety. They may be a valuable first aid measure, or they may be irrelevant. The effectiveness of these devices depends greatly on their location, availability, size, agent type, and the inclination and training of an occupant to use them. Portable fire extinguishers are important devices in "prevent EB" evaluations.

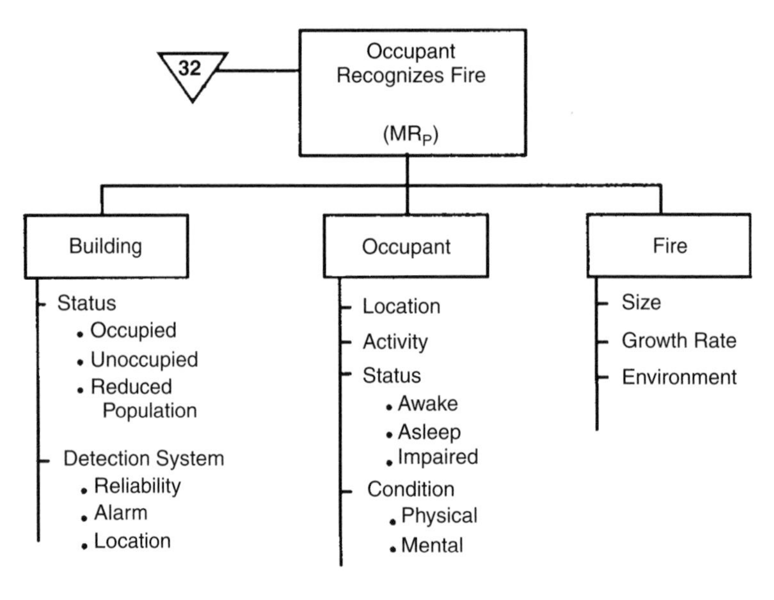

Figure 20.7

The classification of fires is important to understanding types and characteristics of fire extinguishers. Fires are classified according to the type of fuel involved:

- *Class A* fires involve ordinary combustibles of wood, paper, fabric, and plastics. These are the fuel types in most nonindustrial buildings, except for kitchens with cooking greases.
- *Class B* fires involve fires in flammable liquids, such as petroleum products, greases, and gases. These fuels are further categorized with regard to certain flammability characteristics. When the depth of a flammable liquid is greater than 1/4 in, it is usually contained in tanks or containers. When flammable liquids have no real depth (defined as less than 1/4 in) they are not contained and become spill fires or running fires. A third type of class B fire involves flammable liquids or gases that are released from damaged pressurized containers or distribution lines.
- *Class C* fires involve charged electrical equipment. The actual fuels may be class A, class B, or class D. For example, an oil-filled electrical transformer fire involves class B fuels. When the electricity is energized, the fire is classified as class C. Similarly, a class A fire in ordinary combustibles becomes a class C fire when electrical current is present.
- *Class D* fires involve combustible metals such as magnesium, sodium, and potassium.

The extinguisher capacity is important for the size and class of fire. Portable fire extinguishers are rated with regard to their relative extinguishment capacity at a testing laboratory. For example, a class A fire extinguisher is used by an experienced operator on wood panels above excelsior and on wood cribs. A 1-A rated extinguisher has only enough extinguishing agent for this experienced operator to extinguish a fully involved panel and wood crib once. If enough capacity exists for this experienced operator to extinguish this fire six times, the rating is 6-A.

Extinguishers that are rated for class B fires are tested by an experienced operator on burning flammable liquids in a flat pan that contains *n*-heptane floating on water. The flammable liquid surface area that this operator can extinguish with one tank of agent is the rating. For example,

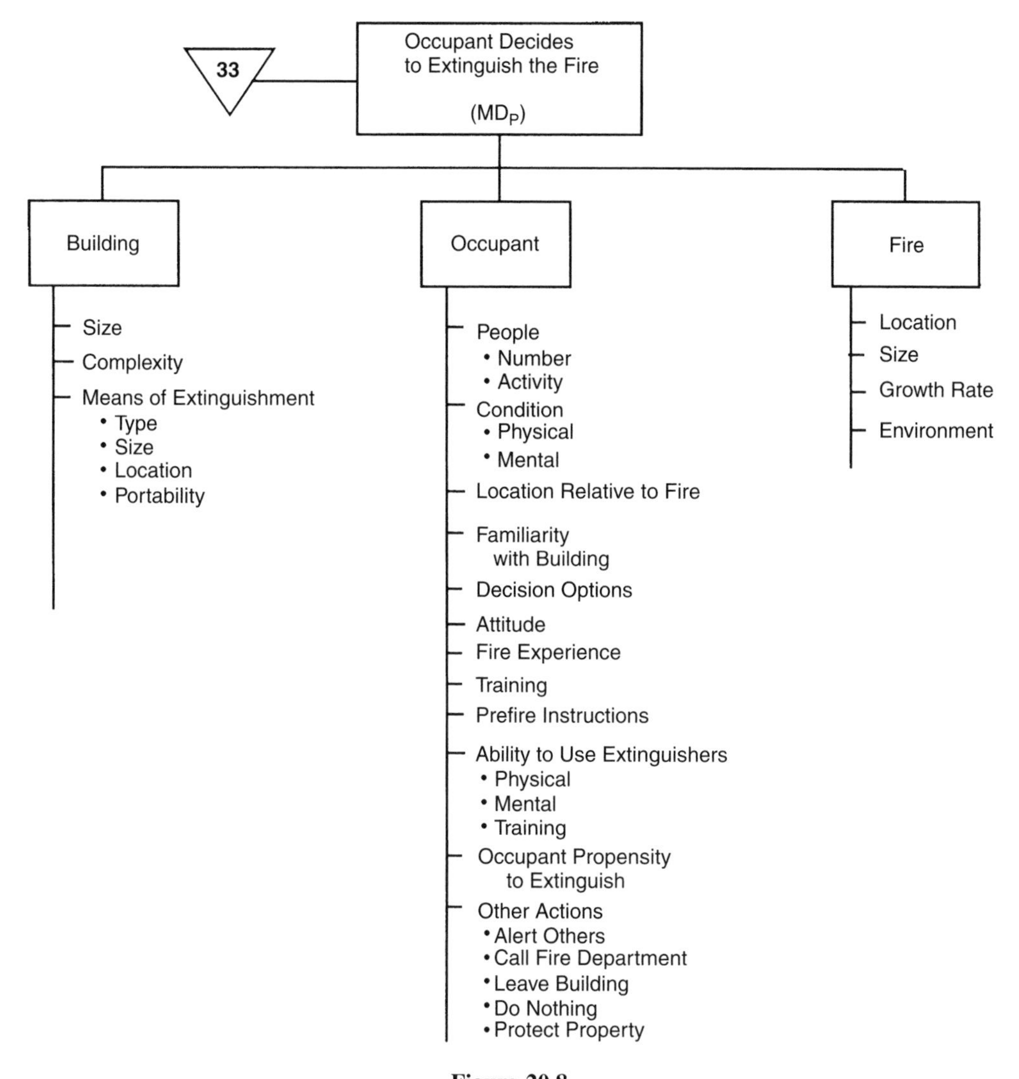

Figure 20.8

a 10-B rating means that an experienced operator is able to extinguish the area of a 1-B rated extinguisher 10 times before the agent in the extinguisher is expended.

The C rating indicates that the agent is suitable for fires in which energized electrical power is present. No rating numerals are used for these fires because they indicate only that the agent can be used with electrical fires. The C rating is used in conjunction with either class A or class B fires, or both. An extinguisher rated 6-A, 20-B:C is suitable for class A and class B fires and also can be used when energized electrical equipment is present. The numerical rating indicates its relative extinguishment capability.

Notice that the rating indicates the capability of an experienced operator who extinguishes these fires on a routine basis. An untrained individual will not apply the agent as efficiently as a trained operator. Nevertheless, this fire extinguisher rating system does provide a numerical comparison for the capacity of different fire extinguishers.

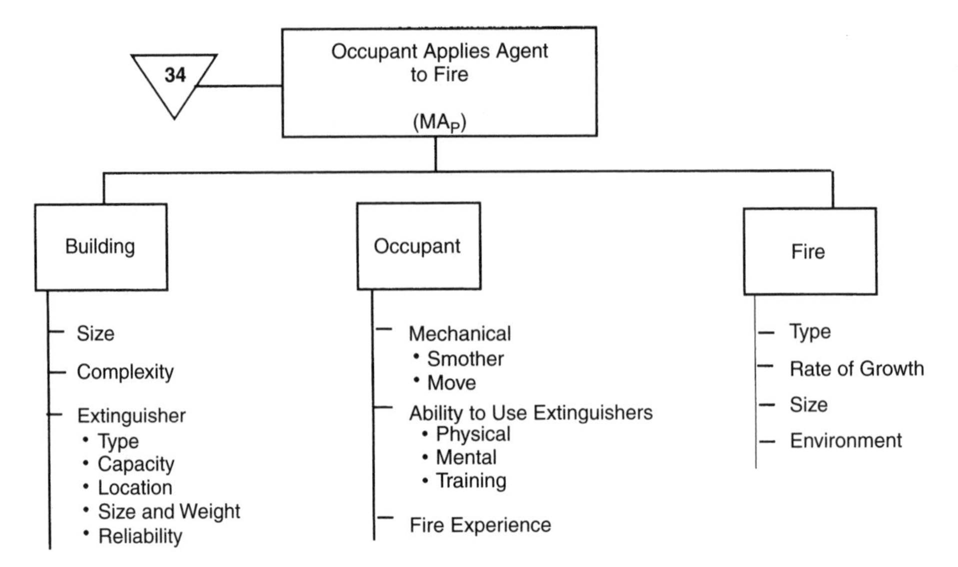

Figure 20.9

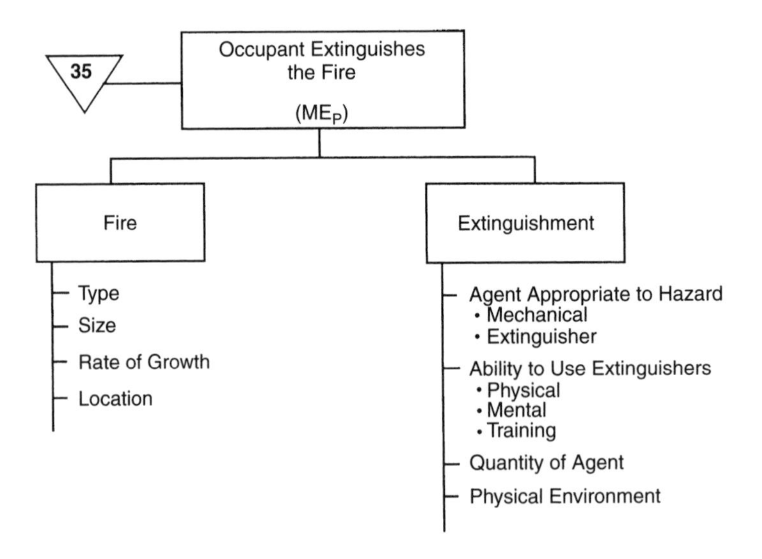

Figure 20.10

Fire extinguishers may be grouped into five different categories based on their extinguishing agent:

- water;
- carbon dioxide;
- halogenated agents (or their replacement);
- dry chemical;
- foam.

Dry powder extinguishers are another category that are used for class D fires.

All portable fire extinguishers have the following common elements:

- a storage container to hold the extinguishing agent and allow the unit to be mounted or moved;
- the extinguishing agent;
- a means to develop internal pressure sufficient to propel the agent toward the fire;
- operating features such as hoses, nozzles, and discharge levers.

20.7 Evaluating occupant extinguishment

Fire extinguishers are often viewed as an important fire safety appliance and a wide variety of different types and sizes are available. Portable fire extinguishers are often given more credit by the layman for effectiveness than may be justified in practice. Nevertheless, occupant extinguishment (M_P) is an important part of a fire prevention program and a realistic understanding of performance must be established. The evaluation must go far beyond a selection of the type and placement of extinguishers.

The manual extinguishment part of a fire prevention evaluation estimates the likelihood that the fire will be extinguished by an occupant before it reaches (or slightly exceeds) established burning. Two factors are the availability of portable extinguishers and the familiarity of occupants with their locations and operation. For example, some designers organize the extinguisher location and manual pull stations near or on the way to the exits. In this arrangement the occupant is moving toward an exit and has the opportunity to pull the manual alarm, get an extinguisher and go back to fight the fire, or leave the building. Occupant extinguishment can be planned and coordinated or it may be disjointed, reducing the likelihood of effective occupant decisions and actions.

A part of the evaluation includes assessment of extinguisher weight and the ability of occupants to reach, secure, and move extinguishers into position before the fire grows too large. For example, a library did not want patrons to touch or walk into the fire extinguishers. To prevent unwanted accidents, extinguishers were placed in rarely used exits with the bottom 8 ft from the floor. The extinguishers were very safe. The ability of occupants to apply agent before a fire grew too large was questionable. On the other hand, extinguishment of small fires in industrial manufacturing plants is often successful. The success of occupant manual suppression with portable fire extinguishers can vary from nonexistent to relatively good. A careful "prevent EB" analysis gives a sense of the relative quality of this part of the fire prevention program.

The final evaluation event estimates extinguishment capability. The class and size of fire have a great deal to do with the effectiveness of extinguishment. The extinguisher's horizontal stream range relates to the ability of an individual to move close enough to the fire to be able to apply the agent effectively. The quantity and the discharge duration are important considerations in an analysis. Some larger extinguishers have discharge durations of about 1 min. Other extinguishers will expend the agent within about 10 s. There is little time for an untrained individual to develop extinguishing skills when the emergency is being confronted. The most important components of portable fire extinguisher effectiveness are the experience and training of the operator.

Example 20.1

A fire prevention program is being evaluated as a part of a complete fire risk analysis. The occupant suppression evaluation was based on the following observations.

Locations for potential ignition scenarios were identified, and several of the more likely were selected for study. Ignition sources, initial kindling fuels, and early fire development were examined. It was estimated that the time frame between ignition and established burning would range between 4 and 15 min, with 9 min the most likely time to reach EB.

The building has substantial time periods when it is occupied by only a small security and cleaning staff. During periods of low population, an individual is unlikely to be near a small developing fire, so this chance event was not considered in the analysis. The detection and alarm system was incorporated in the analysis. The placement of detectors and their sensitivity to a slowly developing fire indicated that a fire would not be detected earlier than 10 kW, and it could be as large as 40 kW. Regardless of when detection occurs, it was estimated that an occupant could not find, move, and apply agent for at least 6 min, and a more reasonable time is 9 min. On this basis, probability estimates were $P(\mathrm{MR_P}) = 0.1$, $P(\mathrm{MD_P}) = 0.8$, $P(\mathrm{MA_P}) = 0.1$, and $P(\mathrm{ME_P}) = 0.2$.

During the periods when the building is in normal operation, many spaces are occupied by humans much of the time. Detection for the occupied condition will be more rapid than instrument fire detection and alarm alone. Many occupants are in the vicinity at detection, and the time duration for locating and moving an extinguisher to the fire and discharging it improves significantly. However, the fire extinguishers are not in clearly recognizable locations, and most individuals did not know the locations. Discussions revealed that most individuals had no training with fire extinguishers, and no prefire training or instructions had been given. Based on analysis of the site-specific situation, the estimates for occupant suppression were $P(\mathrm{MR_P}) = 0.95$, $P(\mathrm{MD_P}) = 0.85$, $P(\mathrm{MA_P}) = 0.4$, and $P(\mathrm{ME_P}) = 0.3$.

Determine the likelihood of occupant manual extinguishment for these two conditions.

Solution

Figures 20.11 and 20.12 show the SVNs that describe these conditions. The results show that during the unoccupied condition, the probability is so low that it can be discounted. In the occupied condition, the probability of occupant manual extinguishment is only about 1 in 10.

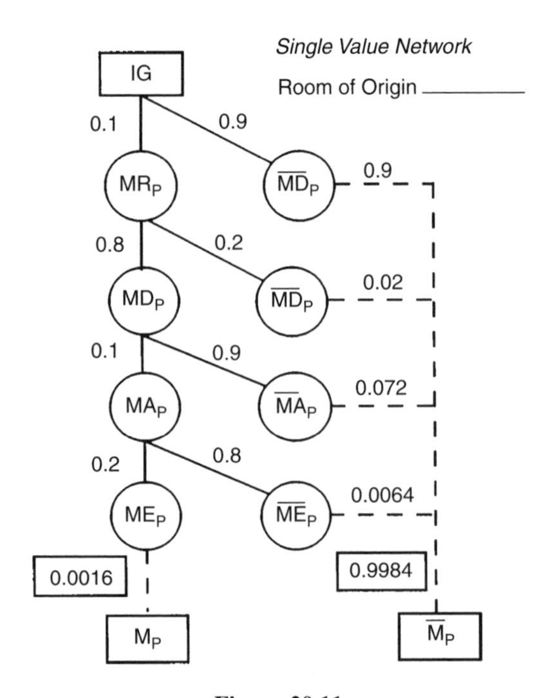

Figure 20.11

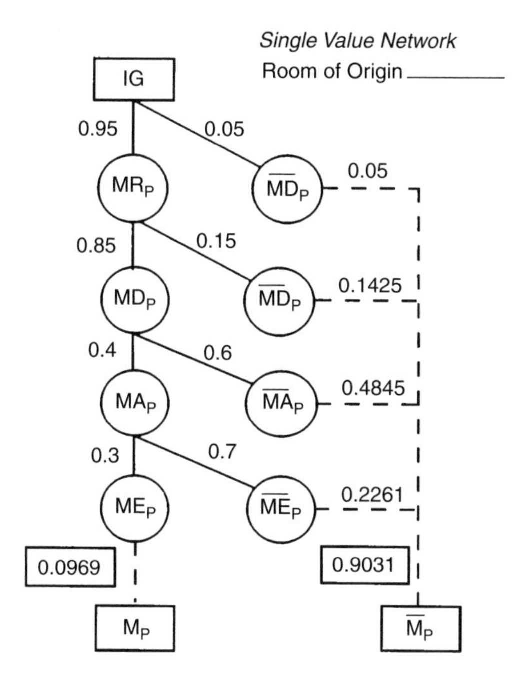

Figure 20.12

20.8 Prevent EB: discussion

A variety of applications can use fire prevention analyses. The building performance will be the same whatever the cause of ignition. A clear separation avoids obscuring building performance with fire prevention effectiveness. This creates a better understanding of the modes of failure and their outcomes. Fire prevention is usually an integral part of a comprehensive risk management program. This separation allows a clearer communication with others regarding the roles and outcomes of failure.

One or a combination of the following ways to describe fire prevention may help to communicate composite building performance.

OPTION 1

Start with a fire-free status and establish a frequency of ignition. This type of analysis is statistically based and indicates the frequency of ignition in a class of buildings similar to the one being evaluated. This information may be useful to inform others about the regularity of fires for comparative purposes. One must interpret the statistics to determine if the term "ignition" really means "established burning." Documentation procedures often unintentionally encourage reporting only ignitions that cannot be ignored. These larger fires may be of the size that established burning is the actual incident being reported.

OPTION 2

Consider selected scenarios in order to tell a story to the building management. This procedure starts with an ignition (i.e., $P(\text{IG}) = 1.0$) in a specified location. It then examines initial fire growth and occupant manual suppression activities. This technique is useful to evaluate the effectiveness of different fire prevention programs under consideration.

OPTION 3

OPTION 2 can also start with a fire-free status. Estimates give a subjective probability of ignition for specified conditions. This technique is useful to compare the relative potential for ignition at various locations that have different conditions of heat energy, kindling fuels, and heat/fuel separations. The estimates give an insight into ignition prevention effectiveness and helps to order possible changes that may reduce the likelihood of ignition. Given ignition, the analysis of initial fire growth and occupant extinguishment can proceed as usual. This probability is not a frequency of ignition, but a relative index to relate critical combinations of ignition factors.

OPTION 4

Rank the effect of ignition locations on the building. This signifies that $P(\text{IG}) = 1.0$ for the building, rather than for a specific location in the building. The analytic hierarchy process (AHP) is a technique for establishing this ranking [1, 2]. The AHP can assist in making multilevel decisions and it can help to translate judgments into quantified priorities. Then, given the probability of ignition in a specific space (i.e., $0.0 < P(\text{IG}) < 1.0$), the initial fire growth and occupant suppression actions can be analyzed for each situation. This procedure has the advantage of being able to combine ignition frequencies for different spaces with dedicated decisions and initial occupant actions in that space. In some ways, this technique can "weight" the response to the problem for an entire building. One can more easily compare a site-specific building with statistical data. The AHP for fire protection applications is more suitable for a computer analysis.

APPLICATIONS

Each of these options has application for corporations or the insurance industry. A detailed fire prevention program could incorporate all of them to get a clearer picture of the effectiveness of specific management recommendations. The purpose of any analysis is to understand performance in a way that enables site-specific alternatives to be compared. A comparison of cost and value helps to establish how much is enough.

20.9 Automatic special hazard suppression

Automatic suppression normally is not perceived as part of a "prevent EB" program. Some explanation is needed to describe why and how we incorporate automatic special hazard suppression into a performance evaluation.

Special hazard suppression equipment is normally installed to protect a spot area within a larger space. Usually these hazards involve the potential for rapid flame spread such as class B fires in industrial buildings or food preparation areas in restaurants. Applications such as flammable liquid storage, dip tanks, oil filled transformers, and deep-fat fryers for cooking are examples of hazards that require special attention. This automatic special hazard protection is not part of the automatic sprinkler system of the flame movement analysis. It is common to have a general area sprinkler system for the entire building, as well as special hazard suppression systems to protect local problem areas of an operation. In industrial or commercial occupancies where this type of automatic protection equipment is installed, an analysis of each component is made for these spaces independently, because of their potential for disaster.

Special hazards equipment includes the following:

carbon dioxide system;
dry chemical system;
halogenated total flooding system;
water-spray system;
foam extinguishing systems;
explosion protection system;

An evaluation of the reliability and design effectiveness of special hazard automatic suppression equipment first considers characteristics of the flammable materials and combustion products and the speed and extent of fire propagation. After these potential fire characteristics are understood, we evaluate the reliability and design effectiveness of the suppression system. This involves the following:

- agent selection;
- detection system;
- release system;
- storage and delivery system.

The examination determines whether the agent is appropriate to the hazard. One may discover a special hazard being protected by an agent that is inappropriate for the fire. Detection and release systems must account for the expected fire propagation speed and the released combustion products. Special hazard equipment is frequently located in corrosive environments, and long-term maintenance becomes an important consideration in the evaluation of agent delivery. Also part of a management program are the cost of cleanup and salvage from fire suppression and to neighboring materials susceptible to damage by the extinguishing agent.

20.10 Carbon dioxide systems

Carbon dioxide (CO_2) is an inert, noncorrosive, and electrically nonconductive gas. Carbon dioxide extinguishes fires by smothering. The CO_2 gas displaces the normal atmosphere and reduces the oxygen available to the fire to a level where it will not support combustion. While CO_2 may be nontoxic, it can have serious physiological effects on humans at concentrations of more than about 6%. Consequently, humans should not be in confined spaces in which CO_2 is discharged.

Carbon dioxide systems are particularly suitable for fires involving flammable liquids and machinery spaces as well as with food preparation equipment, such as deep-fat fryers, hood and duct systems, and ranges. Because CO_2 systems do not leave a residue and are inert, they often are used in spaces that contain electronic equipment, legal documents and records, valuable clothing, and other materials that could be damaged by other types of extinguishing agent.

Carbon dioxide systems are used as a total flooding extinguishment in closed rooms and vaults where openings are small and the gas has an opportunity to remain for a time. In spaces where ventilation is greater or where equipment requires a period of time to shut down, special design features are incorporated to accommodate these needs. The volume and storage requirements depend on the room volume and the design concentration of the CO_2. The storage of the CO_2 may be in high- or low-pressure storage tanks or in banks of cylinders, depending on the amount needed.

Carbon dioxide systems are usually actuated automatically by smoke, heat, or flame detectors. In some spaces, flammable vapor detection or detectors that sense other process abnormalities may be used. All automatic CO_2 systems must also have an independent means for manual actuation. In spaces where humans may be present for extended periods of time, such as electronic computer rooms, manual actuation may be the primary discharge actuation.

In addition to total flooding, there are local application systems that can protect a part of a larger space without flooding the entire area. Food preparation areas, such as deep-fat fryers and hoods and ducts, are one type of local application. Industrial processes, dip tanks, oil-filled transformers, pumps or motors, and access openings in containers of flammable liquids are other types of application. Because the carbon dioxide in local applications is dissipated into the larger volume, special attention is given to direction, duration of discharge, and ventilation. If the heat energy source and the fuels remain, there is the possibility of reignition after initial suppression.

Other types of CO_2 application involve hand hose lines or standpipes with a mobile supply. This use is another form of manual fire fighting. An evaluation for M curve effectiveness would be restricted to the CO_2 application. Detection, notification, agent application, and the final extinguishment are evaluated for the local building fire brigade. Size, experience, equipment, and training of the building occupants become major factors in the evaluation.

20.11 Dry chemical extinguishing systems

Dry chemical extinguishing systems discharge a powder mixture onto flames. The original use, and still a major use for this type of extinguishment, is with portable hand fire extinguishers. However, one can also utilize total flooding, local application, and hand line hose installations. In many ways, these systems use the same types of components as carbon dioxide systems. For our purposes, the evaluation process for the two systems would be similar with adjustments for individual capabilities and limitations.

While dry chemical systems are a powder mixture, the term "dry powder" is usually reserved for the graphite and special compounds used to extinguish class D fires. The dry chemical systems of this discussion are used primarily to extinguish class B fires. They can also be used for class A and class C fires. Extinguishing agents may include chemicals such as sodium bicarbonate, potassium bicarbonate (also known as Purple K or PKP), and monoammonium phosphate (used in ABC portable fire extinguishers). The principal mechanism that contributes to a rapid fire extinguishment is the interference of the dry chemical particles with the combustion chain reaction through the thermal decomposition of their chemical powders.

Some dry chemicals are slightly corrosive. Consequently, the scene should be cleaned up shortly after extinguishment. Also, dry chemicals have insulating properties that can cause electrical contacts and relays to become inoperative. Their use in these types of installation could cause substantial damage and repair expense.

Total flooding dry chemical extinguishing systems are functionally similar to those of carbon dioxide. Total flooding systems are used in enclosed spaces when reignition is not anticipated. Actuation is usually triggered by a detector. Manual operation can also be installed.

Fixed local application systems are similar to carbon dioxide systems. They discharge the agent directly onto the spot area being protected. They are commonly used for flammable liquid fires, such as dip tanks, electrical transformers, kitchen equipment, and storage vessels.

20.12 Halogenated extinguishing systems

Halons are a family of hydrocarbons in which one or more hydrogen atoms have been replaced by fluorine, chlorine, bromine, or iodine. The extinguishing mechanism appears to employ interference with the combustion chain reaction. Used in confined spaces, total flooding halon is very effective in extinguishing flammable liquid and vapor fires. Halons have also been used extensively in rooms that contain delicate equipment or property, such as electronic computers, data storage rooms, and electrical switching gear. The discharge of the heavier-than-air agent is rapid and turbulent. Extinguishment is extremely rapid with little damage to sensitive contents.

Unfortunately, halons have been identified as ozone-depletion agents. Although halon extinguishing systems have many fire protection attributes, halon production was to be phased out by 2000. Existing systems will be replaced eventually by other extinguishing systems.

Halogenated systems have also been designed and installed as total flooding systems or in local application systems. Their use in portable fire extinguishers was common.

20.13 Water-Spray extinguishing systems

A water-spray extinguishing system is another type of fixed special hazard system. Water-spray systems are used primarily to extinguish fires in installations involving flammable liquid and gas storage tanks, electrical transformers, large electric motors, and industrial process piping. Water-spray systems are not used for building protection, but for special hazard installations to extinguish class B fires or to protect these installations from exposure fires.

A water-spray system has fixed piping and specially designed nozzles that provide a defined pattern of water distribution, with specific water particle size, density, and velocity. Water sprays are deluge systems that do not have fusible elements. Water is discharged directly and forcefully on the surface being protected. Because of the hazards of the installations being protected, the water density is large and the entire surface area is covered. A variety of specific considerations, such as detection, actuation, and operating devices, water supply, drainage, and maintenance practices, are incorporated into the design.

20.14 Foam extinguishing systems

Another type of special hazard system for flammable liquid fires involves foam application. Foam is a mixture of water and a solution of specially formulated liquid agents that when mixed create gas-filled bubbles. The bubbles float on the flammable liquids and create a vapor layer that seals the surface, cools the material, and excludes air from the potential combustion process. Foam fire extinguishment is very effective for aircraft fuel fires or for flammable materials that spill and flow along surfaces.

Foam extinguishment can be delivered in three forms. One is a fixed system in an enclosed room. The room may be a tank or vault, or it may be an aircraft hangar. In a fixed system, the delivery system is essentially a deluge-type sprinkler system. The open heads allow a blanket of foam to be spread over the entire area. The second is a local application system. Here the fixed deluge-type sprinkler equipment is positioned around the area to be protected and the foam is applied in a similar manner over the spot area. The third and most common application is by hand hose lines. Fire-fighting foam is particularly effective for flammable liquid fires because the fire fighters can move against the flammable liquids and extinguish the fire in a progressive manner.

Foam is produced by having a separate storage container of foam concentrate that mixes in a predetermined concentration with the water as it flows through the hose line or pipe. The foam may be mechanical (air) foam that is formed by premixing or proportioning the solution with the water. Alternatively, the foam may be a chemical foam that is created by a reaction between an alkaline solution and an acid solution. Foaming agents and equipment can generate expansions ranging from low to high. This involves bubble aggregation expansions from about 20:1 to 1000:1. Much of the technology involves the type of foaming agents, the mixing chamber and process, and the transportation of the equipment to the fire scene.

These special hazard systems can be designed for fixed, local application, or portable hose lines. Their function is to extinguish a flammable liquid fire. Foam extinguishment can be very effective for these types of fire, especially when used as the extinguishing agent by fire-fighting forces.

20.15 Explosion suppression systems

When concentrations of dusts, mists, gases, or flammable vapors are in a confined space the possibility of an explosion is present. An explosion is an extremely rapid combustion that produces a rapid rise in temperature and pressure because of the speed and confinement of the reaction. The speed of the energy release and the resulting pressure rise become the explosion. A wide variety of industrial operations can create explosive conditions.

It is possible to design for an explosion by incorporating pressure-sensitive construction, such as walls that are designed to blow away cleanly with the rapid pressure rise. This releases the pressure and reduces the damage in other parts of the building. Alternatively, explosion suppression systems can be installed. These systems use extremely sensitive detection devices that actuate during the time lag between ignition and the development of the destructive pressure–a time lag measured in milliseconds. The detection device is electronically connected to the suppression agent discharge device. This device is an explosively actuated suppressor or a high discharge rate extinguisher. The suppression agent is discharged faster than the flame is propagated. The system confines and neutralizes the explosion.

20.16 Building evaluations for special hazard installations

Special hazard systems, also called spot systems, are designed and installed to protect a distinct and specific hazard. Most of the time, the hazards involve flammable liquids that are associated with industrial occupancies. Hood and duct systems in kitchens also are protected by these systems. However, industrial plants, restaurants, electrical installations, and aircraft hangars are all buildings, and these procedures can be used to evaluate their performance. Performance evaluations of special hazard systems must have a clear position within the complete analytical system.

The exact position of a special hazard system within the building evaluation depends primarily on its role. These systems may be viewed in three categories. The first is a total flooding system in which the agent covers an entire enclosed space. In this role the system is analogous to a sprinkler system. Although there are substantial differences, the function is to extinguish small fires and prevent extension of the fire throughout the space. If the system is successful, relatively little damage will occur, and an A_f curve (automatic flooding system) can be constructed for this single space. System reliability is an important assessment in an A_f curve. When the system is unsuccessful, full room involvement occurs and succeeding barrier/space modules must be evaluated for fire propagation along other paths. It is common to have automatic sprinkler systems in addition to special hazard systems. If both systems exist in the same enclosed room, the two automatic suppression curves, A and A_f, are evaluated separately. Their combination can be calculated using the SVN of Figure 20.13. Order does not matter because the two components are evaluated independently.

Local application systems pose a different organizational problem because they protect specific zones within a larger room. In addition, the special hazard systems provide rapid initial extinguishment to a special problem. It is typically expected that a company fire brigade will provide additional extinguishing support in the case of insufficient agent capacity or a reignition. We incorporate this as a part of the prevention analysis.

Figure 20.14 shows a plan for a large room that has three local application (spot) protection systems within the space. An ignition can occur within one of the spot protection areas or outside

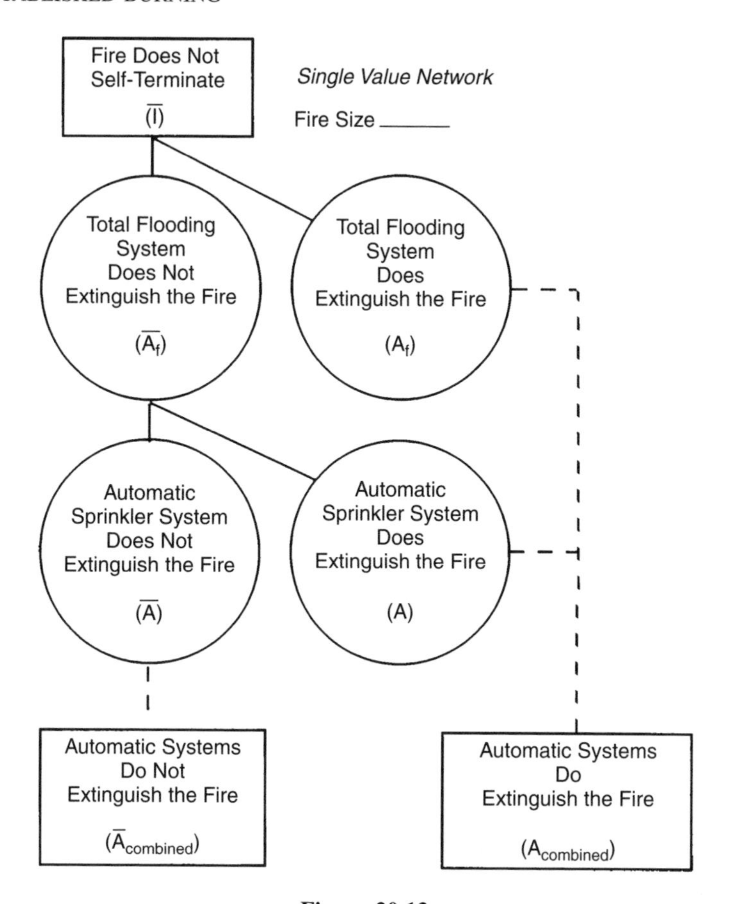

Figure 20.13

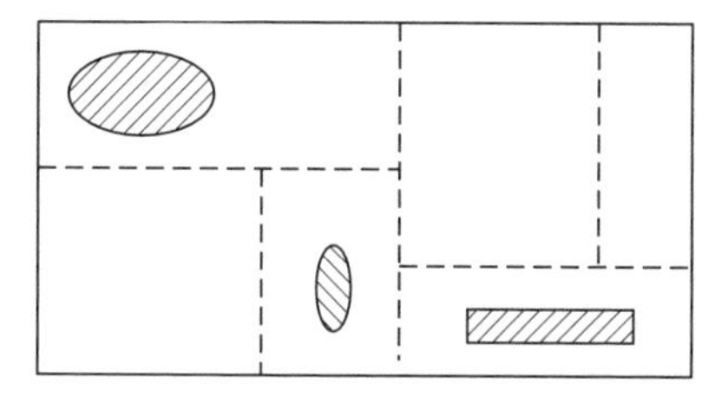

Figure 20.14

of those areas. If the ignition occurs outside of a spot protection area, the "prevent EB" analysis is conducted as described earlier in this chapter. The initial fire growth potential and occupant suppression are evaluated and combined by the network of Figures 20.1 and 20.2. The analysis examines the quality of the fire prevention program outside the special hazard areas. If the program fails and EB occurs, the normal L curve analysis for the room is used.

The room of Figure 20.14 is zoned into analytical spaces, as shown by the dashed lines. Each zone becomes a space described with I, A, and M evaluations. The dashed lines are zero-strength barriers that have no $\overline{T}$ or $\overline{D}$ resistance. However, this technique allows the impact of ignition and EB in the different zones, as well as the flame/heat movement into different zones, to be evaluated and described rationally. Different L curves for the space will reflect the differences in building performance for the ignition locations and conditions that exist.

If the fire occurs in a zone that is protected by a local application special hazard installation, an A_p curve (local application system) or SV estimates for the area is constructed. Section 20.17 describes the evaluation of that A_p curve or SV estimation. In addition to the A_p values, the initial fire growth potential and occupant suppression are part of the protection for that zone. These three components are treated as part of the fire prevention analysis for that zone. Figure 20.15 shows the network that combines these events. If the fire is not controlled by fuel control,

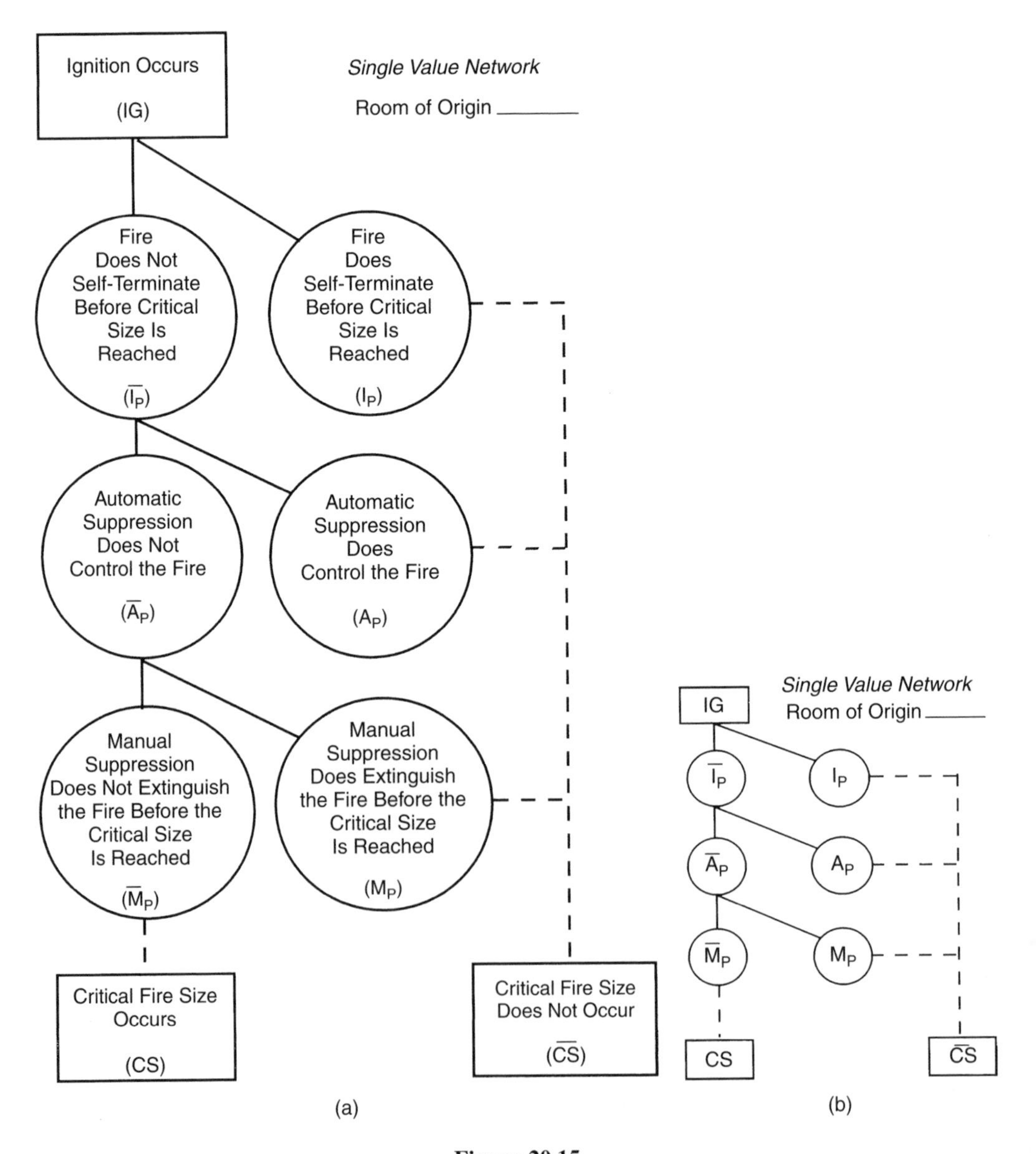

Figure 20.15

occupant extinguishment, and special hazard installation, the building design and the normal L curve analysis become operational. At that time the building fire growth potential, the automatic sprinkler suppression, and the fire department extinguishment define the performance for that zone and the subsequent (zero-strength) barrier/space performance in the room.

In summary, special hazard installations are handled in one of two ways, depending on their function in the building. Total flooding systems are evaluated in a similar way to a sprinkler system because the entire floor area is covered. If these systems are to be augmented by an industrial building fire brigade, the effectiveness is evaluated and incorporated in the usual manner for a space. Subsequent reinforcement by the local fire department may be incorporated using the normal logic for a space evaluation.

When spot protection is part of an analysis, the larger space is divided into zones. The spot protection is initially treated as an A_p curve or an SV component of a complete space analysis. The fire prevention analysis includes that component. Then, if prevention is unsuccessful, the adjacent spaces outside that zone are evaluated for the zero-strength barrier/space module given that the initial zone is fully involved.

20.17 Special hazard evaluations

Performance evaluation of automatic special hazard installations follows procedures similar to those of automatic sprinkler systems. Figure 20.16 shows the events that are used to evaluate a special hazard installation. The analytical events are the reliability of the system for agent discharge (AA_p) and the operational effectiveness for the agent to control the fire (AC_p).

Figure 20.17 shows factors that relate to the system reliability, (AA_p). Figure 20.18 shows factors that influence the operational effectiveness (AC_p).

Example 20.3

A large room of origin in an industrial building is evaluated. A special hazard installation protects a process in one part of the room, as shown in Figure 20.19. The room is divided into two zones to analyze the flame/heat performance of the building and to organize a fire prevention plan for a risk management program. Each zone was evaluated first as a space of origin and then as a subsequent space, given that fire in the other room was not limited. The results of the analyses are as follows.

Zone 1 as a room of origin

Given ignition $P(\text{IG}) = 1.0$, the initial fire growth potential is $P(\text{I}_p) = 0.2$. That is, given an ignition, the probability that the fire would grow to EB was estimated at 80%. Also, the estimate of occupant extinguishment was $P(\text{M}_p) = 0.4$. Given EB in zone 1 as a room of origin, the I, A, and M curve from EB to FRI within the zone are shown in Figure 20.20. The barrier between zones 1 and 2 is zero strength. The $\overline{\text{T}}$ and $\overline{\text{D}}$ curves are shown in Figure 20.21.

Zone 2 as a second room

Given a fire in zone 1, the I, A, and M curves for zone 2 as a subsequent room are shown in Figure 20.22. Two sets of I, A, and M curves would normally have been constructed, one for a $\overline{\text{D}}$ failure and the other for a $\overline{\text{T}}$ failure. However, in this case the cohesiveness requirement indicates that the curves for a $\overline{\text{T}}$ failure would never be used, because $P(\overline{\text{D}}) = 1.0$ for all heat energy applications.

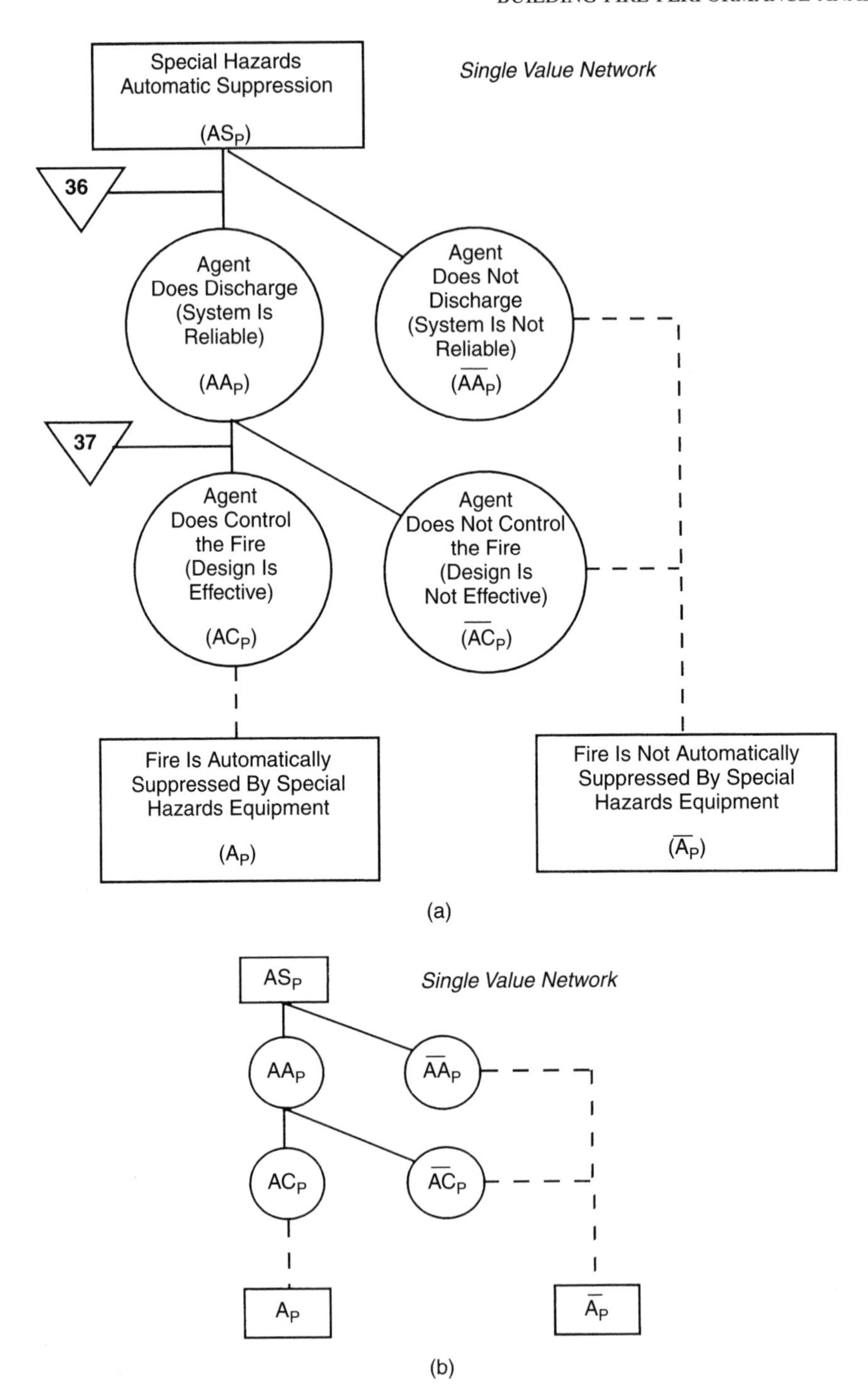
Special Hazards
Automatic Suppression
(AS_P)
Single Value Network
36
Agent
Does Discharge
(System Is
Reliable)
(AA_P)
Agent
Does Not
Discharge
(System Is Not
Reliable)
($\overline{AA}_P$)
37
Agent
Does Control
the Fire
(Design Is
Effective)
(AC_P)
Agent
Does Not Control
the Fire
(Design Is
Not Effective)
($\overline{AC}_P$)
Fire Is Automatically
Suppressed By Special
Hazards Equipment
(A_P)
Fire Is Not Automatically
Suppressed By Special
Hazards Equipment
($\overline{A}_P$)
(a)
AS_P
Single Value Network
AA_P
$\overline{AA}_P$
AC_P
$\overline{AC}_P$
A_P
$\overline{A}_P$
(b)

Figure 20.16

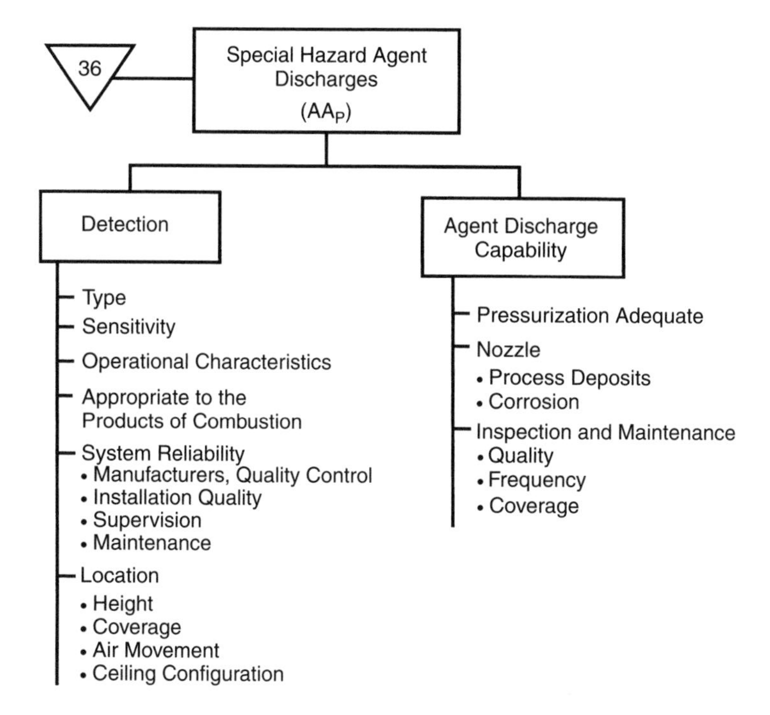

Figure 20.17

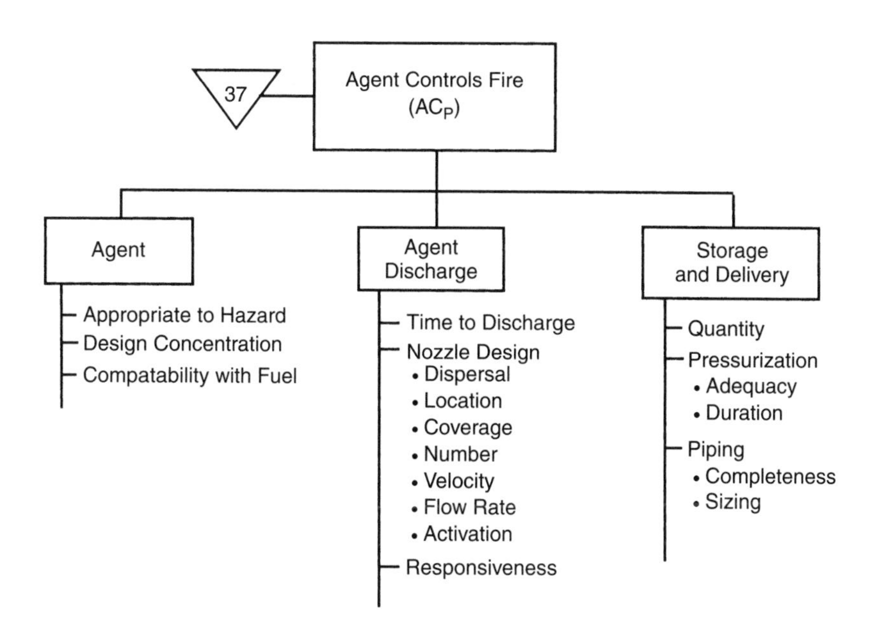

Figure 20.18

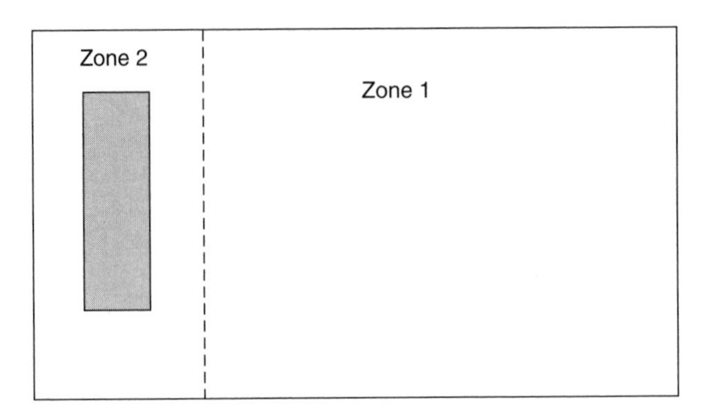

Figure 20.19

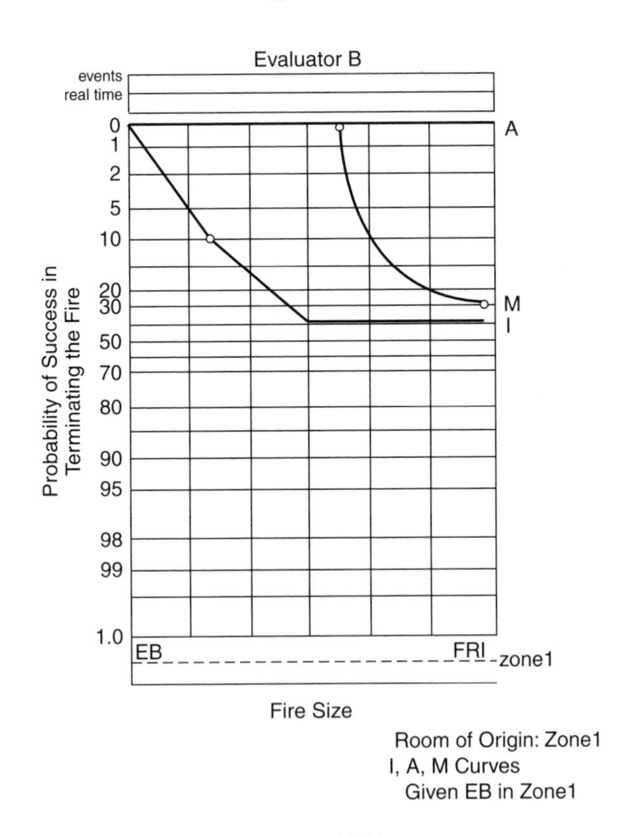

Figure 20.20

Zone 2 as a room of origin

Given ignition in the area protected by the special hazard installation, the value of $P(\mathrm{IG})$ is set to 1.0. The initial fire growth potential is $P(\mathrm{I_p}) = 0.0$. That is, given ignition, the fire will certainly grow to the critical size defined by EB. The speed of the fire in this zone will be so great that the occupant will not be able to apply agent before the entire zone is involved. Therefore $P(\mathrm{M_p}) = 0.0$.

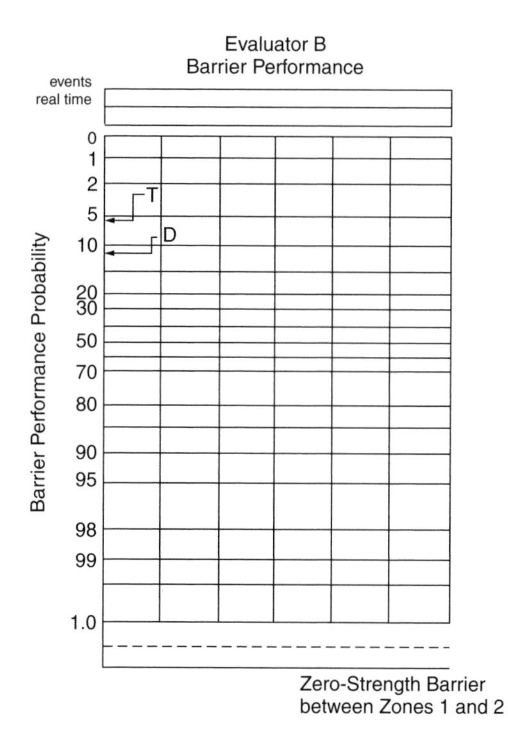

Figure 20.21

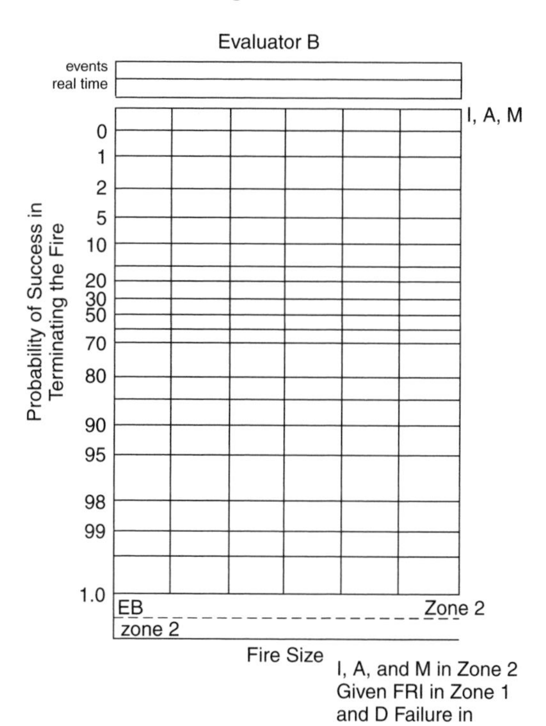

Figure 20.22

The special hazard installation was evaluated for its reliability and design effectiveness. The detection system was inspected for its sensitivity and operational characteristics relating to the hazard being protected. Also, the agent discharge capability was inspected to estimate the likelihood that the system would operate when the detector activated. These were combined and assigned a value for the probability of agent application, $P(\mathrm{AA_P}) = 0.95$.

Given agent application, the fire characteristics, type of agent, agent discharge characteristics, and storage and delivery systems were studied. The probability of success in extinguishing the fire with this equipment, given agent application, was estimated at $P(\mathrm{AC_P}) = 0.90$.

Zone 1 as a second room

Given the fire in zone 2, the I, A, and M curves for zone 1 as a subsequent room are shown in Figure 20.23. Again, two sets of I, A, and M curves would normally have been constructed, one for a $\overline{\mathrm{D}}$ failure and the other for a $\overline{\mathrm{T}}$ failure. However, the evaluations for a $\overline{\mathrm{T}}$ failure cannot be used because $P(\overline{\mathrm{D}}) = 1.0$ for all heat energy applications.

Solution

Here the information enables one to combine fire prevention (prevent EB) and the L curve for the room of origin. The solution illustrates the calculations and their combination. The analysis uncouples each part and then combines the results to produce two different L curves for the room of origin, depending on the zone in which ignition occurred.

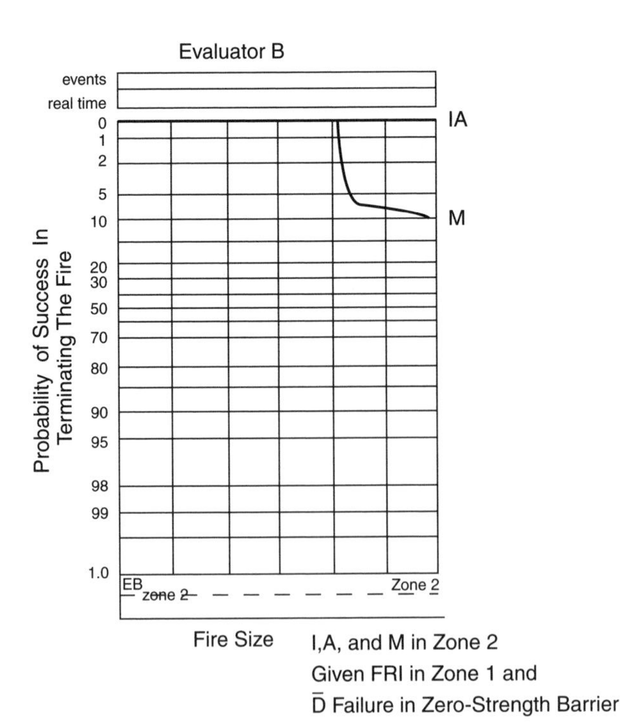

Figure 20.23

Step 1: Calculate EB for an ignition in zone 1

Using the values described above, the probability of reaching CS, given ignition, is shown in Figure 20.24 to be 0.48. The probability of achieving EB (given IG = 1.0) is shown in Figure 20.25 to be $P(\text{EB}) = 0.48$.

Step 2: Construct the L curve for zone 1

The I, A, and M curves for zone 1 given EB are shown in Figure 20.20. The calculation of the L value at FRI of zone 1 is shown in Figure 20.26. The L curve is sketched in Figure 20.27.

Step 3: Calculate the L curve for zones 1 and 2

Given EB in zone 1, the L curve was shown in Figure 20.27. The L curve for zones 1 and 2 is calculated with the network of Figure 20.28. This figure used the $\overline{\text{T}}$ and $\overline{\text{D}}$ curves of Figure 20.21 and the I, A, and M curves of Figure 20.22. The resulting L curve is sketched in Figure 20.29.

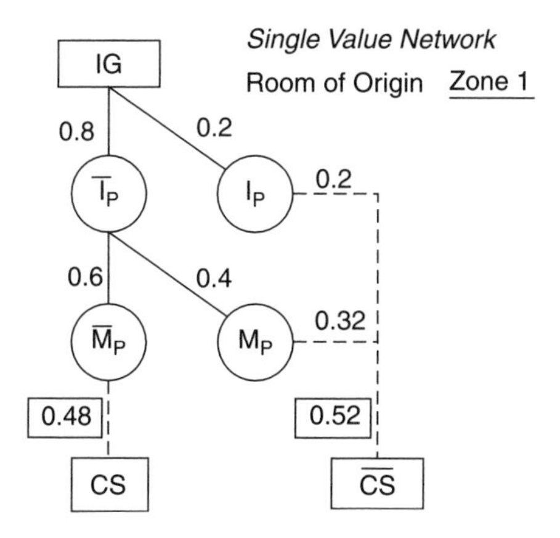

Figure 20.24

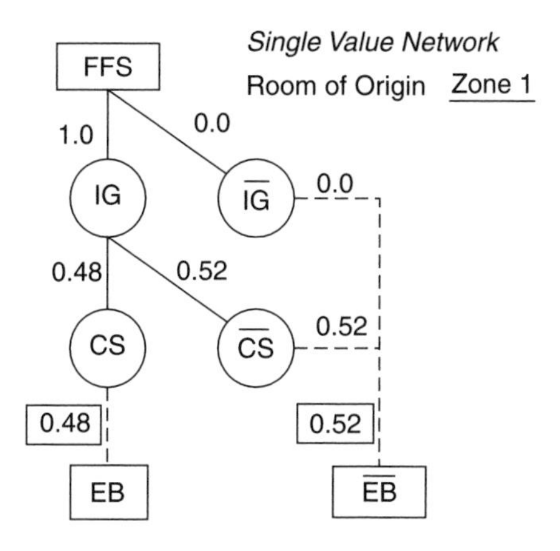

Figure 20.25

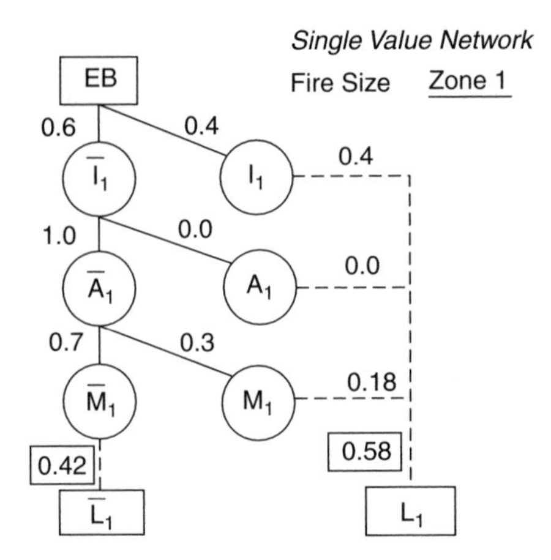

Figure 20.26

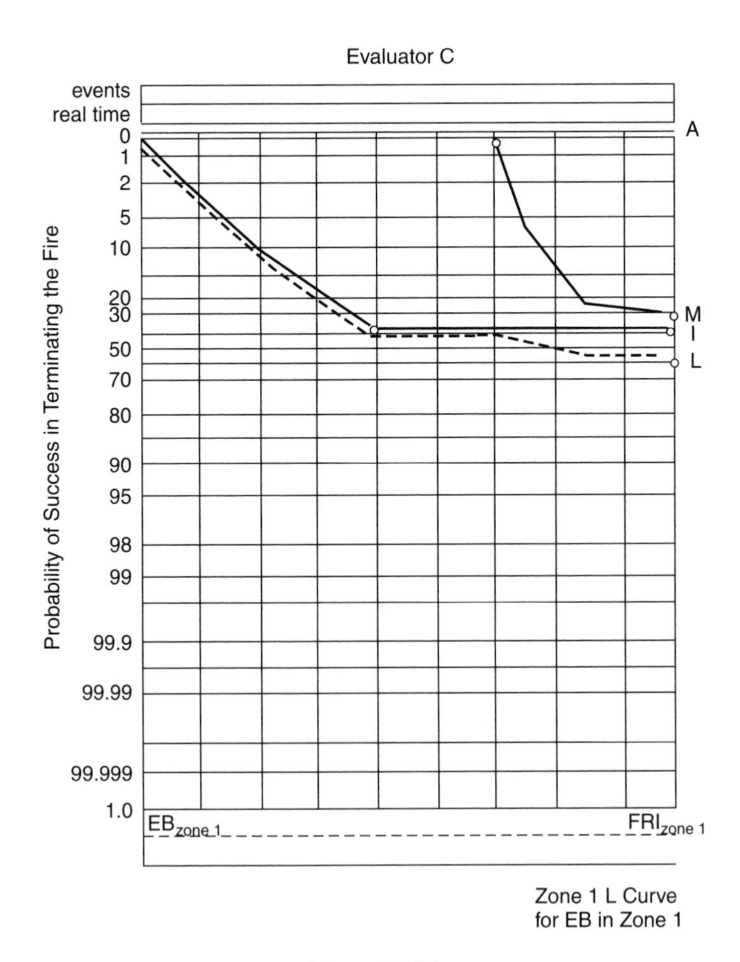

Figure 20.27

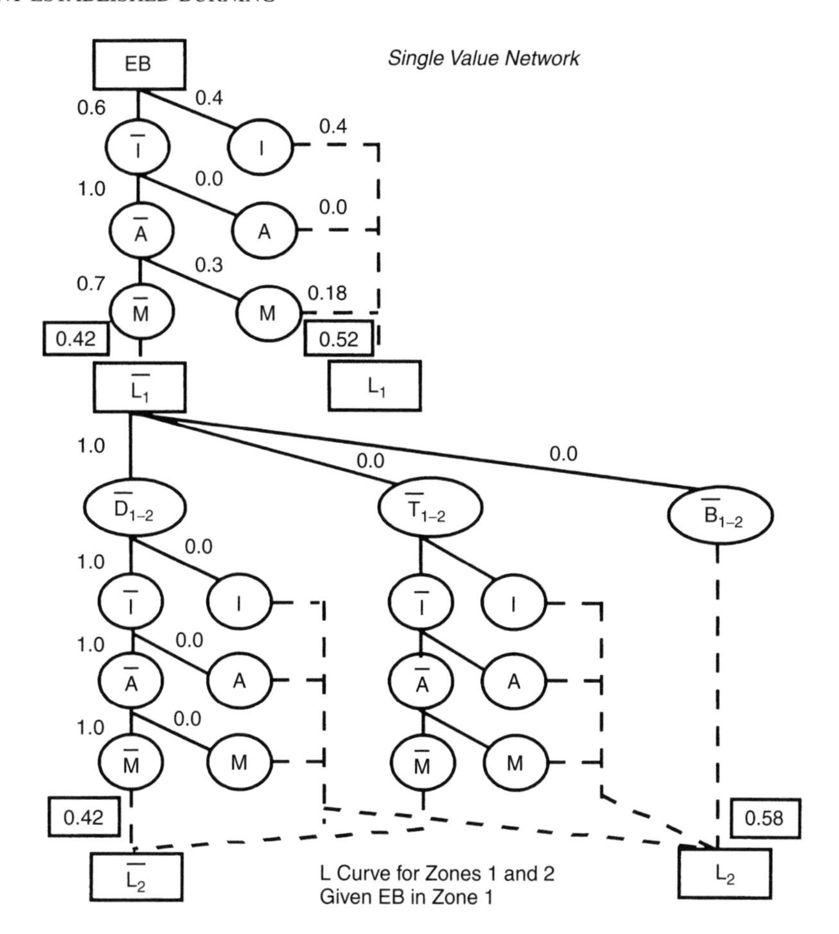

Figure 20.28

Step 4: Calculate L for zones 1 and 2 from a fire-free status

Given an ignition in zone 1, the probability that it would be terminated before reaching EB, $P(\overline{\text{EB}})$, was calculated in Figure 20.25 to be 0.52. Its complement, $P(\text{EB}) = 0.48$. Given EB in zone 1, the probability that FRI would occur, $P(\overline{\text{L}}_1)$, was calculated in Figure 20.26 to be 0.42. Its complement, the probability that the fire will be extinguished, is $P(\text{L}_1) = 0.58$. Given FRI in zone 1 and zero-strength barriers between zones 1 and 2, then $P(\text{L}_2) = 0.0$. This was calculated in Figure 20.28.

Incorporating these values into the SVN of Figure 20.30, we determine that the likelihood of losing both zones 1and 2 is rounded to 0.20. The probability that the fire will be limited before including zones 1 and 2 is rounded to $P(\text{L}) = 0.8$.

Now the process can be repeated for ignition in zone 2.

Step 5: Calculate EB for an ignition in zone 2

The special hazard installation is included in this analysis. Using the values described in the problem statement, the probability of success of the special hazard suppression equipment is shown in Figure 20.31 to be $P(\text{A}_\text{p}) = 0.855$. The probability of an ignition in zone 2 to reach

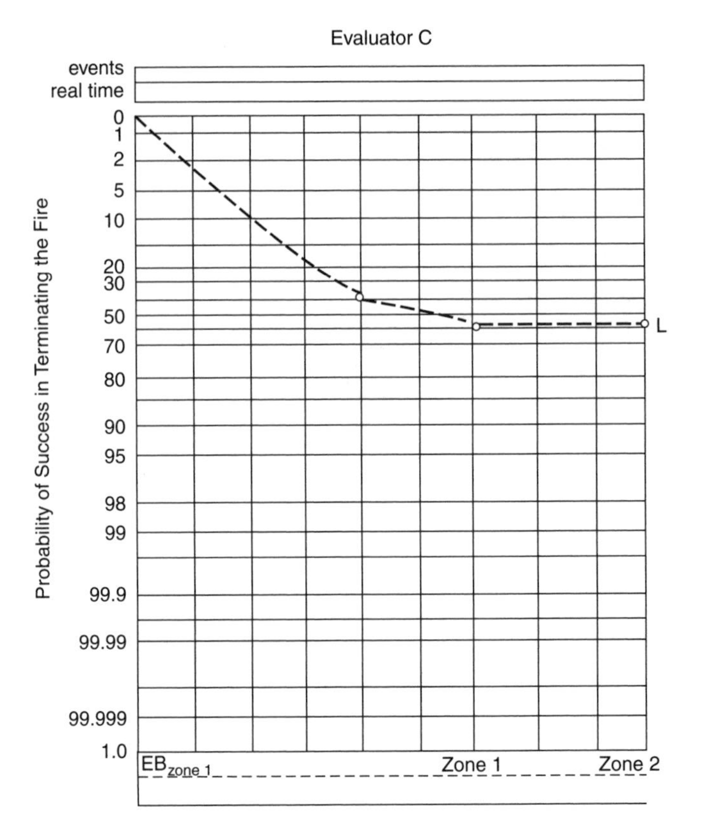

Figure 20.29

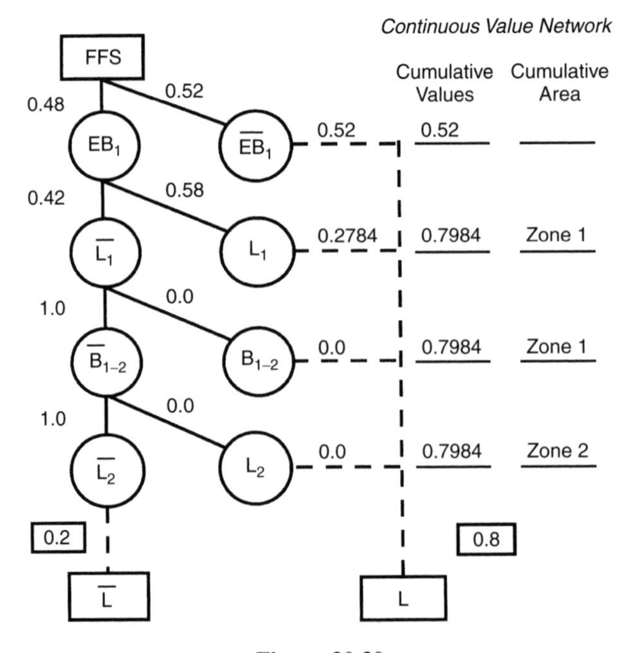

Figure 20.30

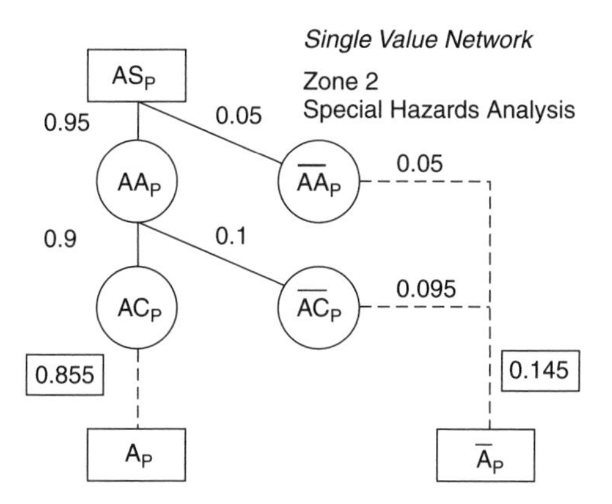

Figure 20.31

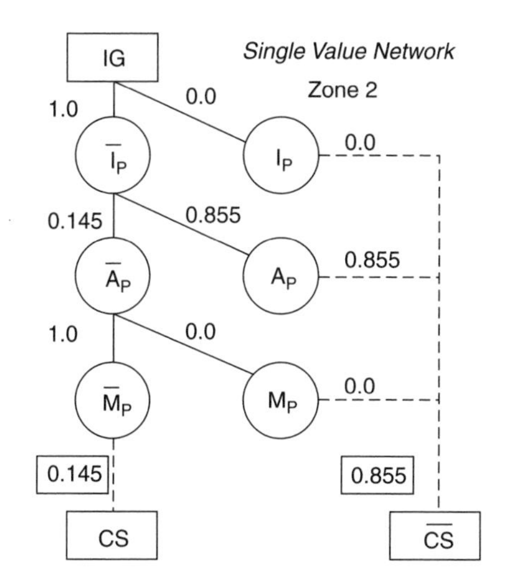

Figure 20.32

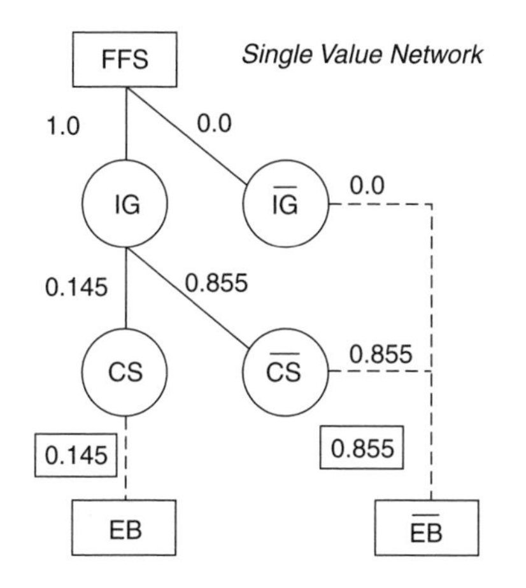

Figure 20.33

the critical size is shown in Figure 20.32 to be 0.145, and the probability of EB is shown in Figure 20.33 to be $P(\text{EB}) = 0.145$.

Step 6: Construct the L curve for zone 2

The I, A, and M curves for zone 2, given EB in zone 2, are not needed as the calculation for the special hazard spot area was known as $P(\text{EB}) = 0.145$. This involves the entire area of zone 2.

Step 7: Calculate the L curve for zones 2 and 1

Given EB in zone 2, the L value is the same number, as described in step 6. Figure 20.23 shows the I, A, and M curves for zone 1 as a second room. The barriers are again zero strength because they merely zone the space. Figure 20.34 shows the calculations for this condition. The resulting L curve is sketched in Figure 20.35.

Step 8: Calculate L for zones 2 and 1 from a fire-fee status

Given an ignition in zone 2, the probability that it would be terminated before reaching EB was calculated in Figure 20.33 to be 0.855. Its complement is $P(\text{EB}) = 0.145$. Given EB in zone 2,

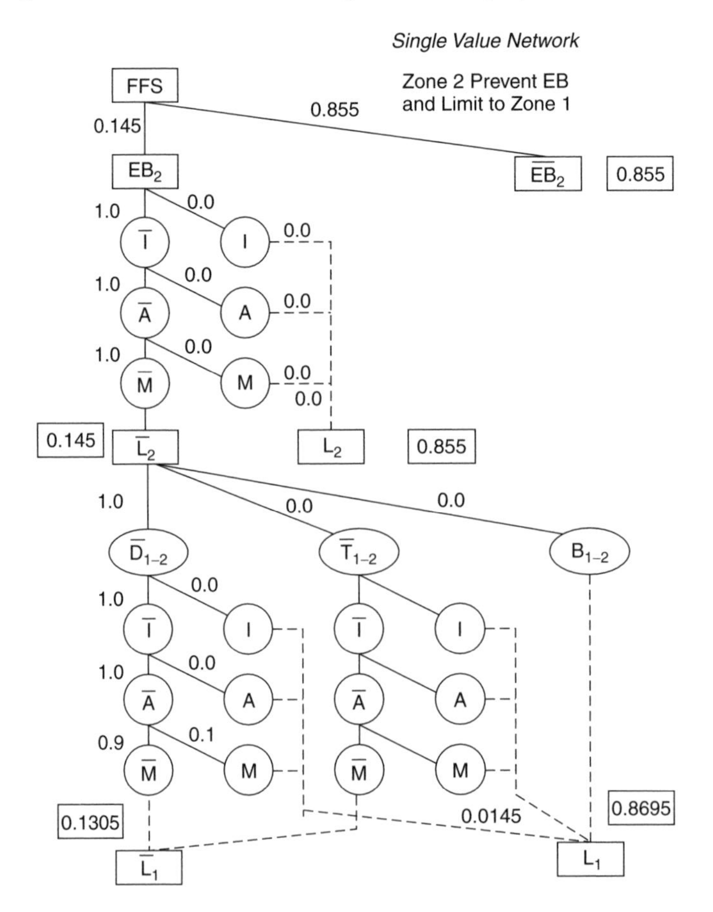

Figure 20.34

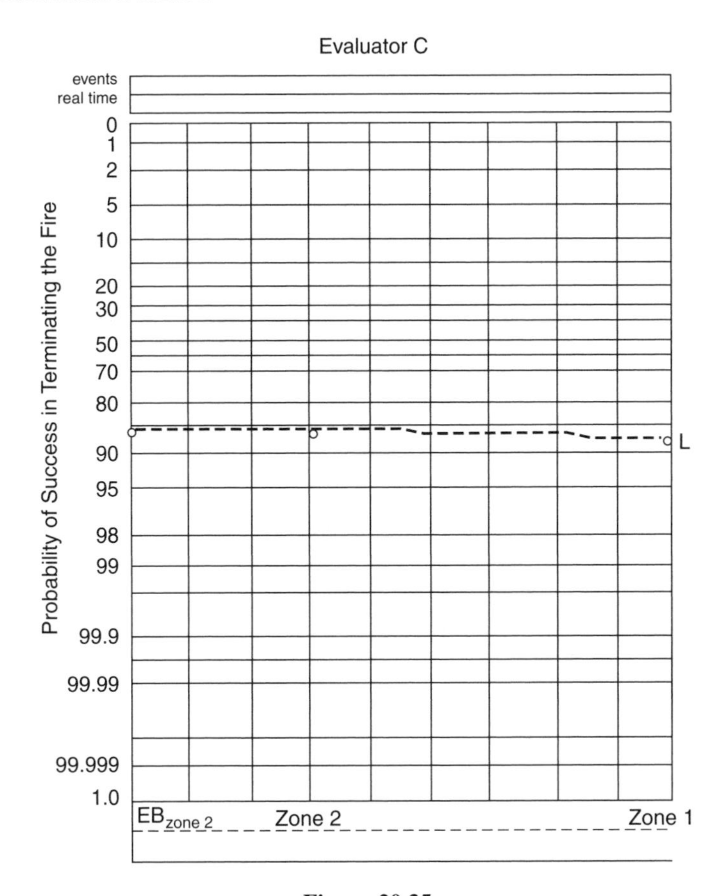

Figure 20.35

the probability that FRI would occur is the same value because EB for zone 2 involves the entire area. Considering fire propagation into zone 1, it was believed that some chance ($P(\mathrm{M}) = 0.1$) of manual suppression existed. Figure 20.34 shows these values for the condition of fire moving from zone 2 to zone 1.

Incorporating these values into Figure 20.36, we determine that given ignition, the likelihood of losing both zone 2 and zone 1 is 0.13. The probability that the fire will be limited before including zone 2 and zone 1 is $P(\mathrm{L}) = 0.87$.

Discussion

The primary objective of this example is to illustrate a way to structure level 2 evaluations for a combined fire prevention and fire propagation situation. Illustrating the technique of using zero-strength barriers in calculations was also a motive.

The room of origin was divided into two zones. The conclusion is that if a fire starts in zone 1, the probability of losing the entire room is 20%. If the fire starts in the area protected by the special hazard equipment, the likelihood of losing the entire room is reduced to 13%. However, more important than the slight difference in numbers is the knowledge of the conditions that produce those values. The problem areas would be sufficiently understood that if improvements

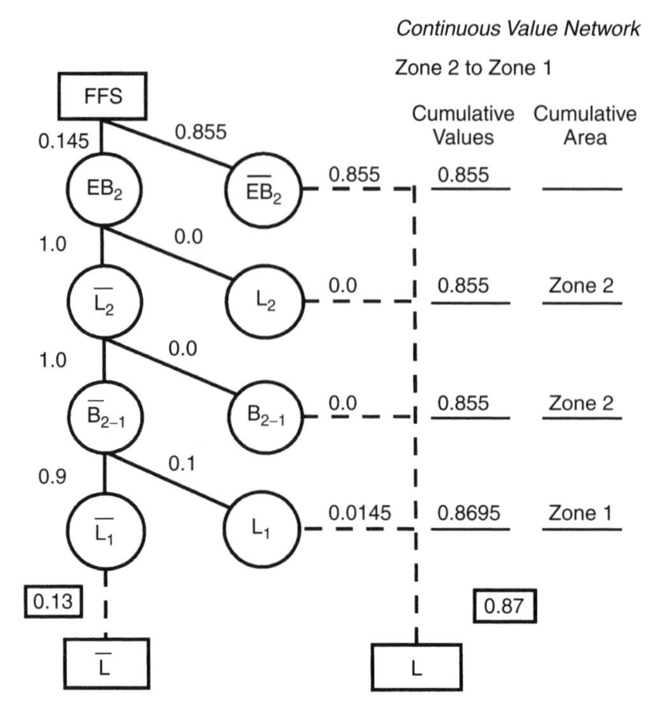

Figure 20.36

were warranted, they would be identified and incorporated into an alternative that would improve the protection.

In this problem the probability of established burning was combined with limiting the fire. This was done for an ignition in each zone. The analysis was based on a probability of ignition equal to 1.0. This enables one to determine the expected outcome independent of the frequency of ignition. If it is not feasible to improve the fire protection, another alternative is to improve the ignition prevention. The relative improvement could be incorporated into the analysis to develop cost-effectiveness analyses for this type of change.

20.18 Closure

Knowledge of a building unfolds logically as it is evaluated in an organized, structured manner. It is important to be able to uncouple and distinguish performance after EB from the conditions that lead to EB. This orderly process enables one to develop a clear understanding of the building's strengths and weaknesses.

Many times, the evaluation will not require a "prevent EB" analysis. Plans approval is one such situation. A risk identification and consequence evaluation is another. However, many buildings survive only on the success of fire prevention. Owners and occupants should be aware of these buildings. If the post-EB protection cannot be improved, attention can be directed to the pre-EB protection.

This chapter attempts to organize the fire prevention component of the complete system. Here fire prevention means ignition prevention and also established burning prevention. The building occupants are incorporated into this part of the system. Within the comprehensive fire safety system, building occupants can be important to the total picture. However, because of the dynamic changes in the history of a fire, the role of the occupant changes dramatically over time.

Reference

1. McGuire, P. J. An analysis of ship fire safety assessment codes and their application to commercial vessels. MS thesis, WPI, 1995.

21 DECISION MAKING AND COMMUNICATION

21.1 Applications

This book organizes building fire performance into a systematic, analytical process. The goal is to provide an understanding so that (1) site specific risks may be characterized, (2) professionals can make better decisions in routine day-to-day activities, and (3) people may communicate more effectively. Here is a summary of this way of thinking:

- *Risk*: the absence of risk in buildings is not feasible. Nevertheless, relative risks associated with alternate fire safety systems can be understood and compared.
- *Purpose*: all fire performance evaluations are intended to satisfy a need or a perceived goal that is initiated by a question from an individual.
- *Understanding*: the same analytical process can be applied for all buildings, ships, tunnels, and transportation vehicles. The way of thinking is easily adapted to any site and structure and involves the following parts:
  a. *The building and site*: the process describes analysis rather than design. Building and site features that affect performance are identified for an existing building or a proposed new building.
  b. *Analytical framework*: the analysis integrates individual component performance into a macrobehavior of the complete system. Network diagrams provide a rigorous, methodical structure to the process.
  c. *Performance estimates*: probabilities express a degree of belief based on all available information and the time available for the analysis. Each of these estimates uses judgment to bridge the gap between performance predictions and information, knowledge, and state-of-the-art fire science and engineering.
  d. *Visualizing component dynamics*: a performance curve provides a visual description of behavior over a continuum of time-related events. One bases the sense of proportion on examining selective events for particular instants of time.
- *Decision making*: understanding the performance and associated risk characterizations provides a way to compare effectiveness, feasibility, costs, and values for options. Comparisons provide a basis for rational decision making.

Building Fire Performance Analysis R. W. Fitzgerald
 ISBN: 0-470-86326-9 (HB)

- *Communication*: the efficiency and clarity with which one tells the story has a major influence in helping others understand the problem so that they can make rational decisions.

21.2 The big picture

Figure 21.1 provides a complete picture of the complex system of building fire performance. This spreadsheet identifies each component and the time during which it is active. Descriptors in spreadsheet cells enable one to visualize component performance curves. Viewing this big picture enables one to understand each component's contribution to performance.

21.3 Prescriptive codes and performance

How much is enough? Risk is inherent in all human activities, including building design. Acceptable risk is a philosophical question that seems appropriate for case-by-case situations and is not addressed here directly. Nevertheless, risk comparisons are helpful for individuals to make informed decisions.

The prescriptive building code is often viewed as society's definition of acceptable risk. Most questions relate to comparisons with contemporary code and standards requirements. This can occur with renovations, rehabilitation, reconstruction, or merely because most of our buildings have been built under a wide variety of code requirements. It may be impractical or unfeasible to change some buildings. One must make a decision to accept, reject, or require certain changes to make the structure acceptable.

This book describes a process for evaluating performance and associated risks for a specific building design. The building may differ from contemporary codes and standards in minor ways or with major variations. One may evaluate all differences, and compare performance and risk with other alternatives.

Existing buildings or proposed new building plans can be compared with contemporary code and standards requirements by comparing the performance with that of a code-compliant building. One may identify specific differences and examine system performance and associated risks. Any procedure acceptable to all parties may be used, but the technique described in Section 21.5 provides a way to organize the process.

21.4 Risk management

Although code compliance is important, the range of other applications involving fire risk management may be much larger. Owners are becoming aware that code compliance or insurance recommendations may not be sufficient for the operations of a particular building. Risk management programs are based on the understanding that evolves from an analysis appropriate to the client's needs. Risk-based decisions involve a mixture of three ingredients:

- accept the risk;
- transfer the risk;
- change the risk.

When a fire occurs, one alternative is to accept the loss and move on with life. Military organizations and individual families often use this form of fire management.

The most common way to transfer the risk is to purchase insurance. Here risk acceptance (as deductible costs) may be combined with a level of insurance to cover catastrophic costs. Management can compare specific building needs with insurance premiums and costs.

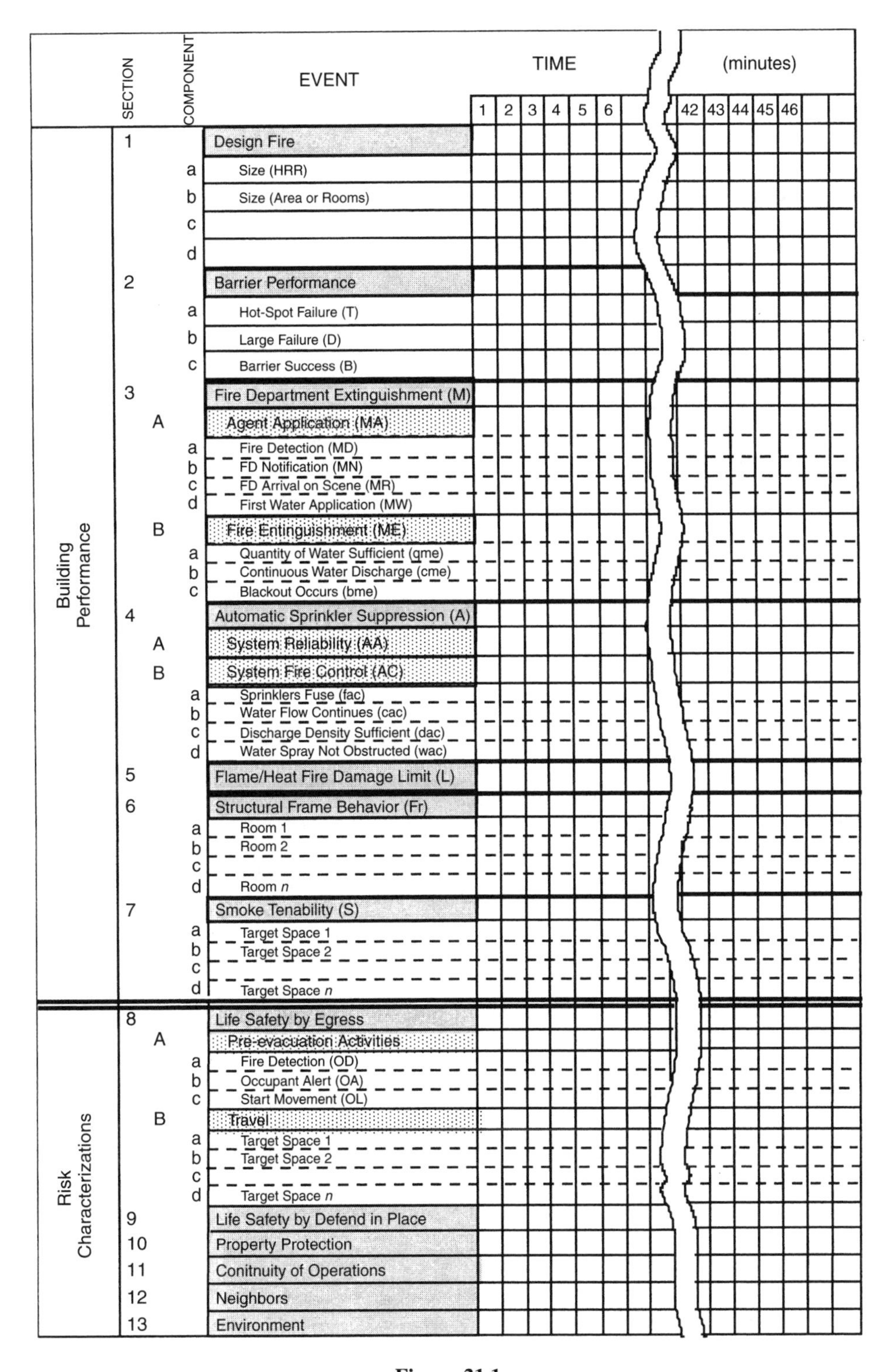
SECTION
COMPONENT
EVENT
TIME (minutes)
1 2 3 4 5 6 42 43 44 45 46
Building Performance
1 Design Fire
a Size (HRR)
b Size (Area or Rooms)
c
d
2 Barrier Performance
a Hot-Spot Failure (T)
b Large Failure (D)
c Barrier Success (B)
3 Fire Department Extinguishment (M)
A Agent Application (MA)
a Fire Detection (MD)
b FD Notification (MN)
c FD Arrival on Scene (MR)
d First Water Application (MW)
B Fire Entinguishment (ME)
a Quantity of Water Sufficient (qme)
b Continuous Water Discharge (cme)
c Blackout Occurs (bme)
4 Automatic Sprinkler Suppression (A)
A System Reliability (AA)
B System Fire Control (AC)
a Sprinklers Fuse (fac)
b Water Flow Continues (cac)
c Discharge Density Sufficient (dac)
d Water Spray Not Obstructed (wac)
5 Flame/Heat Fire Damage Limit (L)
6 Structural Frame Behavior (Fr)
a Room 1
b Room 2
c
d Room n
7 Smoke Tenability (S)
a Target Space 1
b Target Space 2
c
d Target Space n
Risk Characterizations
8 Life Safety by Egress
A Pre-evacuation Activities
a Fire Detection (OD)
b Occupant Alert (OA)
c Start Movement (OL)
B Travel
a Target Space 1
b Target Space 2
c
d Target Space n
9 Life Safety by Defend in Place
10 Property Protection
11 Conitnuity of Operations
12 Neighbors
13 Environment

Figure 21.1

One may change the risk in many ways. Relocation of a certain operation is often a suitable way to change a risk. Installing better fire defenses is another way. Developing appropriate emergency preparedness plans is still another.

Often it is impractical to change a building or to close it down. One may identify a variety of mix 'n match alternatives for any building. Risk management sorts out the problem with possible solutions. A suitable package of alternatives may be developed to help management understand their implications and make better decisions. Knowledge of performance and risk comparisons provides a basis for decisions.

21.5 Performance alternatives

The identification of potential fire defense alternatives is a routine part of the decision-making process. One can evaluate the impact, effectiveness, and costs of possible changes from the base of understanding developed in the initial building analysis. Packaging groups of individual items enables one to organize an immediate action plan and long-range alternative plans.

Performance may be documented by an organized format that can accommodate potential changes and their evaluation. Example 21.1 illustrates the process.

Example 21.1

Show a way of identifying potential changes that will affect the building performance for the example building described in Chapter 4.

Solution

Section 17.4 described the performance of this building. Understanding the performance for this specific building enables one to package alternatives in a cohesive manner.

Column 1 of Figure 21.2 shows factors of the original building that may be considered for change.

Column 2 identifies potential changes that may affect building performance. These features are not a complete list, but illustrate the process. Brainstorming ideas provides an initial base from which potential improvements may be collected and organized. Other performance and risk components can use this format.

Column 3 shows the influence of factors (or groups of combined factors) on the individual component performance. A rating system is often useful to organize the evaluations. For example, a rating of 1 to 4 describes the expected significance of the change on the microsystem performance. Thus, a rating of + is much less important than a rating of + + + or + + ++. A rating of $^{-}$ or $^{--}$ shows a decrease building performance.

A large improvement in one element may not have a significant influence on the macroperformance of a component when viewed in context. For example, changing human fire department notification to an auxiliary fire alarm system may significantly improve that component. However, this effect on M curve performance may be significant or minimal, depending on the sensitivity of notification to the component evaluation. The analysis provides a way to discriminate the significance of element behavior on component functions and holistic building performance.

Column 4 provides a location for recording costs. Initially one may use descriptive cost ranges. Alternatively, one may incorporate numerical values. Column 4(c) identifies functional costs. This may incorporate either monetary expenses or productivity and function costs. This column involves understanding how the building works. Often, functionality and productivity can improve with thoughtful changes in fire performance.

Initial building	Alternative considerations	Change significance	Costs		
			Capital	Operational	Functional
1. Furniture	1. Use less flammable materials in furniture	I curve ++ L curve 1/2 (+)	Not feasible for private homes		
2. Exterior siding (including roof overhang soffit)	2. Change to noncombustible materials	I curve +++ L curve +++	Moderate		
3. Corridor door status	2. Install door closers	B curve +++ L curve + Sm curve ++	Low		
4. Single-station smoke detectors	4. Add heat detectors interconnected to other apartments	Life safety 1/2 (+) MN ++ M curve 1/2 (+)	Low		
5. Single-station smoke detectors	5. Interconnect smoke detectors to other apartments	Life safety 0 to – MN 0 M curve 0	Low		
6. No sprinklers	6. Install sprinklers	A curve ++++	High		
7. Chain-link fence	7. Provide chain-link fence access	M curve 1/2 (+)	Low		
8. Occupant actions	9. Provide occupant training for fire emergency	MN 1/2 (+) Life safety ++ L curve 0 to 1/2 +	Low	Low	Requires constant attention
9. Fire doors at exits	10. Install magnetic holders interconnected with smoke detectors	Sm curve ++ in stairwells and other floors			
10. No automatic fire department notification	10. Provide automatic FD notification with interconnected heat detectors of item 4	MN +++ M curve +++ L curve +	Moderate		
11. Etc.					

Figure 21.2

21.6 Packaging alternatives

Worksheets similar to Figure 21.2 enable one to identify a variety of ideas and their influences on individual components and on the entire system. Similar worksheets may be constructed for emergency preparedness planning. These worksheets allow examination of ideas in isolation and within the context of the system as a whole.

One may evaluate factors from these worksheets separately or in groups to recognize performance changes. When the original building is used as a base for comparison, the cost and effectiveness of individual changes and on packages of features may be described easily. Then two or three thoughtfully constructed alternatives may be developed for presentation to management. Performance curves for alternatives, similar to those of Figure 5.8, can display the effect of proposed changes.

21.7 Summary

This way of thinking enables a variety of applications involving building performance to be considered. All of them involve understanding building performance. Alternate solutions may be considered when risk is too high.

Professional decisions relating the value of information to the cost in acquiring that information are central to applications. Evaluation levels 1, 2, and 3 enable one to tailor the needs of an application to its cost while maintaining an appropriate sense of perspective. The question of how much is enough pervades building performance applications.

APPENDIX A: ANALYTICAL FOUNDATIONS

A.1 Introduction

This method, originally called the interim guide to goals-oriented systems approach to building firesafety, was developed by Harold E. "Bud" Nelson in 1972 as a means to aid management in decision making [1].

Fault tree concepts were the original basis for structuring performance evaluations. Initially this approach enabled individuals to recognize a complete building/fire system. Unfortunately, with all of their strengths, fault trees were never satisfactory for technical analysis and they posed difficulty with practical applications. Although fault trees give a good picture of the logic and interrelationships that lead to failure, they did not generate a clear way of thinking to understand dynamic fire performance. Over the years, an analytical framework evolved by adapting techniques from fields such as systems analysis, risk analysis, psychology, probability and statistics, risk management, operations research, and construction project management. Care was taken to blend theoretical rigor and logic with practical applications for functional performance.

In 1978 network diagrams were first used to track the thought process of performance analysis. Event trees and fault trees were later adapted into a network structure to integrate functional operations and produce the same mathematical results. This network diagram structure magnified technical capabilities and enhanced understanding and communication. This appendix describes the correspondence of the network framework with fault and event trees.

A.2 Logic diagrams

An event tree is a logic diagram that starts with a defined initial event and establishes a forward (inductive) logic that organizes sequences of future events that together describe all possible outcomes. Conditionality and sequencing are usually easy to recognize. The events show states of success or failure that can be represented by probabilistic measures. In many ways, an event tree may be viewed as a way to describe a scenario, analogous to a motion picture. For example, Figure A.1 shows a simplified event tree to describe the likelihood of fire extinguishment by an occupant for a building with no automatic sprinkler protection. The initiating event is ignition, and the event tree describes the sequential events.

Building Fire Performance Analysis R. W. Fitzgerald
 ISBN: 0-470-86326-9 (HB)

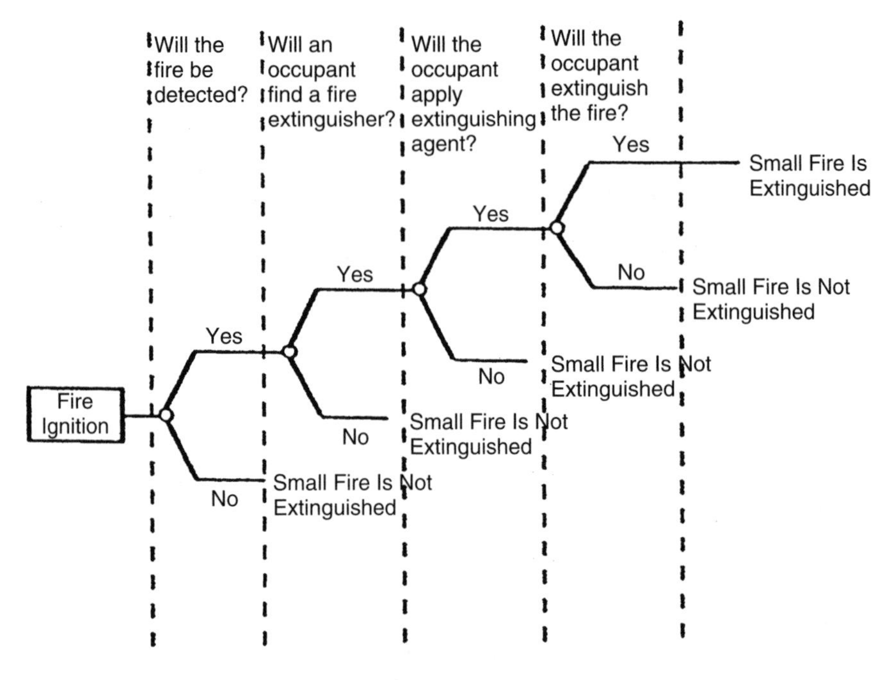

Figure A.1

One can get a sense of other variables such as time, fire sizes, and intervention measures that relate to the event tree sequences. The main advantage of an event tree is the ability to describe a sequence of related events. An event tree enables one to trace a sequence of events from the beginning to the end of a scenario. A major disadvantage of event trees is the inability to decompose events and establish a hierarchy to integrate contributing factors. Also, it is cumbersome to incorporate concurrent paths for redundant activities.

The traditional way of representing an event tree is shown in Figure A.1. Figure A.2 illustrates the same thought process in network form. The differences are inconsequential.

After some experience with fire safety applications, the traditional event trees were adapted to fit routine analysis more efficiently. Risk assessment event trees typically order dissimilar events to sequence a scenario, as shown in Figure A.1. The building/fire system in this book does not need this type of scenario description, because the performance and risk events of Figure 4.7 incorporate all relevant events. Scenarios systematically consider all events that affect performance and sequence their time integration.

Every horizontal row of Figure 4.7 represents a sequence of changes in state for an isolated component. In this book, these component performances are described as continuous value networks (CVNs). They are actually event trees. This organization enables one to focus on a single component, as described in Section 4.5. The results describe changes in state during sequential time increments or building conditions. They become performance curves. These CVNs may also be used to structure cause-consequence analyses and failure modes and effects studies. Figures A.3(a) and (b) illustrate event trees for a fire scenario and the limit of fire damage.

Traditional event trees and CVN diagrams use probabilistic measures of performance, and the calculation rules are the same. That is, probabilities are multiplied along each path, and like outcomes are summed. The CVN diagram structures these calculations to enable their results to be graphed more readily. Examples 6.1 and 6.6 illustrate this process.

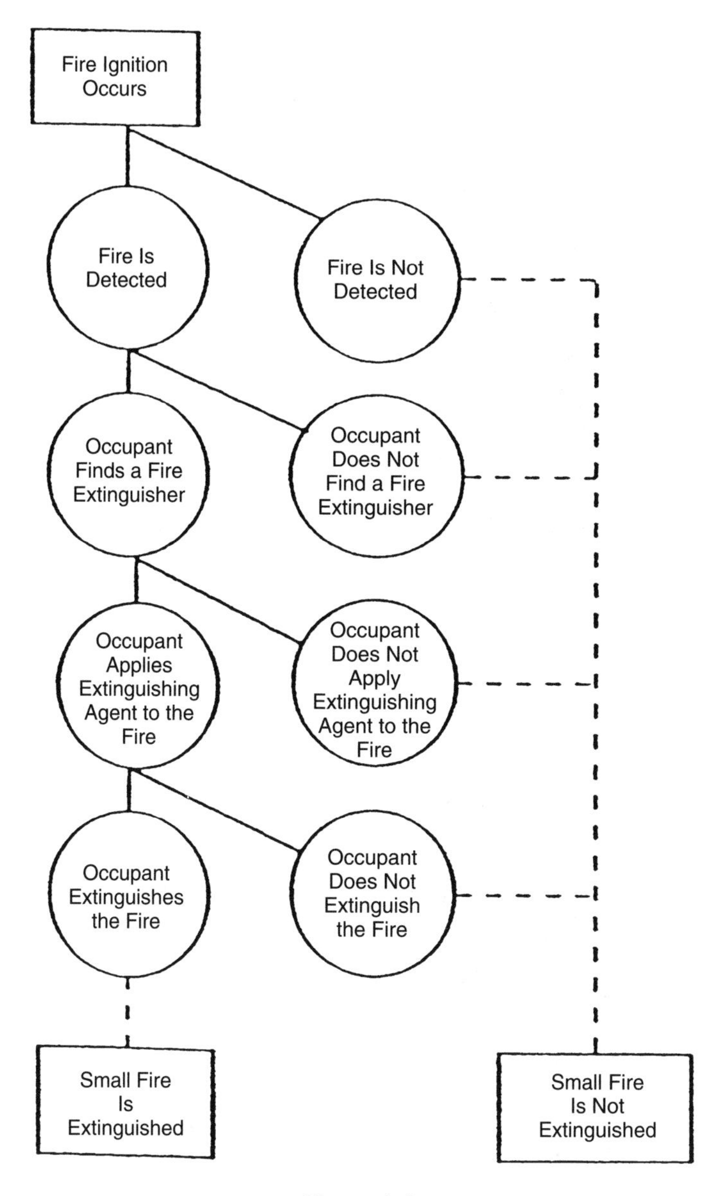

Figure A.2

A.3 Fault and success trees

The fault tree is an important, useful tool that provides a good insight into component behavior. A fault tree supplements event tree weaknesses by identifying causes that can lead to an event failure. The tree organizes the events into a logical framework that enables one to deductively trace the roots of a failure event.

An event is an occurrence. For example, the fire reaching a size of 800 kW; or, the actuation of a sprinkler; or, the notification of the fire department, or the safe egress of an individual are examples of events. Event descriptions are important because of logic gate connections. An AND gate

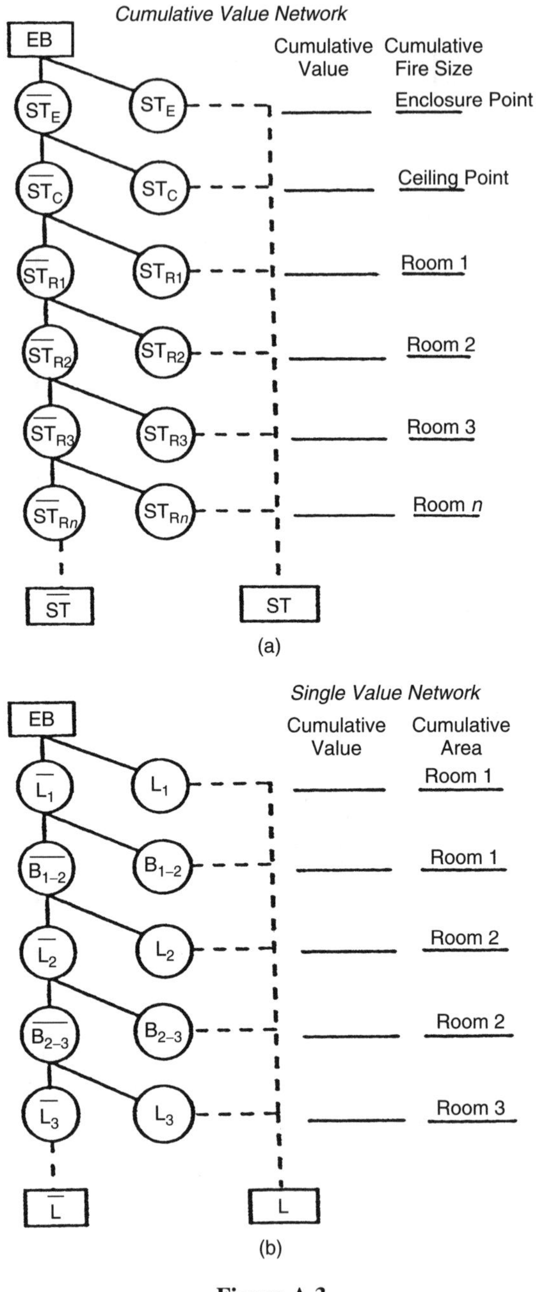

Figure A.3

indicates that all events below that gate are necessary for the event above the gate to occur. An OR gate symbolizes an either/or relationship with the event above the gate. Figure A.4 illustrates the first few layers of a fault tree to examine failure of a sprinkler system to control a fire.

The construction of a fault tree is more of an art than a science. It requires a thorough understanding of the system being studied and the role of the component within the system. All

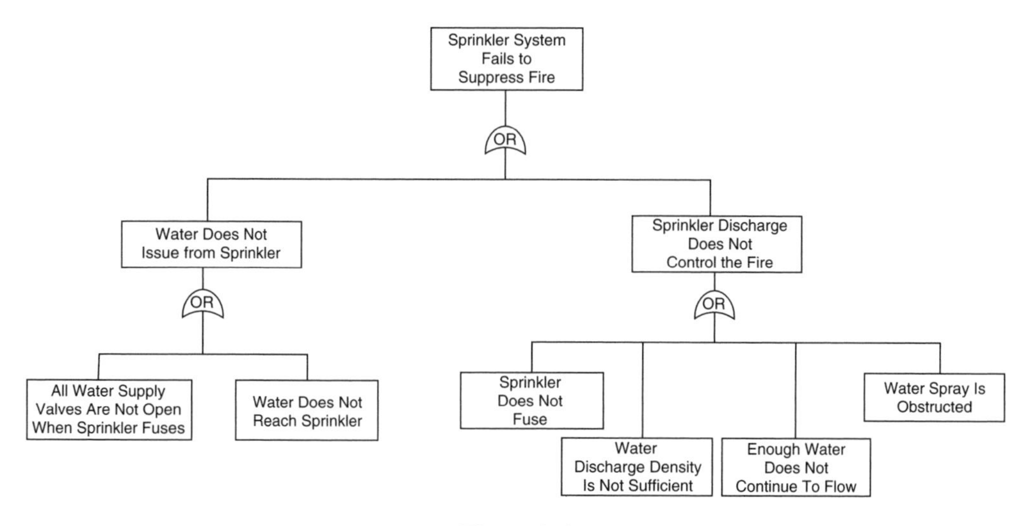

Figure A.4

the factors that contribute to or could cause operational failure of the system are identified. These factors are ordered and grouped into distinct failure events that are synthesized into a logical framework. After initial completion the tree logic is reviewed and tested. The main test for a fault tree is, Will it work? Is it logical?

A fault tree identifies a failure mode and traces back (deductively) the root causes of failure. In design it is often more useful to view the process in terms of success rather than failure. Therefore one can convert a fault tree into a success tree with only two modifications. The first expresses all events in success terms. The second converts the logical OR gates to AND gates and vice versa. Therefore a success tree is the inverse of a fault tree. Figure A.5 shows the success tree that corresponds to the fault tree for sprinkler performance.

Fault and success trees are useful to show all the factors that affect performance on a single diagram. One can observe all important factors and their logical relationships on this single

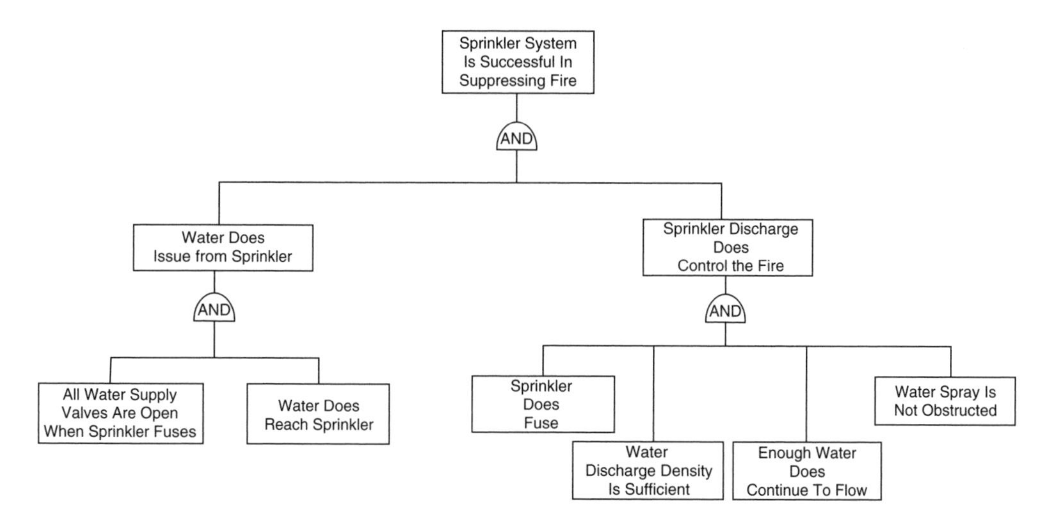

Figure A.5

diagram. There are two cautions to using conventional fault trees. First, one must recognize that evaluation is for one instant of time. One must carefully scrutinize sequencing and coordination of events. The second is that independent, dependent, conditional, and exclusive events are rarely identified. Consequently, one must clearly understand event relationships and their sequencing when making an evaluation.

The conceptual process for fault tree and success tree construction is used to construct networks. Figure A.6(a), (b), and (c) show the network logic for the sprinkler system described above. The network can show success and failure on a single diagram, and the relationship of events is usually easier to recognize. The top event shows the starting condition, and the bottom events show the outcome. Networks have the disadvantage of showing only one evaluation level on a diagram. Consequently, when several levels of causes are present, as in the sprinkler example, one must use a separate diagram for each level. The levels may be linked, as shown in Figure A.6(d). As with fault/success trees, a network evaluation is based on a single instant of time. In this book we designate these as single value networks (SVNs) because they provide an evaluation for a single instant of time or condition.

The network organization uses the logical gates. Events along a continuous path represent AND gates, whereas events that terminate represent OR gates. For example, in Figure A.6(c) successful extinguishment occurs when, given a reliable sprinkler system, the sprinkler(s) fuse AND the water density is sufficient AND continuous water flows AND the water spray is not obstructed. The operational effectiveness of the system fails when the sprinklers do not fuse OR the water density is not sufficient OR water flow is not continuous OR the water spray is obstructed. These may be compared with the event trees of Figures A.4 and A.5.

Although traditional event trees and network diagrams use the same probabilistic measures and obtain the same results, the calculation procedures are different. Events in all these representations are expressed in binary form. That is, an event plus its complement must be 100%. Thus

$$P(\mathrm{E}) + P(\overline{\mathrm{E}}) = 1.0, \tag{A.1}$$

where $P(\mathrm{E})$ is the probability that an event will occur and $P(\overline{\mathrm{E}})$ is its complement, the probability the event will not occur.

Three types of condition are considered by logic gates in fault/success trees:

- the union of two mutually exclusive events, as shown by the Venn diagram of Figure A.7(a);
- the union of two or more independent events that overlap, as shown in Figure A.7(b);
- the intersection of two or more events, as shown in Figure A.7(c).

A union is a combination of events that is represented by an OR gate. When the events are mutually exclusive, as shown by Figure A.7(a), their union is calculated as

$$P(\mathrm{E}) = P(\mathrm{A}) + P(\mathrm{B}). \tag{A.2}$$

The intersection of two independent, mutually exclusive events is shown in Figure A.7(b) as the shaded area of the circle overlap. This represents an AND gate whose value may be calculated as

$$P(\mathrm{E}) = P(\mathrm{A})P(\mathrm{B}). \tag{A.3}$$

The union of independent, mutually exclusive events in Figure A.7(c) is the area enclosed by the circles. This area represents the probability of either OR both events. The shaded area has been included twice and must be deducted. The equation for this union is

$$P(\mathrm{E}) = P(\mathrm{A}) + P(\mathrm{B}) - P(\mathrm{A})P(\mathrm{B}). \tag{A.4}$$

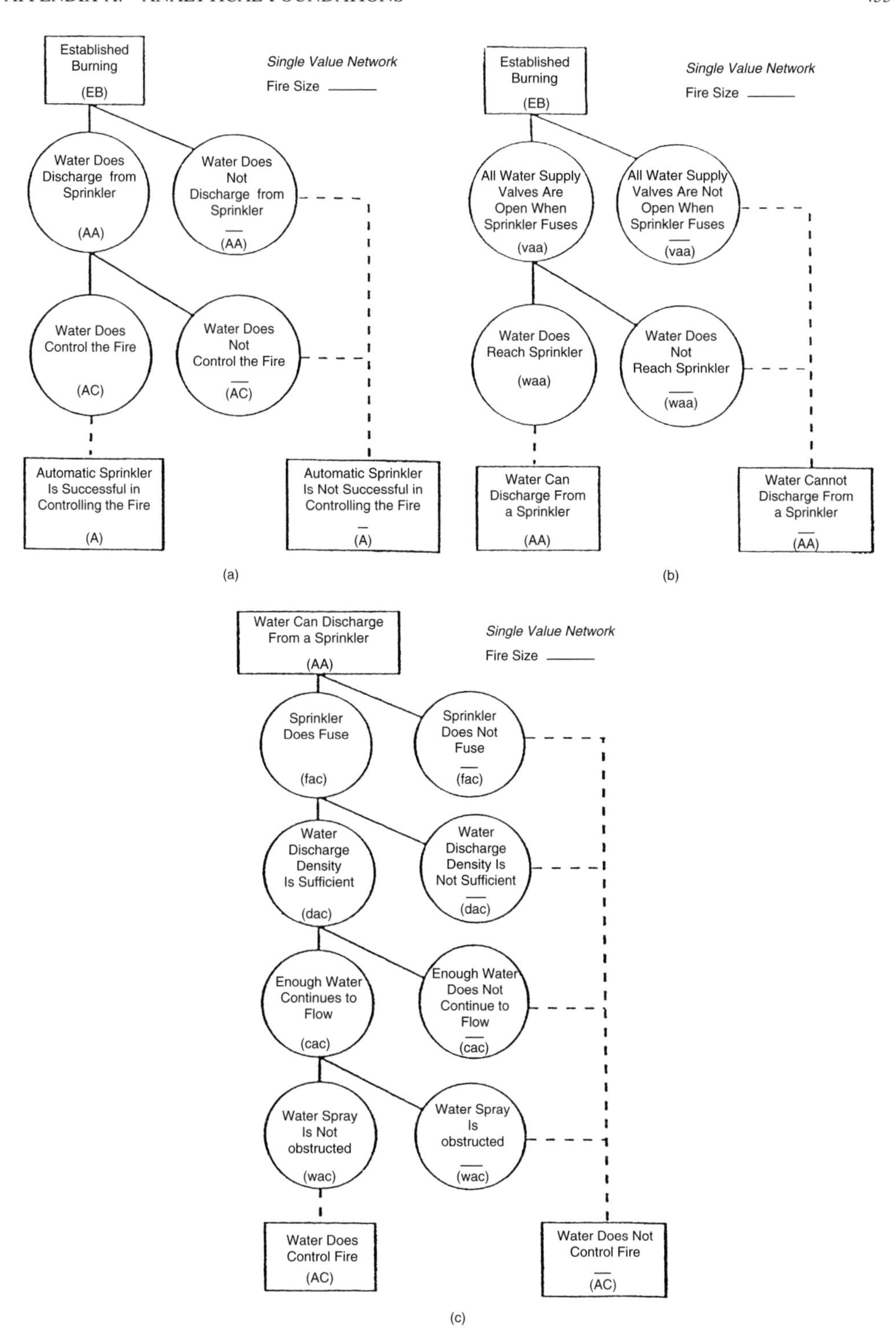
Established Burning (EB)
Single Value Network
Fire Size
Water Does Discharge from Sprinkler (AA)
Water Does Not Discharge from Sprinkler (AA)
Water Does Control the Fire (AC)
Water Does Not Control the Fire (AC)
Automatic Sprinkler Is Successful in Controlling the Fire (A)
Automatic Sprinkler Is Not Successful in Controlling the Fire (A)
(a)
Established Burning (EB)
Single Value Network
Fire Size
All Water Supply Valves Are Open When Sprinkler Fuses (vaa)
All Water Supply Valves Are Not Open When Sprinkler Fuses (vaa)
Water Does Reach Sprinkler (waa)
Water Does Not Reach Sprinkler (waa)
Water Can Discharge From a Sprinkler (AA)
Water Cannot Discharge From a Sprinkler (AA)
(b)
Water Can Discharge From a Sprinkler (AA)
Single Value Network
Fire Size
Sprinkler Does Fuse (fac)
Sprinkler Does Not Fuse (fac)
Water Discharge Density Is Sufficient (dac)
Water Discharge Density Is Not Sufficient (dac)
Enough Water Continues to Flow (cac)
Enough Water Does Not Continue to Flow (cac)
Water Spray Is Not obstructed (wac)
Water Spray Is obstructed (wac)
Water Does Control Fire (AC)
Water Does Not Control Fire (AC)
(c)

Figure A.6

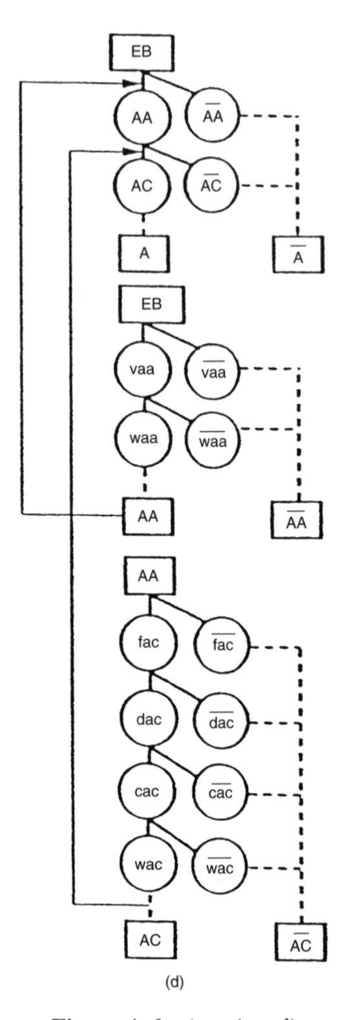

Figure A.6 (*continued*)

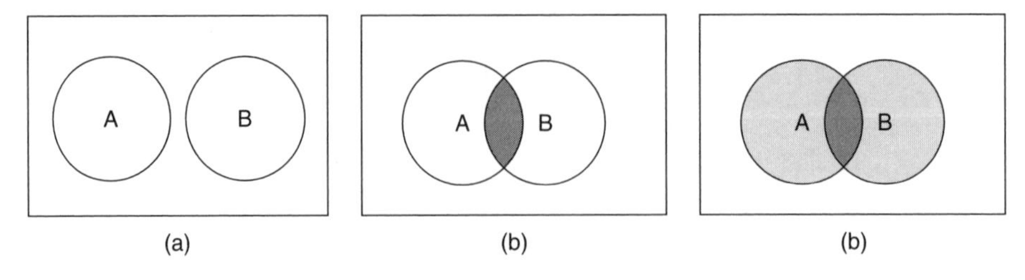

Figure A.7

Assume that $P(\text{A})$ is the probability that the sprinkler system will extinguish a fire and $P(\text{M})$ is the probability that the fire department will extinguish the fire. If each of these probabilities were estimated independently, the probability of fire extinguishment may be calculated as

$$P(\text{extinguishment}) = P(\text{A}) + P(\text{M}) - P(\text{A})P(\text{M}). \tag{A.5}$$

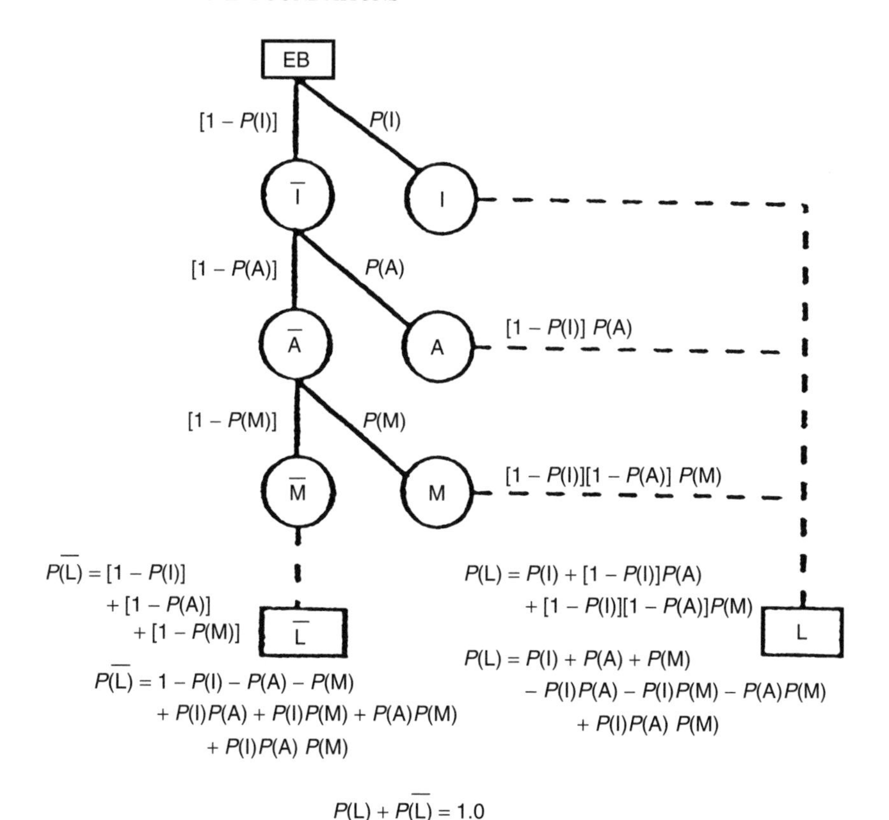

Figure A.8

The complexity increases with the number of events. For example, consider the three ways of limiting a fire. Fire can be limited by self-termination, $P(\mathrm{I})$, OR by automatic sprinklers, $P(\mathrm{A})$, OR by the fire department, $P(\mathrm{M})$. The limit $P(\mathrm{L})$ may be calculated as

$$P(\mathrm{L}) = P(\mathrm{I}) + P(\mathrm{A}) + P(\mathrm{M}) - P(\mathrm{I})P(\mathrm{A}) - P(\mathrm{I})P(\mathrm{M}) - P(\mathrm{A})P(\mathrm{M}) + P(\mathrm{I})P(\mathrm{A})P(\mathrm{M}). \tag{A.6}$$

This equation is cumbersome and may be expressed in alternate form as

$$P(\mathrm{L}) = 1 - [1 - P(\mathrm{I})][1 - P(\mathrm{A})][1 - P(\mathrm{M})]. \tag{A.7}$$

Networks avoid these equations by their organization and structure. One may follow the procedures described in Chapter 6 to get the identical results. If one incorporates the symbolic OR values of Figure A.8 and applies the network calculation procedures to find $P(\mathrm{L})$, equation (A.7) will result. When calculating the outcome for $P(\mathrm{L})$ and using (A.1), equation (A.7) can be obtained.

A.4 Discussion

Traditional risk analysis trees and networks are closely related, and calculations from equations or the network structure are identical. Besides being easier to remember and use mathematically, the networks provide additional attributes. The most important is their ability to track a thought process following functional behavior. Another is that it is somewhat easier to recognize event dependence and independence with networks. Networks are not completely transparent, but fewer problems have been detected in their application to building analyses.

The importance of structuring the thought process for applications cannot be overemphasized. The mathematics of calculation using either event and fault/success trees or network diagrams is trivial. The problem is unimportant computationally because both approaches give the same numerical result. The significance occurs during event evaluations to determine the probabilistic estimate. Over the years, experience with practical building applications has shown that understanding greatly improves with network diagrams. Each event estimate is based on the individual placing himself or herself into the condition and environment of the previous event. Conditionality, dependence, and environmental conditions take on a different perspective in the evaluations.

Understanding a component's performance in isolation and also as a part of a complete building system is the primary goal of an application. Mathematical results are merely a means of comparison. The ability to move back and forth between macro- and microperformance using CVNs and SVNs as guidance makes the process easier.

Reference

1. *Interim Guide to Goal Oriented Systems Approach to Building Firesafety*, PBS 5920.9 Building Fire Safety Criteria, Appendix D, General Services Administration, 1972.

APPENDIX B: WHAT ABOUT THE NUMBERS?

B.1 Introduction

The methods in this book have evolved over 30 years by adapting a variety of techniques from other disciplines. Risk assessment and risk management have been central to the process. Systems analysis, operations research, psychology, and construction project management have contributed much to the procedures. The meaning of probability and techniques of assessment have received the greatest attention.

Appendix A demonstrated that the organization of network diagrams is directly related to fault trees and event trees of risk assessment. The original reason for recasting these tree diagrams was to track a thought process. Calculation simplification and relating numerical results to performance curves was an added benefit.

There are two inseparable constituents to this evaluation method. One is the framework for analysis. After nearly 20 years of development and testing, general acceptance of its organization and structure seemed apparent. The second involves selecting numerical measures of performance. Over the years, the philosophy of probability selection moved from a nonissue to a question of intense scrutiny and debate to a recognition that subjective judgment has great value in understanding performance and documenting rationale.

This appendix describes a rationale for using judgment and subjective probability in assessing event performance. The advocacy for using subjective probability as the measure for uncertainty did not come easily. The path has been difficult and filled with many experiments, much study, and considerable discussion. Out of necessity, some philosophy and observations have been mixed into the discussion.

B.2 Background

The original function of this method was to help individuals view fire and buildings as a complete system and understand risk characterizations for any site-specific building. While this function has remained a primary focus during all of its evolution, the technical and experiential base has changed substantially. Throughout its continuing evolution, questions have been constantly asked concerning the logic, technical basis, application procedures, numerical measures, and

Building Fire Performance Analysis R. W. Fitzgerald
 ISBN: 0-470-86326-9 (HB)

documentation of results. Over the years, the nature of the questions has changed dramatically. They may be grouped into three general categories:

- During the early years, the dominant focus was associated with, What was the thought process used to analyze that component?
- During the middle years, questions about analytical logic subsided and were replaced by questions associated with, How do I get the numbers to evaluate that component?
- In recent years, interest has been directed to, How can I find better state-of-the art information to document my conclusions?

Questions associated with all of these groups have existed throughout the evolution and will probably continue into the future. As answers to many perplexing questions emerged, their results became integrated into the process. Then new, more advanced questions appeared. Much of the evolution can be traced to addressing concerns of students and fire safety professionals in using the procedures for practical applications. The rest was directed toward technical accuracy and validity.

A technical basis for the organization, structure, and mathematics of the system is necessary, but it is not sufficient. The most important attribute of the method is the thought process that it portrays. Much effort has been spent in attempting to capture the way in which a wide variety of fire safety professionals understand and describe building fire performance within their specialty of expertise. The framework has attempted to organize the ideas of these professionals into a structure with which the fire performance of any building can be evaluated in an integrated, consistent, and cohesive manner.

B.3 Meaning of probability

Probability has had a dual meaning almost from the time it was invented. There are two extreme definitions (perhaps we can call them philosophies), as well as many positions between these extremes. One definition, called classical or objective, is that probability is a measure of the frequency of occurrence of one event out of a possible set of the same events. Thus, in a well-shuffled deck of 52 cards, the probability that a random card selection will be a queen is 4 queens / 52 available cards $= 4/52 = 1/13$. The probability that a queen of hearts will be selected is 1/52.

The classical position states that probability is valid only for events that can be repeated under essentially the same conditions. The objectivist extends this idea with limit theorems to include relative frequencies of long-run "experiments." For example, if a sprinkler manufacturer or a detector manufacturer wishes to determine the reliability of the company's product, a sample number of devices is selected from among those that were manufactured. These devices are tested and the number of failures is recorded. The probability of a bad device reaching the construction site would be very nearly the ratio of the number of bad devices out of the total sample that was selected for testing.

The objectivist likes to determine outcomes only from experiments or situations that are repeated under the same conditions. To obtain the expected value of an event, statistical data is needed. The probability is based on the historical record from a set of situations that are essentially similar. This data also enables the variance, confidence limits, and other useful information to be determined. An objective probability is viewed to be a characteristic of an identifiable physical process and is a property of the event. This definition of probability is commonly adopted in the physical and biological sciences and it has wide use in a variety of industries such as medicine, pharmaceuticals, life insurance, and property insurance.

The second definition of probability is called personal or subjective. This interpretation is more commonly used for unique situations where experimental repetition under the same conditions is impossible or unfeasible. For example, the probability that the sprinkler system will extinguish a defined fire with four sprinklers in a site-specific building, or that a detection system installation will actuate before the fire reaches 100 kW is more appropriate for personal assessments. These estimates use subjective probability to describe an expected performance.

The subjectivist can apply probability to all of the problems that an objectivist considers, and to many more. A subjectivist views probability as a measure of the degree of belief that an individual has in a judgment or prediction. Here probability is interpreted as an intellectual process, rather than as a physical property of the objective definition. Some authors [1] view all probability as subjective, and the type and use of actuarial information distinguish the relative position of the application on the scale of interpretations.

When this book describes an estimate calibration, it normally uses the concept of subjective probability. Tribus describes this process as encoding a state of knowledge [2]. The assessments are based on acquired knowledge such as calculated values, computer models, physical relationships, observational facts, experimental information, failure analyses, and personal experience. Assessments use the full spectrum of information that is available and seems relevant to the problem. The existence of uncertainty is clearly understood.

One interprets subjective probability assessments as a judgmental description of expected performance for a site-specific condition. For example, success probabilities of 0.4 and 0.9 should not be interpreted literally as statistical outcomes of 40 times and 90 times out of 100. Rather, they should be recognized as a calibration of a performance estimate by the individual making the assessment. Although it is possible that both outcomes can occur, the event with probability 0.9 is substantially more likely than the event with probability 0.4. Values are normally based on a deterministic evaluation of the situation involved. For example, assume that an individual evaluates a building and describes an estimate for the sprinkler system to control a fire within a floor area of $20\,\text{ft}^2$ as $P(\text{A}) = 0.80$. This means that, based on the information available for the evaluation, the likelihood that the sprinkler system will control the fire before it extends beyond $20\,\text{ft}^2$ is estimated as 80%. The numerical value expresses a best judgment for the specified SV analysis (using the terms of Chapter 6). The calibration is expressed as a probability based on deterministic evidence.

Sometimes this definition of subjective probability is confused with Bayesian theory which involves a combination of subjective judgments and experimental observations. In obtaining Bayesian probabilities, initial subjective judgments form the basis of evaluation. Then those subjective values are upgraded with experimental observations to improve the quality of the assessment. We prefer not to characterize the process advocated for site-specific building performance evaluations as Bayesian. We characterize it as subjective, personal, or estimation calibrations or as encoding a state of knowledge, because acquiring additional statistical data from experimental upgrades is not a usual part of the process. The results should be interpreted as being analogous to estimating construction costs or project completion dates.

B.4 Mathematical theory of probability

There is no disagreement between the frequentist and the subjectivist over the mathematical foundations of probability. The difference of opinion is largely philosophical. Each definition of probability has philosophical extremists at the ends of the spectrum of opinion, extremists who will not concede that the opposite view has any validity. However, in practical evaluations for engineering disciplines, most individuals will adopt the probability view that enables the best decision to be made for the specific problem under consideration.

One of the attributes of the mathematical theory of probability is that it does not consider how the numbers are determined or what they mean. The theory of probability only identifies how to use the numbers in a consistent manner. The analytical framework (i.e., network diagrams) has been carefully constructed to conform to mathematical principles. Consequently, this framework can be used regardless of the interpretation of probability.

Now, let's be a little careful. Just because the analytical framework enables the mathematics to be consistent, it does not relieve the individual of understanding what the numbers mean. After all, the goal of a performance evaluation is to understand the building and its behavior in a fire. Numbers are a means to an end, not the end itself. The goal is not to generate a number or a graph that relieves someone from thinking about its meaning and making a rational decision. Numbers and graphs are used to assist understanding, compare alternatives, and communicate with others.

B.5 Firesafety evaluations

A question arises about which interpretation is appropriate for firesafety applications. The answer depends on the application. If one is dealing with a class of buildings (a population), the objective view is likely to be better. For example, if one wants to compare occupant deaths, property damage, or other characteristics between sprinklered and nonsprinklered buildings, objective probability is the most appropriate. Insurance decision making on which general policy guidelines are based would seem to be more suited to objective procedures.

On the other hand, if one is interested in understanding the performance of a site-specific building in order to develop a risk management program or to develop documentation for a code equivalency proposal, subjective probability is the most appropriate. An insurance inspection highly protected risk (HPR) report would be an appropriate application because subjective judgment is already an integral part of those reports. In cases where one needs documentation of performance and a rationale for the selections, the subjective calibration of estimates is more valuable.

B.6 Accuracy

A group of questions are often posed relating to accuracy of the numbers. In particular, when subjective estimates are used, different people can select different probabilistic values for events. One may ask, Is this a problem? Who is correct? Why isn't it better to have a procedure where everyone gets the same number?

The personalist interpretation of probability accepts that different individuals may differ, even with the same evidence. This may lead to a perception that subjective probabilities are just guesses and not to be used in serious or rigorous applications. This perception is not accurate and, when used appropriately, subjective judgment expressed as a probabilistic measure can be a valuable contribution to analytical rigor. Subjective probability is a valid representation of knowledge and its strengths can be exploited to great advantage.

Many individuals who routinely solve mathematical problems view a numerical answer as being correct or incorrect. Consequently, if all answers using the same deterministic conditions are not the same, which is correct? In using subjective probabilities, a numerical answer is not viewed as of right or wrong. Rather, it is the quality of the assessment that becomes significant. That assessment quality may be viewed as good or bad. The quality of an estimate depends on the amount of information that is used, the skill in translating that knowledge into performance estimates, and the constraints, such as the available time, to make the estimate.

The quality of an assessment requires that the event and its qualifying conditions must be identified clearly. Network evaluations enable this to be structured and described more easily. Nevertheless, humans can have preconceived biases that affect judgment. Here are some of the more common influences:

- Overconfidence in scientific knowledge and computer modeling is often displayed.
- Human error and anticipated human activities or capabilities are often inappropriately portrayed.
- Rare events influence judgement, particularly if they have occurred recently.
- Preconceived assumptions about the effect or value of codes and standards on performance sway opinions.
- Fear of uncertainty can produce excessive conservatism or overadjustment of values.

It is possible to develop scoring models or checklists to enable different individuals to arrive at values that would have numerical consistency. Those models streamline the process and have values of standardized procedures, relative uniformity among evaluators, and a comfort level in use. There can be additional ways of administering the process more clearly. For example, Chief Fire Marshal Harvey developed routine inspection procedures for fire department inspections in Boulder, Colorado, that proved to be very successful [3]. The inspection procedures integrated judgment into an organized form and provided ways of effectively handling the human biases.

Although it is possible to develop standardized procedures based on this method, those procedures are not consistent with the goal of this book. Our goal is to help an individual understand the fire performance of a site-specific building. This understanding can be translated into better communication with others in the building industry. It can also contribute to a better technical documentation of the rationale for the performance expectations. Nevertheless, a collection of decision-aiding tools to address different questions becomes a valuable asset in enabling estimates to be made with greater confidence. Decision-aiding tools based on the framework and evaluation concepts can be developed to handle reasonable variation in personal judgment if there is a sound reason to do so.

The issue of different numerical estimates among individuals should be discussed further. Although numerical differences have occurred during many years of observing different individuals evaluating the same building or academic exercise, rarely have they led to differences in identifying the important features and understanding performance. The framework is constructed with the principle of divide and conquer, which structures the events to enable one to move along an evaluation path in methodical, comfortable steps. If a meaningful numerical difference for an event were to arise, the individuals could focus on a relatively narrow issue and identify sources of information and the bases for the estimate. This discussion usually resolves the issue. Communication becomes enhanced because of the deterministic base for performance evaluations. Because numbers are used primarily to assist in understanding performance, the numerical differences do not seem to be a problem. In other words, while the numerical values may differ (by relatively minor amounts in most cases), the identification of building problems and their relative importance rarely differs among individuals.

B.7 Comfort and confidence

The issue of comfort or confidence in making evaluations among users is real and important. This is especially the case among users first exposed to this way of thinking. It also occurs with individuals who are very knowledgeable in one area of fire safety, but who may not have experience with or knowledge of some components of a complete system. There are a whole series of concerns relating to confidence of an individual in making evaluations. Let's try to address the common issues that cause discomfort.

The usual questions may be grouped into areas involving inadequacy of personal knowledge, weakness in state-of-the-art professional tools of fire science and engineering, constraints relating to available time for evaluation or acquiring building knowledge, and communication needs.

PERSONAL KNOWLEDGE

The building/fire system is very complex compared with other engineering disciplines. In addition, from a historical perspective, the usual building design team does not have individuals whose professional practice is devoted primarily to fire performance decisions. Consequently, the regulatory system (before building acceptance), the fire service (after building acceptance), and to a more limited extent the insurance industry were used to take responsibility for controlling and providing acceptable levels of fire safety for citizens, commerce, and industry. The "code" became the measure of acceptability for design and performance.

The building code and associated standards have improved the level of fire safety tremendously over the years. The fire-related parts of the code have been written in a way that allows individuals with no knowledge of fire safety to apply the code with a degree of comfort. The fact that the level of individual fire risk is indeterminate (and may be very good or very bad) has not been as important as the fact that the building complied with the regulations at the time of construction. Overall the quality of buildings is recognized to have improved during this nearly 80 years of modern building codes. As the codes have improved over the years, the cost of bringing existing buildings to code compliance may not be economical. This results in "grandfathering" the certificate of occupancy. Is this important?

As codes have improved over the years, so has knowledge acquired by experience and augmented by fire science and engineering. Performance expectations for a fire scenario can be predicted at this time much more accurately. This means that individual components and their associated building features or human factors can be evaluated for expected performance with greater confidence than before. The (often justified) criticisms of codes can now be discussed on a more rational basis. However, one must be careful to consider individual factors as a part of a complete system rather than in isolation. This becomes even more important in building fire performance than in other disciplines, because of its complexity and interrelationships.

An individual's personal experience and knowledge become very intimate with confidence in making assessments. Fire science and engineering are often not adequate to provide appropriate justification. Experience may be gained over the years, but what is one to do in the meantime?

The most important ingredient of an evaluation is to understand what constitutes a complete building/fire system and to gain an initial awareness of the function of all components. The macro or global view of the building and its performance evaluation is central to the process. Within this global perspective it is useful to know what you don't know. Even if this is not possible, becoming knowledgeable about the major factors that affect component behavior enables one to seek help with an evaluation. It is difficult for any one individual to know enough about all parts of the complete fire safety system to make detailed evaluations all of the time. Specialists often have a reluctance to stray beyond their field of expertise. Nevertheless, one may make level 1 evaluations to identify those aspects where more detailed knowledge by others is appropriate. Management of knowledge needs and its acquisition is an important part of any evaluation process.

One may not be expected to be all things to all persons (or needs), but a professional should be expected to build up a broad base of knowledge during his or her professional life. The network events (bubbles) can provide a set of pigeonholes to store ad hoc experiential or documented information. They are a place to organize available technical information into a form that can be used for evaluation purposes. Decision-support systems that provide analytical and experiential information and spreadsheets that can be organized with network events help to make evaluations.

PROFESSIONAL TOOLS

The evolution of fire safety engineering is following a developmental emergence similar to that of other, more mature disciplines. During the early stages, analytical capability is limited and technical decisions are based largely on observation and judgment. As greater technical capabilities emerge, some of the judgment is replaced by analytical procedures. However, even in mature disciplines, professional judgment remains a major part of decision making.

Fire safety performance has much informational experience and some good analytical capabilities. However, it has relatively few guidelines for decision making because the analytical tools have not been used widely, and performance is rarely analyzed after the fact. Many technical tools do not describe the necessary range of performance expectations, and there are gaps in the analytical technology. For example, building performance for fire ground operations, barrier behavior, and multiple-sprinkler behavior have deficiencies in analytical capabilities. As research is extended to these topics, the profession will mature rapidly. In the interim, performance evaluations have to incorporate a strong element of judgment into the process.

Judgment is a valuable attribute that should not be mistaken for uninformed guessing. Judgment is based on an understanding that is founded on technical behavior. It has served civilization well for many centuries and is a principal basis for technological advancement of the professions. Judgment should not be feared or flaunted. It is a part of all professional work.

CONSTRAINTS

The quality of all projects is affected by constraints. Some constraints exist because of gaps in professional knowledge, as described above. Many are present because the information is either concealed or too expensive to obtain. Information may be grouped into five categories:

- information that is observable;
- information that may be measured or tested;
- information that may be deduced by calculation or other analytical means;
- information that may be obtained from plans or other data retrieval sources;
- information that is not available.

The economy of information becomes an important factor in professional decision making. When the cost of information becomes greater that its value, alternative, less expensive approaches are used. The cost of information can be thought of in two ways. One is the solution to the client's problem. An individual must use the information available within the time and budget constraints of the employment contract. The three evaluation levels described in this book become useful to consider in this context. The second is that of personal protection. One must ensure that shoddy or inadequate professional services are not a part of the solution. This can be difficult to define in an emerging discipline. A meeting of minds on the type and scope of professional services is important to minimize potential criticism or legal actions. Systematic documentation of the technical basis for decision making can have legal and professional benefits.

COMMUNICATION

The most important benefit that comes with understanding is the enhancement of communication. The ability to provide logical and convincing evidence for judgments and opinions is a great value. The use of curves to convey dynamic performance comparisons can be helpful. The primary purpose for constructing performance curves is to help you understand the behavior of the building or alternative considerations. However, their use to convey this understanding to others may also have a place in some situations.

B.8 Probability based on deterministic analysis

A question that may arise is, If this method is deterministic rather than probabilistic, why are calibration estimates expressed as subjective probabilities? The simple yet unsatisfactory answer is that Nelson [4] used subjective probability as the performance measure when this method was first published 30 years ago. Although Nelson's insightful way of describing performance had a major influence on the method's early evolution, as time progressed, that reasoning had to be examined more carefully.

The focus of the early years of teaching, research, and applications testing with this method was on the organizational framework for thinking. So much effort was concentrated on developing a logical, practical framework that the meaning and ways of determining probability were relegated to the back burner. The need for having reasonable numbers to structure and test the framework required the expediency of continuing to use subjective probabilistic estimates. Because the framework was structured around an analytical thought process rather than available data, an assumption was made that objective probabilities could be developed at a later date.

As the framework for thinking began to take form and its structure was organized to provide analytical rigor with regard to risk analysis and mathematical theory, attention shifted to evaluation methods. During the first 16 years of development, the objectivist view of probability was an implicit expectation. Some techniques were explored, and efforts were made to develop objective probabilistic values. Some were successful and it was recognized that it is possible to develop procedures by which objective probabilistic values could be generated. However, after some testing and more reflection it was concluded that this type of procedure was inappropriate. An evaluation that focuses on understanding specific behavior uses deterministic state-of-the-art technology.

The primary objective of a performance analysis is to understand the building behavior. Experimental and anecdotal situations showed that frequency-based evaluations often give a very different perception of performance than engineering judgment. For example, the basement of a fully sprinklered (wet pipe) office building was assigned a likelihood of success of 0.2 while that same building and sprinkler system was assigned a value of 0.95 for the floors above. Why should the disparate values be chosen by very experienced are knowledgeable firesafety professionals for a system that statistically forecasts a success of about 96%? In another situation a fully sprinklered (wet pipe) warehouse was analyzed with this method by two students who initially did not have any background in fire protection or sprinklers. They selected a probability of success of 0.3 for that sprinkler system. Why would they describe a wet-pipe sprinkler system in this manner? Later examinations by the insurance carrier, a sprinkler design company, and a professional fire protection engineer all concluded that the students had identified significant weaknesses in operational effectiveness that had previously been overlooked. Another example assigned a fully sprinklered commercial building a probability of success of 0.05. This unusual value attracted attention (i.e., told a story) to a condition where the fire pump was not connected.

These examples, as well as many more that could have been described, illustrate that a technical analysis which looks at the important details of a site-specific building can provide a good understanding of performance. That performance can be calibrated by subjective probability estimates. Understanding a site-specific building performance is not a process to generate numbers that can be cited. Rather, the goal is to identify an expected behavior for specific conditions that are present or proposed. This does not imply that statistical data is not important or useful. On the contrary, it can have an important role in the description of buildings and in the development of analytical support structures. However, one must ensure that the role of a performance evaluation is to enhance understanding of each site-specific building. Statistically based analyses are not able to discriminate certain combinations of details that affect installation quality or component performance. The logic of analysis must be transparent, and the basis for performance clearly documented.

For many years this author had a bias against subjective probability values and the somewhat related Bayesian theory. Mathematics, systems analysis, and risk analysis are dominated by objective probability procedures. Death of the frequency concept of probability does not come easily to an individual who has been brought up to identify with a world of accepted practices and scientific objectivity. However, it was objectivity that caused a critical look at these procedures for building evaluations. Gradually, and with some initial difficulty, the subjectivist view was eventually advocated. More comprehensive literature sources associated with subjective probability were eventually uncovered. The distinction between frequency measures and engineering decision making using deterministic information became more clearly understood. Intellectual conversion became more certain with the ability to understand performance for specific applications that was based on deterministic fire science and traditional standards and codes. Although degree-of-belief probabilistic calibration estimates are now advocated, they are viewed as a temporary expedient until well-defined deterministic equations for fire defenses are developed by the profession. After well-defined deterministic procedures for the fire defenses become available, statistically based safety indices can be developed using conventional load and resistance analyses.

Finally, the only reason to use subjective probabilities with CV curves and SV networks is because fire science and engineering are not yet able to provide a sufficient range of analysis and understanding to predict performance. Subjective probability fills the gap between what is known and what may be logically deduced. This technique has been part of all emerging engineering disciplines. Much of the judgment will eventually be replaced by appropriate deterministic procedures. During the interim, the transparency of the method and associated technical documentation provides the basis for decisions.

B.9 Further reading

There are a large number of excellent books on objective probability and statistics. There are a much smaller number on the history of probability and subjective probability. Many of these deal with management science, psychology, and decision making. For those who may wish to investigate this topic further, the following books will provide an interesting perspective on the subject:

- *Degrees of Belief* [5] skillfully integrates the meaning of probability with reliability, risk, judgment and thinking. The examples are drawn from the field of geotechnical engineering. The thought process of geotechnical engineers parallels that of fire safety professionals. This is an enjoyable and informative book on technical decision making. Of the information discovered up to this point, this is the most highly recommended.
- *The Emergence of Probability* [6] gives a fascinating history of the duality of probability philosophies and their applications. Hacking gives an excellent insight into the evolution of the subject.
- *Decision Analysis and Behavioral Research* [1] contains a well-written and very useful mathematical treatment of probability and its applications. The discussions of uncertainty, Bayesian theory, inference, value, and utility are particularly helpful.
- *Decision Analysis* [7] describes the mathematics and structuring of decision making under uncertainty, utility, and ways to organize judgmental probability.
- *Against the Gods* [8] tells the historical development of the mathematics of probability and statistics in a thoroughly entertaining and informative manner. The emphasis is on objective

probability with (eventual) applications to investment risk. The psychological treatment of risk is useful in understanding better ways to communicate with management.

- *Probability and Risk Assessment* [9] provides a useful perspective on subjective probability, risk assessment, and mathematical coherence.
- *Probabilistic Engineering Design* [10] focuses on codifying risk and judgment in design using probability. The distinction between engineering design and engineering modeling using probability and statistics is useful to distinguish applications and philosophies of probability.

References

1. Edwards, W. and von Wintersfeldt, D. *Decision Analysis and Behavioral Research*, Cambridge University Press, 1986.
2. Tribus, M. *Rational Descriptions, Decisions, Designs*. Pergamon Press, 1969.
3. Harvey, C. S. Inspection overload? Not in Boulder. *NFPA Journal*, January/February, 1995.
4. *Interim Guide to Goal Oriented Systems Approach to Building Fire Safety*, PBS 5920.9 Building Fire Safety Criteria, Appendix D, General Services Administration, 1972.
5. Vick, S. G. *Degrees of Belief: Subjective Judgment and Engineering Judgment*. ASCE Press, 2002.
6. Hacking, I. *The Emergence of Probability: A philosophical study of early ideas about probability, intuition, and statistical inference*. Cambridge University Press, 1984.
7. Raiffa, H. *Decision Analysis: Introductory Lectures on Choices under Uncertainty*. Addison-Wesley, 1970.
8. Bernstein, P. *Against the Gods: The Remarkable Story of Risk*. John Wiley & Sons, 1998.
9. Apostolakis, G. Probability and risk assessment: the subjectivistic viewpoint and some suggestions. *Nuclear Safety*, Vol. **19**, No. 3, May/June 1978.
10. Siddall, J. N. *Probabilistic Engineering Design: Principles and Applications*. Marcel Dekker, 1983.

APPENDIX C: THE ROLE OF JUDGMENT IN ENGINEERING EVALUATIONS

C.1 Introduction

Judgment is a part of all engineering evaluations. The amount and role of judgment depend on the maturity of the discipline. When a discipline is in its infancy, judgment has a dominant influence in performance evaluations. As the discipline matures, judgment gradually becomes replaced by quantification methods.

The role of engineering is to make decisions. The decisions may relate to design, to analysis, or to planning. Often the decisions are incorporated into reports that allow others to understand issues and alternatives and make a final selection. The process involves understanding the problem, organizing information, and developing a logical framework for making decisions. Calculations are only tools to assist understanding. The gap between technical knowledge and performance must be filled with what has been called engineering judgment.

This appendix discusses two aspects of evaluation. One looks at the philosophical role of judgment in assessing performance and the other provides an example of level 1, 2, and 3 evaluations. Fire safety engineering is the central theme, although discussion of other disciplines is incorporated when appropriate.

C.2 Building decisions

Many practical questions in fire safety deal with the adequacy of decisions. For example, the density of the sprinkler discharge is *xx* gpm/sq ft; type *yy* photoelectric smoke detectors will be installed in the following locations; fire department connections to the standpipes are located *zz* feet to the west of the main entrance; or the local fire department will respond with two pumps, one ladder, a chief, and a staff of 12. Someone in the design and construction groups or in the community government will make those decisions. Results of those decisions may have a significant influence on the building fire performance, or they may be relatively unimportant to a fire's outcome. A performance evaluation should be able to assess the significance of any detail with respect to building fire performance.

Sometimes decisions that influence fire performance are based on interpretations of prescriptive code or standard requirements. Sometimes they are well formulated by individuals with substantial knowledge and experience in the design, installation, and operation of the particular fire

Building Fire Performance Analysis R. W. Fitzgerald
 ISBN: 0-470-86326-9 (HB)

defense under consideration. Sometimes they are made in an arbitrary and uninformed manner by individuals with no knowledge or training in firesafety. The quality of the decision can have an important effect on the building's performance.

C.3 Firesafety engineering

All engineering disciplines evolve through three stages of growth that may be described as infancy, adolescence, and maturity. Cornell [1] describes the characteristics of these evolutionary stages. Mature engineering disciplines, such as structural, mechanical, and electrical engineering are able to associate loads and resistance in a manner that allows performance to be predicted relatively accurately by calculation methods. At maturity, an analytical framework has been established. Researchers and practitioners use the same framework and vocabulary to work together on the same problems.

Firesafety engineering has not yet reached an analytical level of the mature engineering disciplines. Its evolution may be described more accurately as in an adolescent stage. Among the adolescent characteristics are the lack of a common framework that allows researchers and practitioners to view issues in the same way. Also, loads and resistance are not yet related in a manner that allows a continuum of predictable behavior to be calculated within a clear, practical, and analytically based framework. This evolutionary stage for fire safety is no different from that experienced by the other engineering disciplines as they moved from infancy to maturity.

C.4 Judgment in engineering

Engineers work in a very imperfect world where they must make timely decisions for projects that must be built today. Koen [2] defines an engineering method as "the strategy for causing the best change in a poorly understood or uncertain situation within the available resources." Available resources include knowledge, information, equipment and procedures, confidence, money, and time.

The understanding for any type of engineering performance uses a mix of observation, technology, and judgment to arrive at decisions. During infancy, judgment is a dominant influence. As technology capabilities improve, judgment is gradually supplanted by analytical capabilities that become proven with experience. However, judgment is never eliminated. Judgmental components become integrated with the evolving technological information base.

The history of engineering provides many examples illustrating the transformation between technology and judgment. One discipline, geotechnical engineering, will be briefly noted because the complexity of its problems and the engineering thought process of practitioners is so similar to that of firesafety practitioners. Building and bridge foundations, dams, tunnels and other earthworks have been designed and constructed since ancient times. Although books and papers on the subject have been written for several hundred years, modern geotechnical engineering can be traced to 1923, when Terzaghi published a theory of soil consolidation to compute foundation settlement. This soil consolidation procedure, along with effective stress and earlier theories on earth pressures, established the basis for modern soil mechanics.

Although analytical calculations began to assume a greater role in practice, substantial engineering judgment was still needed to interpret information from theoretical calculations and laboratory experimentation into practical engineering decisions. Peck [3], a highly regarded practitioner and theoretician in soil mechanics, describes clearly this important judgmental component of engineering practice. The evolutionary development of soil mechanics enables one to trace and recognize the ever changing roles of engineering practice, engineering science, and judgment in this field. Similar analogies may be drawn from other engineering areas.

This brief description of another engineering field may appear to be quite dissimilar to firesafety engineering. The nature of the problems and the technology are quite different, but the problem complexity and the fundamental way of thinking are very similar. In both cases a framework is necessary to integrate local and global behavior in a manner that enables the engineer to understand and describe performance. The analytical process allows an engineer to blend theory, calculations, experiments, and experience with judgment to arrive at a performance understanding to create the best solution for available conditions and resources.

Fire science and engineering can provide much information about measures of behavior. Unfortunately, they have not yet reached a maturity to provide the needed scope of computational tools. The gap between computational capabilities and the needs of performance evaluations must be filled by judgment in a manner similar to that experienced by other engineering disciplines.

C.5 Language and culture

The architect, the different design engineers, contractors, code officials, and the fire service all speak different technical languages. Within each group, the technical language, the jargon, and the shades of meaning are clear. Communication within a single group is easy and seldom misinterpreted. However, technical communication barriers between the groups are substantial. Except for a few verbal and conceptual descriptors that establish an interface to conduct necessary business, a common language that ensures understanding is not present. Consequently, a structural engineer and an electrical engineer have as much difficulty in communicating technical details of their respective disciplines as an architect and a fire service officer.

In addition to the language differences between professions, the cultures also are distinct. In many ways, the language and culture between professions involved in the design, construction, operation, and safety of a building are as different as the language and cultures of different nationalities. Even when technical language communication can be established, a cultural distinction and associated acceptable behavior or experiences may inhibit easy flow of ideas and concepts between groups.

This book is about fire performance evaluations. The goal is to help firesafety professionals to understand the performance and to communicate that understanding to others. Engineers normally think and communicate with pictures. The visual thinking of firesafety performance curves provides a universal language between different professional groups. These curves can communicate expected behavior for alternatives efficiently and clearly.

Performance curves could be constructed for the other engineering disciplines. This is unnecessary because the technical experience and understanding are sufficient to make decisions and communicate behavior. Fire safety engineers do not yet have the technical base or a clear, codified description of loading conditions that are characteristic of a mature engineering discipline. The prescriptive code is the perceived basis for fire performance. Unfortunately, all of the building industry professions make decisions and incorporate details that affect fire performance. Some decisions or details can be insignificant to the overall fire performance. Others may have major implications. If fire performance is of interest for a building, its evaluation must be based on understanding that specific building's performance. One must be able to communicate with other disciplines involving their decisions on fire performance.

C.6 Firesafety evaluations

Evaluations provide a way to understand, evaluate, describe, and communicate the expected reliability, quality, and performance of the firesafety subsystems, and a basis for comparing alternatives.

In many ways, the process by which the evaluation is achieved is more revealing than the numerical values that ensue. The understanding that unfolds when the thought process of the framework is augmented by numerical measures of performance enables documentation and communication to be enhanced.

Professional time is a valuable commodity. The professional must conserve that commodity by reducing the time necessary to arrive at better decisions. Time may be reduced by tailoring the analysis to needs. It can also be accomplished by creating or improving decision support systems to enable information to be organized, retrieved, and used more efficiently.

Three levels of evaluation are appropriate to address differing needs:

- *Level 1* observational evaluations provide a rapid, inexpensive assessment of behavior. A level 1 evaluation gives a basic understanding of component behavior within the context of the building's macroperformance. This evaluation level provides a sense of proportion to order risk characterizations, or to make decisions involving additional information.
- *Level 2* detailed evaluations develop a better understanding of component performance. They may assist in approval decisions by authorities having jurisdiction (AHJ) for conventional or performance-based approvals. The analysis augments level 1 information to develop greater knowledge. It may be sufficient for its purpose or it may identify additional information needs.
- *Level 3* evaluations examine component performance sensitivity to expected variations or "what if" conditions. These evaluations extend the knowledge base for decision making.

The next sections illustrate these evaluations for the heat detector actuation in a studio apartment shown in Figure C.1. This apartment was initially described in Chapter 4 and the detector performance was described in Section 11.11.

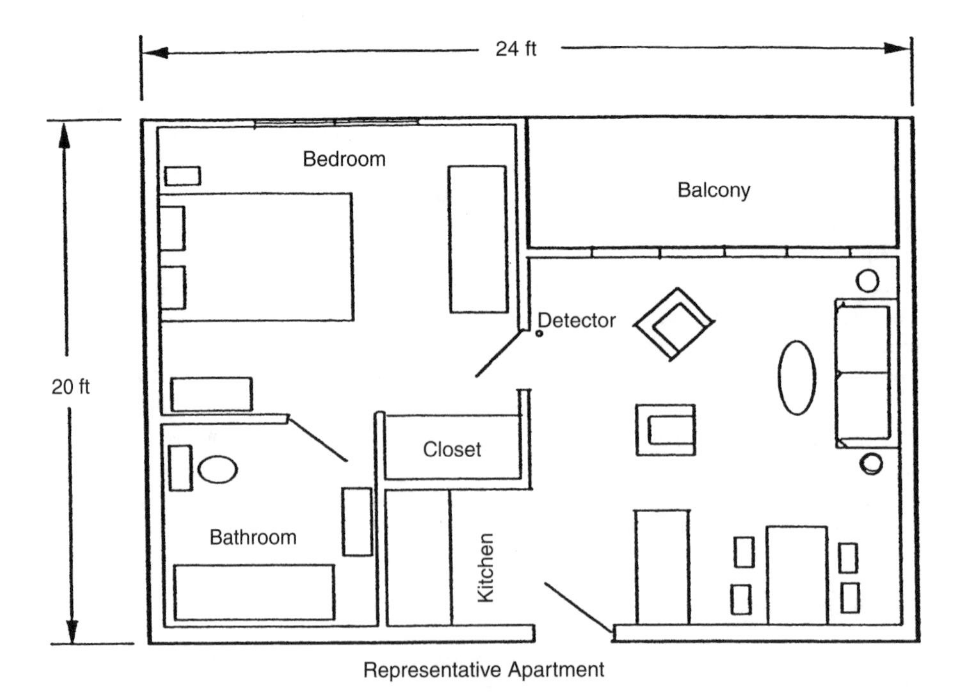

Figure C.1

C.7 Level 1 evaluations

A level 1 evaluation gathers performance information for risk characterizations. It provides a rapid assessment of component performance. One should not form the impression that the speed and simplicity of evaluations relates to careless analysis or an inferior quality product. A level 1 evaluation requires knowledge, experience, and skill. Observational skills for details requires attention and diligence. Rules of thumb and tabular or graphical guidelines can guide estimates.

A component evaluation uses the same framework for thinking at each analytical level. The process envisions a design fire and relates time-related combustion products with a continuous value network (CVN) that describes functional performance. The CVN thought process conceptually identifies performance outcomes associated with a scenario. Single value networks (SVNs) stop the world to examine performance at selected conditions or times. Estimates use whatever informational sources may be available.

Example C.1

Construct a level 1 curve to describe the performance of a 135°F (57.2°C) fixed-temperature heat detector on the ceiling by the bedroom door in Figure C.1. Construct the curve for a fire that involves the wastebasket and sofa.

Solution

Although a good mental picture of the room geometry, fire objects and locations, and detector location is available, other technical information is missing. For example, we do not know specific fire characteristics of the burning objects or the response characteristics of the detector. Nevertheless, a sense of proportion can be ascribed to the detector performance, as shown in Figure C.2. The evaluation considers the factors of Figure 11.17 and is guided by the SVN of Figure 11.3.

Discussion

Construction of this level 1 performance curve requires little time and communication of the results is clear. These attributes are typical for most type 1 evaluations. The price for this speed is confidence in numerical values. The effect of a component within the macro building performance will indicate if more detailed studies are needed.

Experience, rules of thumb, and numerical guidelines or diagrammatic relationships speed the process and give greater confidence and consistency in the results. However, the objective here is merely to describe the process, not to provide decision-support systems to streamline decision making.

It is often useful to estimate limits of certainty as a first step. Bracketing the largest and smallest sizes for actuation (points *b* and *c* in Figure C.2) gives a sense of behavior and identifies a window of uncertainty. Then scenarios involving conditions such as open doors, other ignition locations, wind conditions, and different detector types may be related to the original scenario.

C.8 Level 2 evaluations

One generates a level 2 evaluation to understand performance better. It provides more information by using calculations, if available, to guide the estimates. A greater level of observational or specified details is provided, as illustrated by Example C.2.

Example C.2

Construct a level 2 evaluation for the heat detector of Example C.1.

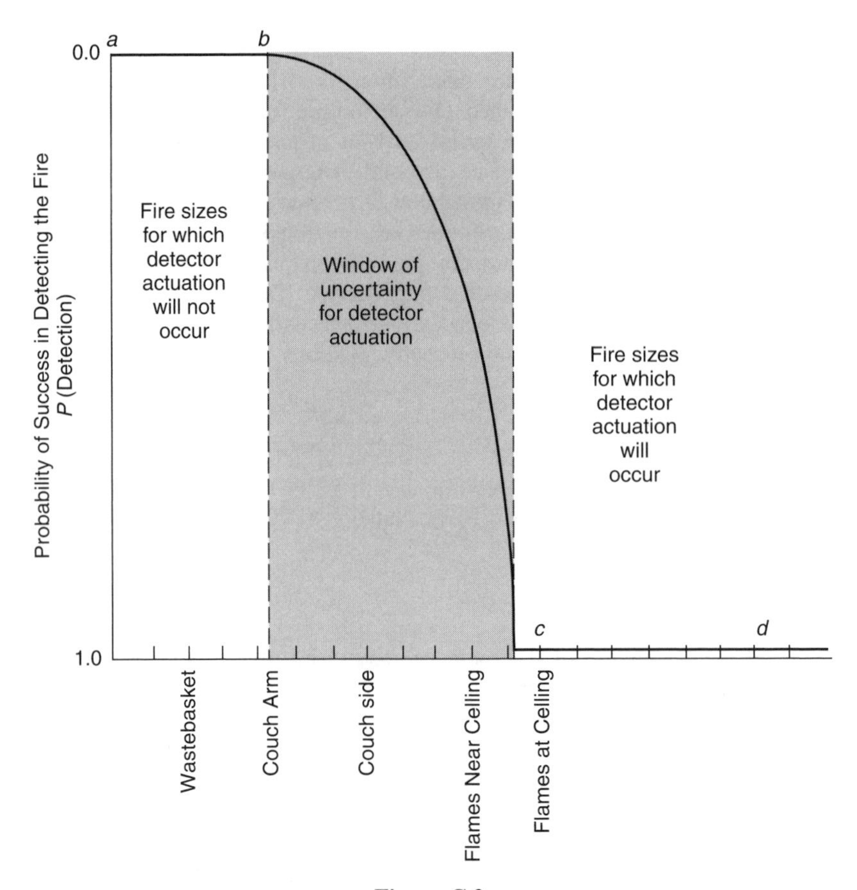

Figure C.2

Solution

This analysis could be addressed in several ways because of the breadth of technical information available for ceiling jets and heat detection. For example, Appendix B of NFPA 72 provides an organized procedure for analyzing heat detectors. Computer programs, such as DETACT - QS or DETACT - T^2, can be used to investigate the influence of variables on detection. These programs are based on large compartments with unconfined ceilings, and the analysis requires adjustment of results to reflect differences between computer model conditions and room conditions. Here we illustrate the process for detector analysis using ceiling jet calculations described in the *SFPE Handbook of Fire Protection Engineering* [4].

Alpert's correlation (C.1) relates a pool fire heat release rate ($\dot{Q}$), its radial distance from the detector (r), and the ceiling height (H) with the detector temperature increase ($T - T_\infty$).

$$T - T_\infty = \frac{5.38(\dot{Q}/r)^{2/3}}{H}. \tag{C.1}$$

For this example, a number of temperature rises could be produced for different heat release rates using spreadsheet calculations. An extensive range of calculations and conditions would produce a better understanding of the detector/fire behavior. However, here only a single condition will be selected to demonstrate a process for organizing the information in a level 2 performance evaluation.

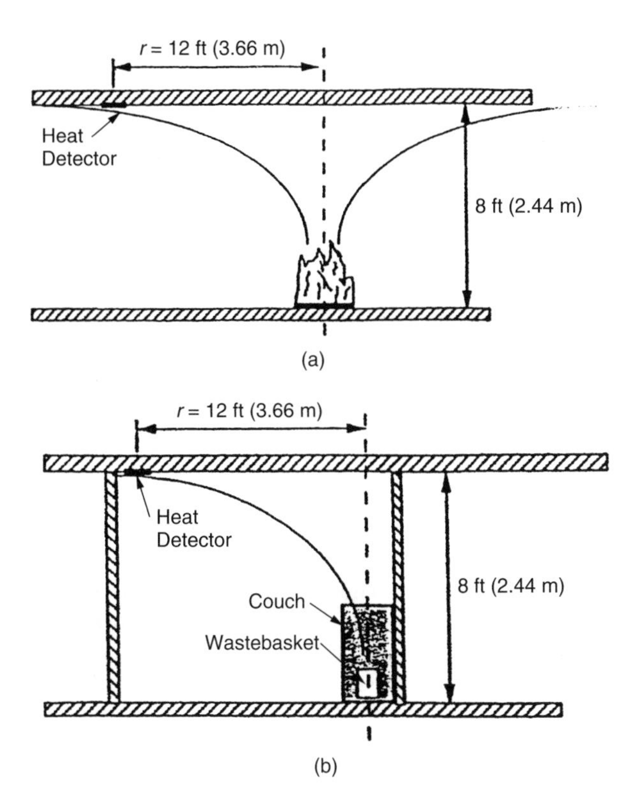

Figure C.3

Figure C.3(a) shows the fire condition for which Alpert developed a correlation to calculate the maximum ceiling jet temperature. Figure C.3(b) shows a section for the room being analyzed. This problem could be addressed by selecting a fire size that is believed to be within the window of uncertainty and calculating the detector temperature. An alternative approach is to select the operating temperature of the detector and calculate the fire size that will cause that temperature to occur. Using this latter approach, with a ceiling height of $H = 2.44\,\mathrm{m}$ (8 ft) and a radial distance from the fire to the detector of $r = 3.66\,\mathrm{m}$ (12 ft), the fire size that will cause a temperature rise of $T - T_\infty = 57.2°\mathrm{C} - 21.1°\mathrm{C} = 36.1°\mathrm{C}$ (65°F) may be calculated as follows:

$$T - T_4 = \frac{5.38(\dot{Q}/r)^{2/3}}{H}$$

$$\Delta T = 36.1 = \frac{5.38(\dot{Q}/3.66)^{2/3}}{2.44}$$

$$\dot{Q} = 240\,\mathrm{kW}.$$

Initially these calculations indicate that the detector will actuate with a fire of 240 kW. A probabilistic performance evaluation examines the conditions inherent in the correlation:

- Fixed-temperature detectors are designed to operate at the rated temperature in a water bath whose temperature rises at a rate of 1° per minute. The environmental conditions of ceiling temperatures rising from hot fire gases will be affected by the mass of the detector response mechanism.

- The ceiling jet is a relatively thin layer of hot gases that moves horizontally away from the fire. Equation (C.1) is based on an unconfined flow of gases (ceiling jet) over the ceiling. Here the confining walls may create zones of lowered flow velocity at walls, corners, and unusual geometric configurations that intercept the otherwise unobstructed ceiling jet flow. A lower flow velocity may lead to longer detector actuation times. This would allow the fire to grow to a larger size than predicted by the calculations.
- The correlation of Figure C.3(a) is based on empirical data in which the enclosure walls were at least 1.8 times the ceiling height (here $1.8 \times 2.44 = 4.4\,\text{m}$) from the fire source. The conditions of Figure C.3(b) show that this is not the situation for this room. When the fire source is in contact with a flat wall surface, some air entrainment will be blocked from the fire plume. If a value of $2\dot{Q}$ were used to consider the effects of the wall and substituted into the correlation above, the resulting fire size would become $\dot{Q} = 120\,\text{kW}$. However, the value of $2\dot{Q}$ is more appropriate for a semicircular burner with the entire flat side against the wall. If a value of $1.05\dot{Q}$ were used, as recommended in the SFPE handbook [4], the fire size would become $Q = 230\,\text{kW}$.
- The model for Alpert's empirical correlations is a steady-state pool fire at a fixed level from the ceiling. The studio apartment involves a growing fire of the wastebasket and sofa. The question arises as to whether a growing fire would have a magnitude greater or smaller than the steady-state fire that is the basis for equation (C.1). If the fire is fast growing, the thermal lag of the sensing element will trigger at an HRR magnitude greater than that of the steady-state fire. On the other hand, a slowly developing fire will allow the sensing element to track the gas temperature. The detector will trigger when the fire heat release rate corresponds nearly to the detector actuation temperature. This HRR would be below that of the steady-state fire. Therefore, for a growing fire, the calculation results relate to the rate of fire development. In this example the fire growth will have a medium to fast development producing a fire size greater than that predicted.

These factors produce conflicting effects. Some indicate larger fires and others smaller fires at detector actuation. Judgment integrates deterministic calculations with theory and field conditions to form the basis for the probabilistic estimate. A useful technique to view differences in condition between calculation results and field conditions is shown in Table C.1. Here one lists the differences (or variables), their relative significance, and the direction of the performance graph movement. The arrows of column 2 show the direction of movement that the performance curve of Figure C.2 would move due to the factor. The asterisks (*) indicate the relative importance of the factor to cause change.

The integrating framework model for this evaluation is Figure 11.3. Based on the pool fire represented by the calculations and judgmental analysis using Table C.1, numerical values for the events were selected and shown in Figure C.4. These values express a judgmental probability for the events, given a fire of 240 kW. The extrapolated performance curve is shown in Figure C.5. The additional information developed by the calculations and analysis enabled the performance description to shift and be compressed slightly. Confidence in the results is better than for the less time-consuming level 1 analysis of Example C.1. The major additional causes of uncertainty relate to content materials, fire growth, and detector (RTI) listing.

C.9 Discussion of Level 2 evaluations

A level 2 evaluation increases the understanding of the component performance and enables more technical documentation to be established. Table C.1 illustrates a way to incorporate conflicting effects of deterministic parameters on site-specific conditions. Noting uncertainties and variables,

Table C.1

Factor	Direction of influence on Q (<– detection with smaller fire and –> detection with larger fire)	Relative significance to the calculated values (**** major to * minor)
Heat confinement by room	<–	**
Heat transfer from the room ($k\rho c$ of bounding surfaces)	–>	*
Fire location with regard to enclosure wall	<–	**
Growing fire rather than pool fire	<– or –>	**
(a) Slow growing	<–	
(b) Fast growing	–>	
Unknown RTI (or listing) of detector	<– or –>	0 to ****
(a) Low RTI	<–	
(b) High RTI	–>	
Unknown materials for sofa affecting rate of fire development, $\dot{Q} = \alpha t^2$	<–	0 to ***
Doors closed rather than open	<–	*
Use of ceiling height, H, rather than height of fire, h_v	<–	**
Location of detector relative to maximum temperature, T, and velocity, v, of ceiling jet	<– or –>	*
Ceiling configuration and obstructions of heat movement to detector	No effect in this example	

along with their influences and relative importance helps one to consider the synergism of the collective influences. This enables one to identify the judgment logic for rational performance estimates. A comparable list of factors and performance influences can be constructed for each deterministic procedure used in component performance evaluations. These listings provide a basis for documentation, if necessary.

Each component evaluation bases performance estimations on state-of-the-art technology for the particular component. Judgment links what is known with estimate predictions. Level 2 evaluations are based on more information and often, but not always, require more time to complete than a level 1 evaluation. Any increased analytical cost would presumably produce a commensurate value for the information. The cost of this information can be reduced substantially by constructing tables, graphs, and other decision-support systems for routine situations.

C.10 Level 3 evaluations

The role of a level 3 evaluation is to extend the knowledge base and better understand the effect of building or component variation on performance. Example C.3 illustrates the process.

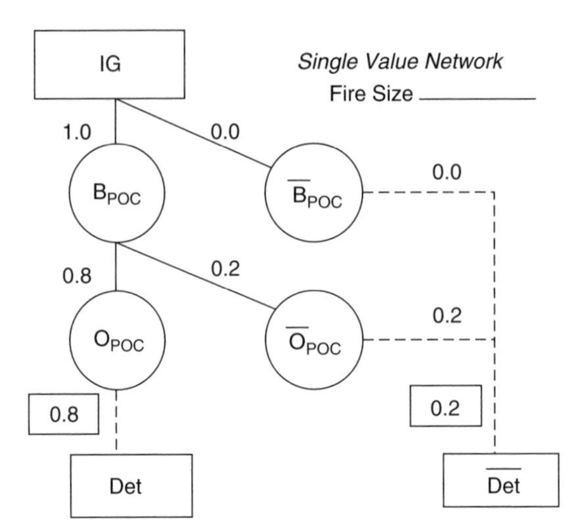

Figure C.4

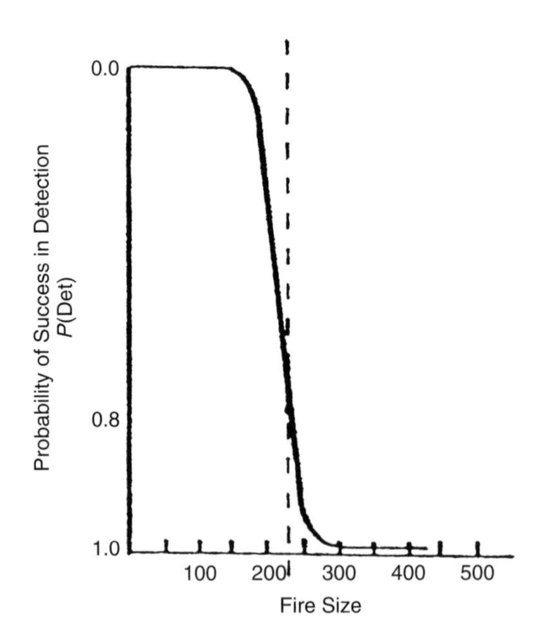

Figure C.5

Example C.3

Develop a level 3 sensitivity evaluation for the 135°F (57.2°C) fixed-temperature heat detector of Example C.1.

Solution

This analysis will develop a better understanding of possible variations in the response time index (RTI) on detector sensitivity. We will assume that this knowledge is important to the building's performance. To illustrate the process, we shall use the Heskestad–Delichatsios [6] relationships for a growing fire and information in NFPA 72, Appendix B.

This analysis is based on an exponentially growing fire under a flat, extensive ceiling having no interior partitions. The fire follows the power-law growth model, $Q = \alpha t^2$. Figure C.6 shows the five model fires that are commonly used as the basis for analysis or design. The time, t_g, for each fire to reach a specified heat release rate of 1000 Btu/s (1055 kW) and the associated values of α are shown in Table C.2.

All 135°F heat detectors do not respond to growing fires in the same way. Detector sensitivity is established by Underwriters Laboratories (UL) or FM Global standard test procedures. The sensitivity of a heat detector is described by a time constant, t_g, which reflects the time that a specific temperature-rated detector will actuate when subjected to a heat flow of defined temperature and velocity. The response time index (RTI) is a function of the time constant and is a characterization of detector sensitivity. The RTI may be calculated from $\mathrm{RTI} = t_g u_o^{1/2}$ where u_o is the velocity of the heated air at which t_g was measured. The range of the time constant, t_g, for

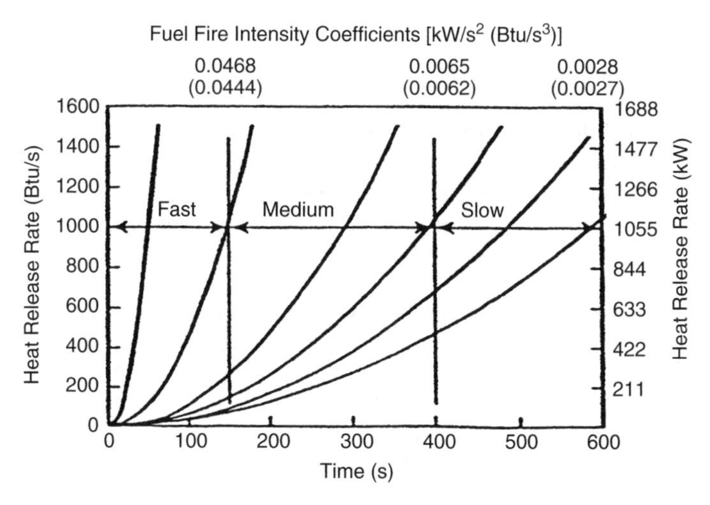

Figure C.6 Reprinted with permission from NFPA 72®, *National Fire Alasm Code* Copyright © 2002 National Fire Protection Association, Quincy, MA 02269. This reprinted matesial is not the complete and official position of the National Fie protection Association, on the referenced subject which is represented only by the standard in its entirety

Table C.2

Model design fires		Threshold fire size, $\dot{Q}_d$ (kW)							
α	t_g	57.2°C (135°F) fixed-temperature heat detector RTI ($\mathrm{m}^{1/2}\,\mathrm{s}^{1/2}$)							
($\mathrm{kW\,s^{-1/2}}$)	(s)	29	54	70	98	122	165	233	404
0.422	50	840	1150	1320	1600	1800	2160	2660	3750
0.044	150	380	510	590	700	790	940	1160	1630
0.011	300	250	330	370	440	490	580	710	990
0.004	500	200	250	270	320	350	410	500	690
0.003	600	190	220	250	290	320	370	440	6

National Fire Alarm Code®, is a registered trademark of the National Fire Protection Association, Inc., Quincy, MA 00269.

Table C.3

Listed spacing (ft)	10	15	20	25	30	40	50	70
Time constant τ_o (s)	330	190	135	100	80	57	44	24
RTI ($m^{1/2}\,s^{1/2}$)	404	233	165	122	98	70	54	29
RTI ($ft^{1/2}\,s^{1/2}$)	738	425	302	224	178	127	98	54

135°F heat detectors and the associated RTI can vary substantially. Unfortunately, the RTI is not being measured at this time. Consequently, if one wishes to estimate an RTI, it is necessary to convert the listed detector spacing in the manner described in NFPA 72, Appendix B. Table C.3 shows the range of RTI estimates for listed 135°F fixed-temperature heat detectors.

The listed detector spacings are based on relatively large, constant HRR fires. The heat detectors are mounted on a smooth, flat ceiling and the listing distance is a measure of the sensitivity of the detector to this fire. The RTI is related to this listed spacing, and the values of Table C.3 show that the smaller the RTI, the more sensitive the detector. Consequently, one must consider more than the temperature rating in detector evaluations.

An additional consideration is the influence of the fire growth rate on the detector response. In fast-developing fires, the thermal lag of heat detectors causes a delay in actuation in developing fires until the fire has grown to a size larger than the size that would initially produce an ambient air temperature equal to the temperature rating of the detector. Consequently, fast-developing fires will reach much larger sizes at the time of heat detector actuation than will slower-developing fires.

Table C.2 gives a sense of proportion for this physical difference between fire growth rate, fire size, and detector sensitivity. The values in Table C.2 show the threshold fire size, $\dot{Q}_d$ (kW), for a radial distance of 3.66 m (12 ft), ceiling height = 2.44 m (8 ft), and a temperature difference between the ambient and the detector temperature rating of $\Delta T = 57.2°C - 21.1°C = 18.3°C$ (65°F). The fire growth values are the standard αt^2 fires shown in Figure C.6, and the RTI values are those associated with the UL listings of Table C.3. The relationship between the fire growth rate and the thermal sensitivity of the detector can make an enormous difference in the threshold fire size at detector actuation. Detection listed spacings may be important during approval inspections.

Another part of this analysis explores the influence of the fire growth characteristics on detector actuation. Figure C.1 shows a love seat and a wastebasket as the objects of the fire scenario. NBS [7] shows experimental heat release rates for a variety of furniture items, including the following information about the love seats tested:

Test 18: chair F33 (trial love seat) 39.2 kg	$\alpha = 0.0066\,kW/s^2$
Test 31: chair F31 love seat 39.6 kg	$\alpha = 0.2931\,kW/s^2$
Test 37: chair F31 love seat 40.4 kg	$\alpha = 0.1648\,kW/s^2$
Test 38: chair F32 sofa 51.5 kg	$\alpha = 0.1055\,kW/s^2$
Test 54: love seat, metal frame, foam cushions	$\alpha = 0.0042\,kW/s^2$
Test 57: love seat, wood frame, foam cushions	$\alpha = 0.0086\,kW/s^2$

A comparison of the heat release rates for these specific furniture items and the RTIs for a 57.2°C (135°F) heat detector yields the results of Table C.4.

The possible variations considering only the range of RTI values for a 135°F heat detector and a love seat that may be purchased to meet an individual's interior design tastes can produce a wide variation in performance. A level 3 analysis provides a better sense of the uncertainty in

Table C.4

Experimental fires			Threshold fire size, Q_d (kW)							
Test	α (kW $s^{-1/2}$)	t_g (s)	7.2°C (135°F) heat detector RTI ($m^{1/2}$ $s^{1/2}$)							
			29	54	70	98	122	65	233	404
31	0.2931	60	730	1000	1150	1380	1570	1870	2310	3260
37	0.1648	80	590	800	920	1110	1260	1500	1850	2620
38	0.1055	100	500	680	780	940	1060	1270	1570	2210
57	0.0086	350	240	300	340	400	440	520	630	880
18	0.0066	400	220	280	310	360	410	480	580	810
54	0.0042	500	200	250	270	320	350	410	500	690

Table C.5

Factor	Direction of influence on $\dot{Q}$ (<– detection with smaller fire and –> detection with larger fire)	Relative significance to the calculated values (**** major to * minor)
Heat confinement by room	<–	**
Heat absorption of the room ($k\rho c$ of bounding surfaces)	–>	*
Fire location with regard to enclosure wall	<–	**
Fire location in the fuel package (i.e., at arm rather than interior)	–>	**
Doors closed rather than open	<–	*
Use of ceiling height, H, rather than height of fire, h_v	<–	**
Location of detector relative to maximum temperature, T, and velocity, v, of ceiling jet	<– or –>	*
Ceiling configuration and obstructions to heat movement to detector	Smooth, flat ceiling in this example	*

a building analysis. This uncertainty may be greater than had been implied by earlier analyses, although it was nonetheless present. This uncertainty is inherent in all levels of analysis and a knowledgeable user is aware of these factors in evaluations. This example indicates that if the RTI (listed spacing) and fire growth characteristics are not known, the fire size at detection has the potential to vary enormously for fixed-temperature heat detectors. Judgment involving these local differences within the context of the more global building performance is inherent in the process.

To narrow the focus for a moment, let us assume that the heat detector in this room is known to have a UL listing of 50 ft spacing. This calculates to an equivalent RTI of 98. Also, assume that we use the specific style of love seat described by test 57. For this situation, a threshold fire size $\dot{Q}_d = 400$ kW is indicated. However, this value corresponds to a growing fire located in a large room with smooth unobstructed ceilings. This value may then be adjusted for the actual room conditions to determine a degree-of-belief performance description. The factors of Table C.5 may be used to adjust the calculated values and establish a better representation for the fire size at detection for the more specific room conditions of Figure C.1.

Based on this additional information, the performance of Figure C.7 may be more appropriate. However, a more conservative set of conditions would indicate a larger fire at detection. Depending on specific conditions, the spread of performance conditions could be quite broad.

C.11 Discussion of level 3 evaluations

The role of a level 3 evaluation is to understand component performance more completely. Often this implies one or a combination of the following: (1) a greater effort to understand component behavior, including sensitivity assessments; (2) a more carefully identified range for the window of uncertainty in the performance description; or (3) more careful specificity of conditions that enable "what if" questions to be interpreted with greater clarity. Example C.3 illustrates the process.

For example, one may state that if the 135°F fixed-temperature heat detector has a UL listing of 50 ft and the fuel package materials conform to those of test 57, the detector performance may be described as that of Figure C.7. If the detector listing is changed to 70 ft and materials in the fuel package are selected to produce a slower-developing fire, the detector performance will improve. On the other hand, if the UL listing is not known and the type of interior contents is not controlled, then uncertainty grows.

Example C.3 was selected to provide an association with the calculations of Example C.2. Detailed analyses and efficient documentation of conclusions often do not require a large professional time expenditure because of computer and spreadsheet analyses. If design fire characteristics were standardized, the time for an analysis would be reduced significantly. A one-time investment with available information technology can produce a wide range of information for developing a sense of proportion for estimates. Level 3 performance knowledge can be systematized when

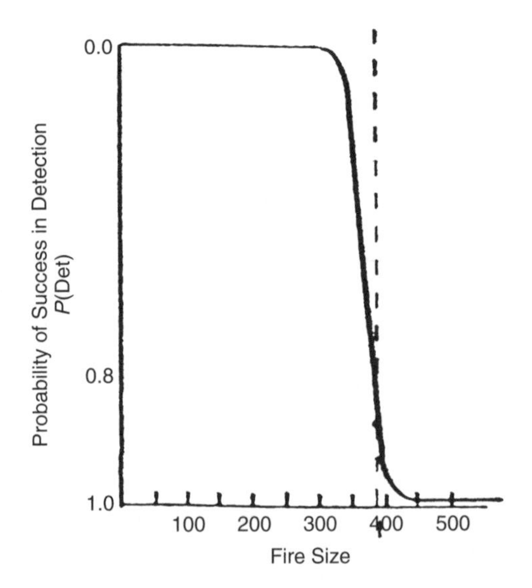

Figure C.7

- fire technology is organized around a performance structure;
- elements of uncertainty and variability of computational methods are cataloged;
- routine building features that influence specific components are identified.

Increased knowledge does not increase the uncertainty of level 1 or level 2 evaluations. The same factors creating the uncertainty are present in all evaluations. Integration of this information into a more complete analysis creates a better understanding of factors that influence building performance. This understanding enables one to recognize when details are important to the overall performance. A greater understanding enables one to discount factors that do not influence the results. Judgmental decisions can be made with greater confidence.

The primary objective of a level 3 evaluation is to *understand* the component performance better so one can communicate more effectively. One does a level 3 evaluation if the added information is important to making decisions.

C.12 Uncertainty and performance

Concepts of classical safety analysis may be adapted to organize the uncertainty information that is associated with a performance evaluation. Variability and uncertainty may be grouped into three categories:

- stability of conditions,
- variability in physical parameters,
- professional practices.

Conditional stability is associated with the relative stability of the situation being evaluated. Stability factors reflect the influence of common arrangement changes to predicted performance. For example, how will shifting the position of the fuel packages or openings affect the fire growth potential used for the room? How will the repositioning of movable partitions influence fire propagation or smoke movement? Conceptually, the conditional stability factor relates geometrical and dimensional variability to evaluation scenarios.

Physical parameter variability incorporates two aspects. The first involves the variability associated with the physical performance of different materials. For example, burning characteristics of a cotton batting upholstered chair will be different from the characteristics of a functionally similar chair constructed with foam plastics. Similarly, operating and extinguishing characteristics of a common sprinkler will differ from those of an early suppression fast response (ESFR) sprinkler. In other words, all chairs and sprinklers are not created equal. The variability of species types introduces a form of uncertainty into performance evaluations.

Consistency of physical behavior is another type of physical variability. For example, the thermal conductivity of many materials varies with temperature and testing procedures. Similarly, the coefficient of linear expansion for materials may change with temperature. Uncertainty arises in the calculation of expected performance with many parameters.

A third category of uncertainty relates to the formulation and use of the deterministic relationships and calculation procedures used in professional practice. For example, what are the limits of validity involving flashover correlations? When does the uncertainty begin to affect the outcomes significantly? Computer programs have default values and limits of applicability incorporated into their codes. The professional practice category identifies and addresses the uncertainty and inaccuracies associated with calculation procedures.

The discussion up to this point has not included the use of statistics. Statistics and classical probabilistic analysis have a role in the complete fire safety system. For certain components that role can be very important. However, one must be aware that a building analysis evaluates a singular, case-specific installation. Distributional information relating to a class of situations may or may not be appropriate. The goal is to understand the details that influence the performance of specific buildings so that the impact of specific alternatives can be understood and communicated clearly. Biases inherent in judgmental decisions are also present in statistical studies. Unfortunately, this is not always evident.

C.13 Summary

The goal of an evaluation is to understand fire performance. Evaluations do not average or weight different conditions. The purpose is to tell a story. Consequently, specific scenarios are selected to represent conditions that are most appropriate to describe the risk characterizations of the building.

Judgment is an integral part of engineering practice. The role of judgment changes as technical knowledge evolves. Calculation procedures are not equally developed for all fire safety components. Even though technology may be inadequate to address the necessary performance functions, an engineering evaluation may be completed using the information that is available and reasonable engineering estimates of the loading and resistance behavior.

Subjective judgment bridges the gap between technical capability and fire performance evaluation needs. Subjective probability is a way to introduce rigor into performance evaluations.

Evaluations and their associated performance curves provide an opportunity to compare functional behavior. The greatest benefit of subjective probability evaluations is that comparisons of expected performance can be made more accurately and consistently with substantially less cost. The framework enables one to understand fire performance and portray risk characterizations.

The quality of an assessment requires that events and their qualifying conditions be identified carefully and clearly. One must understand the knowledge base and construct an interface to access it effectively.

A first-time analysis for a scenario can be time-consuming because it is an initial learning exposure. As experience is gained, comparisons are studied, methods are defined and understood, and decision-support structures are constructed, the time for any level of analysis will be reduced and results will become more consistent.

References

1. Cornell, C. A. Structural safety: some historical evidence that it is a healthy adolescent. *Proceedings of the 3rd International Conference on Structural Safety and Reliability*, Trondheim, Norway, June 23–25, 1981.
2. Koen, B. V. *Definition of the Engineering Method.* American Society of Engineering Education, 1985.
3. Dunnecliffe, J. and Deere, D. Peck, "a man of judgment." In *Judgment in Geotechnical Engineering: The Professional Legacy of Ralph B. Peck* John Wiley & Sons, 1984.
4. Design of detection systems. In *SFPE Handbook of Fire Protection Engineering*, 3d ed. Copyright SFPE 2002, published by NFPA.
5. Alpert, R. L. Ceiling jets. *Fire Technology*, Vol. **8** (1972).
6. Heskestad, G. and Delichatsios, M. *Environments of Fire Detectors – Phase 1*: Vol. 1 *Measurements* (NBS-CGR-77-86); Vol. 2 *Analysis* (NBS-CGR-77-95).
7. Babrauskas, V., Lawson, J. R., Walton, W. D., and Twilley, W. H. *Furniture heat release rates measured with a furniture calorimeter*. National Institute of Standards and Technology, 1982.

APPENDIX D: INSPECTION REPORT FORMS

It is possible to develop inspection forms to organize building evaluations. The following forms illustrate information that may be used in field inspections. These forms may be modified to reflect the type of information and risk collection. The information may be used to construct performance curves for rapid screening and ordering of risks.

Building Fire Performance Analysis R. W. Fitzgerald
 ISBN: 0-470-86326-9 (HB)

Form 1

Inspection Report

Basic Building Data

Building Name -
Address -
- -
- -

Date - - - - - - - - - -
Inspector -

Room(s) of Origin -
- -
- -

Building Size	**Number of Stories**
Small	**One**
Moderate	**2–4**
Large	**5–8**
	Above 8

Occupancy Type -

Unusual Risk Potential

Normal Risk
High Life Safety Risk
High Property Value Risk
One Room
Several Rooms
High Continuity of Operations Risk
High Risk to Neighboring Exposures
High Environmental Risk
Other ______________________

Reason for Room of Origin Selection

Threat to Life Safety
Fire-Fighting Difficulties
Strong Potential for Multiroom Involvement
High Value
Other ______________________

Form 2

Inspection Report

Life Safety

DETECTION

Type	Fire Size q (kW) at Detection	Detection Quality	System
Human Only	Small ($q<20$)	Good	Professional Design
Instrument Only	Moderate ($20<q<500$)	Poor	Authority Approved
Human and Instrument	Large ($500<q<1000$)	Unknown	Unknown
	Very Large (>1000)		

Recognized Detection Deficiencies:

ALERT

Occupant Alert	Delay t (min) between Detection and Occupant Alert	Alerting Quality	System
Human Only	Short ($t<1$)	Good	Professional Design
Audible	Moderate ($1<t<3$)	Poor	Authority Approved
Lighting	Long ($3<t<5$)	Unknown	Unknown
Telephone	Very Long ($t>5$)		
Loudspeakers			
Other			

Recognized Alerting Deficiencies:

EGRESS INFORMATION

Occupant Mobility	Speed S (fps)	Exit Access	Exit Distance d (ft)
Mobile	Fast ($S>4$)	Direct	Short ($d<30$)
Limited (Infirm)	Moderate ($4>S<2$)	Circuitous	Moderate ($30<d<80$)
Limited (Age)	Slow ($2>S>1$)	Dead Ends	Long ($80<d<150$)
Movable	Very Slow ($S<1$)		Very Long ($d>150$)

Egress difficulties:

Form 3

Property Protection

Date - - - - - - - - - - - - - - - - - -

Building - - - - - - - - - - - - - - - - - -

Room -

Range	**Flame/ Heat**	**Flame/ Heat**	**Smoke/ Gas**	**Smoke/ Gas**	**Water**	**Water**
	Building	Contents	Building	Contents	Building	Contents
Up to $1000						
$1000 to $10 000						
$10 000 to $50 000						
$50 000 to $100 000						
$100 000 to $500 000						
$500 000 to $1 000 000						
Greater than $1 000 000						
Other						

Basis **General Comments** -

Owner -
Occupant -
Appraisal -
Inspector Estimate -
Other -

Form 4

Inspection Report

Date - - - - - - - - - - - - - - - - - -

Building - - - - - - - - - - - - - - - - -

Room -

Downtime						
	0–1 K	1 K–10 K	10 K–50 K	50 K–100 K	100 K–1 M	> 1 M
$t < 1$ min						
1 min $< t <$ 1 h						
1 h $< t <$ 1 day						
1 day $< t <$ 1 week						
1 week $< t <$ 1 month						
1 month $< t <$ 3 months						
$t > 3$ months						
Other						

Basis **General Comments** -

Owner -
Occupant -
Appraisal -
Inspector Estimate -
Other -

Form 5

Inspection Report

Fire Growth Hazard Potential to Reach FRI (I Curve)

Room -

Room of Origin		**$\overline{T}$ Failure**		**$\overline{D}$ Failure**	
Certain	[0.0]	Certain	[0.0]	Certain	[0.0]
Very High	[0.1]	Very High	[0.1]	Very High	[0.1]
High	[0.2]	High	[0.2]	High	[0.2]
Moderate	[0.4]	Moderate	[0.4]	Moderate	[0.4]
Low	[0.6]	Low	[0.6]	Low	[0.6]
Very Low	[0.8]	Very Low	[0.8]	Very Low	[0.8]
Optional Inspector Value	_____		- - - - -		- - - - -

Comments -
- -
- -

Room Size	**Ceiling Height**	
Small	< 2.5 m	(8 ft)
Moderate	2.5–3.5 m	(8–12 ft)
Large	> 3.5 m	(> 12 ft)

Dominant Fuel Combustibility	**Dominant Fuel Package Size**	**Interior Finish**
High	Large	Noncombustible
Moderate	Moderate	Combustible
Low	Small	Walls
		Ceiling
		Unknown

Fuel Loading	**Number of Fuel Packages**
< 3 psf	1–3
4–6 psf	4–6
7–10 psf	> 6
11–15 psf	
> 15 psf	

Form 6

Inspection Report

Automatic Sprinkler Suppression in the Room of Origin (A Curve)

None	$P(A) = 0.0$	**Comments** - - - - - - - - - - - - - - - - - -
Very Bad	$P(A) = 0.75$	- -
Bad	$P(A) = 0.85$	- -
Moderate	$P(A) = 0.90$	- -
Good	$P(A) = 0.95$	- -
Very Good	$P(A) = 0.98$	- -

Optional Inspector Value - - - - - - - - -

Building Information

Building - - - - - - - - - - - - - - - - - - - **Year Built** - - - - - - - - -
Room of Origin - - - - - - - - - - - - - - - - - - - **Use** - - - - - - - - -

Fire Growth Classification	αt^2	**Fuel Load**
Very Bad	Ultra Fast	High (< 20 psf)
Bad	Moderate	Moderate (10–20 psf)
Moderate	Small	Low (0–10 psf)
Good	Slow	
Very Good	Very Slow	

Type of Sprinkler System	**Year of Installation**	**Design**
Wet	Before 1960	Professional Design
Dry	1960–1980	Design Authority Review
Preaction	1981–Present	Unknown
Unknown		

Ceiling Height	**Obstructions**	**Overhead Venting**
< 2.5 m	Recognized to	Automatic
2.5–4 m	Be a Problem	Manual
4–8 m	Minor Obstructions	None
> 8 m	None	

Valve Supervision	**Water Supply**	**Maintenance Record**
External Supervisory	Known to Be Adequate	Frequent, Regular
Internal Supervisory	Recognized Deficiencies	Irregular
None	Unknown	None
Unknown		Unknown

Company Attitude	**Company Knowledge of System**
Good	Good
Bad	Weak
Unknown	Unknown

Form 7

Inspection Report

Fire Department Suppression (M Curve)

		Room of Origin	2 Rooms	3 Rooms	4 Rooms
None	$P(\mathrm{M}) = 0.0$				
Bad	$P(\mathrm{M}) = 0.2$				
Moderate	$P(\mathrm{M}) = 0.4$				
Good	$P(\mathrm{M}) = 0.6$				
Very Good	$P(\mathrm{M}) = 0.9$				
Optional Inspector Value		- - - - -	- - - - -	- - - - -	- - - - -

Comments -
- -
- -

Building Information

Building	- - - - - - - - - - - - - -	Year Built	- - - - - - - - - - - - - -
Room of Origin	- - - - - - - - - - - - - -	Use	- - - - - - - - - - - - - -

Ease of Fire Propagation	2 Rooms	3 Rooms	4 Rooms	6 Rooms
Very Easy				
Moderate				
Difficult				

Speed of Propagation	2 Rooms	3 Rooms	4 Rooms	6 Rooms
Very Fast (min)	< 10	< 15	< 20	< 30
Moderate (min)	10–20	15–25	20–30	30–40
Slow (min)	20–30	25–35	30–40	40–50

Detection
- Knee High
- Man High
- Ceiling High
- Room of Origin
- Multiple Rooms

Notification
- Direct FD Connection
- Automatic Third Party
- Human Only
  - Fast (< Man High)
  - Moderate (< Ceiling High)
  - Slow (< FRI_1)
  - Very Slow (< FRI_2)

Arrival After Notification	Response	First Water On After Arrival
Fast (< 4 min)	1 E	Fast (1–4 min)
Moderate (4–8 min)	2 E	Routine (4–8 min)
Slow (>8 min)	2 E, 1 L	Difficult (8–12 m)
	3 E, 2 L	Very Difficult (12–20 min)
	- - - - -	- - - - - - - - - - - - -

Form 8

Inspection Report

Barrier Effectiveness

Room of Origin
Horizontal Failure (After FRI_1) — Type of Failure
Interior Failure — $\overline{D}$ — $\overline{T}$

	$\overline{D}$	$\overline{T}$
One barrier fails (0–5 min)	**Door**	**Hole**
One barrier fails (5–10 min)	**Window**	**Weak Protection**
One barrier failure (< 10 min)	**Opening**	**Grille**
	Other - - -	**Other - - - - -**

Interior Failure — $\overline{D}$ — $\overline{T}$

	$\overline{D}$	$\overline{T}$
Two barrier fails (0–5 min)	**Door**	**Hole**
Two barrier fails (5–10 min)	**Window**	**Weak Protection**
Two barrier failure (< 10 min)	**Opening**	**Grille**
	Other - - -	**Other - - - - -**

Room of Origin
Vertical Failure (After FRI_1) — **Type of Failure**
Interior Failure — $\overline{D}$ — $\overline{T}$

	$\overline{D}$	$\overline{T}$
Fast (0–5 min)	**Large Hole**	**Small Hole**
Moderate (5–15 min)	**Combustible**	**Weak Protection**
Slow (< 15 min)	**Unprotected Steel**	**Heat Transmission**
	Other - - - - -	**Other - - - - -**

Room of Origin
Vertical Failure (After FRI_1) — **Type of Failure**
Exterior Failure — $\overline{D}$ — $\overline{T}$

	$\overline{D}$	$\overline{T}$
Fast (0–5 min)	**Window**	**Small Path**
Moderate (5–15 min)		
Slow (< 15 min)	**Other - - - - -**	**Other - - - - -**

Multiple Room Involvement	**2 Rooms**	**3 Rooms**	**4 Rooms**	**6 Rooms**
Easy				
Moderate				
Difficult				

Comments -
- -
- -

APPENDIX E: FIRE SAFETY EVALUATORS

Two aspects of the framework direct the thought process and enhance communication of fire safety performance. One aspect is the integrated networks that describe the analytical and thought process for evaluating performance. The second aspect is the use of visual means of describing performance using graphical means. The networks are shown in the textbook. This appendix shows a sample of each evaluator (Figures E.1 to E.3). Collectively, the evaluators and networks are so central to performance evaluations that copyright permission is granted for unlimited applications of these aspects of this book.

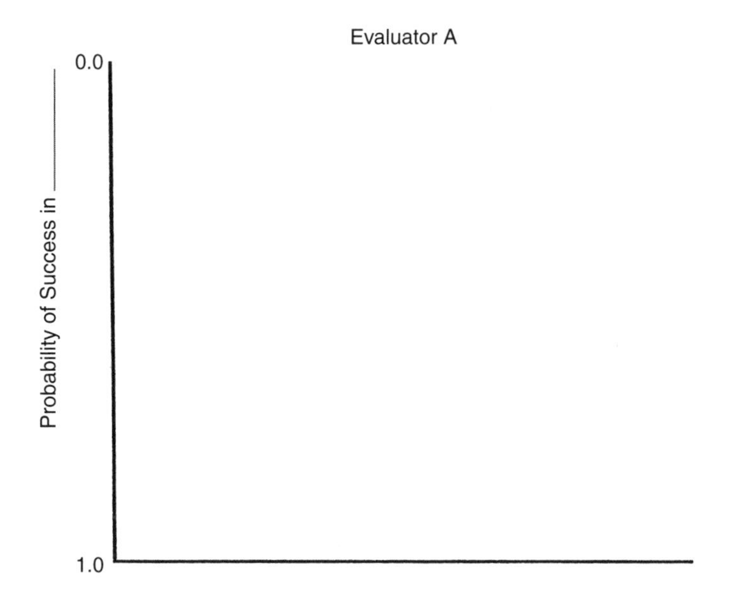

Figure E.1

Building Fire Performance Analysis R. W. Fitzgerald
© 2004 John Wiley & Sons, Ltd ISBN: 0-470-86326-9 (HB)

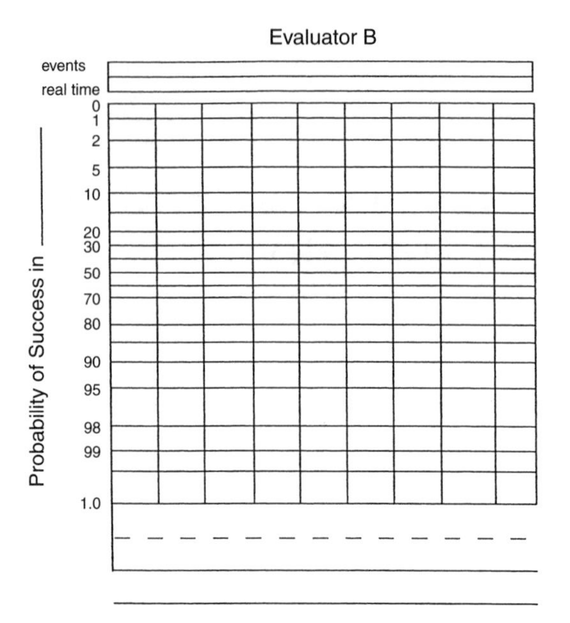

Figure E.2

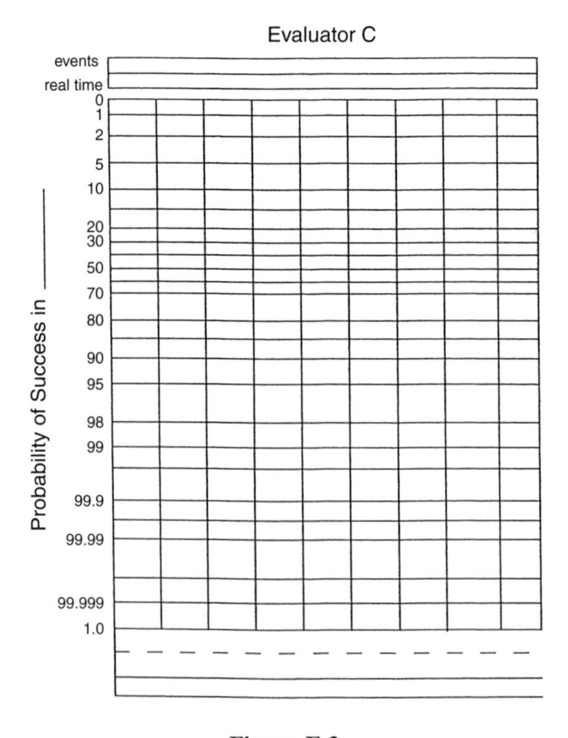

Figure E.3

SYMBOLS AND ABBREVIATIONS

A – automatic sprinkler suppression
AA – agent application (water discharges from a sprinkler)
AA_P – automatic special hazards agent discharges
A curve – continuous curve which describes the probability that a sprinkler system will control (or extinguish) fires of successively larger sizes
AC – automatic sprinkler control (operational effectiveness)
AC_P – automatic special hazards agent controls the fire
adi – detection instrument can actuate (maintenance control)
AE – automatic sprinkler extinguishment (operational effectiveness)
A_F – floor area
A_f – total flooding system extinguishes the fire
ama – agent discharges from nozzle
A_P – automatic special hazards suppression extinguishes the fire
A_t – compartment bounding surface area
A_v – ventilation area
B – barrier success in preventing any ignition in the adjacent space
bme – blackout does occur
B_{POC} – enough products of combustion reach the detector boundaries to cause actuation
B_S – enough smoke can reach the boundary of the target space
c – bounding surface thermal properties
C – ceiling point
cac – enough water continues to flow
C_C – barrier effectiveness factor for influence of construction
C_L – barrier effectiveness factor for influence of applied loading
cme – continuous discharge of agent occurs
C_O – barrier effectiveness factor for influence of openings
C_T – barrier effectiveness factor for influence of thermal restraint
CS – critical size fire is reached
CVN – continuous value network
$\overline{D}$ – condition when a barrier experiences a large failure that can cause a massive influx of fire gases into the adjacent space

Building Fire Performance Analysis R. W. Fitzgerald
 ISBN: 0-470-86326-9 (HB)

dac – water discharge density is sufficient
Det – fire detection
Det_H – human detects the fire
Det_I – instrument detects the fire
DI – detection instrument is operational
E – enclosure point
EB – established burning
E_S – enough smoke accumulates in the target space to maintain tenability
fac – sprinkler fuses
Fr – structural frame behavior
Fr curve – continuous curve which describes the probability that the structural frame will not fail
FRI – full room involvement
fss – signaling system can function (maintenance control)
H_c – heat of combustion
HRR – heat release rate; fire size
h_v – average height of ventilation openings
I – fire can go out itself (originally "interactive burning")
I curve – continuous curve that represents the fire growth for a room interior design
idi – detection instrument is installed properly (construction control)
IG – ignition
Ig_n – fire ignites fuel package n
I_P – initial fire growth potential
iss – signaling system equipment is installed properly (construction control)
L – limit or extent of flame/heat movement and damage before the fire is terminated
L curve – continuous curve that describes the limit of flame/heat damage for successively larger building areas
LA – occupant alert (life safety alert)
LB – leave building
LD – decide to leave room (life safety decision)
LR – leave room
LS – life safety
LS curve – life safety by egress curve
L_{t_n} – occupant does not experience untenable conditions while traversing target space n
M – fire department extinguishes the fire
M curve – continuous curve which describes the probability that the local fire department will extinguish fires of successively larger sizes
MA – first water applied to the fire
MA_P – occupant applies agent to fire
M_B – fire is extinguished by the building fire brigade
MD – fire detection for M curve analysis
MD_P – occupant decides to extinguish the fire
ME – fire is extinguished
ME_P – occupant is successful in extinguishing fire
M_f – fuel mass
M_F – fire is extinguished by the local fire department
MN – fire department is notified
M_P – occupant manual extinguishment
MR – fire department response
MR_P – occupant recognizes the small fire

MW – fire department applies first water
NHL – normalized heat load
nma – nozzle reaches fire area
O_{POC} – detector sensitivity will cause actuation for the products of combustion in the sensing chamber
pmn – decide to notify (process signal) fire department
POC – products of combustion
pss – signaling system is present
Q – fire heat release rate
qme – quantity of agent is sufficient to extinguish the fire
QOD – quick-opening devices
R – room fire point
rdi – detection instrument is reliable (manufacturing quality control)
rmn – message received correctly by the fire department
rss – signaling system is reliable (manufacturing quality control)
RTI – response time index
SCBA – self-contained breathing apparatus
Sm – smoke/gas tenability
Sm curve – continuous curve which describes the probability that the target space will remain tenable
smn – send message to fire department
ST – self-termination
ST_C – the fire self-terminates before reaching the ceiling point
ST_E – the fire self-terminates before reaching the enclosure point
ST_n – the fire self-terminates within fuel package n
ST_R – the fire self-terminates before reaching the room fire point
SS – signaling system is operational
SV – single value
SVN – single value network
T – temperature
$\overline{T}$ – condition when a barrier experiences a small, localized failure that can cause ignition in the adjacent room
vaa – all water supply valves are open
waa – water reaches sprinkler
wac – water spray is not obstructed
wma – water supply is established
α – fire intensity coefficient
τ_e – equivalent standard test time
$\sqrt{k\rho c}$ – thermal absorptivity

$\overline{}$ means "not"
| means "given that"

Index

Note: Figures and Tables are indicated by *italic page numbers*

Building Fire Performance Analysis R. W. Fitzgerald
 ISBN: 0-470-86326-9 (HB)

Index compiled by Paul Nash